The
PREVENTION
of
GENETIC DISEASE
and
MENTAL
RETARDATION

AUBREY MILUNSKY, MB.B.Ch., M.R.C.P., D.C.H.

Assistant Professor of Pediatrics, Harvard Medical School; Director,
Genetics Laboratory, Eunice Kennedy Shriver Center at the Walter E. Fernald
State School; Medical Geneticist, Massachusetts General Hospital and the
Center for Human Genetics, Harvard Medical School, Boston

W. B. SAUNDERS COMPANY / Philadelphia / London / Toronto

W. B. Saunders Company: West Washington Square
Philadelphia, Pa. 19105

1 St. Anne's Road
Eastbourne, East Sussex BN21 3UN, England

833 Oxford Street
Toronto, M8Z 5T9, Canada

Library of Congress Cataloging in Publication Data

Main entry under title:

The prevention of genetic disease and mental retardation.

Includes index.

1. Medical genetics.   2. Mental deficiency – Prevention.
3. Genetic counseling. 4. Medical screening.   I. Milunsky,
Aubrey. [DNLM: 1. Hereditary diseases – Prevention and
control. 2. Mental retardation – Prevention and control.
QZ50 P944]

RB155.P67   616'.042   74–21015

ISBN 0–7216–6395–8

Prevention of Genetic Disease and Mental Retardation   ISBN 0-7216-6395-8

Last digit is the print number: 9   8   7   6   5   4   3   2

To my love, Babette

If, of all words of tongue and pen,
The saddest are, "It might have been,"
More sad are these we daily see:
"It is, but hadn't ought to be."

Francis Bret Harte
in *Mrs. Judge Jenkins*

CONTRIBUTORS

RICHARD D. ASHMORE, Ph.D.
Associate Professor, Livingston College, Rutgers University, New Brunswick, New Jersey.
Societal and Individual Orientations Toward Prevention

LEONARD ATKINS, M.D.
Associate Professor of Pathology at the Massachusetts General Hospital, Harvard Medical School; Associate Pathologist, Massachusetts General Hospital, Boston, Massachusetts.
Prenatal Diagnosis of Genetic Disorders

HELGE BOMAN, M.D.
Acting Assistant Professor of Medicine, University of Washington School of Medicine, Seattle, Washington.
Screening for the Hyperlipidemias

RONALD W. CONLEY, Ph.D.
Research Director, Interdepartmental Task Force on Worker's Compensation, Washington, D.C.
The Economics of Prenatal Genetic Diagnosis

ROBERT J. DESNICK, M.D., Ph.D.
Assistant Professor of Pediatrics, Laboratory Medicine and Pathology, Department of Pediatrics and The Dight Institute for Human Genetics, University of Minnesota Medical School, Minneapolis, Minnesota.
Enzyme Therapy in Genetic Diseases: Progress, Principles, and Prospects

SHIRLEY GRIFFITH DRISCOLL, M.D.
Associate Professor of Pathology, Harvard Medical School; Pathologist, Boston Hospital for Women, Boston, Massachusetts.
Prevention of Prematurity and Perinatal Morbidity

RICHARD W. ERBE, M.D.
Assistant Professor of Pediatrics, Harvard Medical School; Chief, Genetics Unit, Massachusetts General Hospital, Boston, Massachusetts; Medical Geneticist, Boston Sickle Cell Center, Boston, Massachusetts.
Screening for the Hemoglobinopathies

M. B. FIDDLER, B.S.
National Science Foundation Predoctoral Fellow, The Dight Institute for Human Genetics, University of Minnesota, Minneapolis, Minnesota.
Enzyme Therapy in Genetic Diseases: Progress, Principles, and Prospects

JOHN W. GRAEF, M.D.
Instructor in Pediatrics, Harvard Medical School; Associate in Medicine, Children's Hospital Medical Center, Boston, Massachusetts.
The Prevention of Lead Poisoning

YUJEN EDWARD HSIA, B.M., M.R.C.P., D.C.H.
Associate Professor of Human Genetics and Pediatrics, Yale University School of Medicine; Director, Genetics Clinic, Yale University; Attending Pediatrician, Yale–New Haven Hospital, New Haven, Connecticut.
Treatment in Genetic Diseases

MICHAEL M. KABACK, M.D.
Associate Professor, Departments of Pediatrics and Medicine, U.C.L.A. School of Medicine; Associate Chief, Division of Medical Genetics, Harbor General Hospital, Torrance, California.
Heterozygote Screening for the Control of Recessive Genetic Disease

EDWIN H. KOLODNY, M.D.
Assistant Professor of Neurology, Harvard Medical School; Assistant Neurologist, Massachusetts General Hospital; Assistant Attending Neurologist, McLean Hospital; Senior Physician, Walter E. Fernald State School; Associate Biochemist, Eunice Kennedy Shriver Center, Waltham, Massachusetts.
Heterozygote Detection in the Lipidoses

WILLIAM KRIVIT, M.D., Ph.D.
Professor of Pediatrics, Department of Pediatrics, University of Minnesota Medical School, Minneapolis, Minnesota.
Enzyme Therapy in Genetic Diseases: Progress, Principles and Prospects

JEROME E. KURENT, M.S., M.D.
Research Associate, National Institutes of Health, National Institute of Neurological and Communicative Disorders and Stroke, Infectious Diseases Branch, Bethesda, Maryland.
Infectious Diseases and the Prevention of Mental Retardation

MARC ALAN LAPPÉ, Ph.D.
Associate for the Biological Sciences, Institute of Society, Ethics and the Life Sciences, Hastings-on-Hudson, New York.
Can Eugenic Policy Be Just?

JOHN H. MEIER, Ph.D.
Associate Professor of Pediatrics, Psychiatry (Clinical Psychology), and Education, University of Colorado School of Medicine; Director, John F. Kennedy Child Development Center at the University of Colorado Medical Center, Denver, Colorado.

Early Intervention in the Prevention of Mental Retardation

AUBREY MILUNSKY, MB.B.Ch., M.R.C.P., D.C.H.
Assistant Professor of Pediatrics, Harvard Medical School; Medical Geneticist, Massachusetts General Hospital and Center for Human Genetics, Harvard Medical School; Director, Genetics Laboratory, Eunice Kennedy Shriver Center at the Walter E. Fernald State School, Waltham, Massachusetts.

Introduction
The Causes and Prevalence of Mental Retardation
Genetic Counseling: Principles and Practice
Prenatal Diagnosis of Genetic Disorders
The Economics of Prenatal Genetic Diagnosis

ARNO G. MOTULSKY, M.D.
Professor of Medicine and Genetics, University of Washington School of Medicine; Director, Center for Inherited Diseases, Seattle, Washington.

Screening for the Hyperlipidemias

HAROLD M. NITOWSKY, M.D.
Professor of Pediatrics and Genetics, Albert Einstein College of Medicine of Yeshiva University; Attending Pediatrician, Director, Genetic Counseling Program, Bronx Municipal Hospital Center and Albert Einstein College of Medicine, Bronx, New York.

Heterozygote Detection in Autosomal Recessive Biochemical Disorders
Associated with Mental Retardation

LEONARD PINSKY, M.D.
Assistant Professor of Pediatrics and Biology, McGill University; Associate Pediatrician, Montreal Children's Hospital; Staff Investigator, Lady Davis Institute for Medical Research, Jewish General Hospital, Montreal, Quebec, Canada.

Carrier Detection in X-Linked Disease

PHILIP REILLY, J.D.
Adjunct Assistant Professor of Law, University of Houston Law School; Postdoctoral Fellow, Medical Genetics Center, University of Texas at Houston, Houston, Texas.

The Role of Law in the Prevention of Genetic Disease

VINCENT M. RICCARDI, M.D.
Assistant Professor of Biophysics, Genetics, and Medicine, University of Colorado Medical School; Director, Genetics Unit, University of Colorado Medical Center, Denver, Colorado.

Regional Genetic Counseling Programs

JOHN L. SEVER, M.D., Ph.D.
Chief, Infectious Diseases Branch, National Institutes of Health, National Institute of Neurological and Communicative Disorders and Stroke, Bethesda, Maryland; Professor of Pediatrics, Georgetown University Medical School, Washington, D.C.

Infectious Diseases and the Prevention of Mental Retardation

VIVIAN E. SHIH, M.D.
Assistant Professor of Neurology at the Massachusetts General Hospital, Harvard Medical School; Assistant in Neurology, Massachusetts General Hospital, Boston, Massachusetts.

Homozygote Screening in the Disorders of Amino Acid Metabolism

PREFACE

Opportunities to prevent genetic disease reflect in large measure the application of recent and continuing technological advances. Carrier detection and prenatal genetic diagnosis are at the root of this challenging progress. Major advances have also occurred in the prevention of mental retardation in the past 15 years, particularly in the increased understanding of causality and in the role and management of environmental agents. Genetic origins of mental retardation are common and are invariable considerations in differential diagnosis. Recognition of both genetic and environmental contributions to the causes of mental retardation has brought about the compelling need for the consideration of both factors in one volume.

The sequence of chapters follows a logical order, opening with an overall perspective, proceeding to an understanding of the causes and prevalence of mental retardation, followed by the recognition of the crucial attitudinal aspects toward prevention. Principles of genetic counseling are formulated next—perhaps for the first time in a text—to provide the delivery of new information with more meaningful perspectives and options. Not all genetic counselors can be expected to concur with the principles as delineated, but few would argue against the need for recognition of all these factors in the counseling process.

The body of this text is devoted to detailed consideration of the detection of carriers of genetic disease, either by mass screening or by individual testing. Special attention is given in a comprehensive chapter to carrier detection in X-linked disease. Since prenatal diagnosis represents a cogent opportunity to prevent about two thirds of *all* recognized X-linked diseases (roughly 150), especially detailed consideration was warranted. Prenatal genetic diagnosis is considered against the background of what is now known and possible, with particular emphasis on problems and pitfalls. This same chapter on prenatal diagnosis contains a thorough analysis of a large experience in a single laboratory service, including data on over 1000 cases assayed for alpha-fetoprotein. Discussions on screening and therapy for genetic *disease* precede detailed chapters on perinatal and environmental problems and emphasize various approaches to their prevention, including early intervention. The concluding chapters focus on legal, economic, and ethical issues concerned with the prevention of genetic disease.

The essential thrust throughout has been to elucidate those principles that form the basis for any program whose purpose is the prevention of genetic disease and mental retardation. Although every genetic entity has not been described, careful emphasis has been given to the recognition of a wide range of problems and to approaches to their management. The principles delineated should not easily become outdated and will hopefully serve for many years as standard guidelines to anyone concerned with the prevention of genetic disease and mental retardation. Certainly, the background against which these principles have evolved provides a prodigious wealth of information, as exemplified by the bibliography of over 1700 references. In this context, the Appendix will be found to be an additional valuable resource on carrier detection for selected autosomal recessive genetic disorders.

This book has been written for geneticists, pediatricians, obstetricians, and those concerned with the development and application of health care-and-delivery systems. However, in view of the wide range covered by the subject matter, this text should prove to be of value to physicians and students in *all* specialties as well as ethicists, sociologists, psychologists, pathologists, lawyers, and even theologians.

In a heavily populated world beset by increasing, serious problems, parental wishes for healthy "quality" offspring have become paramount. It is my fervent hope that *prospective* parents will realize their wishes through programs based on principles and approaches elaborated upon in this volume.

AUBREY MILUNSKY

ACKNOWLEDGMENTS

I have been singularly fortunate in having had internationally and nationally known colleagues contribute their expertise to this monograph. I am appreciative of their contributions and concerted efforts. Notwithstanding extremely heavy schedules with research, patient care, teaching, and writing, all have made sufficient time to share the quintessence of their experience because of their recognition of the need for an authoritative and critical text.

Despite threatening deadlines, my secretary, Miss Carmela Antonellis, completed all assigned tasks with her customary charm and efficiency. I am grateful to my very able Research Assistant, Miss Judy Heck, for her painstaking valuable efforts. My thanks are due Mrs. Theresa Linietsky for her diligent care in copy-editing. My wife, to whom this text is dedicated, well knows my appreciation for her continuing love, inspiration, and patience.

My studies reported in this volume have been supported in part by USPHS Grant No. 1–P01–HD05515–01 and the NICHHD Contract No. 71–2451, as well as the Maternal and Child Health Project No. 906.

CONTENTS

INTRODUCTION

Aubrey Milunsky, MB.B.Ch., M.R.C.P., D.C.H.

Genetic Disease

Endowed with four to eight deleterious genes, all individuals have reason to be concerned about their own health and that of their loved ones—both present and anticipated. Rare is the family that is entirely free of genetic disorder—however mild or insignificant—of genetic predilection, or of genetic-environmental effects. Almost 2000 autosomal dominant, recessive, and sex-linked disorders have been catalogued.[1] Not included are a large number of polygenically inherited disorders. Admissions to major children's hospitals in the developed countries indicate that 25 to 30 per cent have been for genetic disease, congenital malformations, and/or mental retardation[2, 3] (Table 1).

Coronary artery disease (polygenic or autosomal dominant) is a major scourge; some 30 to 40 per cent of the population suffer from or eventually succumb to its effects. Disorders with genetic components, such as mental retardation, congenital malformations, malignant disease in families, diabetes mellitus, hypertension, schizophrenia, and skin and allergic diseases, are conditions with which we have daily experience. Genetic disorders with ethnic or racial predilection (*e.g.,* Irish, neural tube defects; Jews, Tay-Sachs disease; Blacks, sickle cell anemia), are well known (see Chapters 1 and 5). Critchley has ventured estimates of dyslexia (commonly, autosomal dominant) in the population as high as 5 to 10 per cent.[4] Disorders associated with different HL-A haplotypes are increasingly recognized,[5, 6] whereas those associated with blood groups (Rh, ABO, and others) are well known.

Genetic predisposition to certain disorders has been recognized only in recent years (*e.g.,* α_1-antitrypsin deficiency and chronic obstructive lung disease). Individuals with glucose-6-phosphate dehydrogenase

TABLE 1. CHARACTERISTICS OF ADMISSIONS TO A
PEDIATRIC HOSPITAL IN NORTH AMERICA*†

Class of Admission‡	Number of Admissions		Total	% of All	Male/ Female Ratio	Average Length of Stay (Days)
	MEDICINE	SURGERY				
All			1145	100	1.76	7.31
I. Genetic AR	18	5	23	2.0	1.3	
AD	11	12	23	2.0	1.1	
X	25	6	31	2.7	15.0	
MF§	44	1	45	3.9	1.6	
Chr.	5	0	5	0.4	0.66	
	(103)	(24)	(127)	(11.0)		(7.28)
II. Congenital malformations‖ (excludes group I)	5	205	210	18.4	1.6	8.60
III. Unknown	34	45	79	6.9	1.1	6.30
IV. Nongenetic	278	452	730	63.8	1.65	5.70

*From Clow, C. L., Fraser, F. C., Laberge, C., and Scriver, C. R.: On the application of knowledge to the patient with genetic disease. *In* Steinberg, A. G., and Bearn, A. G. (eds.): Progress in Medical Genetics. Vol. 9. New York, Grune & Stratton, Inc., 1973, pp. 159–213. Reprinted by permission.

†Randomized sample of 1145 patients, which is statistically representative of the 12,800 admissions to the Montreal Children's Hospital during the 12-month period May 1969–May 1970.

‡Abbreviations: AR, autosomal recessive; AD, autosomal dominant; X, X-linked; MF, multifactorial; Chr., chromosomal.

§Multifactorial group includes 24 cases of "atopic hypersensitivity."

‖Group includes 55 cases of hernia, according to classification of Hay and Tonascia. (From Hay, S., and Tonascia, S.: A classification of congenital malformations. San Francisco, U.S. Dept. of Health, Education, and Welfare. Public Health Service, Dental Health Center.)

deficiency may have severe reactions to many different drugs. Anaphylaxis to an antibiotic might be the first indication of a serious genetic predisposition in a family. In addition, patients with certain myopathic diseases[7]—often mild or subclinical—may suffer a fatal outcome with malignant hyperpyrexia[8, 9] during or after general anesthesia.

In 1973 Kellerman and Shaw[10] reported intermediate or high levels of aryl hydrocarbon hydroxylase in association with lung cancer. This inducible enzyme, which converts hydrocarbon carcinogens into the reactive epoxy form, was present in 48 of 50 cases (all smokers) with lung cancer, but was present in only about half of their control subjects. Genetic predisposition may eventually explain why some smokers develop lung cancer while other equally heavily addicted smokers are spared. Indeed, the "metabolic" or biochemical reason for smoking "addiction" may be genetic!

In countries where the problems of malnutrition and infectious disease have been overcome, society has turned its attention to the prevention and management of genetic disease. Recent advances in carrier detection and prenatal genetic diagnosis have undoubtedly heightened public awareness and perception of genetic disease.

A strange tardiness exists, however, in the attitude of society toward preventive action programs. Phlegmatic (if any) responses to programs aimed at preventing smoking, treating obesity, and exhorting the public to use seat belts have been the rule. Programs beamed at the prevention of genetic disease and/or mental retardation are not likely to do much better, unless significant inroads are made in the understanding of factors that effect positive attitudes toward preventive health proposals and programs (see Chapter 2). Society must recognize that certain minimum prerequisites are principal factors in stimulating such attitude development. These prerequisites, modified from Rosenstock,[11] are as follows:

1.   The individual must personally feel and understand the seriousness of the threat to his or her own health.

2.   The individual must perceive that he or she is susceptible to the disorder in question.

3.   The "threatened" individual must perceive that he reaps personal benefit from the program offered.

4.   The individual must believe that no coercive element enters the preventive program.

Participation in screening programs—at least for the Tay-Sachs gene—correlates well with the first two criteria and with perception of benefits from compliance.[12, 13] Poor specific knowledge and lack of motivation among subjects in one Tay-Sachs disease heterozygote screening program were clearly recognizable factors in noncompliance.[13] Conversely, the need to inform and motivate practicing physicians about such a program was also demonstrated.[13]

Mental Retardation

Approximately 3 per cent of the general population have I.Q.'s < 70.[14] Prevalence rates of mental retardation are confounded not only by variations in definitions, but also by differing assessments of coexisting impairment in adaptive behavior.

The causes of mental retardation frequently have genetic components (see Chapter 1). Knowledge of this intertwined relationship prompted consideration of both preventive programs and approaches in one volume. Aside from the many genetic disorders (chromosomal, biochemical) characterized by mental retardation, a significant number of cases with so-called "idiopathic" mental retardation do have, in fact, solely or partly genetic origins (Chapter 1).

In recent years, the perception that mental retardation—regardless of cause—is simply a facet within a spectrum of possible abnormalities, has given rise to the much more realistic and helpful concept of "developmental disability." To a great extent much that is written in this text, in the context of prevention, is best seen as referring to all developmental disabilities. In 1970 the U.S. Congress enacted Public Law 91–517, which defined the term developmental disability as follows:

> ... a disability attributable to mental retardation, cerebral palsy, epilepsy or another neurological condition of an individual found by the Secretary to be closely related to mental retardation or to require treatment similar to that required for mentally retarded individuals, which disability originates before such individual attains age eighteen, which has continued or can be expected to continue indefinitely, and which constitutes a substantial handicap to such individual.

Although lacking in clarity, this definition does serve to recognize disorders of communication, locomotion, adjustment, and intellectual function, while at the same time it emphasizes the irreversibility and generally nonprogressive nature of these disabilities.

Notwithstanding new technologies valuable in prevention, the frequency of developmental disabilities continues to rise (see Chapter 17). To an extent this may reflect greater expertise and sophistication in diagnosis. Not to be ignored, however, are the recoveries made by those who some years ago would have died without the aid of advanced technology. In particular, premature infants have shown improved survival rates, only to have developmental disabilities become apparent with high frequency later[15, 16] (see Chapter 1).

Generally available data have shown a decrease in handicap related to increasing birthweight. Lubchenco et al.[17] unexpectedly observed an approximately similar incidence of handicap in small-for-dates and appropriate-for-dates premature infants. Still unexplained is the observation by Wiener that a high risk of mental subnormality exists for infants with birthweights *more than* 2500 grams and gestational ages of *less than 28 weeks*.[18]

Much has been written about early intervention projects aimed at preventing mental retardation in settings of poverty and social deprivation (see Chapter 17). In a Milwaukee early intervention project a major finding was that maternal intelligence proved to be the best single predictor of intellectual development in poor, deprived children.[19] Data from this study pointed again to the frequently observed phenomenon that mild mental retardation is not randomly distributed among the poor, but is heavily concentrated in clusters of families with intellectual subnormal parents.[19] Sex-linked, polygenic and autosomal recessive modes of inheritance are probable explanations for some of these cases.

Retrospective and Prospective Prevention of Genetic Disease

The short and long-term implications of new techniques for the prevention of genetic disease have been considered in detail.[20-25] Fraser[24] has viewed as eugenic certain preventive approaches, including the following:

1. Genetic counseling leading to reproductive restraint
2. Early diagnosis of the homozygote
3. Heterozygote detection
4. Artificial insemination by donor
5. Prenatal genetic diagnosis and selective abortion
6. Lowering of parental age at reproduction.

Matsunaga, using Japanese data, showed that the incidence of maternal age-dependent chromosome abnormalities has decreased by about 20 per cent over the past 17 years.[26]

Genetic counselors would attest to the ineffectiveness of retrospective prevention, which in addition hardly alters the gene or disease frequency in the population. Lubs and Lubs[27] have calculated (with certain assumptions) that with retrospective prevention of autosomal recessive disease, only 6 per cent of affected cases could be prevented. For X-linked disease 6 per cent of the cases are similarly preventable, but this figure is in contrast to the 16 per cent of cases preventable prospectively. Motulsky et al.[21] calculated that retrospective counseling could prevent only 12.5 to 25.0 per cent of genetic disease. For autosomal dominant disorders, knowledge of the family history, use of artificial insemination, and educated concern and motivation of those at risk, presently serve as the most important methods of disease prevention. Almost invariably, no predictive tests have been taken to determine the likelihood of the occurrence of such conditions.

MAJOR APPROACHES TO THE PREVENTION OF MENTAL RETARDATION AND GENETIC DISEASE

Much of this text is devoted to consideration of the major approaches to the prevention of both mental retardation and/or genetic disease. Some possible strategies for prevention are represented schematically in Figure 1, which is largely self-explanatory.

Central to the theme of prevention is the recognition of causality. As mentioned earlier, programs of prevention are bound to fail if they lack careful attention to development of positive attitudes toward prevention—in both the public and the profession. Closely aligned with attitude development is the simultaneous imperative to educate both doctor and patient—a process which by necessity must be wide-ranging and *continuous.* All may come to naught if mothers-to-be attend their first antenatal visit at 5 months of pregnancy, or later. Educating parents to the need for early and continuous antenatal care may be facilitated by drawing public attention to the known causes of prematurity, as well as those related to perinatal morbidity and mortality (see Chapter 14).

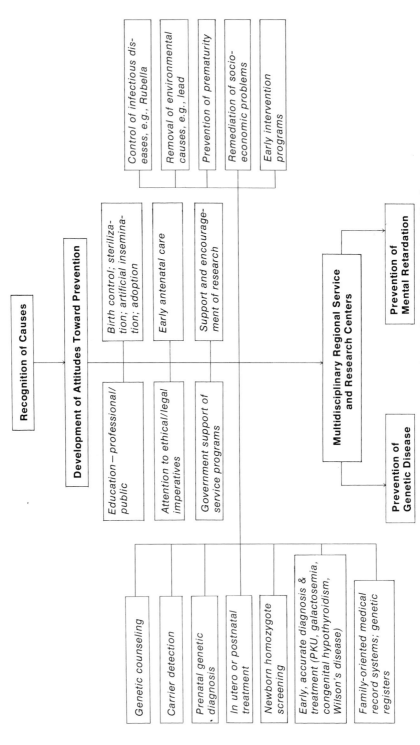

Figure 1. Strategies for the prevention of mental retardation and genetic disease.

The preventive programs under discussion most properly should be funded and implemented by governmental public health authorities. The economic aspects of prenatal genetic diagnosis (Chapters 9 and 20) provide ready evidence for both cost-benefit and cost-effectiveness. Screening programs to detect inborn errors of metabolism and transport in newborn infants have been in progress for years. Except for phenylketonuria, the frequencies of the disorders screened are low (Chapter 10). Much criticism has therefore been leveled against these multiphasic screening programs because of the low yield of cases diagnosed and the alleged high expense incurred. The highly successful Massachusetts screening program, described by Levy,[28] has provided a cost-benefit analysis of its most recent fiscal year of operations.[29] Like others who have reported cost-benefits for phenylketonuria alone,[30] the Massachusetts Department of Public Health now reports the anticipated benefits (Table 2) for their multiphasic screening program.[29]

Programs designed to prevent mental retardation and/or genetic disease are best located within regional medical-school–hospital complexes, where multiple and varied expertise is available and where problems at the interface of many disciplines can be managed simultaneously. The mandatory need for surveillance and the insistence on the highest standards of excellence (aside from the budgetary sense) make such regional centers ideal.

Carrier detection of genetic disorders that currently are not amenable to mass screening is no less important. Observations in carriers of autosomal recessive disorders generally without mental retardation are listed in the Appendix (for observations in carriers of X-linked dis-

TABLE 2. COST-BENEFIT ANALYSIS FOR MASSACHUSETTS NEWBORN METABOLIC DISORDERS SCREENING PROGRAM*

Item	Cost ($)	
Cost of specimen collection	155,730	
Cost of laboratory testing	200,971	
Cost of PKU care	68,187	
Cost of other disorders care	35,750	
Total		460,638
Averted Costs:		
Institutionalization:		
PKU	648,240†	
Other disorders	162,060†	
Evaluation & hospitalization:		
PKU	6,000	
Other disorders	9,000	
Total		825,300
Estimated Savings (1972–73)		364,662

*From Massachusetts Department of Public Health: Cost-benefit analysis of newborn screening for metabolic disorders. N. Engl. J. Med., 291:1414, 1974. Reprinted with permission.

†Based on an average cost of $8,103/yr for institutionalization in Massachusetts.

orders, see Chapter 6). Subjects at risk in these categories should be detected most frequently through family-oriented medical records systems or genetic disease registries (see Chapter 3). While carrier detection for many such disorders may not presently be possible in 100 per cent of cases, technologic refinements may certainly narrow the gap in a relatively short period. Westwood and Raine have discussed the analytic and biologic factors interfering with the confident separation of heterozygotes from normal homozygotes.[31] Indeed, in view of these difficulties, they have recommended that genotype be stated only as a probability.

SCREENING AND DIAGNOSIS IN SELECTED GENETIC DISORDERS

Neonatal Chromosomal Screening

Chromosome surveys were done initially on particular groups of individuals, essentially as part of diagnostic evaluation. Hence, extensive surveys were performed on institutionalized persons with mental retardation or congenital malformations. The effects of X-irradiation (used in the treatment of ankylosing spondylitis[32]) on lymphocyte chromosomes were studied and followed by similar efforts in those exposed to industrial toxic agents, such as benzine and mercurials.[33] All these studies had potential benefit to the individuals tested.

Subsequently, the observation was made that an unusually high incidence of XYY individuals was observed in males confined to maximum security institutions.[34-36] Because of the multiple questions raised by these observations on XYY individuals in the penal institutions, Court-Brown[37] turned to neonatal chromosomal screening aimed at identification at birth of XYY males with continued surveillance, thereafter. Since 1968 a number of groups have initiated newborn chromosomal studies.[38]

From these studies the frequency of structural chromosomal rearrangements, autosomal aneuploidy, and sex chromosome aneuploidy has been determined (Table 3). Much has been learned about human chromosomal variation, the relationship of some karyotypes to the phenotype, and possible associations between behavior and learning problems.

Neonatal chromosomal screening studies have been seen to have definite, though not compelling, benefits to the screenees. Certainly, the detection of structural rearrangements may be of no benefit to the patient so diagnosed, but may be of the greatest significance to the parents—in light of the possibilities for prenatal genetic diagnosis in subsequent pregnancies. Benefit would accrue, then, not only to the

TABLE 3. INCIDENCE OF SIGNIFICANT CHROMOSOMAL
ABNORMALITIES FROM CUMULATIVE SURVEYS OF 22,860
CONSECUTIVELY BORN INFANTS*

Chromosome Disorders	Average Incidence/1000 Live Births (Range)
Structural rearrangements	1.80 (1.7–2.2)
Autosomal aneuploidy	1.00 (0.5–1.6)
Sex-chromosome aneuploidy	2.28 (1.1–4.7)

*From Milunsky, A.: The Prenatal Diagnosis of Hereditary Disorders. Springfield, Ill., Charles C Thomas, Publisher, 1973.

parents or to siblings who may also have balanced translocations, but also to the diagnosed individual in childbearing years, for the same reason. A minimal estimate of 0.5 per cent has been given for the frequency of structural changes in the general population.[39] The most frequent translocation occurring in the general population is t(DqDq), with a frequency of about 1 per 1000.[40] Other Robertsonian translocations have been estimated to occur in the general population with a frequency of about 1 in 20,000.[40]

Other reasons, in addition to prenatal diagnosis and research, can be offered to justify neonatal chromosomal screening programs. Determination of the XYY karyotype in the newborn period may be of debatable value since it is not entirely clear whether criminality,[41, 42] impulsivity,[43] or mental retardation[44] occurs with significant frequency. Prospective screening for the XYY karyotype in the newborn would ultimately allow determination not only of the exact frequency of occurrence, but also of the relationship of this karyotypic abnormality to behavioral and other aberrations. Mellman[45] has argued for the use of sex chromatin studies for such screening because the data accumulated would be comparable to that obtained through chromosomal screening and would also allow for more rapid acquisition of a large data base. Either way, the problems of disclosure, informed consent, counseling, voluntarism, and prolonged follow-up still remain. Mellman has pointed to the value of a noncoercive environment (that is, one not in a hospital setting), but rather in schools and other similar locations.

Detection of the XXY karyotype in the newborn period may be of benefit in providing possible answers to behavioral and learning problems in childhood. Even if such screening studies were delayed until a few years prior to puberty, there would still be adequate time for proper androgen therapy. Children with XXY karyotypes treated with androgens have been the subject of favorable, preliminary reports.[46, 47]

No clear benefit could be anticipated from early detection of XXX individuals in the newborn period since all that is currently known is that a higher rate of incidence of persons with this karyotype is to be found in institutions for the mentally handicapped.[48] Like the XYY syndrome, however, no predictions in a particular case can or should be offered.

Cystic Fibrosis

This disorder is the most common fatal genetic disorder in Caucasian populations, occurring with a frequency of approximately 1 per 2000 live births.[49] Marked heterogeneity occurs in the manifestations of cystic fibrosis and ranges from obvious disease *in utero* to diagnoses made in adulthood. The sweat test is the most reliable diagnostic tool for cystic fibrosis, but it cannot easily be applied to entire populations. For more than 20 years, an increased protein content in meconium ileus has been known,[50, 51] with albumin recognized as the main constituent.[52] More recently, it was shown that the albumin content of meconium was also elevated in cystic fibrosis.[53, 54]

West German[55] and Swedish[56] investigators have developed and clinically evaluated a new, simple reliable test strip (BM-test meconium) and have reported their experience in 68,000 cases studied both in Europe and the U.S.A.[55] They diagnosed 55 cases of cystic fibrosis, later confirmed with sweat tests. Only four false negative tests were obtained — two of which might have been the result of mailing. It should be noted that a *decrease* in albumin concentration was observed in consecutive meconium samples.

Therapeutic advantages aside, only modest benefits to the prevention of disease in other subsequent children could be achieved by diagnosis of the homozygote at birth. The frequency of cystic fibrosis heterozygotes is approximately 1 per 20 in the Caucasian population. Research studies aimed at the detection of the cystic fibrosis gene have progressed favorably,[57] but have not reached the stage of reliability and reproducibility for clinical application.[58] Success with these studies will facilitate the prenatal diagnosis of cystic fibrosis (Chapter 9).

Congenital Hypothyroidism

Klein et al.[59] estimated the frequency of congenital hypothyroidism at between 1 per 5000 and 1 per 10,000 live births. Thus, congenital hypothyroidism is found more frequently than any other congenital metabolic defect associated with mental retardation. Since the clinical diagnosis of congenital hypothyroidism in the neonatal period is difficult and often missed, a newborn screening test would be most valuable. Current evidence points to optimum intellectual attainment when thyroid replacement therapy is initiated prior to 3 months of age.[59, 60]

Extensive Canadian screening studies using dried blood on filter paper obtained on or before the fifth day of life have been done with very promising results.[61-63] A very sensitive radioimmunoassay to detect thyroxine has been utilized, yielding a small percentage of false positive measurements, but no false negative results.[63]

Serum thyroid-stimulating hormone (TSH), however, is the most sensitive test for primary hypothyroidism and is raised in mild hypothyroidism, even when serum thyroxine is still within the normal range.[64, 65] Klein et al. therefore elected to use serum TSH and cord blood for their successful screening studies.[66] A choice may ultimately have to be made between the TSH assay, which may be more reliable, and the thyroxine assay, in which samples can be collected en masse *and* at the same time as phenylketonuria samples.

Although further studies to refine the present assay techniques are still needed, there can be little hesitation in urging the adoption of screening for neonatal hypothyroidism in those centers without such programs.

Alpha₁-antitrypsin Deficiency

Available data indicates that α_1-antitrypsin deficiency is an autosomal disorder with multiple codominant alleles.[67] Affected individuals have a high risk of developing chronic obstructive pulmonary disease, usually with onset during the fourth decade of life. Multiple electrophoretic phenotypes have been recognized and designated as the Pi (protease inhibitor) system.[68] At least 23 such phenotypes have been described — Pi^M being found to be by far the most common allele — with a frequency of approximately 0.9 or higher in all populations tested.[69] Of particular interest are those alleles that lead to lower than normal concentrations of α_1-antitrypsin in serum. For example, Pi^Z homozygotes carry a high risk of developing chronic obstructive pulmonary disease. Cigarette smoking apparently accelerates the deterioration of lung infection, especially in individuals who are Pi^Z homozygotes and heterozygotes. Homozygosity for Pi^Z alleles is associated with neonatal hepatitis and cirrhosis of the liver.[70] Infants with α_1-antitrypsin deficiency have a 20 to 30 per cent risk of developing neonatal hepatitis.[69]

In one large population study, the prevalence of α_1-antitrypsin phenotypes (MZ and NS) was found to be 3.6 and 6.0 per cent, respectively.[71] It is not yet clearly known whether heterozygotes for Pi^Z are predisposed to obstructive pulmonary disease. Nevertheless, Kueppers and Black[69] have suggested that screening of persons at risk in certain occupations — and perhaps even newborns — should be considered.

Huntington's Chorea

In North America the frequency of Huntington's chorea is approximately 1 in 10,000.[72] This severe degenerative brain disorder is inherited as an autosomal dominant trait with complete penetrance. Studies to determine the basic defect have not as yet been successful, although clear progress has been made. Perry et al.[73] have reported that the concentrations of six amino acids – proline, alanine, valine, isoleucine, leucine, and tyrosine – were significantly lower in the plasma of fasting patients with Huntington's chorea than in control subjects studied. No such changes, however, were observable in the plasma of 31 asymptomatic offspring of choreic patients. Studies from the same group revealed a deficiency of γ-aminobutyric acid (GABA) in the brains of choreic patients.[74] These investigators suggested that such a deficiency could result from reduced activity of the enzyme glutamic acid decarboxylase, or it might reflect a decrease in the number of neurons that normally utilize GABA as a transmitter. In confirming the low concentrations of GABA in choreic brains, others[75] indeed have found a highly significant reduction in the activity of the synthetic enzyme glutamic acid decarboxylase in the caudate and the putamen of choreic brains. Bird and Iversen[75] concluded that the key neuropharmacologic feature in Huntington's chorea was the presence of normal dopaminergic systems in association with their reduced availability of GABA.

Various efforts have been made to detect the presymptomatic patient with Huntington's chorea.[76-78] Using L-dopa, 35.7 per cent of subjects at risk developed chorea that always disappeared after discontinuation of the medication.[76] These observations were highly suggestive of the fact that those subjects developing chorea would subsequently develop the full-blown disease. Long-term studies now in progress will eventually test the degree of accuracy of this prediction.

Aminoff et al.[78] also with a predictive test in mind, studied the uptake of dopamine and 5-hydroxytryptamine by platelet-rich plasma from patients with Huntington's chorea. Significantly increased uptake of both dopamine and 5-hydroxytryptamine was observed, and in addition, significant decreases in plasma adrenaline levels were noted in choreic patients.

Extremely serious questions arise when one considers the wisdom of using predictive tests in subjects at risk for developing Huntington's chorea. A false positive test could have disastrous consequences – including a patient's attempt to commit suicide! False negative tests may lead to reproductive decisions which ultimately would bring misery to the offspring and the parents. Obviously, an absolutely accurate predictive test would be helpful in genetic counseling. Nevertheless, the author would have the greatest hesitation in recommending its use until therapy for Huntington's chorea becomes available.

Myotonic Muscular Dystrophy

Harper suggested that almost all heterozygotes for myotonic muscular dystrophy may be detectable by 14 years of age.[79] Close similarities, with regard to age of onset of this autosomal dominant disease, have been observed;[80] Bundey found[81] that this pattern held for families of index cases with onset in infancy or childhood. In the families of adult onset cases, Bundey noted[81] manifestations may become apparent only after 40 years of age.

Somewhat less ominous, perhaps, are predictive tests for myotonic muscular dystrophy, for which slit-lamp examination, followed by electromyography, have been recommended.[81, 82] Measurement of serum immunoglobulins, for the same purpose, have been less successful.[82]

Legal and Ethical Considerations

The ethical and legal aspects of prenatal genetic diagnosis, genetic counseling, and screening programs for carriers have been discussed throughout this text (see, especially Chapters 3, 9, 19, 21) and more extensively, elsewhere.[83-89] Historically, society has not taken a benign stance in the face of preventable disease. The author's recommendation, for example, that couples seeking a marriage license first obtain information about genetic counseling, as well as carrier detection tests,[86] may soon be formalized by laws. In 1973 the Chicago Bar Association proposed that a marriage license may be issued only after presentation of a certificate signed by a physician "setting forth that such person. . . . is free from venereal disease, and *has been advised* [italics added] of abnormalities which may cause birth defects, as nearly as can be determined by a thorough physical examination and such standard laboratory tests as are necessary for discovery of such diseases."[89]

Green and Capron have addressed in some detail the issues of voluntary and mandatory genetic screening, with particular emphasis on the problems of privacy and confidentiality.[89] They concluded that voluntary government screening programs do not appear to raise insurmountable constitutional problems. In contrast, compulsory screening programs are within the powers of state public health authorities if the purposes are to:

1.   provide information about the incidence and severity of the disease

2.   protect members of society from the disease

3.   conserve health resources through the prevention and appropriate treatment of disease

PRINCIPLES FOR THE PREVENTION OF GENETIC DISEASE

Wilson and Jungner recognized ten principles of early disease detection.[87] These points have been elaborated on and modified to refer more specifically to the prevention of genetic disease.

1. The disorder to be prevented should constitute an important health problem.
2. The mode of inheritance and the natural history of the disease should be understood.
3. Cost-effectiveness should first be established and its total be justifiable in relation to expenditure on general health care.
4. Public health education should precede all testing and should continue during and after such programs.
5. Program effectiveness should be continually evaluated against the background of pre-arranged goals to deal with unexpected problems.
6. The rights of individuals and their freedom not to participate in a program must be recognized and safeguarded.
7. Privacy and confidentiality must be assured.
8. The test must be acceptable to the family and the segment of the population studied, be of low or no risk, and be participated in voluntarily.
9. Facilities for diagnosis and treatment should be within reach of the patient.
10. Some recognizable benefit should accrue to the patient or his family.
11. A clear policy should be established regarding whom to treat, when to treat, and how long to treat.
12. Carrier detection programs are best offered prior to marriage or conception.
13. Tests for carriers should be accurate, reproducible, safe, inexpensive, simple, and automated.
14. False negative results in screening tests should not occur or should be so rare as to constitute a very small fraction of the positive cases detected.
15. The option of prenatal genetic diagnosis *ideally* should be available prior to a population screening program for carriers, if no treatment for the disease exists.
16. Strictest attention to the highest legal and ethical standards should be mandatory.
17. The economic responsibilities for preventive programs should be lodged in governmental departments of public health.

These 17 principles are expanded upon in discussions throughout this text and constitute the quintessence of ideal programs for the prevention of genetic disease.

References

1. McKusick, V. A.: Mendelian Inheritance in Man. Catalogs of Autosomal Dominant, Autosomal Recessive, and X-Linked Phenotypes. 3rd ed. Baltimore, The Johns Hopkins Press, 1971.
2. Clow, C. L., Fraser, F. C., Laberge, C. et al.: On the application of knowledge to the patient with genetic disease. *In* Steinberg, A. G., and Bearn, A. G. (eds.): Progress in Medical Genetics. Vol. 9. New York, Grune & Stratton, Inc., 1973, p. 159.

3. Day, N., and Holmes, L. B.: The incidence of genetic disease in a university hospital population. Am. J. Hum. Genet., 25:237, 1973.
4. Critchley, M.: The Dyslexic Child. London, W. Heinemann Medical Books Ltd., 1970.
5. Brewerton, D. A., Caffrey, M., Hart, F. D. et al.: Ankylosing spondylitis and HL-A 27. Lancet, 1:904, 1973.
6. Granditsch, G., Ludwig, H., Polymenidis, Z. et al.: Celiac disease and HL-A28. Lancet, 2:908, 1973.
7. King, J. O., Denborough, M. A., and Zapf, P.: Inheritance of malignant hyperpyrexia. Lancet, 1:365, 1972.
8. Denborough, M. A., and Lovell, R. R. H.: Anaesthetic deaths in a family. Lancet, 2:45, 1960.
9. Isaacs, H., and Barlow, M. B.: Malignant hyperpyrexia: Further muscle studies in asymptomatic carriers identified by creatinine phosphokinase screening. J. Neurol. Neurosurg. Psychiatry, 36:228, 1973.
10. Kellermann, G., and Shaw, C. R.: Aryl hydrocarbon hydroxylase inducibility and bronchogenic carcinoma. Am. J. Hum. Genet., 25:40A, 1973.
11. Rosenstock, I. M.: Why people use health services. Milbank Mem. Fund Q., 44: Suppl.: 94, 1966.
12. Kaback, M. M., Becker, M. H., and Ruth, M. V.: Sociologic studies in human genetics. I. Compliance factors in a voluntary heterozygote screening program. *In* Bergsma, D. (ed.): Ethical, Social and Legal Dimensions of Screening for Human Genetic Disease. Vol. 10. New York, Stratton Intercontinental Medical Book Corp., 1974, p. 145.
13. Beck, E., Blaichman, S., Scriver, C. R. et al.: Advocacy and compliance in genetic screening. Behavior of physicians and clients in a voluntary program of testing for the Tay-Sachs gene. N. Engl. J. Med. 291:1166, 1974.
14. Birch, H., Richardson, S. A., Baird, D. et al.: Mental subnormality: A clinical and epidemiologic study in the community. Baltimore, The Williams & Wilkins Co., 1970.
15. Drillien, C. M.: Causes of handicap in the low-weight infant. *In* Jonxis, J. H. P., Visser, H. K. A., and Troelstra, J. A. (eds.): Nutricia Symposium. Aspects of Prematurity and Dysmaturity. Leiden, H. E. Stenfert Kroese, N., 1968.
16. Lubchenco, L. O., Delivoria-Papadopoulos, M., and Searls, D.: Long-term follow-up studies of prematurely born infants. II. Influence of birth weight and gestational age on sequelae. J. Pediatr., 80:509, 1972.
17. Lubchenco, L. O., Bard, H., Goldman, A. L., et al.: Newborn intensive care and long-term prognosis. Dev. Med. Child Neurol., 16:421, 1974.
18. Wiener, G.: The relationship of birth weight and length of gestation to intellectual development at ages eight to ten years. J. Pediatr., 76:694, 1970.
19. Begab, M. J.: The major dilemma of mental retardation: Shall we prevent it? (Some social implications of research in mental retardation). Am. J. Ment. Defic., 78:519, 1974.
20. Mayo, O.: On the effects of genetic counseling on gene frequencies. Hum. Hered., 20:361, 1970.
21. Motulsky, A. G., Fraser, G. R., and Felsenstein, J.: Public health and long-term genetic implications of intrauterine diagnosis and selective abortion. Birth Defects, 7:22, 1971.
22. Fraser, G. R.: Selective abortion, gametic selection, and the X-chromosome. Am. J. Hum. Genet., 24:359, 1972.
23. Fraser, G. R.: The short-term reduction in birth incidence of recessive diseases as a result of genetic counseling after the birth of an affected child. Hum. Hered., 22:1, 1972.
24. Fraser, G. R.: The implications of prevention and treatment of inherited disease for the genetic future of mankind. J. Genet. Hum., 20:185, 1972.
25. Fraser, G. R.: Genetical implications of ante-natal diagnosis. Ann. Genet. (Paris), 16:5, 1973.
26. Matsunaga, E.: Effect of changing parental age patterns on chromosomal aberrations and mutations. Soc. Biol., 20:82, 1973.
27. Lubs, H. A., and Lubs, M. -L.: Genetic diseases. *In* Burrow, G. N., and Ferris, T. F.

(eds.): Medical Complications of Pregnancy. Philadelphia, W. B. Saunders Company (in press).

28. Levy, H. L.: Newborn screening for metabolic disorders. N. Engl. J. Med., 288:1299, 1973.

29. Massachusetts Department of Public Health: Cost-benefit analysis of newborn screening for metabolic disorders. N. Engl. J. Med., 291:1414, 1974.

30. Bush, J. W., Chen, M. M., and Patrick, D. L.: Health status index in cost-effectiveness; analysis of PKU program. In Berg, R. L. (ed.): Health Status Indexes. Chicago, Hospital Research and Educational Trust, 1973, p. 172.

31. Raine, D. N.: Inherited metabolic disease. Lancet, 2:996, 1974.

32. Buckton, K. E., Jacobs, P. A., Court-Brown, W. M. et al.: A study of the chromosome damage persisting after X-ray therapy for ankylosing spondylitis. Lancet, 2:676, 1962.

33. Smith, P. G. and Jacobs, P. A.: Incidence studies of constitutional chromosomal abnormalities in the post-natal population. Pfizer Med. Monogr., 5:160, 1970.

34. Jacobs, P. A., Brunton, M., Melville, M. M. et al.: Aggressive behaviour, mental subnormality, and the XYY male. Nature, 208:1351, 1965.

35. Jacobs, P. A., Price, W. H., Court-Brown, W. M. et al.: Chromosome studies on men in a maximum security hospital. Ann. Hum. Genet., 31:339, 1968.

36. Casey, M. D., Blank, C. E., Street, D. R. K. et al.: YY chromosomes and antisocial behaviour. Lancet, 2:859, 1966.

37. Court-Brown, W. M.: Males with an XYY sex chromosome complement. J. Med. Genet., 5:341, 1968.

38. Wright, S. W., Crandall, B. F., and Boyer, L.: Perspectives in Cytogenetics. Springfield, Ill., Charles C Thomas, Publisher, 1972.

39. Court-Brown, W. M.: Human Population Cytogenetics. Amsterdam, North-Holland Publishing Co., 1967.

40. Hamerton, J. L.: Human Cytogenetics. Vol. 1. New York Academic Press, Inc., 1971.

41. Hook, E. B.: Behavioral implications of the human XYY genotype. Science, 179:139, 1973.

42. Editorial: What becomes of the XYY male? Lancet, 2:1297, 1974.

43. Money, J.: Behavior genetics: Principles, methods, and examples from XO, XXY, and XYY syndromes. Semin. Psychiatry, 2:11, 1970.

44. Clark, D. F., and Johnston, A. W.: XYY individuals in a special school. Br. J. Psychiatry, 125:390, 1974.

45. Mellman, W. J.: Chromosomal screening of human populations. A bioethical prospectus. In Bergsma, D. (ed.): Ethical, Social and Legal Dimensions of Screening for Human Genetic Disease. Vol. 10. New York, Stratton Intercontinental Medical Book Corp., 1974, p. 123.

46. Myhre, S. A., Ruvalcaba, R. H. A., Johnson, H. R. et al.: The effects of testosterone treatment in Klinefelter's syndrome. J. Pediatr., 76:267, 1970.

47. Caldwell, P. D., and Smith, D. W.: The XXY (Klinefelter's) syndrome in childhood: Detection and treatment. J. Pediatr., 80:250, 1972.

48. Barr, M. L., Sergovich, F. R., Carr, D. H. et al.: The triple-X female: An appraisal based on a study of 12 cases and a review of the literature. Can. Med. Assoc. J., 101:247, 1969.

49. Di Sant'Agnese, P. A., and Talamo, R. C.: Pathogenesis and physiopathology of cystic fibrosis of the pancreas. Fibrocystic disease of the pancreas. N. Engl. J. Med., 277:1287, 1967.

50. Glanzmann, E., and Berger, H.: Uber Meconium-Ileus. Ann. Paediatr., 175:33, 1950.

51. Buchanan, D., and Rapoport, S.: Chemical comparison of normal meconium and meconium from a patient with meconium ileus. Pediatrics, 9:304, 1952.

52. Green, M. N., Clarke, J. T., and Shwachman, H.: Studies in cystic fibrosis of the pancreas: Protein pattern in meconium ileus. Pediatrics, 21:635, 1958.

53. Wiser, W. C., Beier, F. R.: Albumin in the meconium of infants with cystic fibrosis: A preliminary report. Pediatrics, 33:115, 1964.

54. Green, M. A., and Shwachman, H.: Presumptive tests for cystic fibrosis based on serum protein in meconium. Pediatrics, 41:989, 1968.

55. Stephan, U., Busch, E. -W., Kollberg, H. et al.: Cystic fibrosis detection by means of test-strip. Pediatrics, 55:35, 1975.

56. Hellsing, K., and Kollberg, H.: Analysis of albumin in meconium for earlier detection of cystic fibrosis. A methodological study. Scand. J. Clin. Lab. Invest., 33:333, 1974.

57. Lockhart, L. H., and Bowman, B. H.: Assay for detection of the cystic fibrosis gene. Tex. Rep. Biol. Med., 31:631, 1973.

58. Bowman, B. H., Hirschhorn, K., and Bearn, A. G.: Bioassays of cystic-fibrosis factor. Lancet, 1:404, 1974.

59. Klein, A. H., Meltzer, S., and Kenny, F. M.: Improved prognosis in congenital hypothyroidism treated before age three months. J. Pediatr., 81:912, 1972.

60. Raiti, S., and Newns, G. H.: Cretinism: Early diagnosis and its relation to mental prognosis. Arch. Dis. Child., 46:692, 1971.

61. Dussault, J. H., and Laberge, C.: Dosage de la thyroxine (T4) par méthode radioimmunologique dans l'éluat de sang sèche: Nouvelle méthode de dépistage de l'hypothyroide néonatale. L'Union Méd., 102:2061, 1973.

62. O'Donnell, J., Frankl, A., and Walfish, P. G.: Screening for neonatal hypothyroidism using dried capillary blood: Observations on sample collection and hematocrit variation. Abstract, 50th Annual Meeting, American Thyroid Assoc., St. Louis, Missouri, Sept. 1974, p. T-22.

63. Dussault, J. H., Coulombe, P., and Laberge, C.: Preliminary report on a screening program for neonatal hypothyroidism. Abstract, 50th Annual Meeting, American Thyroid Association, St. Louis, Missouri, Sept. 1974, p. T-22.

64. Hershman, J. M., and Pittman, J. A., Jr.: Control of thyrotropin secretion in man. N. Engl. J. Med., 285:997, 1971.

65. Hayek, A., Maloof, F., and Crawford, J. D.: Thyrotropin behavior in thyroid disorders of childhood. Pediatr. Res., 7:28, 1973.

66. Klein, A. H., Agustin, A. V., and Foley, T. P., Jr.: Successful laboratory screening for congenital hypothyroidism. Lancet, 2:77, 1974.

67. Fagerhol, M. K., and Gedde-Dahl, T., Jr.: Genetics of the Pi serum types: Family studies of the inherited variants of serum $alpha_1$-antitrypsin. Hum. Hered., 19:354, 1969.

68. Fagerhol, M. K., and Laurell, C.-B.: The polymorphism of "prealbumins" and α_1-antitrypsin in human sera. Clin. Chim. Acta., 16:199, 1967.

69. Kueppers, F., and Black, L. F.: α_1-antitrypsin and its deficiency. Am. Rev. Resp. Dis., 110:176, 1974.

70. Sharp, H. L., Bridges, R. A., Krivit, W. et al.: Cirrhosis associated with alpha$_1$-antitrypsin deficiency: A previously unrecognized inherited disorder. J. Lab. Clin. Med., 73:934, 1969.

71. Webb, D. R., Hyde, R. W., Schwartz, R. H. et al.: Serum α_1-antitrypsin variants: Prevalence and clinical spirometry. Am. Rev. Resp. Dis., 108:918, 1973.

72. Reed, T. E., and Chandler, J. H.: Huntington's chorea in Michigan. I. Demography and genetics. Am. J. Hum. Genet., 10:201, 1958.

73. Perry, T. L., Hansen, S., and Lesk, D.: Plasma amino acid levels in children of patients with Huntington's chorea. Neurology (Minneap.), 22:68, 1972.

74. Perry, T. L., Hansen, S., and Kloster, M.: Huntington's chorea. Deficiency of γ-aminobutyric acid in brain. N. Engl. J. Med., 288:337, 1973.

75. Bird, E. D., and Iversen, L. L.: Huntington's chorea-post-mortem measurement of glutamic acid decarboxylase, choline acetyltransferase, and dopamine in basal ganglia. Brain, 97:457, 1974.

76. Klawans, H. L., Jr., Paulson, G. W., Ringel, S. P. et al.: Use of L-dopa in the detection of presymptomatic Huntington's chorea. N. Engl. J. Med., 286:1332, 1972.

77. Cawein, M., and Turney, F.: Test for incipient Huntington's chorea. N. Engl. J. Med., 284:504, 1971.

78. Aminoff, M. J., Trenchard, A., Turner, P. et al.: Plasma uptake of dopamine and 5-hydroxytryptamine and plasma-catecholamine levels in patients with Huntington's chorea. Lancet, 2:1115, 1974.

79. Harper, P. S.: Presymptomatic detection and genetic counseling in myotonic dystrophy. Clin. Genet., 4:134, 1973.

80. Bundey, S., and Carter, C. O.: Genetic heterogeneity for dystrophia myotonica. J. Med. Genet., 9:311, 1972.

81. Bundey, S.: Detection of heterozygotes for myotonic muscular dystrophy. Clin. Genet., 5:107, 1974.

82. Bundey, S., Carter, C. O., and Soothill, J. F.: Early recognition of heterozygotes for the gene for dystrophia myotonica. J. Neurol. Neurosurg. Psychiatry, 33:279, 1970.

83. Bergsma, D. (ed.): Ethical, Social and Legal Dimensions of Screening for Human Genetic Disease. Vol. 10. New York, Stratton Intercontinental Medical Book Corp., 1974.

84. Milunsky, A., and Reilly, P.: Medicolegal aspects of prenatal genetic diagnosis. Am. J. Law Med., (in press).

85. Hilton, B., Callahan, D., Harris, M., Condliffe, P., and Berkley, B.: Ethical Issues in Human Genetics. Genetic Counseling and the Use of Genetic Knowledge. Plenum Publishing Corp., New York, 1973.

86. Milunsky, A.: The Prenatal Diagnosis of Hereditary Disorders. Springfield, Ill., Charles C Thomas, Publisher, 1973.

87. Waltz, J. R., and Thigpen, C. R.: Genetic screening and counseling: The legal and ethical issues. Northwestern Univ. Law Rev., 68:696, 1973.

88. Veatch, R. M.: Ethical issues in genetics. In Steinberg, A. G., and Bearn, A. G. (eds.): Progress in Medical Genetics. Vol. 10. New York, Grune & Stratton, Inc., 1974.

89. Green, H. P., and Capron, A. M.: Issues of law and public policy in genetic screening. In Bergsma, D. (ed.): Ethical, Social and Legal Dimensions of Screening for Human Genetic Disease. Vol. 10. New York. Stratton Intercontinental Medical Book Corp., 1974, p. 57.

CHAPTER 1

THE CAUSES AND PREVALENCE OF MENTAL RETARDATION

Aubrey Milunsky, MB.B.Ch., M.R.C.P., D.C.H.

CAUSES

Recognition of the cause must rank as the first step in the prevention of mental retardation. Perusal of the etiologic classification (Table 1-1) of mental retardation confirms that this is no simple task. Moreover, efforts are complicated by the fact that in the past, the etiology was unknown in 60 to 70 per cent of the cases.[1] Berg surveyed 800 *severely* retarded institutionalized patients and noted that a definite cause or distinct syndrome could be recognized in only a third of these carefully studied cases.[1] More recently, in Boston at the Fernald State School, studies of 1077 retarded patients with I.Q.'s <50, showed that no etiologic diagnosis could be made in 34 per cent of the cases.[2] Various types of acquired brain disease were found to be most common (26 per cent) in this institutional study. Chromosome disorders, chiefly Down's syndrome, were next most frequent (23 per cent). Known metabolic and endocrine disorders were noted in only 3.5 per cent of the cases. Crome and Stern have emphasized the likelihood that uncertainty about the cause is even greater in patients *not* as severely affected.[3] Their observations highlight the necessity for an appreciation of the nosology of mental retardation.

Moser and Wolf[2] and others[4] addressed themselves to the relationship between etiology, intelligence, and nosology in mental retardation. They recognized the great importance of adverse sociocultural fac-

TABLE 1-1. ETIOLOGIC CLASSIFICATION OF MENTAL RETARDATION

Genetic

TYPE	EXAMPLE
Chromosomal abnormalities	Down's syndrome, trisomy 18, trisomy 13
Disorders of amino acid metabolism	Phenylketonuria, maple-syrup urine disease
Disorders of mucopolysaccharide metabolism	Hunter's or Hurler's syndrome
Disorders of lipid metabolism	Tay-Sachs disease
Disorders of carbohydrate metabolism	Fucosidosis, galactosemia
Disorders of purine metabolism	Lesch-Nyhan syndrome
Miscellaneous inborn errors of metabolism	I-cell disease
Consanguinity, incest, etc.	
Hereditary degenerative disorders	Schilder's disease, retinal degeneration, etc.
Hormonal deficiency	Congenital hypothyroidism, pseudohypoparathyroidism
Hereditary syndromes or malformations	Primary microcephaly, X-linked hydrocephalus
Neuroectodermatoses	Tuberous sclerosis
Unknown	

Acquired

Prenatal	
Infection	Rubella, toxoplasmosis, cytomegalic inclusion disease
Irradiation	Microcephaly
Toxins	Ethyl alcohol, mercury
Unknown	Malformations, placental insufficiency
Perinatal	
Prematurity	
Anoxia	Birth injuries, hypoglycemia
Cerebral damage	Hemorrhage, trauma, infection
Infection	Meningitis, encephalitis
Postnatal	
Brain injuries	Accidents, hemorrhage from coagulation defects or other cerebrovascular accidents, thrombosis, ruptured aneurism
Infection	Meningitis, encephalitis, brain abscess
Anoxia	Cardiac arrest, hypoglycemia, respiratory distress syndrome
Poisons	Lead, mercury, carbon monoxide
Hormonal deficiency	Hypothyroidism
Metabolic	Hypernatremia, hypoglycemia
Postimmunization encephalopathy	Rabies, pertussis, smallpox
Sociocultural	Deprivation
Kernicterus	
Epilepsy	

tors and the expected variability of human intelligence in the mildly retarded (I.Q. >50) group. Patients with moderate, severe, or profound retardation (I.Q. <50) constitute a second group in whom the prevalence of pathologic conditions increases as the I.Q. decreases. Autopsy

TABLE 1-2. DIAGNOSTIC CLASSIFICATION, BY APPARENT MECHANISM, OF PATIENTS WITH MENTAL RETARDATION*†

	NUMBER	PER CENT
Hereditary Issues	53	5
Single-gene abnormalities with biochemical markers, "inborn errors of metabolism"		
e.g., PKU, Hurler's disease, galactosemia		
Other genetic syndromes with variable retardation aspects		
e.g., Muscular dystrophy, neurofibromatosis		
Special chromosomal abnormalities		
e.g., Translocation Down's syndrome		
Familial retardation of probable polygenic origin		
Early Influences on Embryonic Development	366	34
Sporadic preconception germ cell changes (? mutations)		
e.g., Trisomy 21 (Down's syndrome), other trisomies, and other chromosomal anomaly syndromes		
Aberrations occurring during embryogenesis		
"Prenatal influence" syndromes, with structural changes		
from infection; *e.g.,* congenital rubella, C.I.D.		
from drug ingestion; *e.g.,* thalidomide		
from early uterine bleeding		
from other (obscure) causes; *e.g.,* syndromes of Moebius, Goldenhar, Apert, and others		
Other Pregnancy Problems and Perinatal Morbidity	115	11
Placental anomalies, multiple births, postmaturity, "fetal malnutrition"		
Prematurity, birth trauma, and newborn adjustments, including asphyxia, hypoglycemia, and hyperbilirubinemia		
Acquired Childhood Diseases	42	4
Infection		
e.g., Encephalitis, meningitis, abscess, thrombosis		
Endocrine disorders		
e.g., Hypothyroidism		
Intoxications		
e.g., Lead poisoning		
Head trauma		
Intracranial tumors		
Special sensory handicaps		
e.g., Deafness, blindness		
Cardiac arrest, complications of anesthesia		
Environmental and Social Problems	201	19
Deprivation		
Parental neurosis, psychosis, character disorder		
Childhood neurosis, mother/child conflict		
Childhood psychosis		
Unknown Causes	281	27
Total	1058	100

*Per cent incidence is indicated, based on a representative hospital referral experience of 1058 children with borderline to severe mental retardation at the Developmental Evaluation Clinic, Children's Hospital Medical Center, Boston, 1967–1974.

†Data kindly provided by Dr. Allen C. Crocker, Director.

evidence has indeed shown that most individuals with an I.Q. < 50 have demonstrable brain pathology.[5]

Elucidating the exact etiology of mental retardation in many cases may be extremely difficult and further compounded by such major effects as genetic/environmental interaction, low birth weight, prematurity, and perinatal complications. In addition, the causes of mental retardation, congenital malformations, and other developmental disabilities not only overlap but also are inextricably related. These complex inter-relationships are evidenced by the diagnostic categories used in the Developmental Evaluation Clinic at the Children's Hospital Medical Center in Boston (Table 1–2).

Any attempt to correlate the neuropathologic and clinical features of mental retardation is likely to prove inconclusive because of the multiple variables in ascertainment, age, terminology, extent of involvement, type of disease, heterogeneity, and the presence of different lesions in the same brain.[3] Indeed, epidemiologic and other efforts to determine the cause of mental retardation are beset by a whole host of complicating and confounding factors (Table 1–3).

Collins and Turner surveyed the birth weights of 1345 mentally retarded children (excluding those with Down's syndrome)–75 per cent of whom had I.Q.'s < 50.[6] Small-for-dates babies (i.e., those with a reduced rate of fetal growth) occurred in 14.4 per cent of the cases. In contrast, they noted that the incidence of prematurity alone was only

TABLE 1–3. EPIDEMIOLOGIC AND OTHER FACTORS COMPLICATING DETERMINATION OF THE CAUSES AND PREVALENCE OF MENTAL RETARDATION

Lack of specific knowledge about disease in question
Compounding/confusing environmental factors
Insufficiently developed technology
Failure to obtain/examine autopsy reports, family photographs, X-rays, hospital
    records and medical records
Failure of physician to perform complete laboratory investigations
Absent/unreliable family history
Genetic heterogeneity
Varying criteria for diagnosis
Ascertainment bias
Sampling error
Insufficient cases
Lack of homogeneity of sample
Absence of adequate controls
Statistical artifacts
Unrecognized seasonal disease and variable geographic distribution
Compound genetic disease
Failure to recognize consanguinity
Underestimation of sociocultural deprivation
Occurrence of phenocopies
Lack of awareness of environmental agents, such as drugs and poisons
Unknown role (if any) of cytoplasmic inheritance

marginally higher than that found in the normal population. The association of very low birth weight with severe mental retardation and cerebral palsy is well established.[7] Premature and full-term infants with low birth weight may also have associated malformation syndromes. If mental retardation is observed later, the question arises whether the factor(s) causing the malformation(s) also caused both the low birth weight *and* the retardation, or whether the low birth weight was the factor causing the mental retardation.

Drillien[8] addressed this question in careful prospective studies of nearly 300 infants with birth weights of 2000 grams or less and grouped these infants into three main etiologic categories:

1. *Those affected by adverse factors in early gestation* (Prenatal diagnosis and selective abortion would help diminish the numbers in this group).

2. *Those subjected to adverse factors in late pregnancy.*

3. *Those delivered prematurely without known cause.*

She emphasized that the frequency of mental retardation as a consequence of low birth weight or prematurity can accurately be assessed only by comparative evaluation of infants, and by clear recognition of such etiologic categories.

The many genetic diseases of ethnic origin (Table 1–4) provide cogent reasons for careful pedigree analysis during diagnostic evaluations. On occasion, recognition of the ethnicity of a particular disorder in a patient distant from his or her land of origin may suddenly resolve a diagnostic impasse (*e.g.*, porphyria; acatalasia).

PREVALENCE

A large number of factors have a marked effect on the determination of prevalence rates of mental retardation[11] and include

1. Age of individuals studied
2. I.Q. range
3. Calculation of affected children who have died
4. Definition of mental retardation that includes not only low I.Q. individuals, but also those with impairment of adaptive behavior
5. Socioeconomic class
6. Ethnic group
7. Family history
8. Rural or urban population

In 1953 Tizard concluded from various investigations that in various countries from 1 to 4 per cent of the population is considered educationally subnormal and that an additional 6 to 9 per cent of persons so handicapped require special assistance within the normal school system.[12]

TABLE 1-4. ETHNICALLY RELATED GENETIC DISORDERS*

Ethnic Group	Disorders Found with Relatively High Frequency
Africans	Hemoglobinopathies, especially Hb S, Hb C, α and β-thalassemia, persistent Hb F
	G6PD deficiency, African type
	Adult lactase deficiency
Afrikaners (South Africans)	Variegate porphyria
Armenians	Familial Mediterranean fever
Ashkenazic Jews	Abetalipoproteinemia
	Bloom's syndrome
	Dystonia musculorum deformans (recessive form)
	Familial dysautonomia
	Factor XI (PTA) deficiency
	Gaucher's disease (adult form)
	Iminoglycinuria
	Meckel's syndrome
	Niemann-Pick disease
	Pentosuria
	Spongy degeneration of brain
	Stub thumbs
	Tay-Sachs disease
Chinese	α-thalassemia
	G6PD deficiency, Chinese type
	Adult lactase deficiency
Eskimos	$E_1{}^s$ (pseudocholinesterase deficiency)
Finns	Congenital nephrosis
Irish	Neural tube defects
Japanese (Koreans)	Acatalasia
	Oguchi's disease
	Dyschromatosis universalis hereditaria
Mediterranean peoples	Thalassemia (mainly β)
(Italians, Greeks,	G6PD deficiency, Mediterranean type
Sephardic Jews)	Familial Mediterranean fever
	Type III glycogen storage disease
Norwegians	Cholestasis-lymphedema

*Enlarged from McKusick, V. A.: The ethnic distribution of disease in the United States. J. Chronic Dis., 20:115, 1967, and McKusick, V. A.: Ethnic distribution of disease in non-Jews. Isr. J. Med. Sci., 9:1375, 1973.

More recently, McDonald retrospectively surveyed all severely mentally retarded children (I.Q. <50) born in Quebec province in 1958[13] and living there a decade later. She observed an incidence of 3.8 per 1000 children aged about 10 years in a study of 507 cases with almost complete ascertainment (256 children with severe retardation had died before 10 years of age, thereby yielding a more accurate incidence of 5.4 per 1000 live births in a cohort of 763 cases). These results were consistent with previously reported studies of children in the same I.Q. category[14-25] (Table 1-5).

McDonald estimated the respective contributions of various causes of severe mental retardation in the 507 cases she studied, as follows[13]:

Factor	Percentage
Down's syndrome	23
Single recessive genes	21
Postnatal disease	8–11
Postnatal injury	1
Low birth weight	3
Perinatal damage	6–9
Kernicterus	1–2
	63–70

In about one third of the cases no "clue to the cause" was evident. Costeff *et al.*, in contrast, were unable to determine the cause(s) in 58 per cent of 587 cases with severe mental retardation.[26]

The incidence of major congenital malformations (with or without mental retardation) has been estimated to be between 2.1 and 3.3 per cent.[27-30] By Polani's calculations, single gene disorders occur in 1.3 to 1.7 per cent of all infants. Chromosomal abnormalities affect at least 0.5 per cent of infants.[31] For the U.S.A. alone this means over 100,000 births each year with major birth defects. Neel pointed out that the real frequency of genetic disease becomes more apparent with increasing age;[30] hence, newborn surveys or those taken during the first year of life

TABLE 1–5. PREVALENCE OF SEVERE RETARDATION (I.Q. < 50) IN VARIOUS COUNTRIES

Year	Place	Age (in years)	All Causes (%)
1973	New South Wales, Australia[14]	14–24	3.72
1971	London, England[15]	5–14	3.6–4.0
1966–69	Quebec, Canada[13]	8–12	3.84
1968	Northeast Scotland[25]	all ages	2.16
1968	N. Ireland[16]	10–14	4.11
1964	Wessex Co., Boros[17] Counties, England	15–19	3.54 3.84
1962	Aberdeen, Scotland[18]	8–10	3.7
1961	Salford, England[19]	15–19	3.64
1960	Middlesex, England[20]	10–14	3.61
1959	Rural Sweden[21]	all ages	5.8
1955	Onondaga Co., U.S.A.[22]	5–17	3.6*
1936	Baltimore, U.S.A.[23]	10–14	3.3
1925–29	England & Wales[24]	10–14	4.35

*Estimated

TABLE 1–6. SERIOUS MALFORMATIONS IN 12,000 CONSECUTIVE NEWBORN INFANTS RECOGNIZED IN FIRST FIVE DAYS OF LIFE*

	Total Number	% of Total Malfor- mations	% of Total Number of Newborns
LOCALIZED MALFORMATIONS due to:			
Multifactorial inheritance	70	37.2	0.58
Single mutant genes	41	21.8	0.34
aniridia	1		
brachydactyly, type A₁	1		
microtia, bilateral	2		
polydactyly, postaxial, type A	2		
polydactyly, postaxial, type B	32		
polysyndactyly, postaxial and preaxial type	1		
polysyndactyly, postaxial only	1		
split hand/foot syndrome, type I	1		
Unknown etiology	50	26.6	0.42
MULTIPLE MALFORMATIONS due to:			
Chromosome abnormalities	11	5.8	0.09
Single mutant genes	6	3.3	0.05
Unknown etiology	10	5.3	0.08
Total	188	100	1.56

*Courtesy of Holmes, L. B.: Inborn errors of morphogenesis—a review of localized hereditary malformations. N. Engl. J. Med., 291:763, 1974.

would clearly underestimate the frequency of genetic disease, mental retardation, and even major malformations.

Holmes has thus far examined 12,000 consecutively born infants in the first 5 days of life at the Boston Hospital for Women.[32] His data on serious malformations (Table 1–6) (which included a significant number of disorders where psychomotor retardation is a prominent later feature), indicated an incidence of 1.56 per cent. Two infants born and included in his table were diagnosed by us *in utero* (Trisomy 18; 47 XXY). In four other cases pregnancies were terminated following our antenatal diagnosis of unbalanced D/G translocation, fetal rubella, trisomy 18 and Hunter's syndrome. Another infant subsequently developed the adrenogenital syndrome. Without prenatal diagnosis, consideration of these additional cases would have made an incidence of malformations/genetic disease of 1.6 per cent.

Polani estimated that 6 per cent, or more, of all persons "suffer from developmental disorders manifest at birth or in early life, ranging from mild to severe, from curable through remediable, to untreatable or even lethal."[28] When extrapolated to the U.S. population, an astounding minimum figure of 12 to 15 million individuals is reached. When Meier, Chapter 17, points to the increasing frequency of developmental disabilities, our collective credence must give way to activating anxiety.

GENETICS OR ENVIRONMENT

Almost 2000 different genetic disorders have been recognized and catalogued by McKusick.[33] While a significant percentage of these disorders have associated mental retardation, appreciable difficulties may exist in some instances in differentiating acquired from genetic causes of mental retardation (Table 1–1). Indeed, more causes of mental retardation are being identified as genetic. For example, Davison[34] and Swift[35] have elucidated as genetic the causes of mental retardation in a few families where more than one member was affected.

Infants with Down's syndrome or those affected by rubella, cytomegalic inclusion disease, and toxoplasmosis are often of low birth weight. More commonly, low-birth-weight infants have an excess of both major and minor congenital abnormalities—a fact which suggests that both prematurity and intrauterine growth retardation are due to adverse factors in early pregnancy.[36] The delineation of the degree of the genetic component in such circumstances poses recognizable difficulties. Hereditary components may be involved in prematurity per se, in the congenital anomalies, and possibly in placental abnormalities.

There is little doubt that birth injury to the brain (anoxia, intracranial hemorrhage or thrombosis, and mechanical brain damage) may cause mental retardation. It is equally clear that offspring with congenital cerebral anomalies are more likely to be distressed at birth. Hence, the same basic factors—environmental or genetic—may be responsible for the cerebral defect and the susceptibility to brain damage.

Butler and Alberman,[37] in their perinatal mortality studies, determined those factors most commonly associated with perinatal death:
1. Low income
2. Birth order
3. Maternal age
4. Geographical location
5. Past history of abortion/stillbirths/premature infants
6. Toxemia
7. Antepartum bleeding
8. Malposition
9. Dystocia
10. Instrumental or cesarian section birth

It is a reasonable assumption that these same factors, with their effects perhaps operative to a lesser degree, could be related to the causation of some cases of mental retardation.

PROBLEMS IN DISCERNING CAUSES

Some of the factors complicating the determination of the causes of mental retardation (or more properly, of "developmental disabilities") have been alluded to in Table 1–3. Even in the presence of a known or suspected etiology, complex problems exist in defining or evaluating true causal relationships. A selection of such problems is considered in this section.

Infectious Disease

Intrauterine infection by rubella, cytomegalovirus, or toxoplasmosis occurs with greater frequency than was previously suspected.[38-41] Stern *et al.*,[42] in a study of mentally retarded children living at home, showed that rubella and toxoplasmosis together were probably responsible for about 2 to 3 per cent of their cases with mental deficiency. Further, they observed that cytomegalovirus was associated with 10 per cent of cases with microcephaly and mental retardation. The problem in discerning the causes of mental retardation from microbial agents is compounded by the common occurrence of subclinical or inapparent infection with unrecognized disease and is confused by asymptomatic carrier states. Nevertheless, such inapparent disease can still cause cerebral infection and subsequent mental retardation and microcephaly. In 12 newborns with minor or inapparent cytomegalovirus infection, at least four were diagnosed as mentally retarded—even at the early age of 6 months. In twin pregnancies both toxoplasmosis[43] and cytomegalic inclusion disease[44] commonly affect both members of the pair. Exceptions—with only one twin affected—have been reported for both cytomegalic inclusion disease[45] and rubella.[46]

Cytomegalovirus Infection

This virus is the commonest known microbial cause of mental retardation in childhood.[42] Difficulties arise in discerning microbial causes of mental retardation when the infection is endemic. Sever *et al.*,[47] using serologic methods, demonstrated cytomegalovirus infection in as many as 6 per cent of pregnant women. About 3 to 6 per cent of pregnant women have viruria during pregnancy.[48-51] Cervical excretion has been found to be much commoner, with rates varying from 4 to 28 per cent during pregnancy.[51, 52] Fetal infection follows in only a small fraction of these cases. Viral excretion occurs in 0.5 to 2.0 per cent of newborns,[39-41] of whom only about 1 in 3000 develops the classical syndrome of cytomegalic inclusion disease.[53]

Patient age is important when serologic methods are used during investigations for congenital cytomegalovirus infection as the cause of mental retardation, since the incidence of antibodies increases rapidly with age.[54] Hence, postnatal cytomegalovirus infection is a considerable problem in institutionalized patients and may prevent the accurate determination of congenital infection; *e.g.*, positive serology in a child over 6 years of age may be uninterpretable. Moreover, reinfection or reactivation by cytomegalovirus is known to occur very commonly in pregnancies.[50-52] The presence of circulating antibodies in the maternal and cord serum, therefore, may offer no protection against intrauterine infection. Indeed, the presence of circulating antibodies may modify the disease with gross abnormalities in the first and inapparent infection in the second offspring![55]

Rubella Infection

Mental retardation as a sequel of congenital rubella infection has been estimated to occur in 1.5 to 9.3 per cent of cases.[56-58] Subclinical or unrecognized maternal infection is probably at least as common as clinical illness and can also cause fetal damage.[42] During diagnostic evaluation of a child the absence of a maternal history of infection, therefore, should not interdict serologic rubella studies.

Whereas rubella reinfection after natural disease may be very uncommon, this phenomenon is not at all unusual following immunization. In seven previously immunized women with rubella reinfection, three affected offspring (two aborted) were observed.[59-61] Diagnostic evaluation of a young child with congenital defects and/or mental retardation should include rubella studies despite known maternal immunization.

Toxoplasmosis

As with cytomegalovirus and rubella infection, many infants congenitally infected with toxoplasmosis may appear healthy at birth or for months to years thereafter.[62-64] The incidence of infection in pregnancy has been estimated to be as high as 4 to 6 per 1000.[38, 64] Congenital toxoplasmosis may be as common as 1 per 1000 births; only a small proportion of these cases are likely to be severe. Postnatal infection (especially in patients in institutions) is likely to bedevil efforts at diagnosis using serologic methods.[65, 66] A useful guide is the experience of Couvreur and Desmonts,[62] who have indicated that mental retardation in the absence of chorioretinitis or microcephaly/hydrocephalus is rarely the result of congenital toxoplasmosis. More recently, these investigators[67] have described a number of instructive cases of infants born to mothers with toxoplasmosis. One infant had a normal antibody titer and was physically healthy until 4 months of age, at which time the titer rose and chorioretinitis and cerebral calcifications were found. Two infants had repeatedly examined normal fundi, but both were discovered to have chorioretinitis at 7 months of age. Another infant, thought to have had subclinical infection, was noted to have a retinal scar in his second year. Alford et al.[68] also observed significantly low I.Q. development in children over 2 years of age who had untreated subclinical infection.

Miscellaneous Infections

The role of other known infectious agents in causing congenital defects, including subsequent mental retardation, is much less well delineated than those discussed previously. A recent Finnish report[69] focuses attention on the association of maternal herpesvirus infections during pregnancy and later delivery of infants with congenital defects. Although earlier studies have suggested that the herpesvirus is teratogenic,[70-72] the Finnish workers wisely concluded from their study that there was no clear causal relationship between herpesvirus infection and later congenital malformation. They suggested[69] that reactivation or reinfection with herpesvirus may occur more readily in women bearing a defective child. Further, they proposed that the factor(s) causing malformations may also be responsible for reactivations of latent infection. These cautious insights have direct pertinence to the interpretation of serologic results obtained in seeking the cause of mental retardation or congenital defects.

Studies from both England[57] and the U.S.A.[73] have concluded that no clear causal relationship exists between congenital malformations and maternal chickenpox, measles, mumps, influenza, poliomyelitis, or hepatitis.

Genetic Heterogeneity

One of the central characteristics of autosomal dominant disease is genetic heterogeneity. The child with the full-blown features of tuberous sclerosis, the infant with seizures and achromic spots, the child with seizures and mental retardation only, and the normal asymptomatic mother with cerebral calcifications, represent examples of known heterogeneity in this disease. By implication, no evaluation of a child with mental retardation of unknown cause would be complete without examination of *both* parents with a Woods light, a retinal study, and radiographs of their skulls.

Heterogeneity is not confined to dominant disease, but is often seen in autosomal and sex-linked recessive disease, as well. The hereditary storage disorders are replete with different forms, having variable onset and severity. In the juvenile and adult forms of these diseases (*e.g.*, metachromatic leukodystrophy) acquired causes should be carefully considered to separate them from those diseases of genetic origins before heterogeneity is invoked. The role of heterogeneity in a genetic disease may, on occasion, be particularly difficult to fathom. For example, Herrlin described 21 year old phenylketonuric uniovular monochorionic, but diamniotic, twins who showed significant differences in intelligence, nature of epilepsy, and electroencephalographic patterns, despite common rearing.[74] Woolf observed an even greater difference in intelligence between identical twins with phenylketonuria.[75] The importance of birth weight, among other factors, makes an assessment of genetic heterogeneity virtually impossible in such situations.

Holmes has pointed to the limitations of clinical observations in delineating patterns of inheritance. He offered five examples in which the same malformation — anophthalmia, intestinal atresia at multiple levels, absent radius, split hand-foot deformity, and multiple vertebral fusion defects — may be caused by different mutant genes.[32]

The Hurler and Scheie syndromes[76] provide an excellent illustration of different mutations leading to varying degrees of deficiency of the *same lysosomal enzyme* (α-L-iduronidase) and significantly different clinical phenotypes. In particular, the contrast is striking — with severe psychomotor retardation in Hurler's syndrome and normal intelligence in Scheie's syndrome.

Twins

Both monozygous and dizygous twins have been found to have intelligence quotients or mental scale scores lower than matched singletons at different ages.[77] Principally prematurity or dysmaturity, with or without low birth weight, may account for these observations. The perinatal mortality rate in twins may be over three times higher (10.9 per cent) than the corresponding rate among singletons (3.3 per cent).[37] Hence, diminished intensity of those factors that are operative in causing perinatal mortality are very likely to cause significant morbidity in twins, as well. Berg and Kirman and others found a higher proportion of twins, than singletons, in institutions for the retarded and noted a disproportionately large number of monozygous twins, as well as second-born twins.[77, 78] They remarked on the higher frequency of Down's syndrome and physical abnormalities among these twins. McDonald observed in her study of 507 children with severe mental retardation that the proportion of twins was about twice that expected, if twins had no greater risk of severe mental retardation than singletons.[13] Similar figures have previously been reported from other surveys.[79] Smythies has, however, emphasized that studies of twins in institutions for the retarded are likely to yield higher concordance rates, particularly those of monozygotic twins, than similar studies in the general population.[80]

Low birth weight may be a reflection of nutritional deprivation *in utero,* with the consequent detriment to brain growth and intellectual development. In one large study, twins and triplets were found to score significantly lower in tests of verbal reasoning, but low birth weight was found to be less important a factor than postnatal rearing together.[81]

General agreement exists about the increased frequency of congenital abnormalities in twins.[82-84] For singletons, the observed rates of anomalies were 1.4 per cent; for twins, 2.7 per cent; and for triplets, 6.09 per cent in a large statistical study using World Health Organization (WHO) criteria.[84] Most cases of anencephaly occurring in twins almost invariably affect only one of the pair.[85-87] Chromosomal abnormalities occur with an increased frequency in twins.[31] Both monozygous and dizygous twins concordant for Turner's syndrome,[88, 89] Klinefelter's syndrome,[90] and Down's syndrome[91] have been recorded. Dizygous twins discordant for the cri-du-chat syndrome[92] and for trisomy 18[93] have been observed. Nielsen concluded that there is a higher than expected frequency of twins among patients with Klinefelter's syndrome.[94] Years earlier, Nance and Uchida observed that in sibships with individuals affected with Turner's syndrome, the frequency of twins was several times greater than in the general population.[95] The Yale–New Haven data, showing an increased frequency of chromosomal abnormalities, is consistent with these reports[31] in twins.

Many factors clearly compound the difficulties of elucidating the cause of mental retardation in one or both members of a twin pair. In addition to the critical role of zygosity, birth weight, birth order, chimerism, placentation, prenatal nutrition, and postnatal child rearing all rank as confounding factors. Even when the cause is known—as with thalidomide ingestion—the embryopathic effect on twins may be very different.[96]

Polygenically Determined Disease

Multifactorial inheritance reflects the additive effects of several minor gene abnormalities and environmental factors.[97] The features of polygenic inheritance have been carefully delineated.[98] Roberts ascribed a polygenic mode of inheritance to the group of feebleminded and educationally subnormal persons, suggesting that they represented the tail of the normal curve of the distribution of intelligence.[99] Almost four decades ago Penrose noted that the frequency of mental retardation among the parents and siblings of the propositus was seven to nine times that found in the general population.[100] Furthermore, he noted that familial incidence was greater among the less severely retarded. In contrast, the more severely retarded were more likely to have equally badly affected close relatives.

Although the data of yesteryear should be considered in terms of the significant technologic advances of recent times, there is sufficient information to suggest that at least mild mental retardation of otherwise unknown cause may not rarely be due to polygenic inheritance. The necessary exhortation would therefore be always to explore the pedigree of the propositus with great care. In searching through the family pedigree one should remember that all that is familial is not necessarily ge-

TABLE 1-7. OFFSPRING OF MOTHERS WITH PHENYLKETONURIA*

Offspring	Number	I.Q. Range	Maternal I.Q. Range	Maternal Phenyl- alanine
Phenylketonuric	7	15–50	49–102	12–26
Normal (non PKU)	16	normal– 108	49–102	< 10–27
Subnormal I.Q. (non PKU) and/or speech defect	38	30–88	< 49–96	14–34
Subnormal I.Q. (non PKU) with microcephaly, growth retardation, and/or malformations	49	12–88	27–103	12–34
Died before examination	10		49–82	20–30
Not studied	1		80	14

*From Howell, R. R., and Stevenson, R. E.: The offspring of phenylketonuric women. Social Biol., 18:19, 1971. Reprinted with permission.

netic! For example, virtually all the offspring of a mother with phenylketonuria (and not on an appropriate diet during pregnancy) may have mental retardation and/or congenital malformations (Table 1–7).[101] Similar malformations may be seen in the offspring of mothers to whom trimethadione was administered.[102]

Seasonal clustering is especially characteristic of disease with an etiologic environmental component and may complicate epidemiologic and other studies to determine causes of mental retardation. Virtual "epidemics" of anencephaly have been recorded in New England.[102] Both in England and in Scotland anencephaly has been found to occur most often in the autumn and winter.[104, 105] Some studies have, however, failed to confirm these observations.[106, 107]

Dislocation of the hip,[108] patent ductus arteriosus,[109] coarctation of the aorta,[110] and ventriculoseptal defect[111] may all occur with regular seasonal variation. The seasonal occurrence of rubella may account for the variation in occurrence of patent ductus arteriosus. Various environmental influences, suggested to cause nondisjunction, include infection, radiation, emotional stress, increased maternal androgenic hormones, fluoride content of water, and even atmospheric pollution.[112]

Consanguinity

Morton concluded that homozygosity at any one of 69 loci may result in low-grade mental defect, and that about 8 per cent of such cases are recessive.[112a] He estimated, further, that about a third of normal persons are heterozygous for a gene for low-grade mental defect.[112a] Dewey et al. estimated that there were 114 such loci.[112b]

Certainly, the increased morbidity of children from consanguineous marriages has long been recognized.[113, 114] Not unexpectedly, the incidence of incestuous unions, as reported by Penrose, was higher among parents of severely retarded offspring.[113] The devastating adverse effect of incestuous unions on the frequency of congenital malformations and/or mental retardation was also observed in a group of 161 children of such matings.[115] However, much more significant, numerically, are the issues of first and second-cousin marriages. In one unpublished study by Berg,[3] parents who were first cousins had offspring with a mental defect three times more often than expected for the general population. An increase in first cousin marriages among parents of mentally retarded individuals, as compared to the general population, has been noted by a few investigators.[21, 113, 116]

A strikingly high consanguinity rate was observed in a study of 972 retarded patients in Israel, whose parents seemed of normal intelligence and who had had two or more children with severe retardation.[117] The authors pointed out that this phenomenon was not seen among families in which the parents themselves were retarded, or in families where there were two or more *mildly retarded* children. They concluded in their study that autosomal recessive inheritance played a predominant role in severe familial retardation with normal parents and a significant one in isolated idiopathic retardation of all degrees. In India, Sinclair observed consanguinity in 8.3 per cent of the 470 cases (I.Q. < 90) he studied who were under 16 years of age.[118]

Syndrome Recognition

A significant number of malformation syndromes possibly associated with mental retardation and resulting from single mutant genes are already known (Table 1–8). The high risk of recurrence of these disorders and possibilities for subsequent prenatal diagnosis (Table 1–8) makes autopsy of all deformed stillborns and newborns imperative, so that accurate genetic counseling may be provided. Recognition that a certain constellation of signs constitutes a recognizable syndrome has important implications for both the understanding of the cause and the subsequent genetic counseling. Malformations resulting from environmental teratogens, such as Warfarin[133] and thalidomide, must first be recognized before separation from genetic origins is possible. Mere recognition of the "syndrome," however, does not confer upon the genetic counselor the ability to arrive at a recurrence risk automatically. For example, similar congenital defects may occur despite differences in the mode of inheritance. An autosomal dominant mutant gene may account for aniridia occurring as an isolated malformation, whereas aniridia, combined with Wilm's tumor and other congenital abnormalities,[119] may

TABLE 1–8. SELECTED GENETIC MALFORMATION SYNDROMES*
ASSOCIATED WITH MENTAL RETARDATION AND POTENTIALLY
DIAGNOSABLE *IN UTERO* BY ULTRASOUND, RADIOGRAPHY,
AMNIOGRAPHY, OR FETOSCOPY

Malformation Syndrome	Mode of Inheritance†	Associated Malformations	References
Anophthalmia	X/R	Microcephaly, anophthalmia	134
Apert's syndrome	A/D	Tower-like skull, syndactyly of hands and feet	135
Aqueductal stenosis	X/R	Hydrocephalus	148
Beckwith-Wiedmann Syndrome	A/R or A/D	Macroglossia, omphalocele & visceromegaly	209
Carpenter's syndrome	A/R	Acrocephaly, syndactyly of third to fourth fingers, polysyndactyly of the toes	137
Cephaloskeletal dysplasia	A/R	Microcephaly, skeletal dysplasia	192
Cerebrohepatorenal syndrome	A/R	Glaucoma and other ocular abnormalities, hypotonia, high forehead, flexion contractures of the fingers, stippled patellae and digit epiphyses	136
Chotzen's syndrome	A/D	Wide flat forehead, hypertelorism, ptosis, prognathism, and syndactyly	138
Cockayne's syndrome	A/R	Microcephaly, dwarfism, retinal pigmentation, optic atrophy, and progressive neurologic deterioration	139
Coffin's syndrome	A/D	Hypertelorism, acromegaloid features, abnormalities of the sternum, musculo-skeletal abnormalities	211
COFS syndrome	A/R	Microcephaly, ocular abnormalities, typical facies, and skeletal abnormalities	140
Conradi's disease	A/R	Craniofacial anomalies, short limbs, cataracts, skin manifestations, and stippled epiphyses	141
Craniodigital syndrome	X/R	Unusual facies, syndactyly, skeletal abnormalities	195
Crouzon's disease	A/D	Flat forehead, wide skull, exophthalmos, beaked nose, hypoplastic maxilla	142
Cryptophthalmos syndrome	A/R	Skin-covered eyes, facial genitourinary and skeletal anomalies	143
de Lange syndrome	P/I or A/R	Microcephaly, prominent eyebrows and lashes, syndactyly of the toes, micromelia, hirsutism, and skeletal malformations	144, 156
Dyschondroplasia, facial anomalies, and polysyndactyly	A/R	Keel-shaped skull, macrostomia, micrognathia, polysyndactyly, facial abnormalities	207
Dysmorphogenesis of joints, brain, and palate	A/D	Multiple joint contractures, cleft palate, and Dandy-Walker malformation of the fourth ventricle	208
Ectodermal dysplasia	A/D, A/R, X/R	Frontal bossing, typical facies, under-developed maxilla, skin and hair abnormalities	145
Ectromelia and ichthyosis	A/R	Gross limb anomalies with ipsilateral ichthyosiform erythroderma	205
Ellis–van Creveld syndrome	A/R	Chondrodystrophy, polydactyly, and dysplasia of the hair, teeth, and nails	150
Facial abnormalities and kyphoscoliosis	A/R or X/R	As named	168

(*Table continues on following page.*)

TABLE 1–8. SELECTED GENETIC MALFORMATION SYNDROMES*
ASSOCIATED WITH MENTAL RETARDATION AND POTENTIALLY
DIAGNOSABLE *IN UTERO* BY ULTRASOUND, RADIOGRAPHY,
AMNIOGRAPHY, OR FETOSCOPY (*Continued*)

Malformation Syndrome	Mode of Inheritance†	Associated Malformations	References
Facial dysmorphism and right-sided aortic arch	A/D	Microcephaly, broad forehead, abnormal facies	204
Fanconi's anemia	A/R	Microcephaly, hypogenitalism, brown pigmentation, pancytopenia, skeletal and renal anomalies	146
Goltz's syndrome	X/L/D	Congenital skin defects; abnormal teeth, hair, and nails; microcephaly; ocular abnormalities; skeletal anomalies	147
Hereditary myopathy, oligophrenia, cataract, skeletal abnormalities and hypergonadotrophic hypogonadism	A/R	As named	206
Holotelencephaly with cleft lip and hypotelorism	A/R	Microcephaly, hypotelorism, cleft lip, and upward palpebral slant	160
Hypotonia, obesity, facial, oral, ocular and limb anomalies	?A/R	As named	197
Ichthyosiform erythroderma, hair shaft abnormalities and growth retardation	A/R	As named	201
Incontinentia pigmenti	X/L/D	Microcephaly, skin lesions, ocular abnormalities	149
Laurence-Moon-Biedl syndrome	A/R	Polydactyly, hypogenitalism, and retinitis pigmentosa	151
Leprechaunism	A/R	Elfin facies, hypertelorism, hirsutism, and multiple congenital anomalies	152
Lissencephaly syndrome	A/R	Microcephaly, typical facies, polydactyly	153
Malformed, low-set ears and conductive hearing loss	A/R	Deformities of the pinnae, deafness, genital abnormalities	166
Marinesco-Sjögren syndrome	A/R	Microcephaly, cataracts, limb abnormalities	154
Meckel's syndrome	A/R	Microcephaly, encephalocele, cleft palate, polydactyly, polycystic kidneys	155
Median cleft face syndrome	A/D	Hypertelorism, cleft nose, lip and/or palate	127
Microcephaly	A/R, X/R	Microcephaly	157
Microcephaly, peculiar facies, spasticity, and choreoathetosis	A/R	Microcephaly, odd facies, large ears, spasticity, choreoathetosis	202
Microcephaly, snub nose, livedo reticularis, and dwarfism	A/R	Microcephaly, broad up-turned nose, retinal pigmentation, limb deformities, and livedo reticularis	203
Micromelia and coarse facial features	A/R	Typical facies; cleft palate; short limbs, hands, feet, fingers, and toes	159
Multiple anomalies syndrome	A/R	Cataracts, microcephaly, kyphosis and limited joint involvement	198
Multiple anomalies syndrome	A/R	Unusual facies, cleft palate, and limb abnormalities	196
Multiple anomalies syndrome	A/R	Microcephaly, multiple anomalies	212
Neural tube defects	P/I	Anencephaly, myelomeningocele, encephalocele	158

TABLE 1–8. SELECTED GENETIC MALFORMATION SYNDROMES*
ASSOCIATED WITH MENTAL RETARDATION AND POTENTIALLY
DIAGNOSABLE *IN UTERO* BY ULTRASOUND, RADIOGRAPHY,
AMNIOGRAPHY, OR FETOSCOPY (*Continued*)

Malformation Syndrome	Mode of Inheritance†	Associated Malformations	References
Noonan's syndrome	?AR/?AD/?PI	Short stature, pulmonic stenosis, hypertelorism, ptosis, skeletal abnormalities	213
Oculocerebrorenal syndrome (Lowe)	X/R	High prominent forehead, cataracts, cryptorchidism	161
Oral, cranial, and digital anomalies	A/R	Cleft lip and cleft palate, broad nasal bridge, deformed external nares, hypertelorism, abnormal thumbs	193
Oral-Facial-Digital (OFD) syndrome I	X/L/D	Cleft lip, tongue, and palate, hypoplasia of the nasal cartilages; digital anomalies	214
Oral-Facial-Digital syndrome II	A/R	As in OFD I; and broad, bifid, tip of nose; polysyndactyly	215
Osseous dysplasia	?X/L/D	Microcephaly, peculiar facies, pectus carinatum, hypoplastic genitals, skin lesions, and skeletal deformities	199
Osteopetrosis (infantile)	A/R	Increased bone density, retinal degeneration, hepatosplenomegaly	162
Otopalatodigital syndrome	A/D or X/R	Multiple anomalies, short stature, skeletal abnormalities	216
Phocomelia, flexion deformities, and facial anomalies	A/R	Microcephaly, phocomelia, facial anomalies, cleft lip and palate	217
Pseudotrisomy 18 syndrome	A/R	Prominent occiput or an elongated skull, low-set or malformed ears, neck webbing, flexion deformities of fingers and knees	218
Pycnodysostosis	A/R	Short stature, increased bone density, macrocephaly	164
Robert's syndrome	A/R	Tetraphocomelia, cleft lip and palate, microcephaly, facial anomalies, genital hypertrophy	165
Rothmund-Thomson syndrome	A/R	Lesions of the skin, hair, teeth and nails, microcephaly; ocular and skeletal anomalies	163
Seckel's Bird-Headed Dwarfism	A/R	Microcephaly, short stature, prominent eyes, beaklike nose, narrow face and small chin, skeletal and urogenital anomalies	219
Short stature; ocular and articular anomalies	A/D	As named	200
Smith-Lemli-Opitz syndrome	A/R	Microcephaly, typical facies with bilateral ptosis, genital abnormalities in the male, second and third toe syndactyly	167
Telecanthus-Hypospadias syndrome	X/L/D or A/D	Telecanthus, hypertelorism, hypospadias, cleft lip and palate	194
Treacher Collins syndrome	A/D	Typical facies, ocular abnormalities, external ear deformities	210
Xeroderma pigmentosum	A/R	Skin lesions, microcephaly, hypogonadism	169

*Usually normal karyotypes and non-"metabolic" diseases.
†A/R = Autosomal recessive
 A/D = Autosomal dominant
 X/R = X-linked recessive
 P/I = Polygenic inheritance
 X/L/D = Sex-linked dominant

constitute a nonhereditary syndrome. Holmes has emphasized the difficulty of distinguishing between hereditary and sporadic aniridia, without long-term evaluation to determine the presence or absence of the tumor.[32] There are many other examples where similar quandaries may exist. Varied modes of inheritance may also apply to similar defects. For example, cleft palate, usually due to polygenic inheritance,[98] has been described as being due to an X-linked recessive mutant gene.[120, 121]

A syndrome of pulmonary stenosis, short stature, hypertelorism, ptosis, skeletal abnormalities, and mental retardation was described in 1963.[122] Although these patients were phenotypically similar to Turner's syndrome, their karyotypes were normal. Later it became evident that this described syndrome (Noonan's) could be separated from Turner's syndrome by certain distinguishing physical features.[123, 124] In addition, and in contrast to Turner's syndrome, both polygenic[123-125] and autosomal dominant[126] modes of inheritance have been suggested.

The median cleft face syndrome[127, 128] may be associated with mild to severe mental retardation.[127] While sporadic occurrence has been most usual, in several families autosomal dominant inheritance has been evident.[129, 130] Warkany and associates,[131] however, reported two infant girls with this syndrome who were the offspring of the same 14-year-old mother, but different fathers. None of the parents had the same or similar malformations; the authors, therefore, discounted autosomal dominant inheritance as the cause. Indeed, they were unable to elucidate the etiology of these cases, which nevertheless were probably "genetic." (One wonders about the veracity of the family history obtained from the teen-aged unmarried gravida II mother!) Notwithstanding this parenthetic reservation, Kalter[132] working in mouse genetics, has drawn attention to the occurrence of repeated malformations with a genetic background but not in accord with Mendelian ratios, and in the absence of any recognizable teratogenic factors.

X-Linked Mental Retardation

It has been known for almost 40 years that an excess of males with mental retardation are found in both the hospital and the general population.[21, 113, 170] This male excess is most pronounced among the severely retarded, where the etiology is largely unknown. It should be noted that though birth injury affects predominantly males, this greater incidence is insufficient a reason to account for the disparity between the sexes in subsequent mental retardation. Renpenning *et al.*[171] presented strong evidence for X-linked inheritance based on pedigree data. In the family they studied, males in several generations were severely affected, whereas their mothers were of normal intelligence. Davison meticulously studied 141 families with severe idiopathic mental defect in at least two members.[34] Reliable information was obtained from 98 families. She concluded that there were a remarkable number of pedigrees compatible with an X-linked mode of inheritance in which there were affected males with idiopathic mental defect. Further, she was also able to separate out families where only males were at risk, as well as those where both sexes were equally at risk.

Most recently, evidence from a survey in New South Wales of mentally retarded children with I.Q.'s between 30 and 55 has further suggested the significant frequency of X-linked forms of mental retardation.[172] The author's calculations indicate that in their study 20 per cent of mentally retarded boys in this I.Q. range may be retarded on the basis of X-linked inheritance.[172]

Pedigree analysis, therefore, is very important in determining the X-linked cases that are not usually associated with other clinical abnormalities. Small family size, having only one affected individual, and there being a significant rate of mutation (one third) may well bedevil all efforts at resolving the mode of inheritance. Clearly, this recognition that X-linked retardation occurs with significant frequency is of vital importance, in view of the availability of prenatal sex determination.

The Maternal Milieu

Not surprisingly, the encompassing maternal milieu of the fetus — genetic influences aside — is of critical importance in development. Mothers with both *hypo- and hyperthyroidism* are at increased risk of having offspring with chromosomal abnormalities.[31] Indeed, Lubs[31] has evidence that in maternal hyperthyroidism there may be as much as an eightfold increase in the risk of bearing chromosomally abnormal offspring. The problem of endemic goiter and cretinism — so well discussed by Warkany[96] — is entirely preventable by attention to maternal nutrition.

Undetected *maternal phenylketonuria* still poses problems. Howell and Stevenson reviewed the findings in 121 offspring of 33 women with untreated phenylketonuria[101] (Table 1–7). It should be noted that the records were inadequate to determine the clinical condition of the ten infants who died before this study. Although these authors found 16 children who had been reported as apparently normal, they were able to satisfy themselves in only three cases in which the criteria for a diagnosis of classic phenylketonuria had been met. In only one of these cases had the mother been treated with a low phenylalanine diet during pregnancy. The prevention of almost inevitable mental retardation and/or congenital malformation[101] in the offspring of phenylketonuric mothers is possible only by routine urine testing of prospective mothers.

Maternal diabetes mellitus has a well documented association with the caudal dysplasia syndrome[173, 174] in offspring of women with this condition. Whereas congenital malformations have been reported by some investigators to occur with increased frequency in the children of diabetic mothers,[175, 176] others have noted no significant increase.[177, 178] Clearly, the severity and duration of diabetes and the types of drugs used may easily influence such studies. Recently, Robinson has presented evidence of a significant lowering of intelligence in children of diabetic mothers, especially if the mothers had experienced episodes of acetonuria during pregnancy.[179]

An association between *congenital malformations* and *diphenyl-hydantoin ingestion* by the pregnant mother has been recognized since 1968.[180] Further studies have served to confirm these observations,[181, 182, 183] but not to establish an absolute cause and effect relationship. The numbers of affected children born to such mothers treated during pregnancy is small. The role of the disorder with epilepsy and the importance of actual seizures (single or multiple), still require evaluation before diphenylhydantoin use can be incriminated. Meanwhile, it appears that even though diphenylhydantoin is a folic acid antagonist, the risk of congenital malformations seems independent of maternal folate levels.[184]

Fetal vulnerability to *maternal irradiation* has been recognized for almost 50 years. In studies of 74 infants irradiated *in utero*, Murphy found congenital abnormalities in 25 cases.[185] Seventeen of these infants had microcephaly. Twenty-six children born to mothers irradiated during pregnancy were recorded by Dekaban.[186] The most frequent anomalies were retarded growth, microcephaly, mental retardation, pigmentary degeneration of the retina, cataracts, and genital and skeletal malformations. Effects of background irradiation on the frequency of congenital malformations or mental retardation remain uncertain.[79, 187] Uchida, however, has evidence to suggest that maternal irradiation is more frequent in mothers of children with Down's syndrome and may be causally related.[188]

In 1972 Jones *et al.*[189] reported their observations on the offspring of women with *chronic alcoholism*. In addition to a 17 per cent perinatal mortality rate, they found borderline-to-moderate mental retardation in 44 per cent of the 23 offspring. Although mental deficiency was the most frequent problem encountered among these offspring, 32 per cent had sufficiently abnormal physical features to suggest a "fetal alcohol syndrome." This report was based on chart reviews and not on direct examination of patients. Further, the authors recognized that the intellectual development of these children could have been adversely affected by their environment throughout their early years of life. Indeed, the profound detrimental effects of the disorganized problem family on the psychological and intellectual development of children are well recognized.

Much larger and prospectve studies will be needed to separate out the dysmorphogenic, from environmental, origins of mental deficiency in the children of chronic alcoholic mothers. Until such studies are completed, the case against alcohol as the direct etiologic agent must remain conjectural.

Socioeconomic Factors

The association of poverty and low socioeconomic classes with higher rates of prematurity and perinatal morbidity and mortality has long been recognized. Mental retardation and developmental disabilities as a consequence of prematurity are also only too well known. Since nongenetic factors may contribute as much as one half of the total variance in intelligence scores,[190] the environment plays an important role in the causation of mild retardation. Hence, social deprivation and lack of optimal stimulation — more likely among the poor and disadvantaged — lead, expectedly, to a higher frequency of mild mental retardation in the lower socioeconomic classes.[190] When more than one case of mild mental retardation is observed in one family, major difficulties may ensue from efforts to separate out genetic from environmental (*e.g.*, deprivation, lead poisoning) factors.

As further evidence of environment-gene interaction, a social class gradient in the incidence of neural tube defects has long been recognized.[37]

Diagnostic Problems

It must almost certainly be unusual for a physician to cover every possible diagnostic test in the search for the cause of irreparable mental retardation in any particular case. Economic considerations, and failure to recognize that although the patient may not benefit, genetic counseling for the family may be important, probably rank as two common reasons for less than complete diagnostic studies. The advent of new, more sophisticated diagnostic tests unknown to the physician may be a further reason for failure to determine the etiology of mental retardation. The value of the new chromosome staining techniques and the advances in enzymology in somatic cell genetics exemplify the newer approaches. In a Canadian study, 72 patients with mental retardation and/or congenital anomalies, who were originally reported to have had normal karyotypes with conventional stains, were re-analyzed with the Q-banding technique.[191] Previously unrecognized chromosomal abnormalities were found in four cases. Two of these cases had an unbalanced translocation — new information of potentially great import to their respective families (of use in prenatal diagnosis in subsequent pregnancies).

Swift *et al.*[35] selectively studied 30 families having more than one child with an I.Q. < 60. They observed parental consanguinity in five of these families, suggesting autosomal recessive mutant genes in their retarded offspring. In these families they detected no fewer than four new biochemical disorders associated with severe mental retardation [(deficient L-tyrosine amino transferase (1 case); specific aminoaciduria with excessive quantities of hydroxylysine, β-alanine, and 1-methylhistidine (3 cases)]. A renal membrane transport disorder was considered most likely in view of the normal amino acids in the plasma in the latter three cases. The relationship of these renal tubular defects to mental retardation remains unclear.[35]

These studies raise the inevitable and difficult question of how intensive and extensive an investigation should be to determine the cause of mental retardation. The need for genetic counseling and the advent of prenatal genetic diagnosis now effectively dictate the lengths the physician must go to exclude totally all the known causes of mental retardation before using the term "idiopathic."

ACKNOWLEDGMENT: The very helpful comments on the manuscript made by Professor Hugo W. Moser and Professor Allen C. Crocker are gratefully acknowledged.

References

1. Berg, J. M.: Proceedings of the 2nd International Congress of Mental Retardation. Part 1. Basel, Switzerland, S. Karger, AG, 1963, p. 170.
2. Moser, H. W., and Wolf, P. A.: The nosology of mental retardation: Including the report of a survey of 1378 mentally retarded individuals at the Walter E. Fernald State School. Birth Defects, 7:117, 1971.
3. Crome, L., and Stern, J.: The pathology of mental retardation. Baltimore, The Williams & Wilkins Company, 1972.
4. Clarke, A. D. B., and Clarke, A. M.: Mental retardation and behavioural change. Br. Med. Bull., 30:179, 1974.
5. Crome, L.: The brain and mental retardation. Br. Med. J., 1:897, 1960.
6. Collins, E., and Turner, G.: The importance of the "small-for-dates" baby to the problem of mental retardation. Med. J. Aust., 2:313, 1971.
7. McDonald, A. D.: Children of very low birth weight. London, Heinemann and Spastics Society, 1967.
8. Drillien, C. M.: Aetiology and outcome in low-birthweight infants. Dev. Med. Child Neurol., 14:563, 1972.
9. McKusick, V. A.: The ethnic distribution of disease in the United States. J. Chronic Dis., 20:115, 1967.
10. McKusick, V. A.: Ethnic distribution of disease in non-Jews. Isr. J. Med. Sci., 9:1375, 1973.
11. Heber, R.: Epidemiology of mental retardation. Springfield, Ill., Charles C Thomas, Publisher, 1970.
12. Tizard, J.: Prevalence of mental subnormality. Bull. W.H.O., 9:423, 1953.
13. McDonald, A. D.: Severely retarded children in Quebec: Prevalence, causes, and care. Am. J. Ment. Defic., 78:205, 1973.

14. Kraus, J.: Prevalence of intellectual deficiency in New South Wales. Med. J. Aust., 1:795, 1973.
15. Wing, L.: Severely retarded children in a London area: Prevalence and provision of services. Psychol. Med., 1:405, 1971.
16. MacKay, D. N.: Mental subnormality in Northern Ireland. J. Ment. Defic. Res., 15:12, 1971.
17. Kushlick, A.: The prevalence of recognized mental subnormality of I.Q. under 50 among children in the South of England with reference to the demand for places for residential care. Proc. 3rd Int. Conf. Sci. Study Ment. Retard., Copenhagen, 2:550, 1964.
18. Birch, H. G., Richardson, S. A., Baird, D. et al.: Mental subnormality in the community. A clinical and epidemiologic study. Baltimore, Williams & Wilkins Company, 1970.
19. Kushlick, A.: A community service for the mentally subnormal. Soc. Psychiatr., 1:73, 1966.
20. Goodman, N., and Tizard, J.: Prevalence of imbecility and idiocy among children. Br. Med. J., 5273:216, 1962.
21. Åkesson, H. O.: Epidemiology and genetics of mental deficiency in a southern Swedish population. Uppsala, Almqvist & Wiksell Förlag AB, 1961.
22. Mental Health Research Unit NYS: Technical Report. A Special Census of Suspected Referred Mental Retardates, Onondaga County, N.Y. Syracuse, Syracuse University Press, 1955.
23. Lemakau, P., Tietze, C., and Cooper, M.: Mental hygiene problems in an urban district. Third paper. Ment. Hyg., 26:275, 1942.
24. Lewis, E. O.: The report of the mental deficiency committee: Part IV. London, His Majesty's Stationery Office, 1929.
25. Innes, G., Kidd, C., and Ross, H. S.: Mental subnormality in North-East Scotland. Br. J. Psychiatr., 114:35, 1968.
26. Costeff, H., Cohen, B. E., and Weller, L.: Parental consanguinity among Israel's mental retardates. Acta Paediatr. Scand., 61:452, 1972.
27. Nelson, M. M., and Forfar, J. O.: Congenital abnormalities at birth: Their association in the same patient. Dev. Med. Child Neurol., 11:3, 1969.
28. Polani, P. E.: Incidence of developmental and other genetic abnormalities. Proc. R. Soc. Med., 66:1118, 1973.
29. Ekelund, H., Kullander, S., and Kallen, B.: Major and minor malformations in newborns and infants up to one year of age. Acta Paediatr. Scand., 59:297, 1970.
30. Neel, J. V.: A study of major congenital defects in Japanese infants. Am. J. Hum. Genet., 10:398, 1958.
31. Lubs, H. A., and Ruddle, F. H.: Chromosomal abnormalities in the human population: Estimation of rates based on New Haven newborn study. Science, 169:495, 1970.
32. Holmes, L. B.: Inborn errors of morphogenesis—a review of localized hereditary malformations. N. Engl. J. Med., 291:763, 1974.
33. McKusick, V. A.: Mendelian inheritance in man. Catalogs of autosomal dominant, autosomal recessive, and X-linked phenotypes. 3rd ed. Baltimore, The Johns Hopkins University Press, 1971.
34. Davison, B. C. C.: In Davison, B. C. C., Benson, P. F., Swift, P. N., and Studdy, J. D. (eds.): Genetic Studies in Mental Subnormality. Ashford, British Journal of Psychiatry, Special Publication No. 8, 1973.
35. Swift, P. N., Benson, P. F., and Studdy, J. D.: Genetic Studies in Mental Subnormality. Ashford, British Journal of Psychiatry, Special Publication No. 8, 1973.
36. Drillien, C. M.: Pre- and perinatal factors in the etiology and outcome of low birth weight. In Milunsky, A. (ed.): Clinics in Perinatology. Vol. 2, Philadelphia, W. B. Saunders Company, 1974, p. 197.
37. Butler, N. R., and Alberman, E. D.: Perinatal problems. The 2nd report of the 1958 British Perinatal Mortality Survey. Edinburgh, Churchill Livingston, 1969.
38. Sever, J. L.: In The Prevention of Mental Retardation Through Control of Infectious Diseases: Public Health Service Publication No. 1692, Washington, D.C., Government Printing Office, 1962, p. 37.

39. Stern, H.: Isolation of cytomegalovirus and clinical manifestations of infection at different ages. Br. Med. J., 1:665, 1968.
40. Starr, J. G., and Gold, E.: Screening of newborn infants for cytomegalovirus infection. J. Pediatr., 73:820, 1968.
41. Hanshaw, J. B., Steinfeld, H. J., and White, C. J.: Fluorescent-antibody test for cytomegalovirus. N. Engl. J. Med., 279:566, 1968.
42. Stern, H., Elek, S. D., Booth, J. C. et al.: Microbial causes of mental retardation. The role of prenatal infections with cytomegalovirus, rubella virus, and toxoplasma. Lancet, 2:7618, 1969.
43. Miller, L. H., Reifsnyder, D. N., and Martinez, S. A.: Late onset of disease in congenital toxoplasmosis. Clin. Pediatr., 10:78, 1971.
44. Kopelman, A. E., Halsted, C. C., and Minnefor, A. B.: Osteomalacia and spontaneous fractures in twins with congenital cytomegalic inclusion disease. J. Pediatr., 81:101, 1972.
45. Shearer, W. T., Schreiner, R. I., Marshall, R. E. et al.: Cytomegalovirus infection in a newborn dizygous twin. J. Pediatr., 81:1161, 1972.
46. Forrester, R. M., Lees, V. T., and Watson, G. H.: Rubella syndrome: Escape of a twin. Br. Med. J., 1:1403, 1966.
47. Sever, J. L., Heubner, R. J., Castellano, G. A. et al.: Serologic diagnosis "en masse" with multiple antigens. Am. Rev. Resp. Dis., 88:342, 1963.
48. Hildebrandt, R. J., Sever, J. L., Margileth, A. M. et al.: Cytomegalovirus in the normal pregnant woman. Am. J. Obstet. Gynecol., 98:1125, 1967.
49. Feldman, R. A.: Cytomegalovirus infection during pregnancy. A prospective study and report of six cases. Am. J. Dis. Child., 117:517, 1969.
50. Reynolds, D. W., Stagno, S., Hosty, T. S. et al.: Maternal cytomegalovirus excretion and perinatal infection. N. Engl. J. Med., 289:1, 1973.
51. Montgomery, R., Youngblood, L., and Medearis, D. N.: Recovery of cytomegalovirus from the cervix in pregnancy. Pediatrics, 49:524, 1972.
52. Numazaki, Y., Yano, N., Morizuka, T. et al.: Primary infection with human cytomegaloviruses: Virus isolation from healthy infants and pregnant women. Am. J. Epidemiol., 91:410, 1970.
53. McCracken, G. H., Jr., Shinefield, H. M. R., Cobb, K. et al.: Congenital cytomegalic inclusion disease. A longitudinal study of 20 patients. Am. J. Dis. Child., 117:522, 1969.
54. Stern, H., and Elek, S. D.: The incidence of infection with cytomegalovirus in a normal population. A serological study in greater London. J. Hyg. (Camb.), 63:79, 1965.
55. Stagno, S., Reynolds, D. W., Lakeman, A. et al.: Congenital cytomegalovirus infection: Consecutive occurrence due to viruses with similar antigenic compositions. Pediatrics, 52:6, 1973.
56. Pitt, D. B.: Congenital malformations and maternal rubella: Progress report. Med. J. Aust., 48:881, 1961.
57. Manson, M. M., Logan, W. P. D., and Loy, R. M.: Rubella and other virus infections in pregnancy. Report on Public Health in Medical Subjects, No. 101, London, Ministry of Health, 1960.
58. Swan, C.: Rubella in pregnancy as aetiological factor in congenital malformation, stillbirth, miscarriage, and abortion. J. Obstet. Gynaecol. Br. Commonw., 56:341, 1949.
59. Strannegard, O., Holm, S. E., Hermodsson, S. et al.: Case of apparent reinfection with rubella. Lancet, 1:240, 1970.
60. Northrop, R. L., Gardener, W. M. and Geitmann, W. F.: Rubella reinfection during early pregnancy. A case report. Obstet. Gynecol., 39:524, 1972.
61. Chang, T.-W.: Rubella reinfection and intrauterine involvement. J. Pediatr., 84:617, 1974.
62. Couvreur, J., and Desmonts, G.: Congenital and maternal toxoplasmosis. A review of 300 congenital cases. Dev. Med. Child Neurol., 4:519, 1962.
63. Kraubig, Von H.: Präventive Behandlung der konnatalen Toxoplasmose. In Kirchhoff, H., and Kraubig, H. (eds.): Toxoplasmose, Stuttgart, Georg Thieme Verlag, 1966.
64. Desmonts, G., and Couvreur, J.: Toxoplasmose congénitale. 21e Congrès Pédiatrique de la Langue Française. Vol. 3. Paris, l'Expansion Scientifique Française, 1967, p. 450.

65. Robertson, J. S.: Toxoplasma skin-and-dye-test. Surveys of severely subnormal patients in Lincolnshire. J. Hyg. (Camb.), 63:89, 1965.
66. Robertson, J. S.: Toxoplasmin sensitivity: Subnormality and environment. J. Hyg. (Camb.), 64:405, 1966.
67. Desmonts, G. and Couvreur, J.: Congenital toxoplasmosis. A prospective study of 378 pregnancies. N. Engl. J. Med., 290:1110, 1974.
68. Alford, C. A., Jr., Stagno, S., and Reynolds, D. W.: Congenital toxoplasmosis: Clinical, laboratory, and therapeutic considerations, with special reference to subclinical disease. Bull. N.Y. Acad. Med., 50:160, 1974.
69. Lapinleimu, K., Koskimies, O., Cantell, K. et al.: Association between maternal herpesvirus infections and congenital malformations. Lancet, 1:1127, 1974.
70. Hanshaw, J. B.: Herpesvirus hominis infections in the fetus and the newborn. Am. J. Dis. Child., 126:546, 1973.
71. Fuccillo, D. A., and Sever, J. L.: Viral teratology. Bacteriol. Rev., 37:19, 1973.
72. Savage, M. O., Moosa, A., and Gordon, R. R.: Maternal varicella infection as a cause of fetal malformations. Lancet, 1:352, 1973.
73. Seigel, M.: Congenital malformations following chickenpox, measles, mumps, and hepatitis. Results of a cohort study. J.A.M.A., 226:1521, 1973.
74. Herrlin, K. M.: A clinical and electroencephalographic study of a pair of monozygotic twins with phenylketonuria. Acta Paediatr., (Uppsala) (Suppl. 135), 51:88, 1962.
75. Woolf, L. I.: In Anderson, J. A., and Swaiman, K. F. (eds.): Phenylketonuria and Allied Metabolic Diseases. Washington, D. C., U.S. Dept. H.E.W., 1969.
76. McKusick, V. A.: Heritable Disorders of Connective Tissue. 4th ed., St. Louis, C. V. Mosby Co., 1972.
77. Wilson, R. S.: Twins: Early mental development. Science, 175:914, 1972.
78. Berg, J. M., and Kirman, B. H.: The mentally defective twin. Br. Med. J., 1:1911, 1960.
79. Knobloch, H., and Pasamanick, B.: Mental subnormality. N. Engl. J. Med., 266:1045, 1092, 1155, 1962.
80. Smythies, J. R.: Biological Psychiatry. London, Pergamon Press, Ltd., 1968.
81. Record, R. G., McKeown, T., and Edwards, J. H.: An investigation of the difference in measured intelligence between twins and single births. Ann. Hum. Genet., 34:11, 1970.
82. Allen, G.: Twin research: Problems and prospects. Progr. Med. Genet., 4:242, 1965.
83. Edwards, J. H.: The value of twins in genetic studies. Proc. R. Soc. Med., 61:227, 1968.
84. Onyskowova, Z., Dolezal, A., and Jedlicka, V.: The frequency and the character of malformations in multiple birth (a preliminary report). Teratology, 4:496, 1971.
85. Scott, J. M. and Paterson, L.: Monozygous anencephalic triplets—a case report. J. Obstet. Gynaecol. Br. Commonw., 73:147, 1966.
86. Yen, S., and MacMahon, B.: Genetics of anencephaly and spina bifida? Lancet, 2:623, 1968.
87. Bellefeuille, P. de: Contribution a l'étiologie de l'anencephalie par l'étude des jumeaux: Revue de la littérature. Union Med. Can., 98:437, 1969.
88. Horst, R. van der, Frankel, J., and Grace, J.: Congenital hypertrophic pyloric stenosis in phenotypic female twins with X/XX mosaicism. Arch. Dis. Child., 46:554, 1971.
89. Rickhof, P. L., Horton, W. A., Harris, D. J. et al.: Monozygotic twins with the Turner syndrome. Am. J. Obstet. Gynecol., 112:59, 1972.
90. Nielsen, J.: Twins in sibships with Klinefelter's syndrome and the XYY syndrome. Acta Genet. Med. Gemellol. (Roma), 19:399, 1970.
91. Zellweger, H.: Familial aggregates of the 21-trisomy syndrome. Ann. N.Y. Acad. Sci., 155:784, 1968.
92. Koch, G., Rott, H. D., Schwanitz, G. et al.: Cri-du-chat syndrome bei Zwillingen. Aerztl. Praxis., 22:3221, 1970.
93. Gertzer, D., and Nathenson, G.: Case report. Trisomy 18 in one of fraternal twins. J. Med. Genet., 8:392, 1971.
94. Benirschke, K., and Kim, C. K.: Multiple pregnancy (first of two parts). N. Engl. J. Med., 288:1276, 1973.

95. Nance, W. E., and Uchida, I.: Turner's syndrome, twinning, and an unusual variant of glucose-6-phosphate dehydrogenase. Am. J. Hum. Genet., 16:380, 1964.
96. Warkany, J.: Congenital malformations. Notes and Comments. Chicago, Year Book Medical Publishers, Inc., 1971.
97. Carter, C. O.: Genetics of common disorders. Br. Med. Bull., 25:52, 1969.
98. Carter, C. O.: Genetics of common malformations. In Gairdner, D., and Hull, D. (eds.): Recent Advances in Pediatrics, 4th ed. London, J. & A. Churchill, Ltd., 1971.
99. Roberts, J. A. F.: The genetics of mental deficiency. Eugen. Rev., 44:71, 1952.
100. Hilliard, L. T., and Kirman, B. H.: Mental Deficiency. Boston, Little, Brown and Company, 2nd ed., 1965.
101. Howell, R. R., and Stevenson, R. E.: The offspring of phenylketonuric women. Soc. Biol., 18:19, 1971.
102. German, J., Dowal, A., and Ehlers, K. H.: Trimethadione and human teratogenesis. Teratology, 3:349, 1970.
103. Naggan, L., and MacMahon, B.: Ethnic differences in the prevalence of anencephaly and spina bifida in Boston, Massachusetts. N. Engl. J. Med., 277:1119, 1967.
104. McKeown, T., and Record, R. G.: Seasonal incidence of congenital malformations of the central nervous system. Lancet, 1:192, 1951.
105. Record, R. G.: Anencephalus in Scotland. Br. J. Prev. Soc. Med., 15:93, 1961.
106. Frezal, J., Kelley, J., Fuillemot, M. L. et al.: Anencephaly in France. Am. J. Hum. Genet., 16:336, 1964.
107. Leck, I., and Record, R. G.: Seasonal incidence of anencephalus. Br. J. Prev. Soc. Med., 20:67, 1966.
108. Cohen, P.: Seasonal variations of congenital dislocations of the hip. J. Interdiscipl. Cycle Res., 2:417, 1971.
109. Rutstein, D. D., Nickerson, R. J., and Heald, F. P.: Seasonal incidence of patent ductus arteriosus and maternal rubella. Am. J. Dis. Child., 84:199, 1952.
110. Miettinen, O. S., Reiner, M. L., and Nadas, A. S.: Seasonal incidence of coarctation of the aorta. Br. Heart J., 32:103, 1970.
111. Rothman, K. J., and Fyler, D. C.: Seasonal occurrence of complex ventricular septal defect. Lancet, 2:193, 1974.
112. Penrose, L. S., and Smith, G. F.: Down's Anomaly. Boston, Little, Brown and Company, 1966.
112a. Morton, N. E.: The mutational load due to detrimental genes in man. Am. J. Hum. Genet., 12:348, 1960.
112b. Dewey, W. J., Barrai, I., Morton, N. E. et al.: Recessive genes in severe mental defect. Am. J. Hum. Genet., 17:237, 1965.
113. Penrose, L. S.: Biology of Mental Defect. 3rd ed. London, Sidgwick and Jackson Ltd., 1963.
114. Morton, N. E.: Morbidity of children from consanguineous marriages. Prog. Med. Genet., 1:261, 1961.
115. Seemanova, E.: A study of children of incestuous matings. Hum. Hered., 21:108, 1971.
116. Freire-Maia, N.: Consanguineous marriages in Brazil. III. A note on the relation of inbreeding with deaf-mutism, mental retardation, and speech defects. Rev. Bras. Biol., 18:219, 1958.
117. Chemke, J., Chen, R., Klingberg, M. A. et al.: Some indications for genetic factors in congenital malformations. Isr. J. Med. Sci., 9:1400, 1973.
118. Sinclair, S.: Etiological factors in mental retardation: A study of 470 cases. Indian Pediatr., 9:391, 1972.
119. Fraumeni, J. F., Jr., and Glass, A. G.: Wilm's tumor and congenital aniridia. J.A.M.A., 206:824, 1968.
120. Lowry, R. B.: Sex-linked cleft palate in a British Columbia Indian family. J. Pediatr., 46:123, 1970.
121. Lowry, R. B.: Personal communication, 1974.
122. Noonan, J. A., and Ehmke, D. A.: Associated noncardiac malformation in children with congenital heart disease. J. Pediatr., 63:468, 1963.
123. Kaplan, M. S., Opitz, J. M., and Gosset, F. R.: Noonan's syndrome. Am. J. Dis. Child., 116:359, 1968.

124. Summitt, R. L.: Turner syndrome and Noonan's syndrome. J. Pediatr., 74:155, 1969.
125. Noonan, J. A.: Hypertelorism with Turner phenotype. Am. J. Dis. Child., 116:373, 1968.
126. Baird, P. A., and De Jong, B. P.: Noonan's syndrome (XX and XY Turner phenotype) in three generations of a family. J. Pediatr., 80:110, 1972.
127. De Myer, W.: The median cleft face syndrome. Neurology, 17:961, 1967.
128. Sedano, H. O., Cohen, M. M., Jr., Jirasek, J. et al.: Frontonasal dysplasia. J. Pediatr., 76:906, 1970.
129. Francesconi, G., and Fortunato, G.: Median dysraphia of the face. Plast. Reconstr. Surg., 43:481, 1969.
130. Cohen, M. M., Jr., Sedano, H. O., Gorlin, R. J., et al.: Frontonasal dysplasia (Median cleft face syndrome): Comments on etiology and pathogenesis. Birth Defects, 7:117, 1971.
131. Warkany, J., Bofinger, M. K., and Benton, C.: Median facial cleft syndrome in half-sisters. Dilemmas in genetic counseling. Teratology, 8:273, 1973.
132. Kalter, H.: Sporadic congenital malformations of newborn inbred mice. Teratology, 1:193, 1968.
133. Kerber, I. J., Warr, O. S., and Richardson, C.: Pregnancy in a patient with a prosthetic mitral valve associated with a fetal anomaly attributed to warfarin sodium. J.A.M.A., 203:223, 1968.
134. Hoefnagel, D., Keenan, M. E., and Allen, F. H., Jr.: Heredofamilial bilateral anophthalmia. Arch. Ophthalmol., 69:760, 1963.
135. Blank, C. E.: Apert's syndrome (a type of acrocephalosyndactyly). Observations on a British series of thirty-nine cases. Ann. Hum. Genet., 24:151, 1960.
136. Jan, J. E., Hardwick, D. F., Lowry, R. B., et al.: Cerebrohepato-renal syndrome of Zellweger. Am. J. Dis. Child., 119:274, 1970.
137. Temtamy, S. A.: Carpenter's syndrome: Acrocephalopolysyndactyly. An autosomal recessive syndrome. J. Pediatr., 69:111, 1966.
138. Bartsocas, C. S., Weber, A. L., and Crawford, J. D.: Acrocephalosyndactyly type III: Chotzen's syndrome. J. Pediatr., 77:267, 1970.
139. Cockayne, E. A.: Dwarfism with retinal atrophy and deafness. Arch. Dis. Child., 11:1, 1936.
140. Pena, S. D. J., and Shokeir, M. H. K.: Autosomal recessive cerebro-oculo-facio-skeletal (COFS) syndrome. Clin. Genet., 5:285, 1974.
141. Comings, D. E., Papazian, C., and Schoene, H. R.: Conradi's disease. Chondrodystrophia calcificans congenita, congenital stippled epiphyses. J. Pediatr., 72:63, 1968.
142. Vulliamy, D. G., and Normandale, P. A.: Cranio-facial dysostosis in a Dorset family. Arch. Dis. Child., 41:375, 1966.
143. Ide, C. H., and Wollschlaeger, P. B.: Multiple congenital abnormalities associated with cryptophthalmia. Arch. Ophthalmol., 81:638, 1969.
144. Pashayan, H., Wehlan, D., Guttman, S. et al.: Variability of the de Lange syndrome: Report of 3 cases and genetic analysis of 54 families. J. Pediatr., 75:853, 1969.
145. Reed, W. B., Lopez, D. A., and Landing, B.: Clinical spectrum of anhidrotic ectodermal dysplasia. Arch. Dermatol., 102:134, 1970.
146. Gmyrek, D. and Syllm-Rapoport, I.: Zur Fanconi-anamie (F.A.). Analyse von 129 beschriebenen fallen. Z. Kinderheilkd., 91:297, 1964.
147. Goltz, R. W., Henderson, R. R., Hitch, J. M. et al.: Focal dermal hypoplasia syndrome. A review of the literature and report of two cases. Arch. Dermatol., 101:1, 1970.
148. Holmes, L. B., Nash, A., ZuRhein, G. M. et al.: X-linked aqueductal stenosis: Clinical and neuropathological findings in two families. Pediatrics, 51:697, 1973.
149. Morgan, J. D.: Incontinentia pigmenti (Bloch-Sulzberger syndrome). Am. J. Dis. Child., 122:294, 1971.
150. McKusick, V. A., Egeland, J. A., Eldridge, R. et al.: Dwarfism in the amish. 1. The Ellis-van Creveld syndrome. Johns Hopkins Med. J., 115:306, 1964.
151. Bell, J.: The Laurence-Moon syndrome. In The Treasury of Human Inheritance. Vol. 5, Part 3. Cambridge, England, Cambridge University Press, 1958, p. 51.
152. Donohue, W. L., and Uchida, I.: Leprechaunism. A euphemism for a rare familial disorder. J. Pediatr., 45:505, 1954.

153. Dieker, H., Edwards, R. H., Zurhein, G. et al.: The lissencephaly syndrome. Birth Defects, 5:53, 1969.
154. Alter, M., and Kennedy, W.: The Marinesco-Sjögren syndrome. Hereditary cerebello-lental degeneration with mental retardation. Minn. Med., 51:901, 1968.
155. Hsia, Y. E., Bratu, M., and Herbordt, A.: Genetics of the Meckel syndrome (dysencephalia splanchnocystica). Pediatrics, 48:237, 1971.
156. Motl, M. L., and Opitz, J. M.: Studies of malformation syndromes XXVA. Phenotypic and genetic studies of the Brachmann-deLange syndrome. Hum. Hered., 21:1, 1971.
157. Kloepfer, H. W., Platou, R. V., and Hansche, W. J.: Manifestations of a recessive gene for microcephaly in a population isolate. J. Genet. Hum., 13:52, 1964.
158. Carter, C. O., David, P. A., and Laurence, K. M.: A family study of major central nervous system malformations in South Wales. J. Med. Genet., 5:81, 1968.
159. Rudiger, R. A., Schmidt, W., Loose, D. A. et al.: Severe developmental failure with coarse facial features, distal limb hypoplasia, thickened palmar creases, bifid uvula and ureteral stenosis: A previously unidentified familial disorder with lethal outcome. J. Pediatr., 79:977, 1971.
160. Khan, M., Rozdilsky, B., and Gerrard, J. W.: Familial holoprosencephaly. Dev. Med. Child Neurol., 12:71, 1970.
161. Lowe, C. U., Terrey, M., and MacLachlan, E. A.: Organic aciduria, decreased renal ammonia production, hydrophthalmos, and mental retardation. A clinical entity. Am. J. Dis. Child., 83:164, 1952.
162. Johnston, C. C., Jr., Lavy, N., Lord, T. et al.: Osteopetrosis. A clinical, genetic, metabolic, and morphologic study of the dominantly inherited, benign form. Medicine, 47:149, 1968.
163. Silver, H. K.: Rothmund-Thomson syndrome: An oculocutaneous disorder. Am. J. Dis. Child., 111:182, 1966.
164. Elmore, S. M.: Pycnodysostosis: A review. J. Bone Joint Surg. 49-A:153, 1967.
165. Roberts, J. B.: A child with double cleft of lip and palate, protrusion of the intermaxillary portion of the upper jaw, and imperfect development of the bones of the four extremities. Ann. Surg., 70:252, 1919.
166. Mengel, M. C., Konigsmark, B. W., Berlin, C. I., et al.: Conductive hearing loss and malformed low-set ears, as a possible recessive syndrome. J. Med. Genet., 6:14, 1969.
167. Dallaire, L.: Syndrome of retardation with urogenital and skeletal anomalies (Smith-Lemli-Opitz syndrome): Clinical features and mode of inheritance. J. Med. Genet., 6:113, 1969.
168. Jammes, J., Mirhosseini, S. A., and Holmes, L. B.: Syndrome of facial abnormalities, kyphoscoliosis, and severe mental retardation. Clin. Genet., 4:203, 1973.
169. Reed, W. B., May, S. B., and Nickel, W. R.: Xeroderma pigmentosum with neurological complications. Arch. Dermatol., 91:224, 1965.
170. Dewey, W. J., Barrai, I., Morton, N. E., and Mi, M. P.: Recessive genes in severe mental defect. Am. J. Hum. Genet., 17:237, 1965.
171. Renpenning, H., Gerrard, J. W., Zaleski, W. A., and Tabata, T.: Familial sex-linked mental retardation. Can. Med. Assoc. J., 87:954, 1962.
172. Turner, G., and Turner, B.: X-linked mental retardation. J. Med. Genet., 11:109, 1974.
173. Passarge, E.: Congenital malformations and maternal diabetes. Lancet, 1:324, 1965.
174. Thalhammer, O., Lachmann, D., and Scheibenreiter, S.: "Caudale regression" beim kind einer 18 jahrigen frau mit pradiabetes. Z. Kinderheilkd., 102:346, 1968.
175. Hagbard, L.: Pregnancy and diabetes. Springfield, Ill., Charles C Thomas, Publisher, 1961.
176. Pedersen, J.: The Pregnant Diabetic and Her Newborn. Baltimore, Md., The Williams & Wilkins Company, 1967.
177. Rubin, A., and Murphy, D. P.: Studies in human reproduction. III. The frequency of congenital malformations in the offspring of nondiabetic and diabetic individuals. J. Pediatr., 53:579, 1958.
178. Farquhar, J. W.: The influence of maternal diabetes on fetus and child. In Gairdner, D. (ed.): Recent Advances in Pediatrics, 3rd ed. Boston, Little, Brown & Company, 1965.

179. Robinson, R.: Neurological development in infants of diabetic mothers. Dev. Med. Child. Neurol., 12:227, 1970.
180. Meadow, S. R.: Anticonvulsant drugs and congenital abnormalities. Lancet, 2:1296, 1968.
181. Speidel, B. D., and Meadow, S. R.: Maternal epilepsy and abnormalities of the fetus and newborn. Lancet, 2:839, 1972.
182. Lowe, C. R.: Congenital malformations among infants born to epileptic women. Lancet, 1:9, 1973.
183. Monson, R. R., Rosenberg, L., Hartz, S. C. et al.: Diphenylhydantoin and selected congenital malformations. N. Engl. J. Med., 289:1049, 1973.
184. Hall, M. H.: Folic acid deficiency and congenital malformations. J. Obstet. Gynaecol. Br. Commonw., 79:159, 1972.
185. Murphy, D. P.: Congenital Malformations. A Study of Parental Characteristics with Special Reference to the Reproductive Process. Philadelphia, University of Pennsylvania Press, 1947.
186. Dekaban, A. S.: Abnormalities in children exposed to X-radiation during various stages of gestation: Tentative time table of radiation injury to the human fetus. Part I. J. Nucl. Med., 9:471, 1968.
187. Gentry, J. T., Parkhurst, E., and Bulin, G. V.: An epidemiological study of congenital malformations in New York state. Am. J. Public Health, 49:497, 1959.
188. Uchida, I. A., Holunga, R., and Lawler, C.: Maternal radiation and chromosomal aberrations. Lancet, 2:1045, 1968.
189. Jones, K. L., Smith, D. W., Streissguth, A. P. et al.: Outcome in offspring of chronic alcoholic women. Lancet, 1:1076, 1974.
190. Motulsky, A. G.: Population genetics of mental retardation. *In* Jervis, G. A. (ed.): Expanding Concepts in Mental Retardation. A Symposium. Springfield, Ill., Charles C Thomas, Publisher, 1968, p. 13.
191. Solar C. del, and Uchida, I. A.: Identification of chromosomal abnormalities by quinacrine-staining technique in patients with normal karyotypes by conventional analysis. J. Pediatr., 84:534, 1974.
192. Taybi, H., and Linder, D.: Congenital familial dwarfism with cephaloskeletal dysplasia. Radiology, 89:275, 1967.
193. Juberg, R. C., and Hayward, J. R.: A new familial syndrome of oral, cranial, and digital anomalies. J. Pediatr., 74:755, 1969.
194. Opitz, J. M., Summitt, R. L., and Smith, D. W.: The BBB Syndrome. Familial telechanthus with associated congenital anomalies. Birth Defects, 5:86, 1969.
195. Scott, C. R., Bryant, J. I., and Graham, C. B.: A new craniodigital syndrome with mental retardation. J. Pediatr., 78:658, 1971.
196. Palant, D. I., Feingold, M., and Berkman, M. D.: Unusual facies, cleft palate, mental retardation, and limb abnormalities in siblings–a new syndrome. J. Pediatr., 78:686, 1971.
197. Cohen, M. M., Jr., Hall, B. D., Smith, D. W. et al.: A new syndrome with hypotonia, obesity, mental deficiency, and facial, oral, ocular, and limb anomalies. J. Pediatr., 83:280, 1973.
198. Lowry, R. B., MacLean, R., McLean, D. M. et al.: Cataracts, microcephaly, kyphosis, and limited joint movement in two siblings: A new syndrome. J. Pediatr., 79:282, 1971.
199. Ruvalcaba, R. H. A., Reichert, A., and Smith, D. W.: A new familial syndrome with osseous dysplasia and mental deficiency. J. Pediatr., 79:450, 1971.
200. Matsoukas, J., Liarikos, S., Giannikas, A. et al.: A newly recognized dominantly inherited syndrome: Short stature, ocular and articular anomalies, mental retardation. Helv. Paediatr., Acta, 28:383, 1973.
201. Tay, C. H.: Ichthyosiform erythroderma, hair shaft abnormalities, and mental and growth retardation. A new recessive disorder. Arch. Dermatol., 104:4, 1971.
202. Hooft, C., de Hauwere, R., and van Acker, K. J.: Familial non-congenital microcephaly, peculiar appearance, mental and motor retardation, progressive evolution to spasticity, and choreo-athetosis. Helv. Paediatr., Acta, 23:1, 1968.
203. Christian, J. C., Johnson, V. P., Biegel, A. A., et al.: Sisters with low birth weight, dwarfism, congenital anomalies, and dysgammaglobulinemia. Am. J. Dis. Child., 122:529, 1971.

204. Strong, W. B.: Familial syndrome of right-sided aortic arch, mental deficiency, and facial dysmorphism. J. Pediatr., 73:882, 1968.
205. Falek, A., Heath, C. W., Jr., Ebbin, A. J. et al.: Unilateral limb and skin deformities with congenital heart disease in two siblings: A lethal syndrome. J. Pediatr., 73:910, 1968.
206. Lundberg, P. O.: Hereditary myopathy, oligophrenia, cataract, skeletal abnormalities, and hypergonadotropic hypogonadism; A new syndrome. Eur. Neurol., 10:261, 1973.
207. Opitz, J. M., Johnson, R. C., McCreadie, S. R. et al.: The C syndrome of multiple congenital anomalies. Birth Defects, 5:161, 1969.
208. Aase, J. M. and Smith, D. W.: Dysmorphogenesis of joints, brain, and palate: A new dominantly inherited syndrome. J. Pediatr., 73:606, 1968.
209. Cohen, M. M., Jr.: Macroglossia, omphalocele, visceromegaly, cytomegaly of the adrenal cortex and neonatal hypoglycemia. *In* Bergsma, D. (ed.): Birth Defects Original Article Series. Part XI. Orofacial structures. Vol. 7. The National Foundation – March of Dimes. White Plains, New York. Baltimore, Md., Williams & Wilkins Co., 1971, p. 226.
210. Rogers, B. O.: Berry-Treacher Collins Syndrome: A review of 200 cases. Br. J. Plast. Surg., 17:109, 1964.
211. Procopis, P. G., and Turner, B.: Mental retardation, abnormal fingers, and skeletal anomalies: Coffin's syndrome. Am. J. Dis. Child., 124:258, 1972.
212. Neu, R. L., Kajii, T., Gardner, L. I. et al.: A lethal syndrome of microcephaly with multiple congenital anomalies in three siblings. Pediatrics, 47:610, 1971.
213. Baird, P. A., and DeJong, B. P.: Noonan's syndrome (XX and XY Turner phenotype) in three generations of a family. J. Pediatr., 80:110, 1972.
214. Reisner, S. H., Kott, E., and Bornstein, B.: Oculodentodigital dysplasia. Am. J. Dis. Child., 118:600, 1969.
215. Rimoin, D. L., and Edgerton, M. T.: Genetic and clinical heterogeneity in the oral-facial-digital syndromes. J. Pediatr., 71:94, 1967.
216. Gall, J. C., Jr., Stern, A. M., Poznanski, A. K. et al.: Oto-palato digital syndrome: Comparison of clinical and radiographic manifestations in males and females. Am. J. Hum. Genet., 24:24, 1972.
217. Herrmann, J., Feingold, M., Tuffli, G. A. et al.: A familial dysmorphogenetic syndrome of limb deformities, characteristic facial appearance and associated anomalies: The "pseudothalidomide" or "SC-syndrome." Birth Defects, 5:81, 1969.
218. Simpson, J. L., and German, J.: Developmental anomaly resembling the trisomy 18 syndrome. Ann. Genet. (Paris), 12:107, 1969.
219. McKusick, V. A., Mahloudji, M., Abbott, M. H. et al.: Seckel's bird-headed dwarfism. N. Engl. J. Med., 277:279, 1967.

CHAPTER 2

SOCIETAL AND INDIVIDUAL ORIENTATIONS TOWARD PREVENTION

Richard D. Ashmore, Ph.D.

"Mental retardation" refers to a complex set of phenomena that are influenced by a wide variety of factors (see Chapter 1). Any campaign to prevent mental retardation must include an equally diverse range of programs. For such programs to be successful, however, they must be accepted and supported by the public. This chapter will outline the factors that influence the public's acceptance and support of prevention programs, particularly those programs concerned with mental retardation.

Understanding these factors requires the application of sociology and social psychology to medical behavior. Unfortunately, there is not a great deal of extant research and theory on this topic,[1] and the work that has been done is rarely directed at the sociopsychological factors influencing preventative behavior regarding mental retardation and genetic disease.[2] This chapter, therefore, will not be simply a review of what is already known about the causes of pro-prevention thinking, but rather an attempt to relate general "principles" of attitude and value formation and to discuss the topic of preventative health behavior. By "preventative health behavior," or "health behavior," I refer to "any activity undertaken by a person believing himself to be healthy, for the purpose of preventing disease or detecting it in an asymptomatic stage."[3]

"Attitude" is the individual-level variable most frequently used by social psychologists to explain behavior. Although there are many definitions of attitude, most share three elements: (1) that an attitude is an internal psychological response (i.e., attitudes are not directly observable); (2) that an attitude is an evaluative response (i.e., a good-bad, approach-avoidance response); and (3) that attitudes, in conjunction with other individual and situational variables, influence behavior toward the attitude object.[4]

Regarding the prevention of mental retardation, it would be desirable to identify and work to strengthen individuals' general attitudes toward prevention. That such a generalized attitude may exist is suggested by research indicating a rather strong general attitude in another realm of behavior, i.e., the Liberalism-Conservatism factor in political and economic affairs.[5] Unfortunately, however, there are empirical and conceptual reasons to believe that such a general attitude toward prevention may not exist. Haefner and his associates found moderate intercorrelations among four self-report measures of personal preventative health behaviors: checkup visit to physician, checkup visit to dentist, brushing teeth after meals, and taking a tuberculosis test.[6] Green also obtained moderate intercorrelations among a set of measures of preventative health behaviors, but the sizable amount of unique variance led him to treat each measure separately.[7] Although these studies provide some support for the idea of a general attitude toward prevention, a subsequent pair of studies, which were specifically designed to test this idea, do not.[8] In the first study a sample of suburban American women were contacted by telephone and asked the degree to which they had engaged in each of 22 preventative behaviors, e.g., having a Pap test, toothbrushing, carrying flares in their car, and exercising. In the second study, eight of these same behaviors were included in a questionnaire mailed to fathers of ninth grade students attending a suburban school. For both studies the matrix of intercorrelations among the preventative behaviors was factor analyzed. In neither study did the first principal-components factor account for a substantial amount of total test variance, thereby providing evidence against a unidimensional orientation toward prevention. Neither study indicated any evidence of a general factor of preventative health behavior.

To date, then, there is no compelling empirical evidence for the existence of a general attitude toward prevention, or even a general attitude toward health behavior. There is also a conceptual reason to believe that there are several prevention attitudes, at least some of which are relatively uncorrelated. Attitudes have symbolic referents: either words (e.g., "America") or images (e.g., the flag) that serve as the focus of the attitude. Neither preventative behavior, in general, nor preventative health behavior, in particular, has a clear symbolic referent.

Although the concept of attitude alone does not seem to offer a promising framework for understanding the orientation of individuals toward health behavior as a general phenomenon, Rokeach's model of a hierarchical, integrated belief system may be useful.[9] Rokeach argues that individuals have a belief system which is composed of Beliefs about

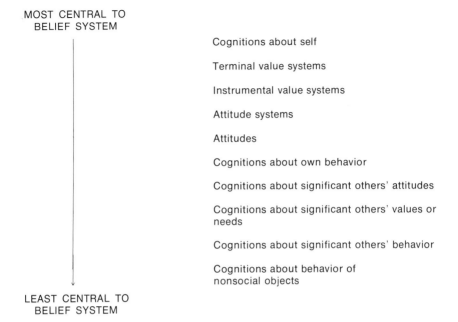

MOST CENTRAL TO
BELIEF SYSTEM

Cognitions about self

Terminal value systems

Instrumental value systems

Attitude systems

Attitudes

Cognitions about own behavior

Cognitions about significant others' attitudes

Cognitions about significant others' values or needs

Cognitions about significant others' behavior

Cognitions about behavior of nonsocial objects

LEAST CENTRAL TO
BELIEF SYSTEM

Figure 2–1. Hierarchical Organization of Elements in the Belief System. (Adapted from Rokeach, M.: The Nature of Human Values. New York, The Free Press, 1973, pp. 220–221.)

Self (taken together, they are roughly equivalent to the "self-concept"), Terminal Values (desired end-states, such as "a world at peace"), Instrumental Values (favored ways of behaving, such as "honesty"), Attitude Systems (organized sets of attitudes, such as Liberalism-Conservatism), and Attitudes and Beliefs about Others. He argues, further, that the elements within the system are hierarchically arranged, with the elements near the top of the hierarchy being wider in scope (i.e., more general), more stable, and serving to shape those elements lower in the system. For example, Terminal and Instrumental Values are higher in the belief system hierarchy than Attitudes. Thus, many individual attitudes may be related to one value, and although these individual attitudes may be subjected to pressure for change, no long-term, large-scale change can occur if the related values are not changed. For the present analysis, two aspects of the Rokeach schema are significant: (1) the hierarchical organization suggests that generality in health behavior orientations might be found at the level of values or self-concept; and (2) the interrelated and hierarchically arranged components of the belief system provide a useful vehicle for organizing discussion of how to create and maintain positive orientations to preventative health behavior.

A comprehensive program to prevent mental retardation and genetic disease must influence the individual's thinking about his or her own health, the health of his or her child, and that individual's values and attitudes as they serve to support the present socioeconomic system—a system which places certain groups at a severe health disadvantage. Although certain organic types of mental retardation are spread rather evenly throughout the population, the persons at the bottom of the socioeconomic status ladder are the most likely victims of mental retardation.[10] In part, this victimization springs from certain attitudes and behavior patterns which are not conducive to maintaining health for self and family. Suchman terms this "medical deprivation"[11] and below are discussed several ideas for altering these behaviors and attitudes. Most of the victimization of the poor, however, is, simply, due to being poor. That is, the poor have less access to good hospitals at the time of birth, less access to a proper diet for mother and child, less access to an environment free of poison and accident, and less access to home, neighborhood, and school environments that are intellectually stimulating. All of these conditions have been implicated in the development of mental retardation, and any overall attack on this problem must seek to reorganize society so that the poor have more nearly equal access to services and environments which foster health, rather than disease and deficit.

"PRINCIPLES" FOR INCREASING PREVENTATIVE THINKING

The previous analysis suggests two basic considerations in developing principles for increasing "preventative thinking." First, rather than conceiving of preventative behavior as the result of some general attitudinal factor, it is best to see such behavior within the context of an individual's entire system of beliefs—from the higher-order, more general self-concepts and values, to perceptions about other people. Second, preventative thinking involves an individual's beliefs and actions relative not only to his or her own health but also to that of his or her offspring. More broadly, such thinking refers to beliefs concerning the health of others in the society at large, particularly those at the bottom of the socioeconomic status hierarchy.

These considerations provide the basis for this section, which details a number of "principles" for increasing preventative thinking. The Rokeach schema for an individual's belief system is used to organize this discussion.

Cognitions About Self

Since self-concept is the highest component in an individual's belief system, proposals for increasing preventative thinking either must be congruent with the existing self-perceptions of the target group or must seek to change these perceptions. Unfortunately, at present there is little research on the role of self-perception in preventative health behavior. There are, however, two lines of research and theory that are suggestive.

The only systematic attempt to explain why and when individuals take preventative health measures is the "health belief model," as articulated by Hochbaum, Rosenstock, and their collaborators.[3,12] Their model explains health behavior in terms of (1) the perception of threat from possible illness, (2) the perception of the costs and benefits of various courses of preventative action, and (3) cues for taking action. The first class of variables, which determines the *motivation* for action, is composed of "perceived susceptibility" and "perceived seriousness." The former is essentially the individual's *subjective* probability of becoming ill, whereas the latter refers to the individual's belief as to how much harm the illness will do him. Although these motivational factors make the individual more or less disposed to action, the individual's actual behavior can be predicted only by also considering what he sees as the most attractive alternative, that is, the one with the highest perceived possibility of success and the lowest perceived cost—"low" not only in monetary terms but also in terms of time, effort, and other factors. The final element of the model—cues—is necessary, because even if there are high perceived threat and one clearly attractive behavior, some "trigger" to action seems necessary.

In this model, perceived susceptibility and probability of success are the two components most relevant to self-perceptions. Although the data are not completely confirming, there is evidence that individuals who see themselves as vulnerable to a particular disease are most likely to be willing to take preventative action.[3] Thus, increasing preventative thinking must begin by making individuals aware that they are susceptible. With respect to genetic disease, members of various ethnic groups who are especially likely to have particular diseases (e.g., sickle-cell anemia, among Blacks) should be made aware of their susceptibilities. Also, individuals living in certain "dangerous" environments (e.g., those with high levels of various toxic substances) should be informed of ways in which this increases susceptibility to disease. Finally, populations (e.g., non-white poor) in which women tend to have babies at biologically dangerous times (e.g., when the mother is very young) should be warned of the danger this entails for both mother and child. In sum, use of the "perceived susceptibility" factor involves identifying the personal and sociostructural factors which make various groups differentially susceptible to disease and disability, and making the members of the group aware of their susceptibility. Since it is the individual's feeling that "*I* am susceptible" that motivates action, it is important that very personal and specific messages be employed. Individuals should be informed why and how *they* are particularly susceptible—not just that "*Everyone* is at risk of having a child with genetic disease and/or mental retardation."

The second aspect of the "health belief" model relevant to self-perception is the perceived probability of success of preventative action. Although the perception that a particular preventative action will be successful is dependent on a number of factors, one minimum requirement is that the individual see himself as able to exert some control over his fate through personal action. The individual must see himself as having some personal control over his future. There is some evidence that the poor—those who engage in preventative behavior least often—are also those who see themselves as incapable of influencing their future through their own behavior.[13] Thus, it may be that the poor do not undertake preventative actions, in part, because they do not believe such actions will do any good. Although more research is needed on this point, it seems advisable at the present time to suggest that educational programs be undertaken to strengthen the perception of personal control among the poor. One such program has been suggested by de Charms; its preliminary results, at least, are positive.[14] (Of course, such programs may prove to be counterproductive, if they are not accompanied by sociostructural changes that increase the real power of the poor.)

Values

Although Americans highly value their own health and that of their families,[15] they do not seem to place a high value on preventing health problems before they occur.[16] Actually, this low value placed on preventative health behavior reflects a more general American orientation to the present, rather than to the future, and the belief that the existing state of affairs is "good" and that when it is not "good" somehow things will take care of themselves. Two examples help make this point clear. Even though most Americans were acutely aware of the gasoline shortage over the winter of 1973–1974, survey data indicated that they also felt that the situation would improve in the future, and that drastic changes in personal and/or societal behavior were not necessary to prevent such shortages in the future.[17] Research on political socialization indicates that elementary school curricula encourage a vague, positive attachment to the American political and economic status quo—that is, that "Things are all right."[13] Thus, there seems to be a basic value, receiving the tacit support of the American educational system, in which the present is represented as good and preventative thinking is not encouraged. To increase preventative thinking about matters of health, educational programs are needed that are more realistic and that support the idea of planning for the future.

Research on political socialization suggests yet another aspect of values that is relevant to health behavior. A number of studies indicate that schools teach the principles of our democratic system without conveying the ways to carry out these principles. For example, the overwhelming majority of Americans endorse the principle of "free speech," but sizable majorities are also in favor of abridging this freedom when the speaker holds a nontraditional viewpoint, such as the support of atheism, communism, or homosexuality. The same condition seems to exist in regard to preventative health care. Friedson and Feldman found that 88 per cent of a sample of American adults agreed with the principle that regular visits to the dentist are desirable, but only 32 per cent reported that they, or members of their family, had had such a visit recently.[18] Thus, programs designed to teach people how to convert preventative health care principles into actual preventative health behavior seem necessary.

Attitudes and Opinions

In the belief-system hierarchy of Rokeach, Attitudes (attitude systems and individual attitudes) fall below Values. Opinions (beliefs about the way people and things are), such as "Most people do not know what is good for them," fall below Attitudes.

In several chapters of this book screening and testing programs such as the screening of infants for phenylketonuria (PKU) are discussed. Such programs must be designed and presented so as not to clash with the attitude systems either of those who must pay for and enact into law such programs, or of those who are the intended beneficiaries of the programs. Resistance from the middle and upper middle class is to be expected, to the degree that a particular program is seen as a free service. This is particularly true if the program is specifically intended for one group of people, e.g., a program to supplement the diets of pregnant women whose income falls below a predetermined "poverty line." Acceptance of such programs is heightened, not so much by appealing to humanitarian ideals (although such appeals may be useful with some audiences), but by stressing long-term cost-effectiveness of the program, as in the case of the PKU screening program in Massachusetts.[2]

Resistance to government prevention programs can also come from low-income minority groups. Although minority-group persons, even those at the bottom of the socioeconomic ladder, believe in "America," their experiences with various government-supported agencies and institutions (e.g., welfare, public housing, police, and public schools) are often abrasive and competitive. Thus, there exists in the ghetto or *barrio* an ideology (a system of attitudes and opinions) of dislike for and distrust of most governmental institutions. Even programs that are well intentioned and generally well designed will be resisted if they arouse the suspicions of the community. Health screening programs are most likely to be resisted if (1) they require more data from respondents than is necessary ("They are using us as guinea pigs again") or (2) there is any reason to believe the data might be used against the individual or community ("They are spies for the welfare or police department"). The best way to avoid such resistance is to include members of the target community in program-planning discussions, beginning at the earliest possible stages.

At the level of specific attitudes, the most obvious possible determinant of preventative health behavior is attitude toward the medical profession and individual physicians. Suchman provides evidence that those lower in socioeconomic status are less favorably disposed toward medicine and its practitioners than those higher in status.[11] According to other research, it is known that the poor are also least likely to take preventative health actions. Kasl and Cobb, on the other hand, report that those taking advantage of free medical examinations do not have a more positive attitude toward medicine than those who do not get such examinations.[3] This apparent inconsistency may be resolved by looking at attitudes toward medicine in the same way that Cloward and Jones looked at parental attitudes toward the public education system.[19] They found that lower-income parents did not devalue education in general or have a more negative, overall attitude toward the public education system than did middle-income parents. The lower-income parents did, however, feel that the schools did not do enough to meet the special needs of the child from a poor family. With regard to health, low-income individuals' preventative behavior may be inhibited not by a negative attitude toward medicine and doctors in general but by negativity toward specific health care systems they have experienced. There is certainly evidence that the health care available to the poor is in need of significant improvement.[20] At least temporarily, though, those interested in developing preventative health programs should not assume that any group has generalized negative attitudes toward medicine.

A number of specific opinions must also be considered when one attempts to increase the public's "preventative thinking." First, many people believe that genetic diseases and deficits are the *fault* of the parents. Such a belief fosters guilt in those parents who have, or might have, genetically-impaired offspring and increases resistance to prevention programs on the part of others who see the problem as the burden of the individual and not of society at large. Second, many people also believe that genetically-determined conditions are not treatable or preventable. Research is needed to determine how pervasive these beliefs are, and to what extent they impede acceptance and use of preventative health care facilities.

Cognitions About the Attitudes and Behavior of Others

A very consistent finding regarding health behavior is that individuals are highly influenced by the attitudes of their family and friends.[16, 21]

Those persons who undertake preventative actions are more likely to have family and friends who approve of such action and who undertake such action themselves. Those who develop negative attitudes toward a particular type of prevention have often been influenced by the attitudes of persons around them.

These findings are congruent with much accumulated research indicating the persuasiveness of face-to-face contacts. In forming attitudes toward politicians or deciding whether to adopt a new product or technology, people are much more often influenced by individuals in face-to-face interactions than by mass media campaigns.[22] The mass media make people aware of new ideas, but persuasion by individuals, particularly those termed "opinion leaders," is required for acceptance of such ideas. Opinion leaders attend more to the mass media than their peers (i.e., they read more newspapers and magazines, watch more TV), have many personal contacts, and are regarded by their own peers as having expertise. Any program to increase "preventative thinking" must capitalize on opinion leaders. Prior to mass media campaigns, therefore, such people must be identified and convinced of the efficacy of the program being developed. This approach is particularly important in low-income and minority communities where there is less attention paid to the mass media and greater importance placed on face-to-face communication.

TOWARD AN OVERALL PROGRAM TO INCREASE POSITIVE ORIENTATIONS REGARDING HEALTH BEHAVIOR

A comprehensive attempt to increase acceptance of preventative health practices should be aimed at three target groups: (1) the adult general public, (2) medical personnel, and (3) children.

Influencing the General Public

The public needs to be convinced to undertake a program of preventative action for themselves and to support programs to help all people—particularly the poor, who are at the greatest health disadvantage. Public support for governmental action is most likely when this action is presented as cost effective, and when the action is not seen as infringing upon the individual's freedom of action in any way.

Although the most effort should be devoted to enlisting the support of opinion leaders, campaigns to influence individual health practices or acceptance of programs should also use the media to convey information. In using mass communication the following point should be recognized: Americans tend to see many dramatic shows as educational[23]—

that is, many Americans regard television shows like *Marcus Welby* as presenting the facts of medical practice and knowledge. Thus, any media campaign to increase public acceptance of measures to prevent mental retardation and genetic disease should include efforts to influence the producers of *Marcus Welby* and related programs to portray a favorable view of such measures. In addition, these programs should be designed to influence the individual viewer's belief system regarding health behavior — that is, the viewer should be made to feel susceptible and, at the same time, should see himself as able to reduce threat through his own actions (self-concept); he should be made to feel that preventative behavior is highly desirable and does not clash with other important values (values). The television viewer should be convinced that genetic disease — or disease, in general — is not a sign of personal failure and is not a condition that is unalterable (attitudes and opinions); he should be shown that people like himself take preventative actions (attitudes and behavior of others). These points are more likely to be accepted by the viewer when he is watching a dramatic series (assuming the viewer sees it as educational) than when watching a documentary program or a commercial. A commercial, especially, is seen as an attempt to influence, and people may resist the message to protect their self-image as independent, thinking agents, free to make their own decisions. When watching a dramatic series, on the other hand, the viewer does not have such a defensive attitude.

Individuals acquire new information and adopt new attitudes most readily when they are most personally relevant and useful. Attempts to prevent genetic disease, therefore, should concentrate on those times in an adult's life when he or she is most likely to be thinking of children. Milunsky offers a promising idea: when a couple obtains a marriage license, provide them with a packet of information about genetic disease and its prevention.[24] Such a packet should be short and easy to read. To increase comprehension and acceptance of the message it might be possible to construct the packet according to principles of programmed learning. Another advantageous time to discuss preventative action is during a family planning session. Doctors and personnel of family planning clinics have access to adults when questions of preventative health are most salient.

Influencing Medical Professionals

As noted in the previous section, people are most open to new information when it is salient to them. With regard to health matters, therefore, medical professionals (particularly obstetricians and pediatricians, in the case of genetic disease and mental retardation) are in a position to be potent shapers of positive orientations toward preventative health behavior. To effectively carry out such a role, however, these professionals must be trained to recognize how attitudes and beliefs fit within the individual's total belief system. Such training is most logically done in medical schools, and some medical schools have added this type of training to their curricula. These educational efforts should be expanded and more attention should be paid to overcoming the problems involved in expanding the physician's role.[25]

Medical school curricula should also be expanded to include training in how to treat effectively and influence patients of various ethnic and social class backgrounds, particularly those of lower class and minority groups. Brown, as well as Coe and Wessen, has discussed the problems that occur when doctor and patient have different cultural perspectives.[26] This clash of cultures impairs the effectiveness of the physician not only as a healer but also as a teacher. The recently developed "culture assimilator" provides a relatively quick and easy way of teaching about another culture and may be a useful addition to medical school curricula.[27]

Influencing Children

In the long run, "preventative thinking" can be increased only if today's younger generation can be positively influenced. Four points must be kept in mind in developing educational programs to influence children: (1) While it is essential to teach the principles of preventative health, it is equally — if not more — important to teach how these principles may be easily and effectively converted into action. For example, a program designed to influence nutritional practices not only should inculcate the value of balanced meals, but also should teach children of varying cultural backgrounds and economic status how to achieve such balance, given their purchasing ability and traditional food preferences.

(2) Educational curricula and the personnel who present these curricula to students must be carefully selected. It is particularly important to avoid presenting preventative thinking and behavior about health in a moralistic, "preachy" fashion. Part of the failure of some school anti-smoking and anti-drug programs can be traced directly to the dogmatic way in which they were presented to students.[28, 29]

(3) Effectiveness of educational programs will be maximized if they are made directly relevant to the students' lives. For example, sickle cell anemia not only can serve to exemplify the relationship between genetics and disease but also can serve to make a discussion highly personally relevant to Black children, especially if the African cultural heritage of the children is woven into the lesson.

(4) Positive orientations toward health behavior will be increased if children are taught to place a high value on prevention in general. In

part, this requires that schools teach a more tempered view of the present state of affairs, since blanket acceptance of the status quo works against planning for the future. Increased preventative thinking also requires that preventative actions be seen as a positive aspect of one's self-concept.

A PHILOSOPHY FOR IMPLEMENTING AND ALTERING PROGRAMS TO INCREASE PREVENTATIVE THINKING

The major portion of this paper has been devoted to presenting ideas which might help in formulating programs to increase positive orientations toward health behavior. As noted at the beginning of the chapter, these ideas are not well grounded in empirical research and many of the ideas are extrapolations from areas quite remote from health behavior (e.g., political socialization). Given these cautions, programs based on the ideas presented should be advanced in the spirit of what Campbell calls the "experimenting society."[30] By this he means the following: (1) programs which are implemented are those which seem most likely to succeed, given our present knowledge (i.e., we do not delay program implementation until "all the data are in"); (2) evaluation is an *integral* part of each program, so that effectiveness or ineffectiveness of various techniques can be clearly demonstrated; and (3) techniques that do not produce positive results are abandoned or, preferably, modified according to information gained from evaluative research. Although this philosophy of social action is much more difficult to achieve than to describe, Ashmore and McConahay have provided some guidelines to facilitate its implementation.[13] If the programs outlined above can be made operational within a social context which at least approximates the "experimenting society," then it would be possible to move more systematically toward the time when the frequency of mental retardation and genetic disease is sharply reduced.

References

1. Coe, R. M., and Wessen, A. F.: Social-psychological factors influencing the use of community health resources. Am. J. Public Health, 55:1024, 1965.
2. Sorenson, J. R.: Social Aspects of Applied Human Genetics. New York, Russell Sage Foundation, 1971.
3. Kasl, S. V., and Cobb, S.: Health behavior, illness behavior, and sick role behavior. Arch. Environ. Health, 12:246, 1966.
4. Cook, S. W., and Selltiz, C.: A multiple-indicator approach to attitude measurement. Psychol. Bull., 62:36, 1964.
5. Wilson, G. D. (ed.): The Psychology of Conservatism. New York, Academic Press, Inc., 1973.
6. Haefner, D. P., Kegeles, S. S., Kirscht, J., and Rosenstock, I. M.: Preventative actions in dental disease, tuberculosis, and cancer. Public Health Rep., 82:451, 1967.
7. Green, L. W.: Status identity and preventative health behavior. Pacific Health Education Reports: Schools of Public Health. Report No. 1. Berkeley, University of California, and Honolulu, University of Hawaii, 1970.

8. Williams, A. F., and Wechsler, H.: Dimensions of preventative behavior. J. Consult. Clin. Psychol., 40:420, 1973.
9. Rokeach, M.: The Nature of Human Values. New York, The Free Press, 1973.
10. Tarjan, G.: Studies of organic etiologic factors. *In* Caplan, G. (ed.): Prevention of Mental Disorders in Children. New York, Basic Books, Inc., 1961, pp. 31–51.
11. Suchman, E. A.: Social factors in medical deprivation. Am. J. Public Health, 55:1725, 1965.
12. Rosenstock, I. M.: Why people use health services. Milbank Mem. Fund Q., 64: Part 2, pp. 94–127, 1966.
13. Ashmore, R. D., and McConahay, J. B.: Psychology and America's Urban Dilemmas. New York, McGraw-Hill Book Company, 1975, Chapters 3 and 8.
14. de Charms, R.: Personal causation training in schools. J. Appl. Soc. Psych., 2:95, 1972.
15. Cantril, A. H., and Roll. C. W., Jr.: Hopes and Fears of the American People. New York, Universe Books, 1971.
16. Russell, R. D.: Motivational factors as related to health behavioral change. *In* Veenker, C. H. (ed.): Synthesis of Research in Selected Areas of Health Instruction. New York, The School Health Education Study, 1963, pp. 92–110.
17. Murray, J. F., Minor, M. J., Bradburn, U. M., Cotlerman, R. F., Frankel, M., and Pisarski, A. E.: Evolution of public response to the energy crisis. Science, 184:257, 1974.
18. Friedson, E., and Feldman, J. J.: The Public Looks at Dental Care. New York, Health Information Foundation, Research Series 6, 1958.
19. Cloward, R. A., and Jones, J. A.: Social class: Educational attitudes and participation. *In* Passaw, A. H. (ed.): Education in Depressed Areas. New York, Teachers College Press, 1963.
20. Bullough, B., and Bullough, V. L.: Poverty, Ethnic Identity, and Health. New York, Appleton-Century-Crofts, 1972.
21. Merrill, M. H., Hollister, A. C., Gibbens, S. L., and Haynes, A. W.: Attitudes of Californians toward poliomyelitis vaccination. Am. J. Public Health, 48:146, 1958.
22. McGuire, W. J.: The nature of attitudes and attitude change. *In* Lindzey, G., and Aronson, E. (eds.): The Handbook of Social Psychology. 2nd ed., Vol. 3, Reading, Massachusetts, Addison-Wesley Publishing Co., Inc., 1969, pp. 227–235.
23. Weiss, W.: Effects of the mass media of communication. *In* Lindzey, G., and Aronson, E. (eds.): The Handbook of Social Psychology. 2nd ed., Vol. 5, Reading, Massachusetts, Addison-Wesley Publishing Co., Inc., 1969, pp. 77–195.
24. Milunsky, A.: The Prenatal Diagnosis of Hereditary Disorders. Springfield, Illinois, Charles C Thomas, Publisher, 1973.
25. Richmond, J. B., and Lipton, E. L.: Studies of health of children with specific implications for pediatricians. *In* Caplan, G. (ed.): Prevention of Mental Disorders in Children. New York, Basic Books, Inc., 1961, pp. 95–121.
26. Brown, E. L.: New Dimensions of Patient Care. New York, Russell Sage Foundation, 1965.
27. Fiedler, F. E., Mitchell, T. R., and Triandis, H.: The culture assimilator: An approach to cross-cultural training. J. Appl. Soc. Psych., 55:95, 1971.
28. James, W. G.: Teenage smoking and lung cancer. J. Hlth. Phys. Ed. Recr., 31:25, 1960.
29. Nowlis, H. H.: Student drug use. *In* Korten, F. F., Cook, S. W., and Lacey, J. I. (eds.): Psychology and the Problems of Society. Washington, D.C., American Psychological Association, 1970, pp. 417–418.
30. Campbell, D. T.: The experimenting society. Paper presented at the 1969 meeting of the American Psychological Association, Washington, D.C.

CHAPTER 3

GENETIC COUNSELING: PRINCIPLES AND PRACTICE

Aubrey Milunsky, MB.B.Ch., M.R.C.P., D.C.H.

Genetic counseling is a communication process concerning the occurrence and the risks of recurrence of genetic disorders within a family. The aim of such counseling is to provide the consultand(s)* with the fullest comprehension of all the implications of the disease in question and all the possible available options. A further goal, and intrinsic to the expressed aims of counseling, is helping families through their problems, their decision-making, and their adjustments. While the primary strategy of counseling is to achieve understanding that leads to rational decisions, there is also the obvious hope that such decisions will indeed lead to a decrease in the incidence of serious genetic disease and will prevent the suffering of both patient and family. Most counselors would probably agree with the broad definition and goals of counseling as stated above. It has already become apparent, however, that opinions about how this information should be conveyed and whether or not it should be withheld vary among counselors.

Inherent differences among counselors in moral attitudes, race, religion, age, training, discipline, and the like make it impossible to obtain any unanimity of approach to the delivery of genetic counseling. Nevertheless, there is a need to provide basic guidelines in genetic counseling to people of widely different cultures. Some of the principles outlined below, which are culled from and refer primarily to peoples of Western culture, will inevitably meet with disagreement. Moreover, they

---

*The term "consultand" denotes the person or persons receiving counseling.[1]

reflect an intensely personal view based upon the premise that there *are* human rights of personal inviolability. Two such fundamental rights are (1) the right to life with a self-determined destiny, and (2) the right to marry and procreate.

These rights delineate the province of individual privacy and moral autonomy in making decisions about procreation. Although parents may seek genetic, or even moral, information and guidance, genetic counseling is seen as enhancing their freedom of choice by clearly defining reproductive options and enriching their decision-making process.

Genetic counseling will be discussed under four headings:
1. Guiding Principles in Genetic Counseling
2. Prerequisites for Genetic Counseling
3. Problems in Genetic Counseling
4. Major Genetic Counseling Programs

GUIDING PRINCIPLES IN GENETIC COUNSELING

The increasing, though belated, prominence of genetic counseling in general health care delivery requires careful attention to its guiding principles. Despite this need, recent texts in genetics have made little or no effort to recognize such principles.[2-8] Hecht and Lovrien[9] did address themselves to certain principles of genetic counseling. Some authors may not agree with *all* the principles and prerequisites outlined in the following sections, but few would disagree with the need to enunciate clearly such guiding principles.

Accurate Diagnosis

Genetic counseling simply cannot begin without the establishment of an accurate diagnosis. It is a common experience for clinical geneticists to see patients referred for counseling for whom the alleged diagnosis is clearly incorrect. Great care and attention must be exercised either in confirming the diagnosis or in obtaining sufficient data to ensure the certainty of the diagnosis in question. Earlier photographs of a previously deceased offspring, autopsy reports, hospital records of other affected family members, results of carrier detection tests elsewhere, and other similar information may be crucial in the confirmation of the exact diagnosis. Failure to reach a proper diagnosis is not rare, however.

Data may be insufficient, tests may still be inconclusive, the syndrome may not have been described before, or other complicated considerations may induce hesitation in diagnosis. In such instances counseling could be very misleading. Since all statements made during counseling would be prefaced by the word "if," none of the information offered the consultand may be correct or accurate. Not infrequently, patients in these uncertain diagnostic categories leave the counseling center and promptly forget the "if," immediately applying the counseling information to themselves and their families. Some counselors in these situations even provide percentage risk figures. In such cases it is not unusual to hear the consultand referring to a recurrence risk by a specific figure which he has come to believe applies accurately to him or her. It would seem more prudent, in the absence of an exact diagnosis, to refrain from guessing the recurrence risk. An unrecognized syndrome may be recessive (similar to Meckel's) and may carry unexpected high risks. An acute awareness of the patient's quandary in such situations is extremely important.

Noncoercive and Nondirective Counseling

The role of the counselor is to assist consultands in recognizing *their personal* major priorities crucial to their decision-making. The counselor's expertise is his or her special perspective which allows discernment of problems often not immediately obvious to the consultand(s). Hence, the previously unrecognized, highly developed distaste of one parent for living with a retarded child in the house may become apparent to the other spouse only during counseling. Their acknowledgment of this facet may have a compelling effect on their decision-making. Failure to consider the *short- and long-term effects* of a child with genetic disease on the siblings is common. Indeed, with their feelings and concerns about *themselves* often foremost in their minds, prospective parents often give little attention to the possible effects of the particular genetic disease on a *future child*. Hence, the counselor should *not* be seen in the role of advising or telling consultands what they ought to do, or how they should act, but in helping them recognize and anticipate issues and problems before they occur.

The World Health Organization Expert Committee on genetic counseling endorsed this nondirective approach,[10] and both I and others have recommended such practice.[11-13] In line with this thinking, the majority of counselors have indeed perceived themselves as being nondirective.[14] Sorenson's data, for example, indicate that 54 per cent of counselors tend to leave all decisions to the parents, and only 7 per cent either always tell the parents what they would do in the same circumstance or give outright advice as to what parents "ought" to do.[14] However, the data further suggest that 64 per cent of counselors report that it has always been appropriate to inform consultands in a way that would guide them toward an appropriate decision. This apparent contraindication in the data highlights the difficult dilemma and possible paradox confronting counselors. In the main, counselors try to remain impartial and objective as they communicate all the information consultands require for rational decision-making. Paradox invariably arises when counselors recognize their parallel desire to decrease the frequency of serious genetic disease and to maintain their objectivity. Indeed, Childs has questioned how the physician who participates so intimately in the afairs of his or her patients in every other way "can maintain an almost passive indifference to the outcome in genetic counseling" (personal communication). Emery has felt that it would be proper to suggest—in at least serious and high-risk cases—that "family limitation would be advisable."[15] When parents already manifest established tendencies or attitudes, an encouraging supportive approach has been advocated by Pearn.[16]

The intrinsic danger of the directive approach is the insinuation by the counselor of his or her own religious, racial, eugenic, or other dictates into the counseling offered. Repeated recommendations by certain physicians to patients to avoid prenatal genetic studies on religious grounds serves as a common example. Some of these physicians have even argued that their actions were in the best interests of the family! (In this author's experience these individuals are often the first to resist any directive approach for themselves or their families.) Such an approach should be condemned not only on the grounds of counselor prejudice, but because it constitutes a moral affront to individual privacy in reproduction and other matters of decision-making.

Some geneticists relate that patients often ask of them, "What would *you* do?" This question, usually coming at the end of a session, may signal failure in counseling. The patient, of course, is *not* interested in what the physician would do in like circumstances, but rather, is asking for a decision to be made for her (or him). If every effort is expended in ensuring recognition by the patient of his or her own priorities, then this question generally does not arise. Failure to spend sufficient time in counseling—with the frequently required repeat visits—may make for greater indecision and unhappiness among consultands and a less effective service.

I believe that in the proper practice of anticipatory genetic counseling this question need not arise if the approach by the counselor has been comprehensive and has been beamed at elucidating the major priorities, beliefs, fears, and other concerns of the family being counseled. Accustomed to the paternal medical approach of receiving therapeutic directives, many individuals may expect or hope that the physician will make their decisions for them.

Every effort, therefore, should be expended in helping the consultand and family to recognize the issues of greatest import in their lives. By thus helping to place the concerns of each family in relative perspective, their decision-making process is clarified, enhanced, and indeed enriched. A decision so reached makes the family as happy as is possible, and the likelihood of possible recriminations, guilt, and fear of culpability is held to a minimum. Further, the need for the counselor to inject his "guidance" is minimized.

Concern for the Individual

Among the many issues that some feel should be raised at the genetic counseling session are those that pertain to the economic and other burdens of the disorder on society. In the assessment of the priorities recognized by the family (or individual), this issue should be relegated to a position very low on the scale of values used in the decision-making process. At least 53 per cent of the counselors studied by Sorenson[14] felt that the question of the possible economic burden on society should not even be raised in counseling. The rest of the counselors were equally split between those who felt that this issue should be raised by the counselor and those who felt that the question should be raised by the consultand. In today's world, I do not believe that responsible parents can totally ignore the results on society of their procreative activity. As part of the comprehensive view of all the issues under consideration, this societal question could receive passing attention. However, the major emphasis should be on concern for the individual, the available priorities, and personal or family needs and choices. Concern for the individual must always override consideration of the needs of society. In the ultimate analysis society has been seen to have ways of influencing or guiding its citizens. The role of the counselor is not seen to be that of an advocate for society.

Truth in Counseling

It might seem strange to some to have included truth in counseling as a general principle. There are, however, many situations in genetic counseling where facts may be purposely distorted, de-emphasized, or hidden. Statements may even be made with the clear intent to mislead. For example, the consultand with neurofibromatosis may be concerned not only about his or her own future, but also with the type of involvement that might occur in 50 per cent of the offspring. While wishing to provide information on the full spectrum of disease in neurofibromatosis with future offspring in mind, the counselor might consider communicating information in a drastically curtailed form to prevent unnecessary anxiety in the affected consultand.

To avoid the obligatory culpability engendered in the mother after the diagnosis of X-linked disease in her child, some counselors may try to conceal the true mode of inheritance. In view of the implications for other members of the family—for example, sisters of the consultand—this approach should not be countenanced. It is not unusual for the geneticist to have to state his or her inability to make, or exclude, a certain diagnosis. There should be no hesitation in admitting the limits of personal ability (a recommendation especially pertinent to practitioners away from major centers) and in seeking additional consultations with specialists. Similarly, the unexpected—but nevertheless inconclusive—nature of a certain test for carrier detection should be communicated to the patient together with its implications. It is equally important to acknowledge both ignorance and the dilemma of an unexpected prenatal diagnosis.

Occasional situations arise where genetic counseling is sought by a close friend of the counselor. As in the treatment of one's own family and close friends, genetic counseling is best not initiated. The need for complete disclosure of highly personal matters concerning reproduction, contraception, family planning, and family defects, provides sufficient emotional fire to mar or destroy any friendly relationship—besides compromising the need for objectivity.

Mass screening for carriers of certain hereditary disorders has unmasked a host of new problems. For example, in screening for sickle cell trait, many families have been found in which paternity could be clearly disputed. We have referred to the legal morass[17] that may bedevil such situations and cannot easily recommend a *modus operandi* in these cases. As far as truth in counseling is concerned, regard for the continuing survival of the family unit might constitute the only reason such a principle might have to be forfeited.

Confidentiality and Trust

As in every other physician-patient relationship, trust and confidentiality are paramount considerations. It is not unusual for persons holding a Ph.D.—rather than an M.D.—to provide counseling, since they are not bound by any Hippocratic oath or peer professional ethical imperatives, this stricture *may* be especially applicable to them. The counselor may discover that his or her intent for confidentiality may be readily assailed. If the diagnosis of an obligate carrier state for an X-linked disease is made in a mother, she may expressly forbid the counselor from communicating this information to her sisters at risk. These particular—and hopefully rare—situations call for very careful decision-making by the counselor to avoid any of the legal pitfalls that have been discussed involving the transmission of privileged communications and breach of medical ethics.[17]

Additional problems have recently arisen as a direct result of advances in genetic technology. Do physicians have a right to communicate all genetic information (including carrier status or likelihood of development of genetic disease, such as Huntington's chorea) to insurance companies? Can this be done only with the express consent of the patient or is this information part and parcel of the usual medical communication from doctor to insurance company?

Yet another situation in which confidentiality and trust in counseling may come under heavy pressure is represented by the following case: One parent, at 45 years of age, was diagnosed as having Huntington's chorea. One of her sons is a commercial jumbo jet pilot—and it is he who seeks counseling. Although the consultand is devoid of any sign of the disease at the time of counseling, he is cognizant of his 50 per cent risk of actually having the disease. Does the physician have the obligation to inform the consultand's employers without his consent if the young man refuses to make this communication himself? There are probably many such examples in genetic counseling in which there is a need to balance the benefit to the individual against the danger to society. Appropriate decisions will invariably be made on the basis of the merits of each individual case. Hopefully, the geneticist will bear in mind the principles of truth in counseling, coupled with awareness of the ethical and moral principles of confidentiality and trust.

Parental Counseling

Capron has argued eloquently that parents have a legal right to be fully informed in their decision-making regarding procreation.[18] At the same time, the counselor has a *duty to convey* to such parents the various options, risks, benefits, and consequences available or foreseeable.[18]

Hence, and notwithstanding this legal exhortation, it is a mandatory principle that genetic counseling be provided to *both* parents at the same time. The complex issues of guilt, culpability, family prejudices, serious differences of opinion between spouses, and ignorance and fear—to mention only a few—form the basis of this important principle. Too often the more concerned parent comes alone to counseling with resulting emotional chaos, incorrect interpretation to the spouse, or lack of appreciation of the true risk situation. No substitute—including letters—can take the place of face-to-face discussions, with opportunities for questions and interchange of unacknowledged ignorance or issues.

Every genetic counselor recognizes the need for follow-up visits by the consultand(s). The follow-up interviews allow for reinforcement and repetition of the counseling. Failure to comprehend the information provided, correction of misapprehensions, and communication of medical advances, options, and other factors are all adequate justification for such visits. Moreover, the follow-up visits may provide valuable insights into the counselors' practice and may lead to improvement of services offered. In this context, Clow et al.[13] in a small study of counseled families, noted that 50 per cent felt that a single counseling session was insufficient and that all expressed the need for some form of continuous counseling.

The Timing of Genetic Counseling

In the past (hopefully) the tradition of medicine has been to initiate preventive actions after the occurrence of tragedy. Hence, genetic counseling in the past has invariably been offered after the birth of a child with genetic disease or congenital malformation. Recent advances in genetic technology — mainly prenatal diagnosis and carrier detection — now enable the physician to counsel prospectively many patients at risk. Today, in a variety of situations, prospective advice is both critical and important. Many can now benefit — the prospective mother found to be a carrier of hemophilia because of the family history; the Black couple discovered to be carriers of sickle cell anemia through a population screening program; an Ashkenazic Jewish couple found to be carriers of Tay-Sachs disease through an ethnic population screening program.

Ideally, counseling should be offered prior to conception or marriage. Those of eugenic inclination hold that genetic counseling should be offered early enough to provide options for those choosing their mates — especially if the option of abortion is odious to them.

Certainly, the time to initiate counseling is *not* mid-pregnancy when, other considerations aside, some carrier tests may be unreliable (e.g., serum hexosaminidase).[19] It is also crucial to bear in mind the inability of patients anguished by a recent loss to assimilate even the essence of counseling offered. While parental counseling immediately after such loss is important in the prevention of unplanned further pregnancies, there should be an awareness of the need to repeat the counseling offered some months later.

PREREQUISITES FOR GENETIC COUNSELING

Knowledge of the Disease

A large body of factual knowledge must be considered a vital prerequisite for counseling. Required knowledge would include information as to how to make and confirm the diagnosis, the recurrence risks, the mode of inheritance, the tests available for carrier detection, the variable onset forms of the disease (e.g., infantile, juvenile, adult), the causes of death, the prognosis for survival. Other considerations are the specific and important complications (e.g., neoplastic change in neurofibromatosis or the likelihood of mental retardation or seizures in tuberous sclerosis), the available treatment and its efficacy, the likely burden on the family (including the effect on other siblings), the family economics, and the marital sex life. Basic to such considerations and knowledge would be an understanding of the suffering and involvement of the affected individual.

Counselor Qualifications

"Know thyself" would be an apt minimum prerequisite for any counselor. A true awareness of personal bias and attitude is important in the counseling process, regardless of which principles enumerated above the counselor perceives as correct or reasonable. No one has accurately delineated the body of knowledge required for proper genetic counseling. The American Society of Human Genetics has begun to wrestle with the problem of the necessary qualifications of counselors and has taken up the question—among others—concerning the place of physicians, basic scientists (with Ph.D.'s), and para-counselors.[20-21] (See later discussion on Training Programs in this chapter.)

I share Epstein's[20] view that optimum genetic counseling is provided by a team constituted of physicians, basic scientists, para-counselors, social workers, psychologists, and administrators. Such teams are best located within medical school–hospital complexes since they provide a wide variety of specialty resources. Extremely few areas in the world are able to muster such extensive in-depth genetic counseling teams. However, it would be foolish to believe that all genetic counseling should occur in major centers (although genetic counseling so provided would surely be the best possible). Practitioners or scientists in many disciplines can and do provide good genetic counseling away from major medical centers. Cognizant of the principles enumerated above and the prerequisites and problems just discussed, such counselors could enrich their efforts by seeking additional consultation or testing at a time when they are allied to major medical centers.

Ability to Communicate

Genetic counseling is both an art and a science. While virtually anyone can communicate the risk figures for autosomal recessive disorders and other genetic diseases, much care, sympathy, understanding, and insight into human nature are necessary to communicate effectively all the required information in a counseling session. The use of simple language, the recognition of anxiety, and the ability to anticipate and answer the unspoken fears and questions help immeasurably to make the counseling experience most beneficial and valuable.

Notwithstanding excellence in face-to-face counseling, many teams have recognized the importance of summarizing the counseling provided in a letter not only to the family physician but also to the consultand(s). Hsia (at Yale) has already shown that almost all patients receiving such letters do indeed keep them in a safe place for further reference or because of the relevance of the information to their children.[22]

Reference to the need for the family to remain in contact with the genetic counseling team because of potential advances in carrier detection, prenatal diagnosis, and other new procedures is as important as reassurance about continuing emotional support to the family in question. This kind of comprehensive approach will assure effective communication and prevent failure of the patient to understand the information being presented.

Knowledge of Ancillary Needs

In the presentation of alternatives and options open to the consultands, the counselor must be familiar with a variety of other important information, such as the possibilities for prenatal diagnosis, the procedures and problems of adoption, the difficulties and intricacies of artificial insemination by a donor, the applicability and problems of the use of contraceptives, and morbidity and even mortality from anticipated therapeutic abortion; all of which constitute aspects of the problem that the counselor can reasonably be expected to discuss.

Humanity

The need for this prerequisite may be regarded by some as superfluous and taken for granted by others. Hopefully, however, there are none who believe that genetic counseling is simply the communication of risk and recurrence figures. Sensitivity and awareness of the patient's plight and predicament are paramount. The consultands, for example, may have lost their only child from a chronic genetic disease. This might have been a child born after years of effort to conceive. Inevitably, there will be fears and anxieties to deal with. The obligate maternal carrier of an X-linked disease may have severe problems with guilt. Similar problems of personal culpability may need airing in situations where both partners are carriers. The economic and emotional burden resulting from having a genetically defective child in the family may lead to endless and often irreparable problems within the family. Marital disharmony, problems in otherwise normal, marital sexual relations, and complicated behavioral disturbances may ensue some time after the birth of a child with serious genetic disease. The overriding anxiety inevitably accompanying other feelings after the loss of a newborn or young child from genetic disease will almost invariably block full assimilation by the parents of any genetic counseling provided. The experienced counselor should verbalize to the parents in this situation his recognition that such anxiety is normal and expected. He should explain, furthermore, that the anxiety frequently interferes with the complete assimilation of all the information provided in counseling. The counselor should recommend—automatically—that the couple return again within three to six months for another complete counseling session. It is important that the consultant display a sympathetic attitude and be a good listener. Frequently, the counselor may perceive that the main questions, fears, and anxieties are *not* in fact related to genetic disease.

Efficacy of Genetic Counseling

To measure the effectiveness of genetic counseling, the degree to which the goals of counseling have been achieved must be studied. The expressed primary goal is to achieve complete consultand understanding, thereby allowing rational decision-making. A secondary aim (but a primary hope) is the prevention of genetic disease.

Carter and Roberts followed up 450 couples who had been studied in their clinic three to ten years earlier.[23] They found—as have other investigators[23-26]—that when there was a high risk of recurrence of genetic disease, as well as a significant long-term handicap, the parents were generally deterred from further reproduction. Their reported deterrence rates, however, refer to changes in *attitudes;* actual changes in reproductive behavior were not as good. Their optimistic results may also have been a reflection of patients in high socioeconomic classes. In contrast, Sibinga and Friedman noted that parents of children with phenylketonuria often failed to comprehend information communicated to them by their physicians.[27] In a similar vein, Taylor and Merrill observed that 12 out of 21 couples who had previously had a child with Duchenne muscular dystrophy did not know their risks for having future affected offspring.[28] Reynolds et al.[29] found that 84 per cent of their consultands had adequate understanding of counseling provided; however, 16 per cent had distinct distortions or rejection of the information given, and two consultands "repressed the entire counseling experience."

Failure to comprehend the information offered in counseling may have sad consequences. In one case, because of the erroneous impression of a high risk, the wife underwent a tubal ligation following counseling.[29] Efficacy of counseling is best measured by studying the rationality of decisions made by couples at risk. In so doing, careful assessment need be made of the degree or extent of understanding by the consultand.[30, 31] Rational decisions usually, but not always, reflect action toward the prevention of high risk disease. Cases in which the couple chooses to conceive another child cannot necessarily be viewed as failures in genetic counseling. Pearn[16] described an unusual, though striking, example of a mother with two children with celiac disease who "wanted" her next child to have the same disease! (The woman had the ease of dietary management in mind for *all* her children.) Four major obstacles (religion, denial, lack of grasp of probabilities, and lack of knowledge of genetics and biology) that interfere with rational decision-making will be elaborated on in the following section.

The value and efficacy of any medical service should be constantly evaluated. It is patently evident that genetic counseling services require careful and extensive monitoring, as well.

Legal and Ethical Problems

See Chapter 19.

PROBLEMS IN GENETIC COUNSELING

The Counselor

Virtually all counselors would concur with the primary goals of counseling just emphasized. Much less of a consensus is likely, however, when counselors consider what they hope their counseling will achieve. Almost all would agree to achieving the prevention of disease, the happiness of the consultand(s), and the tempering of guilt or anxiety. Other counseling achievements, such as the prevention of an increase of deleterious genes in the population, improvement of the general health and vigor of the population, or reduction in the number of carriers of genetic disorders are likely to be ranked of less importance by the majority of counselors. Studies of the efficacy of genetic counseling may reflect, to a significant degree, the unexpressed goals of the counselor. Indeed, Sorenson's data[14] indicate clearly that 64 per cent of counselors do in fact influence decisions.

While, intellectually, the majority of counselors would agree that impartiality and objectivity are extremely important facets of counseling, many of those counselors in an actual consultation are seen to be providing directive counseling. Sorenson noted (Table 3–1)[14] much variability in the counselors' perception of their obligation in genetic counseling. Remarkably, counselors felt that consultands could raise the question of the economic burden of the particular disorder, possible social stigma attached to the disorder, the disease as a reason for limiting family size, sterilization, and sex life. Whereas, 53 per cent of the counselors felt that the question of possible economic burden of the disorder on society should not be raised in counseling, 23 per cent and 24 per cent, respectively, thought that either the counselor or the consultand should raise that question. Clearly, perception by counselors of their duty to raise specific questions in counseling is highly variable.

Counselor attitudes on factors concerning the decision to reproduce are also highly variable. For example, Table 3–2 shows that the economic situation of the family (29 per cent), the desire for children (10 per cent), possible societal burden (34 per cent), and the risk of children being carriers (40 per cent) were factors considered to be not too important in reproductive decision-making.[14]

While most counselors would intellectually agree about the rights of parents to have children, there are those who might feel that mothers whose offspring would virtually all be retarded (phenylketonuric mothers not on diet)[32] should not, by law, be allowed to procreate. It is also likely that perception of the burden of a particular disease by the counselor may differ appreciably from that of the consultand.

TABLE 3-1.   COUNSELORS' PERCEPTIONS OF OBLIGATIONS
IN GENETIC COUNSELING.*

*Question: Which of the following topics do you consider to be primarily
your professional obligation to raise and discuss, and which do you consider
to be the obligation of the counselee?†*

	Normally I Should (%)	Normally They Should (%)	Should Not Be Raised (%)
Religious views of counselees	37	55	8
Ability of counselees to handle economic burden of disorder	58	36	4
Possible effects on family life	88	12	0
Social stigma attached to disorder	38	42	20
Possible economic burden of disorder on society	23	24	53
Alternate forms of parenthood, including adoption, artificial insemination	79	21	0
Notification to extended family at risk for carrier state or disease	77	21	1
Counselee's view of the disease as a reason for limiting family size	47	51	2
Methods of contraception	65	31	4
Counselee attitude on sterilization	40	54	6
Effect on counselee's sex life	33	59	8

*Data courtesy of J. R. Sorenson: Counselor: Self portrait. Genetic Counseling, 1:31, 1973.

†"Counselee" = "Consultand."

Many factors are, therefore, seen to color or influence the counselor and include his or her eugenic views, religion, age, qualifications, training, and personal prejudices, as well as the state of his or her physical and mental health. Any evaluation of counselor attitude, however, will be bedeviled by the lack of specific training in genetic counseling—in the case of the vast majority of counselors.

The Consultand

Why do people come for counseling? What do they seek? For the most part most individuals are clear about their reasons for coming, and—at least in a general way—are conscious of the questions for which they require answers. On occasion they have no idea why they came or were referred.[29]

The various reasons for seeking genetic counseling most often fall into one of five categories:

1. Do they have a genetic disease?
2. Are they at risk of having an (or another) affected child?
3. What are the implications of the disease, its prognosis, and treatment?
4. What help is available to them in decision-making?

5. Where can they obtain help with an affected child?

Although, in general, the consultand and the counselor have the same clear perception of the reason for the consultation, frequently, there are discordant views. The consultand may, for example, be seeking guidance in reproductive decision-making only to find the counselor telling him or her what "ought" to be done. At times the consultand may not vocalize the real question(s) causing much anxiety. There may be guilt about infidelity, problems about previous or recent drug ingestion, or even previous, treated venereal disease. A finely honed sensitivity and an empathetic ear are critical tools needed in understanding, supporting, and guiding the consultand.

While the initial question or response at the beginning of the consultation will invariably be concerned with the reasons for coming, the counselor would be well advised to spend some time getting to know both the consultand and his background. The consultand (as well as the counselor described earlier) often brings to the consultation definite religious or eugenic views, as well as emotional tone which may be colored by fear, ignorance, guilt, denial, despair, depression, poverty, and possibly the tragedy of a previously affected child. A careful appraisal of the educational, economic, emotional, and family status of the consultand would be a *sine qua non* of good counseling. All these factors can be expected to influence counseling.

The appreciation of risk and the interpretation of odds is known to vary with the basic personality type.[33] A pessimistic consultand may, for example, reach different conclusions, with the same odds in mind, than

TABLE 3–2. COUNSELORS' PERCEPTIONS OF FACTORS RELATIVE TO THE DECISION TO REPRODUCE WHEN THE RISK OF GENETIC DISORDER EXISTS.*

Question: How important do you feel the following considerations should be in shaping decisions concerning reproduction where there is a risk of a genetic disorder in future offspring?

	Very Important (%)	Important (%)	Not too Important (%)	Not at all Important (%)
Estimated recurrence risk	71	28	1	0
Clinical manifestations of disease	75	23	2	0
Economic situation of family	17	49	29	5
Desire for children	33	51	10	2
Impact of defective child on marriage	47	45	7	0
Possible societal burden	19	42	34	6
Manageability of disease	48	49	3	0
Impact of disorder on child's mental state	59	38	3	0
Impact of disorder on child's physical state	41	55	4	0
Risk of children being carriers	16	34	40	10

*Data courtesy of J. R. Sorenson: Counselor: Self portrait. Genetic Counseling, 1:31, 1973.

an optimistic individual. Alker[34] has also shown a basic distinction in the attitude toward risk between an achievement-orientated and a failure-threatened subject. Moreover, Steiner indicates that attitudes to risk vary with changing moods.[35] Prior discussion of a risk situation may lead to an increased willingness by an individual to take greater risks.[16, 36-38] It remains unclear if this "risky-shift" phenomenon could occur as a consequence of genetic counseling.[38]

In an important study, Leonard et al. documented various aspects of the views and attitudes of consultands which before had been known or suspected only anecdotally.[30] They confirmed the commonly observed anxiety block to the assimilation of new information and observed that denial was sufficiently strong in some families as to have mothers refuting prognoses or holding that the disease was nongenetic. They reported one father, unable to accept his obligatory carrier status, who denied paternity. Their observations suggested imperfect reception of the genetic counseling in 44 per cent of families. Not unexpectedly, they found that the principal obstacle to the use of information was religion.

Much attention need be paid to three of their major observations. They suggested that reproductive attitudes are determined more by the perception of the burden of the disease, than by knowledge of exact risks of recurrence. They noted a significant inability, or lack of grasp of, probability. Finally, they concluded that lack of knowledge of genetics and human biology was perhaps the greatest obstacle to effective transmission of genetic information.

Family Problems

It is a sad commentary that intrafamily problems may compound the various difficulties normally encountered in genetic counseling. Communication between families, disappointingly, is often either absent or minimal and has made family history-taking an occasionally unrewarding exercise. We have seen families where the parents elected not to tell their living children about their first deceased or institutionalized child with mental retardation and a genetic disease. The new possibilities for carrier detection and antenatal diagnosis now make such parental actions morally and legally questionable. A mother may choose not to inform her sisters of their risk after she has had a child with a serious X-linked disease (e.g., Lesch-Nyhan syndrome). In a recent case of ours, the patient came for amniocentesis at 16 weeks gestation. She thought that her sister's child had Hunter's syndrome, which had not been proven, as it eventuated. An atypical mucopolysaccharidosis was suspected in that child. We suggested that an immediate skin biopsy (the first attempt having failed in culture) be obtained on the child. The mother's sister refused to cooperate, thereby preventing *specific* studies of the amniotic fluid cells (as opposed to the nonspecific radioactive sulfate incorporation studies).

Heterozygosity – Heterogeneity – Expressivity – Mutation

In general, case management is not problematic when there is an accurate test for the heterozygote, or even when there is no such test available. Difficulties arise when there is uncertainty about the heterozygote status – the test being inconclusive. For the Tay-Sachs carrier state, inconclusive results obtained from using serum for hexosaminidase estimations can usually be resolved by studying peripheral blood leukocytes from the same patient. For most of the lipidoses and other inborn errors of metabolism, a similar approach is yet to be proven. In cases where difficulties are being experienced in making a definitive statement about the heterozygote status of the consultand, our recommendation has been to offer definitive prenatal genetic studies when indicated. Many of the hereditary storage diseases still fall into this category of management.

The need to be aware of genetic heterogeneity is self-evident. The lipid storage disorders are typical with variable ages of onset – namely infantile, juvenile, and adult forms of disease. The evidence – at least for metachromatic leukodystrophy – shows no difference between the deficient arylsulfatase A in the infantile and in the adult onset forms of the disease.[39] In Farber's disease, ceramidase, active only at an acid pH (and not at an alkaline pH), is deficient.[40] The implications for such observations are in prenatal diagnosis, where it is crucial to seek a *specific* enzymatic deficiency.

Serious difficulties may result if there is lack of appreciation of genetic heterogeneity and variable expressivity. Take, for example, the firstborn infant of a young couple who is found to have ash-leaf achromic spots with seizures and who is diagnosed as having tuberous sclerosis. The question of recurrence risks posed by the parents may be difficult to answer. Since roughly 50 per cent of these cases are transmitted to offspring through an autosomal dominant mode of inheritance, it is critical to exclude such disease in either of the parents. Extremely careful physical examination of *both* parents is crucial. This includes examination of the fundi following dilatation of the pupils, as well as examination under the Woods light, and x-rays of the skull and abdomen (calcification) – all of which represent minimum absolute requirements for excluding the hereditary form of tuberous sclerosis. Even then, a small percentage of cases may be missed. Pneumoencephalography of a parent would seem to be justified only in the event of a grossly abnormal electroencephalogram or neurologic symptoms or signs.

Problems are also posed by highly significant rates of mutation for different diseases. When roughly 50 per cent of cases result from mutation (e.g., tuberous sclerosis), the recurrence risk would be remote. The difficulty is that a 100 per cent certainty of hereditary type exclusion is usually not possible. When a definitive test for the actual disease, (e.g., Marfan's syndrome and Huntington's chorea), is not available, the difficulties of counseling are further compounded.

MAJOR GENETIC COUNSELING PROGRAMS

From 4 to 5 per cent of liveborn infants suffer from genetic or partly genetic conditions.[10, 41] Furthermore, as the years pass, other genetic disorders become evident,[42] increasing the incidence of hereditary disease. Although genetic counseling should be an integral part of primary medical care[43] it has long been neglected in health care delivery and insurance schemes. Hsia has carefully described all the essential elements involved in an actual genetic counseling service[22] (Fig. 3–1).

While large numbers of people require genetic counseling, Sly has estimated that probably 90 per cent of such families will not be seen by a genetic counselor.[12] Wendt and Theile believe that less than 1 per cent of families with a genetic risk have been reached by genetic counseling in Germany.[44] The personnel and services available worldwide are grossly inadequate. Clow et al. have detailed the organization of medical genetics programs.[13] Emphasis was placed on the need to coordinate efficiently all resources so that defined objectives could be achieved, and patient benefits, maximized.

For major genetic counseling programs to develop, many counselor training programs, as well as massive public (starting in school) and professional education, will be required. Counseling would need to be provided at the most propitious time (before marriage) and should be, preferably, state supported. Regionalization of centers would be critical. The development of a National Genetic Disease Register is appealing, but is riddled with almost insurmountable problems.

All these and other programs would, hopefully, develop in an atmosphere sensitized to basic human rights and to accepted, defined principles of genetic counseling.

Training in Genetic Counseling

Today, the overwhelming majority of individuals who provide genetic counseling have not been specifically trained for the task. Much counseling is offered by general practitioners or specialists in fields other than genetics. In fact, it has been quite unclear as to what constitutes essential and optimal training and who should counsel: a physician, a biomedical scientist (Ph.D.), or a "para-counselor."

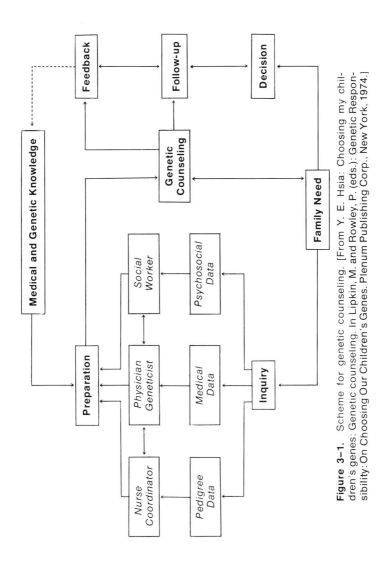

Figure 3-1. Scheme for genetic counseling. [From Y. E. Hsia: Choosing my children's genes: Genetic counseling. In Lipkin, M. and Rowley, P. (eds.): Genetic Responsibility: On Choosing Our Children's Genes. Plenum Publishing Corp., New York, 1974.]

The Committee on Genetic Counseling of the American Society of Human Genetics has given these questions careful consideration. In their report[21] to the Society, and following consideration of the range of problems encountered in counseling and the "tools" that must be employed, the committee members suggested that comprehensive counseling services be provided by a team of trained individuals. Furthermore, they believe that a physician with clinical experience and specific training in medical genetics be a member of the team, and that biomedical scientists continue to play a vital role in such services. Medical social workers and public health nurses were regarded as having important functions in communicating with consultands, as liaisons, in managing social problems, and in obtaining needed medical information. Para-counselors or genetic associates (being trained more recently) function in a similar manner.

The Committee made specific comments and recommendations concerning training for genetic counselors.

Training in either basic genetics or medicine alone is not sufficient to qualify an individual as a genetic counselor. A physician without training specifically in medical genetics, or a scientist without training within a specialized medical genetics group, does not have the medical, genetic, communication and laboratory skills necessary for appreciation and handling of the types of problems listed above. While the training program is not precisely defined in either time or content, it should include for medical graduates board eligibility in a branch of clinical medicine, training for which should include *at least* one and preferably two years within a specialized medical genetics group; for a scientist, the Ph.D. degree in human or medical genetics, and *at least* one year's training within a specialized medical genetics group.[21]

The demand for genetic counseling is very likely to continue at an accelerated pace, especially as the new advances in prenatal diagnosis and carrier detection become more widely known. This increasing demand should not, however, diminish the need for the development and maintenance of standards of excellence for genetic counseling practice.

Human Genetics in the School Curriculum

Good health habits—including the development of attitudes to prevention—can and should be inculcated throughout the school curriculum, starting in the earliest years. Clear documentation has already been provided of the degree of confusion and ignorance about specific genetic diseases in affected families already counseled.[27-30, 45] Recent progress, however, in the development of broad-based science courses in schools[46-49] promises a more enlightened society in the future. Childs[50] has also made a plea that "teaching in school be more specifically oriented to providing that basis upon which specific details of knowledge of disease and health may later be imposed." Others, too,[43, 51, 52] have recognized the need for greater efforts in teaching genetics and human biology in school.

Attention to five guiding principles should help in the development of teaching curricula in human biology and genetics in school.

Extent and Duration. From the kindergarten years through high school, steadily advancing degrees of information and knowledge should be provided and demanded. The interest in bodily function, expressed in early childhood, should be encouraged and stimulated and continually related to health care programs, with an aim toward preventive goals.

Repetition. As an essential pedagogic principle, repetition should find specific application in the human biology school curriculum.

Extent and Need. The frequency of genetic disease and the extent of the problem should be communicated by careful, informed, and sensitive teachers during the high school years. The need for each person to be aware of his or her family history (hereditary disease) should be emphasized.

Individual Perceptions. To motivate young people to seek genetic counseling voluntarily will require great educational effort. Their perception of personal threat or risk is likely to be a critical point in achieving motivation.

Recognition of Benefit. Young people, especially, should recognize personal health benefits that might accrue if they pursue action programs available. Good examples of such programs are carrier detection and options for prenatal diagnosis. Recognition of other alternatives, such as adoption and artificial insemination by donor, is also important.

Regional Genetic Programs:

Government and privately sponsored public health programs aimed at control, treatment, and prevention of genetic disease, have been alluded to by Clow et al.[13] The use of allied health personnel in such programs has also been reported.[15] The role of the regional center—located most optimally in a medical school-hospital setting—could be both central and peripheral to individual care.

"Central" involvement would be exemplified by a team manned by experts who would provide diagnostic evaluation and treatment, as well as complex counseling. A "peripheral" role would be that of support to a practitioner trained by the center in his or her straightforward counseling experiences.

My perception of a regional genetics program is one that combines administration with patient care and counseling, teaching (public and professional), and research. Riccardi describes in detail one successful regional genetic counseling program in Chapter 18.

Premarital and/or Preconception Genetic Counseling

Some evidence exists to show that persons receiving genetic counseling make rational and responsible decisions regarding reproduction. Sadly, however, many of these decisions are made *after* the family has already suffered tragedy. To my knowledge, no state or country has embarked upon a large-scale preventive genetic disease program aimed at young adults of marriageable age. Systematic provision of genetic counseling to young people at the time they obtain their marriage licenses could, potentially, have an important impact on the prevention of genetic disease. Ideally, such counseling should be offered prior to conception (in or out of wedlock) if realistic goals of prevention are to be achieved. Conception out of wedlock, one would guess, is rarely planned—even today. Nevertheless, an educational program, as just outlined, may convey a sufficiently strong signal of awareness even for some of these couples to exercise caution or to seek counseling.

Principles

Certain essential principles (in addition to those elucidated earlier for genetic counseling) form the basis of any premarital/preconception genetic counseling program.

Education.   Extensive and continuing education programs from the earliest grades in school are crucial to the success of any preventive program. (See the discussion on Human Genetics in the School Curriculum.)

Costs.   To have maximal utilization, the state would be expected to carry the costs of the program. Young persons with little money can scarcely be expected to seek and pay for counseling.

Voluntary Program.   A noncoercive voluntary program of prevention would be ideal. Two facets of the program, however, should be mandatory. The first is the educational background required to create awareness of the problem and the indications and need for genetic counseling. The second imperative of the program would be the provision—to couples planning marriage—of information about the need for genetic counseling, the options that exist, and the location of genetic counseling centers. (All such information would be furnished by the state.) It would seem reasonable for the state to insist that couples at risk for having defective children would attempt procreation only after rational, informed decision-making.

Timing of Counseling.   Premarital counseling is desirable only because it is predicated on the philosophy of preventive action. The anxiety, tensions, and other problems that may complicate the days and months immediately prior to marriage, are matters of common experience. To compound the problems of this period by the complex emotions, possibly engendered by genetic counseling, is obviously undesirable. The possible consequences of such eleventh hour counseling can also easily be envisaged, and include immediate dissolution of the relationship, or sowing of the seeds for later marital discord or divorce. There is, however, no more timely an occasion to offer "useful" genetic counseling. The alternative to be remembered is counseling after the occurrence of tragedy, and the oft-associated emotional and economic family chaos[54] — including divorce and suicide.

Counseling of Couples. Only couples concerned with genetic questions about themselves or their family history would be counseled in such a program. It should be emphasized that effective counseling is best provided to *couples* planning to marry, so that *full* and meaningful pedigree analysis is possible.

On occasion, single young persons, motivated by hereditary problems in the family or by eugenic considerations, seek genetic counseling. They may, for example, wish to know their heterozygote status for Tay-Sachs disease, sickle cell anemia or other genetic disorders. Armed with this information, they might seek a noncarrier spouse — for religious, or other reasons. This would be a reasonable approach, and such single persons could benefit from genetic counseling. Single persons at high risk of having a genetic disease themselves, such as Huntington's chorea, may elect not to marry or not to have offspring after they have received genetic counseling.

National Genetic Disease Registry

The vast majority of genetic units around the world have no systematic scheme to follow up entire families after the propositus has been diagnosed. Some years ago, McKusick urged that a system of medical records and follow-up be instituted, in which the family (as opposed to the individual) was the unit under study.[55] The objective of his system, like that of others, was the prevention of genetic disease through proper diagnosis and management. He indicated that the system could be expected to accomplish the following (slight modifications added):

1. Prompt and correct diagnosis (e.g., endocrine adenomatosis); 2. Proper treatment of a rare disorder (e.g., hypophosphatemic rickets; presymptomatic Wilson's disease); 3. Early detection and eradication of life-threatening complications of genetic disease (e.g., intestinal malignancy in polyposis); 4. Proper prophylactic practices (e.g., avoidance of certain drugs in porphyria and in glucose-6-phosphate dehydrogenase deficiency); and 5. Selective prenatal diagnosis (e.g., translocation mongolism, Tay-Sachs disease, and others).

Notwithstanding the few units that pursue the entire geographically dispersed family at risk, it would seem most timely and responsible to build some systematic organization. To some extent, this would call for the reversal of roles, with the physician seeking the patient. This approach would demand a highly responsible organization. That there is a pressing need to reach the unwitting patient is abundantly clear. In Scotland,[15] the United States,[12] and Germany[44] evidence has been accumulated to show that there are a significant number of persons in the population who are unaware of their high risk of having a child with a serious hereditary disorder. Stevenson and Davison[56] years ago made the point that the best organized genetics unit is likely to be unable to fulfill all the requirements of an optimal genetic counseling service (e.g., data storage and retrieval for 25 years; tracking down of widely separated families).

Many persons have, therefore, used or argued the need for some kind of genetic disease register.[55, 57-67] Indeed, registers of handicapped children or those with congenital malformations are kept in many locations around the world.[55, 57-59, 62-64, 67] Some units have operated through voluntary registration,[59, 60] consequently making them less useful. Newcombe and others[59, 60, 68] have also described the use of rapid, automated, computerized methods of "linking" together data (e.g., vital records) from different information sources on the same family. The salutary observation has been made that no human teratogen has, thus far, been identified by a variety of different registries, including those in hospitals, in prepaid medical programs, in cities, regions or nations.[66] Further, no registry has yet confirmed the teratogenicity of a suspected environmental agent.[66]

Goals of a Genetic Disease Registry

The essential goals of a local or national genetic disease registry are the ascertainment and prevention of genetic disease.[55, 62] The principles of any such registry would, naturally, incorporate those already discussed for genetic counseling. Intrinsic to the concept of a genetic registry system is the approach to individuals who have not requested information about genetic disease. This problem of seeking out and telling individuals of their risks of either developing genetic disease themselves, or of having affected children, is not easily resolved. Lappé et al.[69] have addressed themselves to certain ethical aspects of the problem.

The view that parents have a right to know is advocated by the Edinburgh center.[67] Emery et al.[67] have described in some detail the basic organization of their computerized genetic register system, referred to by the acronym RAPID (Register for Ascertainment and Prevention of Inherited Disease). A critical aspect of their system is that, even in approaching newly determined individuals at risk, the assistance of the individual's own practitioner is the key access route. Subsequent access to the data in the computer can be achieved again only through the patients' own doctor. It should be noted that this system aims to ascertain mainly those individuals at high risk (greater than 10 per cent) of having a child with a serious genetic disease about which counseling would be appropriate.

Organization of their register pays particular attention to the problems of privacy and confidentiality. It is essentially a voluntary, as opposed to a mandatory, system. Data storage and retrieval are secured; data-linking, trend-identification, and other research technologies can be easily utilized. Moreover, it is out of the hands of government. It should serve as an excellent model for the development of such registers elsewhere.

Major and continuous efforts would be necessary in the development of similar registers to maintain absolute privacy and confidentiality, to adhere to medical ethics, as well as to assure accuracy of diagnoses made and educating data sources, e.g., new physicians. Not the least of any such plans should be careful, necessary, and on-going public education.

ACKNOWLEDGMENT: I acknowledge with appreciation the constructive comments of Prof. Barton Childs (Johns Hopkins University School of Medicine) and Prof. James R. Sorenson (Boston University School of Medicine).

References*

1. Murphy, E. A.: The rationale of genetic counseling. J. Pediatr., 72:121, 1968.
2. Nora, J. J., and Fraser, F. C. (eds.): Medical Genetics: Principles and Practice. Philadelphia, Lea & Febiger, 1974.
3. Thompson, J. S., and Thompson, M. W. (eds.): Genetics in Medicine. Philadelphia, W. B. Saunders Company, 1973.
4. Goodman, R. M. (ed.): Genetic Disorders of Man. Boston, Little, Brown and Company, 1970.
5. Levitan, M., and Montagu, A. (eds.): Textbook of Human Genetics. New York, Oxford University Press, 1971.
6. Porter, I. H.: Heredity and Disease. New York, McGraw-Hill Book Company, 1968.
7. Lynch, H. T. (ed.): Dynamic Genetic Counseling for Clinicians. Springfield, Ill., Charles C Thomas, Publisher, 1969.
8. Stern, C. (ed.): Principles of Human Genetics. 3rd ed. San Francisco, W. H. Freeman and Company, 1973.
9. Hecht, F., and Lovrien, E. W.: Genetic diagnosis in the newborn: A part of preventive medicine. Pediatr. Clin. North Am., 17:1039, 1970.
10. World Health Organization Expert Committee: Genetic Counseling. WHO Tech. Rep., 416:1, 1969.
11. Milunsky, A.: The Prenatal Diagnosis of Hereditary Disorders. Springfield, Ill., Charles C Thomas, Publisher, 1973.
12. Sly, W. S.: What is genetic counseling? Birth Defects, 9:5, 1973.
13. Clow, C. L., Fraser, C., Laberge, C., and Scriver, C. R.: On the application of knowledge to the patient with genetic disease. *In* Steinberg, A. G. and Bearn, A. G.

---

*For an extensive bibliography, see Sorenson, J. R.: Social and Psychological Aspects of Applied Human Genetics: A Bibliography. Fogarty International Center, DHEW Publication No. 73–412, (NIH).

(eds.): Progress in Medical Genetics, Vol. 9, New York, Grune & Stratton, Inc., pp. 159–213, 1973.

14. Sorenson, J. R.: Counselors: Self portrait. Genetic Counseling, 1:31, 1973.

15. Emery, A. E. H.: Genetic counseling. Scot. Med. J., 14:335, 1969.

16. Pearn, J. H.: Patients' subjective interpretation of risks offered in genetic counseling. J. Med. Genet., 10:129, 1972.

17. Milunsky, A., and Reilly, P.: Medico-legal issues in the prenatal diagnosis of hereditary disorders Am. J. Law & Med. (In press).

18. Capron, A. M.: Informed decision-making in genetic counseling: A dissent to the "wrongful life" debate. Indiana Law J., 48:581, 1973.

19. O'Brien, J. S., Okada, S., Chen, A., and Fillerup, D. L.: Tay-Sachs disease: Detection of heterozygotes and homozygotes by serum hexosaminidase assay. N. Engl. J. Med., 283:15, 1970.

20. Epstein, C. J.: Who should do genetic counseling, and under what circumstances? Birth Defects, 9:39, 1973.

21. Committee on Genetic Counseling. Report to the American Society of Human Genetics. May 15, 1973.

22. Hsia, Y. E.: Choosing my children's genes: Genetic counseling. In: Lipkin, M., and Rowley, P. (eds.): On Choosing Our Children's Genes. New York, Plenum Publishing Corp., 1974, p. 43.

23. Carter, C. O., Roberts, J. A., Evans, K. A. and Buck, A. R.: Genetic clinic: A follow-up. Lancet, 1:281, 1971.

24. Roberts, J. A. F.: Genetic prognosis. Br. Med. J., 1:587, 1962.

25. Carter, C. O.: Comments on genetic counseling. In Crow, J. F., and Neel, J. V. (eds.): Proceedings of the Third International Congress of Human Genetics. Baltimore, The Johns Hopkins University Press, 1967, pp. 97–100.

26. Smith, C., Holloways, S., and Emery, A. E. H.: Individuals at risk in families with genetic disease. J. Med. Genet., 8:453, 1971.

27. Sibinga, M. S., and Friedman, C. J.: Complexities of parental understanding of phenylketonuria. Pediatrics, 48:216, 1971.

28. Taylor, K. and Merrill, R. E.: Progress in the delivery of health care: Genetic counseling. Am. J. Dis. Child, 119:209, 1970.

29. Reynolds, B. DeV., Puck, M. A., and Robinson, A.: Genetic counseling: An appraisal. Clin. Genet., 5:177, 1974.

30. Leonard, C. O., Chase, G. A., and Childs, B.: Genetic counseling: A consumers' view. N. Engl. J. Med., 287:433, 1972.

31. Murphy, E. A.: The effects of genetic counseling. In Porter, I. H., and Skalko, R. G. (eds.): Heredity and Society. New York, Academic Press, Inc., pp. 119–142, 1973.

32. MacCready, R. A., and Levy, H. L.: The problem of maternal phenylketonuria. Am. J. Obstet. Gynecol., 113:121, 1972.

33. Moran, E.: Clinical and social aspects of risk-taking. Proc. R. Soc. Med., 63:1273, 1970.

34. Alker, H. A.: Rationality and achievement: A comparison of the Atkinson-McClelland and Kogan-Wallach formulations. J. Pers., 37:207, 1969.

35. Steiner, J.: An experimental study of risk-taking. Proc. R. Soc. Med., 63:1271, 1970.

36. Rettig, S.: Group discussion and predicted ethical risk-taking. J. Pers. Soc. Psychol., 3:629, 1966.

37. Chandler, S. and Rabow, J.: Ethnicity and acquaintance as variables in risk-taking. J. Soc. Psychol., 77:221, 1969.

38. Horne, W. C.: Group influence on ethical risk-taking: The inadequacy of two hypotheses. J. Soc. Psychol., 80:237, 1970.

39. Percy, A. K., and Kaback, M. M.: Infantile and adult-onset metachromatic leukodystrophy-biochemical comparisons and predictive diagnosis. N. Engl. J. Med., 285:785, 1971.

40. Sugita, M., Dulaney, J., and Moser, H. W.: Ceramidase deficiency in Farber's disease. Science, 178:1100, 1972.

41. Stevenson, A. C.: Frequency of congenital and hereditary disease, with special reference to mutation. Br. Med. Bull., 17:254, 1961.

42. Neel, J. V.: A study of major congenital defects in Japanese infants. Amer. J. Hum. Genet., 10:398, 1958.

43. World Health Organization Expert Committee: Human Genetics and Public Health. WHO Tech. Rep. Ser., No. 282, 1964.

44. Wendt, G. G., and Theile, U.: A pilot scheme for a genetic clinic. Humangenetik, 21:145, 1974.

45. Pearn, J. H., and Wilson, J.: Acute Werdnig-Hoffman disease: Acute infantile spinal muscular atrophy. Arch. Dis. Child., 48:425, 1973.

46. Brandwein, P. F., Cooper, E. K., Blackwood, P. E., and Hone, E. B.: Concepts in Science. 2nd ed. Vols. 1–6, New York, Harcourt, Brace & World, Inc., 1970.

47. Mallinson, G. G., Mallinson, J. B., Ellwood, E. P., Zeiger, L., Feravolo, R., Trexler, C. R., Brown, D. G., and Smallwood, W. L.: Science. Vols. 1–6. Morristown, New Jersey, Silver Burdett Co., 1968.

48. Novak, J. D., Meister, M., Knox, W. W., and Sullivan, D. W.: World of Science Series. Vols. 1–6. Indianapolis, The Bobbs-Merrill Co., Inc., 1966.

49. Marshall, J. S., and Beauchamp, W. L.: The basic science program. Vols. 1–6, Chicago, Scott, Foresman & Co., 1965.

50. Childs, B.: A place for genetics in health education and vice versa. Am. J. Hum. Genet., 26:120, 1974.

51. Steinberg, A. G. (ed.): Conference on Genetic Disease Control. Washington, D.C., December 5, 1970.

52. Chase, G. A.: The background of couples who request genetic counselling. *In* Porter, I. H., and Skalko, R. G. (ed.): Heredity and Society. New York, Academic Press, Inc., pp. 143–157, 1973.

53. Clow, C., Reade, T., and Scriver, C. R.: Management of hereditary metabolic disease. The role of allied health personnel. N. Engl. J. Med., 284:1292, 1971.

54. Farber, B.: The effects of a severely retarded child on family integration. Monograph of the Society for Research in Child Development. Series 71, Vol. 24, No. 2, Lafayette, Indiana, Society for Research on Child Development.

55. McKusick, V. A.: Family-oriented follow-up. J. Chronic. Dis., 22:1, 1969.

56. Stevenson, A. C. and Davison, B. C. C.: Families referred for genetic advice. Br. Med. J., 2:1060, 1966.

57. Wallace, H. M., and Rich, H.: The use of a registry in a community for handicapped children. Except. Child., 24:198, 1958.

58. Miller, J. R.: The use of registries and vital statistics in the study of congenital malformations. *In* Fishbein, M. (ed.): Proc. 2nd Int. Conf. on Cong. Malfs. New York, International Medical Congress Limited, 1964, p. 334.

59. Newcombe, H. B.: Familial tendencies in disease of children. Brit. J. Prevent. Med. 20:49, 1966.

60. Renwick, D. H.: The combined use of a central registry and vital records for incidence studies of congenital defects. Br. J. Prev. Soc. Med., 22:61, 1968.

61. Wertelecki, W., Lawton, T., and Gerald, P. S.: Computer-assisted interview of families with genetic diseases. Excerpt Med. Int. Cong. Ser., No. 191, pp. 87–88, 1969.

62. Emery, A. E. H., and Smith, C.: Ascertainment and prevention of genetic disease. Br. Med. J., 3:637, 1970.

63. Oliver, J. E.: Huntington's chorea in Northamptonshire. Br. J. Psychiatry 116:241, 1970.

64. Kimberling, W. J.: Computers and gene localization. *In* Wright, S. W., Crandall, B. F., and Boyer, L. (eds.): Perspectives in Cytogenetics: The Next Decade. Springfield, Charles C Thomas, Publisher, pp. 131–147, 1972.

65. Welch, J. P.: Some public-health aspects of research in mental retardation. Med. Bull., 51:137, 1972.

66. Miller, R. W.: How environmental effects on child health are recognized. Pediatrics, Suppl., 53:792, 1974.

67. Emery, A. E. H., Elliott, D., Moores, M., and Smith, C.: A genetic register system (RAPID). J. Med. Genet., 11:145, 1974.

68. United Nations–World Health Organization Seminar: The use of vital and health statistics for genetic and radiation studies. New York, United Nations, 1962.

69. Lappé, M., Gustafson, J. M., and Roblin, R.: Ethical and social issues in screening for genetic disease. N. Engl. J. Med., 286:1129, 1972.

CHAPTER 4

HETEROZYGOTE SCREENING FOR THE CONTROL OF RECESSIVE GENETIC DISEASE

Michael M. Kaback, M.D.*

INTRODUCTION

In the past two decades, dramatic advances have been made in the science and technology of human genetics. Since the elucidation of the normal human chromosomal complement in 1956, a wealth of new information concerning cytogenetic mechanisms and chromosomal aberrations in man has become established.[1] Advances in human biochemical genetics have been equally as dramatic. Nearly 2000 single gene alterations in man are now recognized—many associated with specific, and often serious disease.[2] In conjunction with this important scientific progress, major changes have occurred in the social, legal, and ethical attitudes of Western society. Attitudes toward population control, "quality of life," individual freedoms, and abortion (to name only a few major areas) are vastly different today from what they were just a decade ago. It is in this context of scientific and cultural evolution, coupled with the traditional medical philosophy to minimize the frequency and severity of disease, that the effective control of certain inherited disorders of man becomes a realistic and current consideration.

*This work was supported by a grant from the National Foundation–March of Dimes, NIH Graduate Research Training Grant (HD-00-417) and by a contract from the California State Department of Health.

This chapter considers a number of past and present developments in human genetics, from the perspective of how these innovations might be used as effective mechanisms for the control of certain serious inherited disorders. Several modes of disease control are identified, but the discussion is focused primarily on the prevention of recessive genetic disease through programs emphasizing public education, voluntary heterozygote screening, and effective genetic counseling. The predilection of certain genetic disorders for defined subpopulations enhances the feasibility and potential effectiveness of such an approach. With simple, accurate, and inexpensive methods for heterozygote detection of certain of these mutant alleles, and the availability of intrauterine fetal diagnosis for the associated genetic disease, a positive rationale for mass screening becomes more evident.

The goals, strategy, and design of such an undertaking are explored, as are the roles of professionals, community leaders, and volunteers in the organization and delivery of such programs. Important genetic, economic, social, and ethical questions can be and have been raised by and about such an approach.[3, 4, 5, 6] Each of these areas is addressed and the critical importance of ongoing scientific and psychosocial evaluations in programs of this type is emphasized.

BACKGROUND

The basic principles of medical genetics are not new; applications of Mendelian laws to certain diseases of man have long been recognized. Human disorders inherited as dominant or recessive genetic traits have been described in the medical literature throughout the twentieth century. In some instances, these concepts have even had social and legal recognition, such as the prohibition, by law, of close consanguineous matings in many Western countries.

In addition to a large number of chromosomal aberrations and the more common multifactorial genetic disorders, nearly 2000 single gene alterations in man are now recognized. In more than 100 of the recessively inherited disorders (both X-linked or autosomal), the specific inborn metabolic error has been identified.[7] Such discoveries provide a means not only for accurate genetic disease diagnosis, but also for expanded research and the development of effective treatment modes. In certain instances, this already has been accomplished.[8] These advances also have led to a greatly improved capability in genetic counseling, and most recently, to prevention of certain of these serious and untreatable conditions through intrauterine fetal diagnosis and selective abortion.[9, 10]

In more than 40 of the recessively inherited inborn metabolic disorders, techniques are now available to identify accurately healthy individuals who are heterozygous (carriers) for the specific deleterious alleles involved.[10, 11, 12] In the past, many of these conditions led to severe disability, retardation, or death in affected individuals. Financial disaster, family disruption, and life-long stigmatization often developed in families in whom such disorders occurred. Now, in light of these recent discoveries, the situation has changed dramatically; in at least several of these serious inherited disorders, a potential to control is at hand.

APPROACHES TO THE CONTROL OF HUMAN GENETIC DISEASE

The control of human genetic disease might be considered according to the following modes: cure, effective therapy, and prevention.[13]

Cure

Cure of genetic disease, *i.e.,* replacement or correction of the mutant gene or genes involved, is still a futuristic method of control. Selective gene correction, replacement, and "manipulation" of genetic control mechanisms in diploid organisms (*i.e.,* man) through "genetic engineering" techniques recently have had considerable coverage in the lay press and other media. Suggestions, in these publications, that such approaches soon will be applicable to human genetic disorders are misleading, premature, and unfounded at present or in the immediate future.

Therapy

The effective treatment of genetic disease, on the other hand, is far from unrealistic. The pathogenetic mechanisms of a number of recessive single gene disorders are now well understood, and effective modes of therapy have evolved (see Chapter 11). Early diagnosis, followed by immediate therapy, can either totally eliminate the abnormalities associated with the disease (if otherwise left untreated) or greatly reduce the deleterious phenotypic effects of the disorder. In such instances, few would question the benefits to the affected individuals, to their families, or to society-at-large derived from early therapy.

Where effective therapy for an inborn metabolic error is at hand, the availability of an accurate, inexpensive, and simple means of disease diagnosis makes disease control possible. Diagnostic screening, followed by early initiation of therapy for individuals found to be affected, represents a most desirable means of disease control. Obviously, such screening should be directed toward the identification of affected individuals prior to the onset of symptoms, or at least prior to a time when irreversible effects of the abnormality have occurred. Screening of this type should be implemented at a time when the maximum "catchment" would be achieved—probably during the early neonatal period (since the majority of births occur in hospitals) or at certain optimal times, such as upon entry to school (when a relatively "captive" audience is available).

The subject of *disease* screening is discussed elsewhere in this volume (see Introduction, Chapter 8, and Chapter 10). Screening for genetic *disease* raises a different set of questions from those related to screening for genetic *heterozygosity*. In general, the issues, principles, vagaries, and dilemmas of each of these two "screening" approaches are entirely different. The most obvious contrast between these two types of screening is in the goals or purposes. On the one hand, screening for disease is directed at diagnosis, treatment, and normalization of the phenotype in the genetically abnormal individual, to achieve disease control. On the other hand, carrier screening is directed at prevention, attained by either cessation of, reduction in, or selection in reproduction (through "appropriate" mate selection, artificial insemination, or intrauterine diagnosis and selective abortion), to achieve disease control. Here, the mechanism of control is prevention-attainable through avoidance of the genetic illness. From many perspectives, unique and complex issues totally different from those encountered in disease screening are involved. These issues will be addressed more thoroughly in a later section.

Prevention

Historically, prevention of serious genetic disease has been (and remains) the most important instrument of disease control. The recognition and application of Mendelian principles to specific human disorders in the past provided a basis for genetic counseling of families where individuals affected with such conditions were identified. Counseling was extended to parents and close relatives concerning the nature, course, and prognosis of the condition, as well as the recurrence frequency and risks for future affected offspring. In many instances, genetic disease was prevented, since many families either refrained entirely from further reproduction or had substantially fewer children than they had hoped for.[14] This was particularly the case where the affected individual's illness led to a considerable emotional and/or financial burden for the family.[15]

Genetic disease control was achieved through the prevention of (or reduction in) further occurrences of that disorder in families in which the disease had already appeared. This approach was of limited value, however, in significantly reducing the overall incidence of the disorder. For most of the more common autosomal recessive disorders (e.g., sickle cell anemia, thalassemia, cystic fibrosis, and Tay-Sachs disease) the substantial majority of cases tend to occur in families with no known previous history of the disorder. The index case, therefore, most frequently is not preceded by any other known cases in the pedigree.[16] Prevention of additional cases within such families clearly is of major clinical importance to the family, but is of only minimal significance in reducing the overall incidence of the disease.

Postproband genetic counseling, therefore, can be an effective, albeit a limited, instrument in genetic disease control. Control by this mechanism—if achieved—is done so by limiting the reproductive freedoms of the couple involved and possibly the freedoms of close relatives, as well. In some families where a serious and untreatable genetic disorder has occurred, healthy siblings, aunts, uncles, and even first and second cousins have expressed fears and have even taken steps to avoid reproduction for fear of bearing children with this condition. It should be pointed out that often these measures have been the result of inappropriate and inaccurate medical advice.[17]

In many cases recent achievements have substantively changed this situation. The availability of techniques for accurate ascertainment of the heterozygous state in many of these serious recessive conditions has greatly altered what was previously a purely statistical approach to genetic counseling. Heterozygote testing of relatives in such families now provides a much more accurate means for counseling specific family members and their spouses as to the realistic presence (or absence) of risk for the defined condition in their offspring.

The very recent application of amniocentesis and amniotic fluid cell studies to the accurate fetal detection *in utero,* of most chromosomal, and many recessive genetic disorders, in early gestation provides an extraordinary and powerful new instrument in genetic counseling. Evolving social attitudes and the advent of liberalized abortion, coupled with the technical advances in somatic cell genetics, human cytogenetics, and biochemical genetics (all of which make prenatal diagnosis possible), have dramatically changed the less-than-optimal situation that existed just a few years ago. Now, for many families at a defined risk for genetic disease in their offspring, prenatal monitoring for the condition (if it is one of the over 60 inborn metabolic errors now detectable by this means)[9, 10, 12] (see Chapter 9) can provide a means for such couples to have unaffected offspring selectively. This implies, of course, that such families would elect to monitor each pregnancy by amniocentesis and that they would choose to abort those pregnancies in which the fetus was found to be affected (a 25 per cent risk with each pregnancy—for recessive conditions). In this way, the couple having had such an illness in one of their children, and therefore having been made aware of their risks, could now be aided in bearing additional children, all of whom would be free of the disease.

Clearly, this is not a perfect solution, and it is one that might not appeal to everyone. For many, however, this is preferable to not risking pregnancies at all or even to artificial insemination (in cases where that is an alternative consideration). If prenatal detection is possible, disease prevention can be achieved *without* restriction of a couple's reproductive opportunities (assuming such an alternative would be acceptable). However, if the proband case is necessary to identify the risk status of the family, only limited reduction in the overall postnatal incidence of the disease could be expected.

To achieve disease control more effectively by prenatal diagnosis and selective abortion, it would be necessary to ascertain the at-risk status of a family prior to the birth of the first affected individual. Once the at-risk status is determined, two important goals could be achieved—enhancement of disease control (substantial reduction of the postnatal disease incidence) and provision of a realistic and "positive" reproductive alternative—at least for most couples at risk.

PRE-PROBAND IDENTIFICATION OF THE AT-RISK INDIVIDUAL OR FAMILY

The critical prerequisite for implementation of such a preventive control mechanism is a means to identify the at-risk couple before the birth of offspring affected with genetic disease. If a specific genetic disorder tends to occur predominantly in a defined subpopulation, this occurrence greatly facilitates the identification of prospective at-risk couples. This serves to reduce the population base in which the at-risk couple "search" need be made and enhances the possibilities for effective control.

For chromosomal disorders, such a situation is at hand, since pregnancies in women above 35 years of age have long been recognized to be at substantially greater risk (relative to younger women) for producing offspring with nondisjunctional types of chromosomal aberrations (particularly Down's syndrome). If appropriately informed of this information and of the availability of prenatal testing, many pregnant women over 35 years of age might choose to have their pregnancies monitored and to abort those in which a chromosomally abnormal fetus is identified. A substantial reduction in the incidence of these serious disorders could result.[18] Another important consequence of this knowledge would be that many other families could be relieved of the anxiety that they might otherwise have suffered through pregnancies in which they feared the birth of a chromosomally abnormal child. Moreover, in certain instances, pregnancies which might have been electively aborted, perhaps because of such anxieties, could be continued to term.

Single Gene Recessive Disorders

For autosomal and X-linked recessive disorders, prospective identification of the at-risk couple is a more complex matter. Here, accurate
and simple heterozygote detection methods, applicable on a mass scale,
become the essential requirement if disease prevention is to be achieved
through "at-risk family identification" prior to the occurrence of index
cases. With X-linked disorders, carrier detection need be directed only
to women (see Chapter 6). However, in light of the Lyon phenomenon
of random X chromosome inactivation in the somatic cells of the female,
it is not yet possible or practical to identify definitively all obligate carriers of any X-linked recessive gene. Moreover, even the most accurate
methods for carrier detection (90 per cent of obligate carriers for hemophilia A can now be so identified)[19] are not technically or economically
applicable to population screening at this time. Biochemical "handles"
are available for relatively few of the X-linked recessive conditions.
Only in Fabry's disease, Lesch-Nyhan syndrome, G6PD deficiency, hemophilia A, and, perhaps most recently, Hunter's syndrome[20] have specific biochemical methods for carrier detection been described. Again,
however, simple, straightforward, and accurate means for clear heterozygote identification have not yet been described. Complex techniques,
such as skin fibroblast cloning, may be required for the greatest accuracy,[21, 22] and this would certainly be prohibitive as a mass screening tool.

In other X-linked conditions such as Duchenne's and Becker's
muscular dystrophy, nonspecific elevations of serum creatine phosphokinase have been described in some heterozygotes.[23, 24] Histologic evaluation of muscle biopsies may further aid in dystrophy carrier detection,
but it could hardly be envisioned as a method applicable to mass screening. Other methods to delineate female carriers of X-linked recessive
genes (like the quantitative nitro-blue tetrazoleum reduction test in
chronic granulomatous disease)[25] are useful in women related to proband
males, but they are impractical for large-scale screening. These practical,
technical, and genetic limitations, as well as a lack of any clear predilection of a particular subpopulation for most serious X-linked disorders
(with the important exception of the G6PD deficiencies), lead one to
conclude that extensive carrier screening of women for such traits would
be extraordinarily expensive, complex, and insufficiently accurate to justify such efforts at this time. Carrier evaluation in women with affected
males in their families, on the other hand, would seem far more appropriate and advisable. The possibility that as many as one third of the
males affected with X-linked recessive conditions represent new spontaneous mutations[26] adds another difficult aspect in considering effective
means for the prospective prevention of X-linked recessive disorders.

With many autosomal recessive disorders identification of the heterozygous carrier, as well as prenatal diagnosis, is possible.[10, 11, 12] These achievements have dramatically altered the capacity for genetic counseling in families where affected individuals have been detected. As with the X-linked recessives, prospective identification of at-risk couples is another matter. Because of the complexity, expense, and less than sufficient accuracy of most of these carrier detection methods, as well as the lack of a defined higher-risk population for most of these conditions, mass screening for identification of at-risk couples does not seem feasible at this time. Within pedigrees where such disorders have occurred, however, carrier screening and counseling – including discussion of prenatal diagnosis – are certainly worthwhile.

In the relatively few instances where mass carrier screening might be considered (e.g., Tay-Sachs disease (TSD), sickle cell anemia, and β-thalassemia) certain unique criteria (see below) have encouraged some to initiate such efforts.

For conditions where prenatal detection is not possible as yet, but where heterozygote identification can be achieved, another consideration is relevant. The identification of an individual as a heterozygote for an autosomal recessive allele might be used as pertinent information in genetic counseling of that person. Accordingly, individuals found to carry a specific autosomal recessive trait might be counseled thoroughly as to the implications of this finding and the risks they would incur if they chose to mate with another individual carrying the same trait. Implicit in such a program is that such information could and would be utilized in mating choices, and that prevention (and thereby control of the disease) would be achieved. Although straightforward in design, such an effort entails enormous complexities and possible psychological subtleties.[27]

ETHNIC DISTRIBUTIONS OF GENETIC DISEASE

The variation in prevalence of certain genetic traits and, therefore, the incidence of specific genetic diseases in different human population groups have been recognized for some time (see Chapter 1, Table 1–4, Chapter 5, and Table 2–5). Why certain deleterious recessive genes (which result in dramatically reduced reproductive fitness in the homozygous or hemizygous state) persist and remain stable in frequency among heterozygotes in the population remains problematic. Whether heterozygosity imparts some biological or reproductive advantage to the carrier individual, or the high carrier frequency can be accounted for by populational genetic phenomenona, such as spontaneous mutation, founder effect, or genetic drift, is uncertain.

Although each human may carry as many as four to eight recessive deleterious alleles, the frequencies of certain of these genes are dramatically different in different population groups. This is probably best explained by the marital and reproductive patterns of population groups throughout history. People with particular social, geographic, religious, or racial backgrounds have tended to marry and reproduce with other persons of similar backgrounds. Accordingly, as such reproductive patterns have been maintained for significant periods of time, it is understandable how certain recessive traits, and, in turn, specific disorders associated with them, could have segregated differentially.

Alternatively certain environmental factors may have imparted a selective (biological or reproductive) advantage to certain heterozygous states at certain points in history in particular geographic areas. This would result in an enrichment among the surviving population in that area in individuals heterozygous for that particular allele. The relative resistance to falciparum malaria in the early years of life found in individuals heterozygous for sickle cell hemoglobin and for G6PD deficiency (the Negro variant) is one of the better known examples of this mechanism.[28, 29] Similarly, it has been suggested that heterozygosity for the Tay-Sachs disease gene may have conferred a relative resistance to pulmonary tuberculosis on such heterozygous individuals; this might account for the high frequency of this particular recessive allele in Jewish individuals of Central and Eastern European ancestry.[30] Whatever the mechanisms, it is abundantly evident that certain recessive traits do have higher frequencies in certain ethnic groups. Some representative examples, with corresponding frequency estimates, are listed in Table 4–1 (see also Chapter 5 and Table 5–2). Where a defined recessive disorder tends to occur in a specific ethnic group, and where techniques are available to identify the heterozygote for the recessive mutation associated with that disease, an optimal situation for disease control exists.

TABLE 4–1. FREQUENCY AND INCIDENCE ESTIMATES FOR SELECTED AUTOSOMAL RECESSIVE DISORDERS IN DEFINED ETHNIC GROUPS

Disease	Ethnic Group	Gene Frequency	Carrier Frequency	"At-Risk" Couple Frequency*	Disease Incidence In Newborns
Sickle cell anemia	Blacks	.040	.080	1 in 150	1 in 600
Tay-Sachs disease	Ashkenazic Jews	.016	.032	1 in 900	1 in 3600
β-Thalassemia	Greeks, Italians	.016	.032	1 in 900	1 in 3600
α-Thalassemia	S.E. Asians and Chinese	.020	.040	1 in 625	1 in 2500
Cystic fibrosis	N. Europeans	.020	.040	1 in 625	1 in 2500
Phenylketonuria	Europeans	.008	.016	1 in 4000	1 in 16,000

*Likelihood that both members of a couple are heterozygous for the same recessive allele (assuming nonconsanguinity and that both are of the same ethnic group).

CONTROL OF RECESSIVE GENETIC DISEASE THROUGH HETEROZYGOTE SCREENING IN LARGE POPULATIONS

Preliminary Considerations

Before the details of any carrier screening program are considered, it is appropriate to address several simple and very practical questions:

1. For which recessive genetic disorders are simple, accurate, and relatively inexpensive heterozygote detection methods available?

2. What is the nature, severity, and frequency of the associated disease in the general population? In a defined subpopulation?

3. What benefits or alternative plans of action will be available to those individuals or families identified by the screening test to have such genes?

4. To whom should the screening test be directed—children, teenagers, married couples?

Whether heterozygote screening programs are to be implemented, if at all, should be determined by the answers to these "what and why" questions. In certain instances it might be justifiable to screen an entire population; in others, a subpopulation group, in which a specific genetic disease tends to be more prevalent, would be more appropriate as the target for large-scale screening. In still another instance, because of the complexity and nature of the screening test, the rarity of the disease, or its lack of predilection for any specific subpopulation, it might be best to screen only those individuals within pedigrees where proband cases have previously been identified.

The nature and severity of disease, the means for heterozygote detection, and, perhaps most importantly, the alternatives available for those individuals or couples identified to carry such alleles dictate when and to whom screening programs should be directed. Where heterozygote identification for a particular disease-related allele can be coupled with prenatal detection of the homozygous disorder, carrier screening might best be directed only to married couples considering reproduction. Where such "positive" reproductive alternatives would not be available, it might seem more appropriate to screen individuals prior to the time of marriage. This implies that heterozygotes, once identified, would be appropriately informed and counseled so that they could utilize this information in selecting their reproductive partners. Therefore it is the "benefit" of heterozygote ascertainment for the individual that defines when and whom to screen.

Screening for heterozygosity in young school-age children or newborns seems hardly justifiable.[31] Some might argue that such identification could be followed by appropriate screening among the parents of such children; however, if this screening is the goal, then primary screening of adults would be more attractive.

Legislated vs. Voluntary Screening

Many of the complexities of genetic education design and screening program strategy would be obviated simply by making such screening tests mandatory. For example, legislative requirement that all couples at the time of marriage be required to take certain carrier detection tests would greatly simplify many of the issues discussed here. Alternatively, carrier screening might be required before entry into school (as had been invoked in several states for sickle cell screening in the early seventies). Although mandatory carrier screening would greatly simplify many of the complex matters discussed, several substantive issues remain, which more than offset the benefits of legislated testing. None of the genetic disorders considered would be regarded as epidemic conditions of major public health importance. In addition, without a means for total correction (cure) or effective therapy for such conditions, carrier screening has implications relative to only reproductive alternatives. Since it remains to be established whether reproductive counseling can be effectively delivered and integrated into reproductive decisions to reduce significantly the incidence of genetic disease, it would seem unwise to assume that this would be the case.

One might consider the assessment of an individual's genetic constitution as the ultimate invasion of that person's privacy. Accordingly, before this intrusion via the genetic detection test, each individual ought to understand the purpose, significance, and "benefits" to be derived from such an evaluation and have the right not to have such information if he so chose. The concept of mandatory genetic screening among healthy individuals for reproductive counseling has strong "Orwellian" overtones, and in the author's opinion cannot be justified.

Legislated screening for serious genetic *disease,* in which effective therapy is available, is a very different matter. Here, the early identification of disease can be followed by effective treatment to offset the deleterious effects of the abnormal genetic constitution in such individuals. It is much more reasonable that required screening be done for disease detection where clear benefits (and few, if any, liabilities) would be involved.

Specific Program Goals

If it is concluded that carrier detection is worthy of consideration, then the particular program to be implemented must be addressed. The first consideration is to identify the goals of the screening program. Several reasonable and ethically acceptable goals might be (1) to reduce the tragedy and incidence of serious and untreatable genetic disease, (2) to impart health benefits to individuals and families, (3) to explore and establish gene frequencies of selected alleles in particular populations, and

(4) to evaluate the possible heterogeneity of certain genetic markers within that population. (These last two goals are academically motivated.)

To perceive of a carrier screening program as existing to identify heterozygotes of certain deleterious genes to "improve" the genetic constitution of future populations seems totally reprehensible and unacceptable as a goal for such efforts.

Only after careful consideration is given to defining program goals, can a detailed design for the program be established. The goals determine the design of the program and define the means by which they may be achieved.

Technical and Economic Considerations

Once the program goals have been defined, the carrier detection method to be used must be critically assessed before any action is taken. The optimal means of conducting this evaluation is to screen relatives within pedigrees where specific genetic disease has occurred. Accordingly, the accuracy of the method employed can be determined. For example, if one tests a substantial number of individuals who are first cousins to a proband case (with an autosomal recessive disorder) one would expect that 25 per cent of such individuals would be identified as carriers. Similarly, 67 per cent of the nonaffected siblings of affected individuals with such disorders would be carriers, and all parents would be heterozygotes (assuming paternity is accurate). By conducting such intrapedigree assessments, a critical determination of the accuracy (both in a positive and negative direction) of the detection method can be established. An example of such an evaluation of the serum hexosaminidase A assay for TSD heterozygosity is presented in Table 4–2.

TABLE 4–2. RESULTS OF SERUM TESTING IN RELATIVES OF PATIENTS WITH TAY-SACHS DISEASE*

Relationship of TSD Child	Expected Carrier Frequency	# Tested	Noncarriers	Carriers	Observed Carrier Frequency
Offspring	1.000	116	0	116	1.000
Sibling	0.670	59	21	38	0.640
Grandchild, niece, nephew	0.500	64	32	32	0.500
Uncle, aunt	0.330	47	32	15	0.310
1st cousin, grandniece, grandnephew	0.250	73	52	21	0.280
2nd cousin	0.125	77	66	11	0.143
3rd cousin	0.063	36	34	2	0.056

*Determined by quantitative assay of hexosaminidase A. For details, see Kaback, M. M., Zeiger, R. S., Reynolds, L. W., and Sonneborn, M.: Approaches to the control and prevention of Tay-Sachs disease. *In* Steinberg, A., and Bearn, A. (eds.): Progress in Medical Genetics. Vol. 10. New York, Grune & Stratton, Inc., 1974.

With any screening tool, it is preferable that if inaccuracies exist at all, they be in the direction of over-identification of positives. False negatives occurring with any significant frequency would be unacceptable. The screening test should serve as a filtering instrument that either identifies only all positives, or identifies all positives and an additional group of individuals who can later be further evaluated by a follow-up system. This points out the importance of having some corroborative back-up method for further assessment of those individuals where a possible positive result is indicated by the initial screening test.

With regard to testing accuracy, one must be alert to heterogeneity for the parameter to be tested. In fact, it should be anticipated that in screening large numbers of individuals some heterogeneity will exist. Similarly, the possible effects of certain medical, biological, or environmental factors on the screening test should be considered. Certain illnesses, pregnancy, drugs, and other factors might affect the accuracy of the screening method used (as in the case with hexosaminidase A quantitation for TSD carrier detection).[32]

Not only is the accuracy of the method a critical issue but also the applicability on a large scale. Wherever possible, automated systems that allow for large-volume screening are desirable.[33] In only a few recessive disorders is the method sufficiently accurate and the material for carrier assessment sufficiently available to make mass screening a feasible consideration at this time. In many of these conditions, specific loading tests, or the use of cultivated cells *in vitro* (such as skin fibroblasts or established lymphocyte lines) are required for carrier identification.[10, 12] Importantly, such techniques would be economically and technically impractical to apply on a large scale. It is only in those few instances where material such as urine, plasma, serum, and whole blood (or possibly tears)[34] is readily available that carrier detection could be easily applied. In a number of recessive disorders, purified leukocytes have been used successfully for heterozygote identification. However, the current techniques for isolation of leukocytes from peripheral blood samples are complex, time consuming, and relatively cumbersome.[35] Accordingly, it would be difficult to envision a mass screening program based on leukocyte evaluation.

Once it is established that an accurate carrier detection method is available and that it can be delivered in high volume, the next consideration must be an economic one. If the method can be applied at reasonable cost, large-scale screening is more readily justified. Cost-benefit analyses are often misleading, however, One can never assess the human costs that might be saved through prevention of certain genetic diseases. Since the majority of screening efforts are or will be supported by public funding, however, it is appropriate that a detailed cost-effectiveness analysis be considered and utilized in the final decision as to whether to conduct such programs.

PROFESSIONAL COMMITMENTS

Before embarking on a mass heterozygote screening program one must define the responsibilities and commitments required of certain professional personnel. These considerations must be carefully assessed in the early planning stages; to ascertain that certain key commitments and personnel are not available would be unjustifiable after the program has been initiated. Each segment of the screening program in which professional skills may be required should be identified and planned for in advance of the program. One can subdivide a large carrier screening program into several major phases: (1) education of the community, (2) organization of the program within the community and the development of manpower, (3) allocation of laboratory responsibilities, (4) communication of testing results to individuals tested, and (5) provision of counseling services to individuals and families in need of such services as the result of screening tests. The professional responsibilities in each of these main segments may be met by different individuals. Certain nonprofessionals may also be appropriately trained to provide effective service in selected segments of such a program.

Community Role in Subpopulation Screening

Where a defined subpopulation is to be the target of a genetic screening effort, the community-at-large must be involved at various levels in such a program. Not only are the medical personnel in the community important in this regard, but also other community leaders should be informed of and should understand the screening program prior to its general initiation. From the early planning stages of such a program, key leadership of the community — medical, religious, organizational, and others — should be brought into the decision-making and the planning process so that a true perspective of community attitudes can be ascertained.[36] Such an approach will substantially strengthen the program. It is only from the community itself that the dynamics of the community can best be established. Medical researchers rarely have a sensitivity to or awareness of the requirements of any given community. With input from community leaders from the very beginning, program personnel can begin to shape the educational aspects of the program and its delivery to meet the specific needs of that community. In addition, the involvement of critical community personnel in the delivery process can greatly strengthen the effectiveness of such an effort.[37]

Educational Requirements of Mass Screening Programs

If carrier screening is to be conducted on a voluntary basis, an effective educational program should be devised. Ultimately, if healthy adults are to comply with a voluntary test of this sort, they should understand its rationale.[38] Care should be taken not to generate fear with the educational program. Subtle forms of individual or community coercion should be avoided.

Education is required at more than the level of the general public. In the early stages of planning, detailed educational sessions might be carried out with various community leaders, including medical practitioners and religious and organizational leaders, as well as other lay persons in positions of authority. One might speculate that with an effective and lucid educational program, lay individuals in a community would appreciate the basic scientific information required to understand the importance and relevance of any genetic screening program.

Counseling Requirements

In striking contrast to screening programs for genetic disease, where the incidence of the homozygous state is quite small, screening programs for carriers in a defined population identify relatively enormous numbers of individuals who are heterozygous for these alleles. It is therefore essential that any program that conducts such a screening effort have the necessary personnel to meet this potentially enormous counseling need. Not every individual will perceive the genetic information obtained from the screening in exactly the same fashion. Each person's ability to comprehend and understand some of the implications of heterozygosity will unquestionably vary as a function of his own education, background, and general orientation. A carrier screening program, therefore, must have the flexibility and counseling capability to deal with each person in an individualized fashion.

Counseling will be needed not only for individual carriers but also for couples who are identified to be at-risk for disease by virtue of their mutual carrier status. In the case of such couples, the need for ongoing counseling is critical. Obviously, the at-risk couple is the target of mass screening and must demand highest priority in terms of counseling services. Other areas in which important counseling needs can be identified concern those individuals whose tests must be repeated or further clarified. An optimal screening test would not miss any carriers, but might over-identify individuals as possible carriers, who then could be further evaluated. Such individuals require a careful explanation if their initial test is "inconclusive." Often, these persons may be more anxious and concerned than individuals whose tests show them to be carriers. Clearly, special counseling requirements are generated by this group.[32]

Another important area in which specific counseling needs are evident concerns relatives of individuals who are found to be carriers. Does the screening program have the right or responsibility to contact directly close relatives of individuals identified as carriers? The issue of invasion of privacy can be raised if this question is answered affirmatively. Alternatively, the program directors might take the position (as has been done by personnel in most TSD screening programs)[32] that identified carriers be carefully counseled and educated as to the importance of their carrier status to close relatives. In turn, simple and clearly written educational material should be provided to the identified carriers to give or send to appropriate relatives in their family. Relatives become informed in this way and are given the option to contact the program if they desire further information or want to be screened themselves. Experience from ongoing programs indicates that not only do relatives come from the immediate vicinity of the primary screening program, but in many instances relatives living at great distance from a genetic center come for counseling service. This situation requires the use of telephone or mail communications — procedures that can become quite complex. Nonetheless, directors of screening programs must carefully consider the extent of their services and what responsibilities they will take with regard to close relatives of individuals identified as carriers.

When carriers are identified, realistic and ethically acceptable efforts should be made to screen further close relatives of such individuals. The yield of such an effort, in terms of disease prevention, will be greatly expanded since the proband heterozygote in a pedigree has already been identified.

The Role of Public Health and Governmental Agencies

It is highly desirable, in the author's opinion, that heterozygote screening, if carried out at all, be done so on a voluntary basis. However, if effective public education programs and mass screening efforts are to be efficiently delivered, this voluntary aspect may require important input from public health or similar community agencies. Financial support for such efforts, optimally, would come from public agencies (state or local health departments, and other similar agencies). Such agencies may play a critical role in locating potential screenees, especially when the testing has not been mandated by law. With appropriate fiscal and administrative support from public health agencies, and with close alliance with qualified genetic personnel, the educational and service provisions of such programs can be readily achieved.

Ongoing Program Evaluations and Pilot Studies

Screening of the population for mutant gene carriers should be regarded, at this time, as an experimental health service. Accordingly, with no real precedent for this type of health service program, it is im-

TABLE 4–3. RESULTS OF SERUM TESTING IN RELATIVES OF PROGRAM-IDENTIFIED TAY-SACHS DISEASE HETEROZYGOTES*

Relationship To Carrier	Expected Carrier Frequency	# Tested	Noncarriers	Carriers	Observed Carrier Frequency
Parent	0.50	167	79	88	0.53
Sibling	0.50	108	57	51	0.47

*Individuals identified as TSD carriers in the John F. Kennedy Institute TSD Prevention Program (Baltimore, Md.). For further details see Kaback, M. M., Zeiger, R. S., Reynolds, L. W., and Sonneborn, M.: Approaches to the control and prevention of Tay-Sachs disease. In Steinberg, A., and Bearn, A. (eds.): Progress in Medical Genetics. Vol. 10. New York, Grune & Stratton, Inc., 1974.

perative that the design and formulation of such an effort provide for detailed ongoing assessments of the effectiveness and impact of such programs. Scientific and technical matters should certainly be assessed; more importantly, perhaps, the psychosocial implications of carrier screening and genetic counseling should also be evaluated.

The accuracy of the carrier detection test itself can be addressed in several ways. Careful statistical analysis of early testing results should indicate whether peculiar patterns of positive results are occurring in individuals taking certain medications, having certain illnesses (e.g., the common cold), or exposed to certain environmental factors. The sexual distribution of carriers can also be helpful in identifying whether hormonal factors (i.e., use of contraceptive drugs or pregnancy) influence the test. One can statistically assess the accuracy of the carrier detection test by evaluating blood relatives of individuals identified to be carriers (see Table 4–3). Accordingly, one would expect that at least one parent of each carrier would be a carrier and that appropriate ratios would be found among cousins, siblings, and other close relatives. If alternative methods for carrier detection are at hand, then these techniques (even if more complex and expensive) should be applied, at least in the initial phases, to all identified carriers to corroborate the initial screening results (see Table 4–4).

Critical evaluations of the effectiveness and overall "impact" of such programs must be made, particularly in prototype efforts of this kind. Whether effective genetic education and carrier detection can be appropriately delivered to a lay population without generating major anxieties, stigmatization, and misunderstandings is a question of primary importance.[38, 39] If mass screening programs lead to mass anxieties, then one may seriously question the advisability of such programs. These questions will be answered only if the issues are recognized and if careful psychosocial and behavioral research protocols are incorporated into early screening programs.

Before any massive effort to screen a sizable population is attempted, it is particularly valuable that at least one or preferably several small pilot efforts be conducted and evaluated. Many hidden or unanticipated difficulties may surface with such a venture. Technical,

TABLE 4–4. CONFIRMATION OF CARRIER STATUS IN
SERUM-DEFINED TAY-SACHS DISEASE HETEROZYGOTES*

| | | Confirmatory Test | |
Serum Carriers†	# Studied	LEUKOCYTE HEX A†	PARENTAL TESTING‡
Males	59	59/59	43/43
Females§	45	45/45	38/38

*For further details, see Kaback, M. M., Zeiger, R. S., Reynolds, L. W., and Sonneborn, M.: Approaches to the control and prevention of Tay-Sachs disease. *In* Steinberg, A., and Bearn, A. (eds.): Progress in Medical Genetics. Vol. 10. New York, Grune & Stratton, Inc., 1974.

†Carrier state defined as Hex A level (on 3 consecutive determinations) within 1.7 standard deviations of the mean of Hex A levels determined in serum and WBC from 65 obligate heterozygotes.

‡Either one or both parents shown to be carriers by serum and leukocyte Hex A determination.

§Excludes pregnant women or those on birth control pills.

educational, or counseling operations may require slight, or even major, changes or revisions.

The impact of an anticipated screening program on individuals, families, and the community-at-large can best be assessed by pilot projects. Certainly, if major deficiencies are present they readily become apparent. Conversely, projected "benefits" of the program also become obvious. Perhaps many of the difficulties that arose in the early seventies with sickle trait screening in the United States might have been avoided if small pilot programs had first been established and critically evaluated.[40, 41]

PROGRAM PRINCIPLES AND ADDITIONAL ETHICAL CONSIDERATIONS

Carrier screening, both in selected populations and on a panpopulational basis, raises several important issues in addition to the sensitive areas already mentioned. For many of the reasons previously cited, voluntary carrier screening should require the informed consent of the screenee. Individuals "solicited" for a carrier screening test should be informed of the purpose of the test and should be reasonably knowledgeable about the nature of the condition involved. Obviously, this does not mean that all individuals be informed to the level of a physician or geneticist. However, a comprehensive effort should be made to inform each person prior to testing, in a simple and understandable fashion, about the nature of the illness, the meaning of "carrier," the implications of being a carrier, and the alternatives available to individuals found to carry the gene in question. Certainly, this information would constitute a reasonable requirement for informed consent in carrier detection programs among healthy individuals. Whether each individual should be informed of the *theoretical possibility* of psychological complications, even when there is no clear evidence of this effect, is more open to debate. This possibility should be carefully evaluated in early screening programs. If it becomes apparent that carrier identification does, in fact, cause certain behavioral or psychological burdens for significant numbers of those so identified, this issue then should be incorporated into the information explored with potential screenees prior to subjecting them to the carrier detection test.

Every means must be employed to maintain the absolute confidentiality of test results. This information should be communicated on an individual basis (either through personal letter, direct telephone contact, or private interview). Under no circumstances should public access to test results be permitted. The population to be screened must be assured of the preservation of confidentiality prior to becoming involved in the screening process.

Whether or not a specific recessive allele is predominant in a given population, the carrier detection test should be available and accessible to any individuals who wish to have such information about themselves. Although the educational program might be targeted justifiably to that higher-risk population, the program should not be restrictive in the sense that only individuals from that population can be tested. Since any person may carry any recessive allele, anyone who desires such a test should have the opportunity to be tested.

In a more ethical vein, carrier screening with the goal of reproductive counseling (whether directed toward mate selection, adoption, artificial insemination, or prenatal diagnosis and possible abortion) touches upon a number of very basic and human questions: Is it a basic human freedom that each individual reproduce if he or she chooses and what are the "rights" of the fetus? Certainly, with its implicit restriction on an individual's freedom of choice in selecting a mate or in reproducing, carrier screening might be interpreted as an infringement on such freedoms. Where prenatal diagnosis and possible abortion are involved, the delicate question of the "rights" of the fetus in such decisions has been raised by some. This is a substantial ethical question, as is the matter of abortion itself. If carrier screening can identify couples at-risk for not-too-obviously severe diseases (sickle cell anemia or cystic fibrosis, as judged by some individuals) how and where does one draw the line with regard to selective termination of pregnancy?

The information provided an individual or couple through carrier screening is directed at providing that individual or couple with an informed basis upon which to make reproductive decisions. The ultimate questions, in all or many of these matters, are *Who* has the right to make such decisions? and Is the making of such decisions, indeed, a *right*? As with many such ethical dilemmas, no clear unanimity exists. The reader is encouraged to explore several recent texts which have been devoted entirely to such matters.[3, 4, 6, 42]

Genetic Considerations

The widespread application of certain recently acquired genetic technology (and that anticipated in the near future) could dramatically alter the capability to control certain serious inherited disorders. Heterozygote screening in defined populations, as discussed here, is considered primarily as an instrument for genetic disease prevention. What genetic effects might such programs have? If at-risk couples (so identified through screening) could be enabled to have unaffected offspring selectively, by prenatal diagnosis and selective abortion of homozygous fetuses, would this not lead to an increased frequency for that deleterious gene in the population? For autosomal recessive alleles like those associated with sickle cell anemia, TSD, β-thalassemia, and (perhaps in the near future) cystic fibrosis, screening and prenatal disease prevention would have only a negligible effect in this regard.[43] Since the great majority of heterozygotes in each generation result from carrier–noncarrier matings (whose reproduction should be unaffected by screening), only a small increase in the number of carriers would result from the relatively rare carrier-x-carrier couples who could then successfully reproduce. Of course, prenatal diagnosis is not yet available for the hemoglobinopathies or for cystic fibrosis—the former awaits only a reliable and safe means for obtaining minute fetal blood samples;[44, 45] the latter, an accurate and reliable "biochemical handle." Both of these technical achievements are anticipated in the near future.[46, 47, 48]

Without the prenatal alternative, carrier screening for autosomal recessive traits (premaritally), in association with "reproductive" counseling, could lead to complete stabilization in frequency of the recessive allele—if it is assumed that carriers would choose to reproduce only with noncarriers.[49] In either case, "effective" heterozygote screening would result in a dramatic reduction in the incidence of disease, with only a negligible effect on gene frequency.

CONCLUSIONS AND SUMMARY

Advances in the science and technology of human genetics, coupled with important changes in the social and legal attitudes of Western society, provide the foundations for the recent interest in approaches to the control of certain serious inherited disorders. In this chapter, the prevention of autosomal recessive disease through heterozygote screening and genetic counseling has been considered. The prevalence of certain recessive alleles in defined populations, the availability of simple, accurate, and inexpensive methods to identify heterozygotes, and (for certain disorders) the capability to detect the homozygous condition in the fetus *in utero* make such an approach a worthy consideration. Without prenatal diagnosis, carrier detection followed by genetic counseling might lead to disease prevention if "appropriate" reproductive decisions then were made by identified heterozygotes.

Obviously, unique, substantive, and almost global issues are raised by such considerations. Important ethical, social, economic, and genetic issues must be recognized and addressed before such programs are initiated. The nature, design, goals, and feasibility of any screening effort first must be considered thoroughly. These issues interrelate and thereby dictate the scope and structure of the program itself. Whom to screen, when to screen, how to screen, for what to screen and whether to screen are best answered before the program is begun and only after all these matters are critically assessed.

Heterozygote screening in healthy adult populations raises new and important questions for the medical geneticist, the social scientist, the individuals in the "target" community, and, in many ways, the society at large. At the present time genetic medicine and the simplest concepts of human genetics are abstractions for the public. The delicate and possibly threatening aura of this "new health message" and the complex social and ethical issues involved render voluntary heterozygote screening and prospective genetic disease prevention substantially different from other "public health programs."

In light of these innovations, the traditional role of the medical geneticist has changed. In the past, medical genetics has been a respondent discipline, reacting primarily to the proband-related needs of patients and families referred for services and counseling. With the advent of large-scale heterozygote screening programs, the geneticist adopts a new role—that of initiator, or health advocate—one who seeks out the public at large, informs and educates, and then deals with a more abstract type of counseling (non–proband-related) on a relatively large scale.

The need for detailed psychosocial and sociological evaluations of early screening efforts of this type is essential. Although very few recessive disorders are amenable to such an approach at this time, the further development of carrier detection methods and prenatal diagnostic techniques for several important disorders is anticipated. Accordingly, detailed evaluations of prototype programs become even more critical. From a purely theoretical perspective there is little question that such efforts could have a major impact on the incidence of recessive genetic disease. The feasibilities, realities, and liabilities, however, need to be carefully assessed prior to mass introduction or expansion of such programs.

Ultimately, carrier screening and prevention may provide only an interim approach to recessive disease control. With the continuation of basic medical research, effective therapies or even cures for many of these inborn errors of metabolism may be developed. At some future date a more attractive and generally more appealing means of disease control might then be implemented.

ACKNOWLEDGMENT: The author wishes to express his appreciation to Miss Laetitia Nicoll for her assistance in the preparation of this manuscript.

References

1. Hamerton, J. L.: Human Cytogenetics. Vol. 1. Academic Press, Inc., New York, 1971.
2. McKusick, V. A.: Mendelian Inheritance in Man. Baltimore, Md., Johns Hopkins Press, 1971.
3. Hastings Institute: Ethical and social issues in screening for genetic disease. N. Engl. J. Med., 286:1129, 1972.
4. Hilton, B., Callahan, D., Harris, M., Condliffe, P., and Berkley, B. (eds.): Ethical Issues in Human Genetics. New York, Plenum Publishing Corp., 1973.
5. Kaback, M. M.: Heterozygote screening: A social challenge. N. Engl. J. Med., 289:1090, 1973.
6. Bergsma, D., Lappé, M., Roblin, R., and Gustafson, J. (eds.): Ethical, Social, and Legal Dimensions of Screening for Human Genetic Disease. In Birth Defects: Orig. Art. Ser., Vol. X. New York, Stratton Intercontinental Medical Book Corp., 1974.
7. Stanbury, J., Wyngaarden, J., and Fredrickson, D.: The Metabolic Basis of Inherited Disease. New York, McGraw-Hill Book Co., 1966.
8. Howell, R. R.: Genetic disease: The present status of treatment. In McKusick, V., and Claiborne, R. (eds.): Medical Genetics. New York, HP Publishing Co., 1973.
9. Burton, B. K., Gerbie, A. B., and Nadler, H. L.: Present status of intrauterine diagnosis of genetic defects. Am. J. Obstet. Gynecol., 118:718, 1974.
10. Milunsky, A.: The Prenatal Diagnosis of Hereditary Disorders. Springfield, Ill., Charles C Thomas Publishers, 1973.
11. Raine, D. N.: Management of inherited metabolic disease. Br. Med. J., 2:329, 1972.
12. Kaback, M. M., and Howell, R. R.: Heterozygote detection and prenatal diagnosis of lysosomal diseases. In Hers, H. G., and Van Hoof, F. (eds.): New York, Academic Press, Inc., 1973.
13. Kaback, M. M.: Perspectives in the control of human genetic disease. In Genetics and the Perinatal Patient. Evansville, Ind., Mead Johnson & Co., 1973.
14. Carter, C. O., Roberts, J. A. F., Evans, K. A., and Buck, A. R.: Genetic clinic—a follow-up. Lancet, 1:281, 1971.
15. Leonard, C. O., Chase, G. A., and Childs, B.: Genetic counseling: A consumer's view. N. Engl. J. Med., 287:433, 1972.
16. Myrianthopoulos, N. C., and Aronson, S. M.: Population dynamics of Tay-Sachs disease. Am. J. Hum. Gen., 18:313, 1966.
17. Kaback, M. M.: Unpublished observations.
18. Stein, Z., Susser, M., and Guterman, A.: Screening programme for prevention of Down's syndrome. Lancet, 1:305, 1973.
19. Zimmerman, T. S., Ratnoff, O. D., and Littell, A. S.: Detection of carriers of classic

hemophilia using an immunologic assay for antihemophiliac factor. J. Clin. Invest., 50:255, 1971.

20. Sjöberg, I., Franssom, L., Matalon, R., and Dorfman, A.: Hunter's syndrome: A deficiency of L-iduronosulfate-sulfatase. Biochem. Biophys. Res. Commun., 54:1125, 1973.

21. Davidson, R., Nitowsky, H., and Childs, B.: Demonstration of two populations of cells in the human female heterozygous for glucose-6-phosphate dehydrogenase variants. Proc. Natl. Acad. Sci. U.S.A., 50:481, 1963.

22. Migeon, B. R., der Kalovstian, V. M., Nyhan, W. L., Young, W. J., and Childs, B.: X-linked HGPRT deficiency: Heterozygote has two clonal populations. Science, 160:425, 1968.

23. Goodman, R. M. (ed.): Genetic Disorders of Man. Boston, Little, Brown and Co., 1970.

24. Becker, P. E.: Two new families of benign sex-linked recessive muscular dystrophy. Rev. Can. Biol., 21:551, 1962.

25. Baehner, R. L., and Nathan, D. G.: Quantitative nitro blue tetrazoleum test in chronic granulomatous disease. N. Engl. J. Med., 278:971, 1968.

26. McKusick, V. A.: Human Genetics. Englewood Cliffs, N.J., Prentice-Hall Inc., 1969.

27. Stamatoyannopoulos, G.: Problems of screening and counseling in the hemoglobinopathies. *In* Motulsky, A., and Lenz, W. (eds.): Birth Defects. Amsterdam, Excerpta Medica, 1974.

28. Motulsky, A.: Hereditary red cell traits and malaria. Am. J. Trop. Med. Hyg., 13:147, 1964.

29. Bienzle, U., Ayeni, O., Lucas, A. O., and Luzzatto, L.: G6PD and malaria. Lancet, 1:107, 1972.

30. Myrianthopoulos, N. C., and Aronson, S. M.: Population dynamics of Tay-Sachs disease II. What confers the selective advantage upon the Jewish heterozygote? Adv. Exp. Med. Biol., 19:561, 1972.

31. Fost, N., and Kaback, M. M.: Why do sickle screening in children? The trait is the issue. Pediatrics, 51:742, 1973.

32. Kaback, M. M., Zieger, R. S., Reynolds, L. W., and Sonneborn, M.: Approaches to the control and prevention of Tay-Sachs disease. *In* Steinberg, A., and Bearn, A. (eds.): Progress in Medical Genetics. Vol. 10. New York, Grune & Stratton, Inc., 1974.

33. Lowden, J. A., Skomorowski, M. A., Henderson, F., and Kaback, M. M.: Automated assay of hexosaminidases. Serum. Clin. Chem., 19:1345, 1973.

34. Carmody, P. J., Rattazzi, M. C., and Davidson, R. G.: Tay-Sachs disease—use of tears for detection of heterozygotes. N. Engl. J. Med., 289:1072, 1973.

35. Skoog, W. A., and Beck, W. S.: Studies on the fibrinogen, dextran, and phytohemagglutinin methods of isolating leukocytes. Blood, 11:436, 1956.

36. Kaback, M. M., and Zeiger, R. S.: Practical and ethical issues in an adult genetic screening program. *In* Hilton, B., Callahan, D., Harris, M., Condliffe, P., and Berkley, B. (eds.): Ethical Issues in Human Genetics. New York, Plenum Publishing Corp., 1973.

37. Kaback, M. M., and O'Brien, J. S.: The prevention of recessive genetic disease: Feasibility, costs, and genetic impact. (Abs.) Proc. Soc. Pediatr. Res., p. 283, 1971.

38. Kaback, M. M., Becker, M. H., and Ruth, M. V.: Sociologic studies in human genetics: I. Compliance factors in a voluntary heterozygote screening program. Birth Defects, 10:165, 1974.

39. Beck, E. S., Blaichman, S., Scriver, C. R., and Clow, C. L.: Advocacy and compliance in genetic screening. N. Engl. J. Med., 291:1166, 1974.

40. Bowman, J. E.: Mass screening programs for sickle cell hemoglobins: A sickle cell crisis. J.A.M.A., 222:1650, 1972.

41. Beutler, E., Boggs, D., Heller, P., Miumaver, A., Motulsky, A., and Sheehy, T.: Hazards of indiscriminate screening for sickling. N. Engl. J. Med., 285:1985, 1971.

42. Veatch, R. M.: Ethical issues in genetics. *In* Steinberg, A., and Bearn, A. (eds.): Progress in Medical Genetics. Vol. 10. New York, Grune & Stratton, Inc., 1974.

43. Motulsky, A. G., Fraser, G. R., and Felsenstein, J.: Public health and long-term

genetic implications of intrauterine diagnosis and selective abortion. Birth Defects, 7:22, 1971.

44. Hollenberg, M. D., Kaback, M. M., Kazazian, H. H.: Synthesis of adult hemoglobin by reticulocytes from the human fetus at midtrimester. Science, 174:698, 1971.

45. Kan, Y. W., Dozy, A. M., Alter, B. P., Frigoletto, F. D., and Nathan, D. G.: Detection of the sickle gene in the human fetus. Potential for intrauterine diagnosis of sickle cell anemia. N. Engl. J. Med., 287:1, 1972.

46. Kan, Y. W., Valenti, C., Guidotti, R., Carnazza, V., and Rieder, R. F.: Fetal blood sampling in utero. Lancet, 1:79, 1974.

47. Chang, H., Hobbins, J. C., Cividalli, G., Frigoletto, F. D., Mahoney, M. J., Kan, Y. W., and Nathan, D. G.: In utero diagnosis of hemoglobinopathies. N. Engl. J. Med., 290:1067, 1974.

48. Rao, G. H., and Nadler, H. L.: Arginine esterase in cystic fibrosis of the pancreas. Pediatr. Res., 8:684, 1974.

49. Rucknagel, D. L.: Genetic basis of sickle cell disease. *In* Abramson, H., Bertles, J., and Wethers, D. (eds.): Sickle Cell Disease. St. Louis, C. V. Mosby Co., 1973.

CHAPTER 5

HETEROZYGOTE DETECTION IN AUTOSOMAL RECESSIVE BIOCHEMICAL DISORDERS ASSOCIATED WITH MENTAL RETARDATION

Harold M. Nitowsky, M.D.

GENETIC DISEASE AND MENTAL RETARDATION

The causes of mental retardation are diverse, ranging from certain well-defined genetic disorders, such as Down's syndrome and phenylke-tonuria, to disorders with a clear-cut environmental etiology, including prenatal infection and birth trauma. The remarkable advances in knowledge during the past few decades have broadened our understanding of the role of various etiological factors leading to mental defect. Nevertheless, the fact that neither a specific genetic origin nor a definable environmental factor can be implicated as the basis for the intellectual defect in the majority of patients seen with severe mental retardation attests to our continued lack of adequate understanding of the problem.

Recent surveys have shown that 20 per cent or more of all hospitalizations on pediatric services are due to gene-related diseases.[1]

Similarly, about 10 per cent of all institutionalized mental retardates suffer from diseases that are clearly genetic in origin.[2] If one considers only the institutionalized persons with I.Q.'s less than 50, the latter figure is even higher—perhaps reaching 20 per cent.[3] Although the majority of such patients have trisomy 21, significant contributions to the overall total are made by patients with other chromosomal abnormalities, including aberrations of the sex chromosomes, and by patients with single-gene disorders, such as phenylketonuria and tuberous sclerosis.

These figures do not provide a true indication of the problem, because many inborn errors of metabolism associated with severe developmental retardation cause deaths in infancy or early childhood (*e.g.,* Tay-Sachs disease or metachromatic leukodystrophy) and therefore are generally not represented in institutionalized populations of mental retardates. Table 5–1 lists some of the rare single-gene defects causing serious metabolic disorders associated with mental retardation or early death. Virtually all these disorders can be recognized early in life by appropriate laboratory tests. In an increasing number of instances, prenatal diagnosis is possible by amniocentesis and studies of amniotic fluid or cell cultures.[4] This table is indicative of the rapidity with which these disorders are being defined, since most of the conditions listed were unknown as recently as ten years ago.

Although individually rare, in the aggregate the number of patients with these and other inborn errors of metabolism associated with mental retardation is impressive. If patients survive with retardation due to a disorder for which no treatment is available, or if therapy is effective but for some reason the onset of treatment is delayed or poorly maintained, institutionalization may become necessary as the ultimate recourse. If one considers current costs of institutional care, one can readily appreciate the significant public health impact of this problem. For this as well as other reasons, the importance of early diagnosis, treatment, or prevention is readily apparent.

THE PURPOSE OF HETEROZYGOTE DETECTION

Although there has been impressive progress in recent years in the development of biochemical and other techniques for the detection of carriers of a single dose of a mutant gene,[5] until a few years ago heterozygote identification was a procedure of strictly limited usefulness. Generally, the first indication of heterozygous status in a person or couple was the appearance of a disorder, known or assumed to be recessively inherited, in one of their progeny. At this point, there may be no compelling reason for the precise establishment of heterozygosity for the disorder in the parents, aside from the aim to determine or to confirm the mode of inheritance. On the other hand, identification of carriers among

TABLE 5-1. SOME RARE SINGLE-GENE DEFECTS ASSOCIATED WITH MENTAL RETARDATION OR EARLY DEATH*

Disorder	Biochemical Defect	Clinical Features
Hyperammonemia	Carbamyl-PO_4 synthetase	Episodic vomiting and lethargy exacerbated by protein intake
	Ornithine transcarbamylase	As above
Citrullinemia	Argininosuccinic acid synthetase	Episodic vomiting, mental retardation
Argininosuccinic aciduria	Argininosuccinic acid lyase	Variable: from severe neonatal disease to seizures and mental retardation
Argininemia	Arginase	Seizures, mental retardation
Maple syrup urine disease, classic	Branched-chain ketoacid decarboxylase	Neonatal acidosis, lethargy, feeding problems, urine odor
Intermittent	Same; different mutant	Episodic acidemia and CNS symptoms associated with infection
Variant	Same; different mutant	Mental retardation
Thiamine-response	Same; different mutant	Mental retardation
Hypervalinemia	Valine, α-ketoglutarate transaminase	Neonatal vomiting, lethargy, feeding problems
β-Alaninemia	β-Alanine, α-ketoglutarate transaminase (?)	Neonatal lethargy and seizures
Nonketotic hyperglycinemia	Glycine-methylene TH_4 interconversion	Seizures and mental retardation
Homocystinuria	Cystathionine synthase	Variable retardation, lens dislocations, Marfanoid appearance
	Remethylation defect	Variable: from severe neonatal disease to almost no symptoms
Hyperlysinemia	Lysine, α-ketoglutarate reductase	(?) Mental retardation
Saccharopinuria	Saccharopine cleavage enzyme	Mental retardation
Sarcosinemia	Sarcosine reductase	Mental retardation
Hartnup's disease	Gut and renal tubular transport of neutral amino acids	Feeding problems, failure to thrive, mental retardation
		Ataxia, photosensitive rash, mental retardation
Isovaleric aciduria	Isovaleryl-CoA dehydrogenase	Metabolic acidemia, odor of sweaty feet, neurologic signs
Propionic acidemia	Propionyl-CoA carboxylase	Episodic vomiting, metabolic acidemia, ketonuria
Methylmalonic acidemia	Methylmalonyl-CoA carbonylmutase	Same as above
	Defective synthesis of deoxyadenosyl-B_{12}	Same as above
Lactic-pyruvic acidosis	Pyruvate decarboxylase	Variable: from neonatal acidemia and retardation to ataxia and mild retardation
β-Methylcrotonylglycinuria	Methylcrotonyl-CoA carboxylase	Feeding difficulties, mental retardation
α-Methyl, β-hydroxybutyric aciduria	α-Methylacetoacetyl-CoA thiolase (?)	Intermittent metabolic acidemia
Sulfite oxidase deficiency	Sulfite oxidase	Severe mental retardation, lens dislocation
Butyric and hexanoic aciduria	Green acyl dehydrogenase	Neonatal convulsions, odor, hepatomegaly

*Modified from Goodman, S. I.: Some advances in prevention of mental retardation. *In* Schulman, I. (ed.): Advances in Pediatrics. Vol. 19. Chicago. Yearbook Medical Publishers, 1972. p. 257.

close relatives can provide valuable information for use in counseling family members in regard to family planning or choice of a marriage partner.[6] With advances in our knowledge of inherited metabolic disorders, and with the possibility of therapeutic intervention for some of these conditions, carrier detection, and particularly identification of couples at-risk in a pedigree may aid in the prevention or amelioration of a seriously incapacitating or lethal disorder by means of early postnatal diagnosis and treatment. Alternatively, and particularly when no therapy is available, the possibility of prenatal diagnosis and selective abortion of an affected fetus may permit carrier couples to have children free of the disease in question.

Although few inborn errors of metabolism are sufficiently common to be accompanied by high heterozygote frequencies, this is not true for certain population or ethnic groups (Table 5–2). Moreover, the development of quantitative biochemical or immunological assays that are sufficiently sensitive, inexpensive, and easy to apply at the population level has stimulated the implementation of mass screening programs for carrier detection.[7] Although some of these programs have engendered much controversy, there are compelling reasons for heterozygote detection among populations in which a disorder occurs with relatively high frequency, which is lethal or seriously incapacitating, for which there is no effective therapy, and perhaps most important, for which prenatal diagnosis by means of amniocentesis is feasible. In this situation, the outlook for couples at-risk has been completely altered. Each pregnancy can be monitored in the early stages and terminated if the fetus is shown to be affected—thereby providing a ready solution to the problem of heterozygous parents who desire their own normal children.

There is additional justification, with direct clinical relevance, for

TABLE 5–2. INBORN ERRORS OF METABOLISM AT HIGH FREQUENCIES IN PARTICULAR POPULATIONS*

Condition	Population	Frequency (per 10,000)
Phenylketonuria	Scotland	2.0–2.8
Tyrosinemia	Chicoutimi region, Canada	1.5–1.8
Tay-Sachs disease	Ashkenazic Jews, U.S.A.	2.7
Gaucher's disease (adult type)	Ashkenazic Jews, Israel	4.0
Duchenne muscular dystrophy	N.E. England	3.3
Myotonic dystrophy	New Zealand	9.0
Cystic fibrosis of the pancreas	Victoria, Australia	4.8
Huntington's chorea	Tasmania	1.7
Porphyria variegata	South African Whites	30.0
α_1-Antitrypsin deficiency	Scandinavia	8.0
Adrenogenital syndrome	Yupik Eskimos	20.0

*Modified from Brock, D. J. H.: Inborn errors of metabolism. *In* Brock, D. J. H., and Mayo, O. (eds.): The Biochemical Genetics of Man. New York, Academic Press, Inc., 1972.

heterozygote detection. Thus, in some recessively inherited inborn errors of metabolism, clinical manifestations or abnormalities may be observed in the heterozygote. For example, persons heterozygous for cystinuria type II may excrete enough cystine in the urine to form occasional renal calculi if their urine becomes very concentrated—a complication that can be prevented by maintaining a large urine output.[8] Another example of clinical relevance, which has a more direct bearing on the problem of mental retardation, occurs in galactosemia. The finding of cataracts and other abnormalities at the time of birth of infants with this inherited abnormality has been attributed to ingestion by their heterozygous carrier mothers during pregnancy of excessive quantities of milk or other foods which are sources of dietary galactose.[19] In this situation, documentation of heterozygosity in the mother, regulation of her diet, and monitoring of erythrocyte galactose-1-phosphate levels may help to avert noxious effects on a susceptible galactosemic fetus.[10]

Finally, heterozygote identification is useful because it contributes in a number of ways to our knowledge of the inborn errors of metabolism. For example, identification of the heterozygote may provide convincing evidence for a recessive mode of inheritance of a particular disorder, whether the mutation is autosomal or X-linked. In this context, the concept of dominance or recessiveness is defined operationally, and such a distinction is useful for medical and counseling purposes. However, in the strict genetic sense, one may argue that most conditions are not transmitted as completely dominant or recessive, but as relatively so, depending upon the ease with which the character can be detected in the heterozygote. Since the heterozygote has one mutant and one normal gene, the product of the mutant may be distinguishable qualitatively from the normal gene product within the same individual, or there may be simply a reduced amount of activity of the normal gene product. Thus, although the heterozygote carrier for a biochemical abnormality usually is clinically asymptomatic, careful biochemical, immunological, or physicochemical studies may reveal departures from the norm, so that these individuals can be distinguished from the noncarrier population.

Identification and determination of the prevalence of heterozygotes also may be useful in calculating gene frequencies and in estimating more precisely the incidence of an autosomal recessive genetic disorder in a population. Since many recessive inborn errors of metabolism occur very infrequently, estimates of incidence may be unavailable or only rough approximations. Moreover, there may be problems related to ascertainment of cases because of early death prior to recognition of the disorder or difficulties in establishing a correct diagnosis. Although the incidence of an autosomal recessive disorder may be quite rare, the heterozygote frequency is, of course, substantially higher. If carrier detection is feasible and data relating to carrier prevalence are available, then the Hardy-Weinberg law can be applied to estimate the frequency of the mutant allele and the incidence of the disease in the population.

In X-linked recessive disorders, heterozygote frequency is only twice the disease frequency. However, because a carrier female transmits the disease to the same proportion (one half) of her male progeny regardless of the genetic make-up of the male parent, heterozygote detection may be of even greater practical benefit. Thus, if the genotype of the daughters of a known carrier of a disease such as hemophilia can be established, then for practical purposes the daughters' risk of having affected children can be stated quite independently of their future mating patterns.

METHODS OF HETEROZYGOTE DETECTION

Methods for identification of heterozygotes may involve the detection of qualitative or quantitative alterations in a specific gene product, or the demonstration of an abnormality in a metabolic function as the result of the presence of a mutant gene. In disorders transmitted as autosomal recessive, heterozygote detection most often is dependent on a dosage effect. The assumption is made that a single dose of a mutant gene will produce half as much of the abnormal protein as a double dose. If the abnormal protein is an enzyme that has lost its catalytic function, then the expectation is of almost no activity in the affected homozygote, full activity in the unaffected homozygote, and approximately half activity in the heterozygote. Thus, in the case of the carrier of the mutant galactosemia gene, the activity of galactose-1-phosphate uridyl transferase in erythrocyte lysates is approximately half that of normal controls, with only a small degree of overlap between the homozygous-deficient and homozygous-normal subjects.[11]

Dosage effects have been found with some regularity in attempts to establish heterozygous status by assay of enzyme activity in red or white blood cells, cell cultures from skin, amniotic fluid and other sources, tissue samples, and serum or urine.[5] Indeed, heterozygotes for disorders due to enzyme deficiency have been discovered accidentally in the course of quantitative enzymological surveys carried out for other purposes.[12] The hazards of drawing facile conclusions about genetic constitution from enzyme assays should not, however, be underemphasized. Activities are influenced by polymorphisms,[13] genotype interactions,[14] and environmental effects.[15] Nevertheless, the various lists that have been compiled of enzyme tests for heterozygosity[5] suggest that gene dosage is one of the more powerful determinants of enzyme levels.

Although dosage effects usually ensure that the average level of enzyme activity in heterozygotes is significantly less than in normal homozygotes, in some instances there may be considerable variation in both groups, and the activity distributions may overlap. The mean values generally are useful in confirming a postulated recessive mode of inheritance, but may not be helpful if an individual's status as a carrier has to be established. One possibly way of improving the discriminatory power of enzyme assays has been demonstrated with cultured lymphocytes stimulated by phytohemagglutinin. For example, the α-glucosidase activities of lymphocyte extracts of persons heterozygous for Pompe's disease overlap the activities of controls. When lymphocyte transformation is induced with phytohemagglutinin (PHA), α-glucosidase activity in controls increases sharply, but remains constant in heterozygotes so that there is no longer overlap between the two groups.[16] Stimulation of lymphocytes by PHA leads to morphological changes, increased protein synthesis, and mitosis, and it is known to be associated with enhanced activities of several enzymes.[17] If the difference in response is somehow related to a dosage effect, it is not clear why the lymphocytes of a person heterozygous for α-glucosidase deficiency show no increase in enzyme activity after PHA stimulation. A similar result has been obtained with respect to lysosomal acid phosphatase in persons heterozygous for deficiency of this enzyme[18] and for cystathionine synthase in carriers for homocystinuria.[19] Further studies are required to determine whether this phenomenon may have wider applicability for other inborn errors of metabolism in which enzyme deficiency is expressed in circulating lymphocytes.

The failure to demonstrate a gene-dose effect in the obligate heterozygote may furnish insight into the nature of the molecular structure and abnormality in an inherited metabolic disorder. For example, it has been shown that the native molecule of liver cystathionine synthase contains four subunits, composed of two nonidentical polypeptide chains, each represented twice (*i.e.*, the structure is $\alpha_2\beta_2$).[20] The extensive evidence supporting the generalization that one gene determines the primary structure of one polypeptide chain indicates that there should be two structural gene loci involved in the determination of this enzyme (wild type genotype AA, BB). The simplest possible genetic variant would then be heterozygous at one of these loci (genotype AA', BB). Since more than one α chain is present in cystathionine synthase, such a heterozygote potentially forms three types of enzyme molecules: $\alpha_2\beta_2$, $\alpha\alpha'\beta_2$, and $\alpha'_2\beta_2$. The molecule $\alpha\alpha'\beta_2$ is of interest because such a hybrid molecule cannot be present in individuals homozygous for either the A, or the mutant A' allele, and it thus represents a special molecular form found only in the heterozygous state. The possible existence of such hybrid molecules provides one possible explanation for the observation that obligate heterozygotes for cystathionine synthase deficiency tend to have less than 50 per cent of the mean specific activity of this enzyme in extracts of liver,[21] cultured fibroblasts,[21, 22] or phytohemagglutinin-stimulated lymphocytes.[19] Making the simplifying assumptions that each allelic gene locus is responsible for the formation of the same number of polypeptide chains, and that these chains are assembled at random into enzyme tetramers, some have calculated that, in order to account for the mean cystathionine synthase activities of 17 and 38 per cent observed in extracts of stimulated lymphocytes or liver, the $\alpha\alpha'\beta_2$ molecule must have a specific activity of 0.00 to 0.25 (compared to 1.0 for the wild type and negligible activity for the homozygous mutant molecules). In effect, this model, proposed by Mudd and his colleagues,[21] presupposes a negative interaction between the associated normal and the mutant polypeptide chains, such that the catalytic activity of the former is adversely affected. Such interactions have not been detected enzymatically in the few instances studied of enzyme hybrids formed between chemically modified and unmodified subunits.[23] These results, together with the evidence that in most genetic diseases the heterozygote has roughly 50 per cent of the normal activity of the affected enzyme,[24] suggest that negative interactions of the type postulated for cystathionine synthase are not common. Other explanations for the low activity in heterozygotes for cystathionine synthase deficiency may be proposed. However, the model suggested is an indication of the type of study of subunit structure and function that will probably be required to extend present understanding of the relationship between genotype and phenotype in this and other inborn errors of metabolism.

There is ample evidence for genetic heterogeneity in patients with severe cystathionine synthase deficiency.[21, 25, 26] Thus, there may be multiple alleles at both the A and B loci, and the population is likely to contain not only homozygous mutant individuals (*e.g.*, A'A', BB or AA, B'B') but also genetic compounds or mixed heterozygotes for two different mutant alleles at the same locus (*e.g.*, A'A″, BB), as well as double heterozygotes (AA', BB'). Identification of subjects with these various possible genotypes and clarification of the structural and functional changes in the enzymes they form represent important areas for future investigation.

Other examples have been observed recently of anomalous gene-dose effects in heterozygotes which may be explained by the existence of genetic compounds. Tay-Sachs disease is an inborn error of glycosphingolipid metabolism, characterized by the accumulation of Gm_2 ganglioside in nerve cells and other tissues of affected individuals.[27] The defect in this disorder has been associated with the absence of one (the A form) of the two major isozymes of N-acetyl-β-D hexosaminidase (Hex A and Hex B).[28] Hex A can be differentiated from Hex B in serum, leukocytes, cultured skin and amniotic fluid cells, urine, and other sources by the use of various physical characteristics of these isozymes.[29] The reduction of the proportion of Hex A relative to Hex B to about half normal values is the basis of screening for heterozygous carriers of Tay-Sachs disease.[30]

An apparent absence of Hex A activity has been demonstrated in obligate heterozygotes and in other healthy members of two families in which infants have been found to have Tay-Sachs disease.[31, 32] It has been suggested that the anomalous detection of a lack of Hex A activity in these adults by the use of a synthetic substrate (4-methyl-umbelliferyl β-N-acetylglucosamine) reflects the presence of two mutant alleles (A', A'') at the locus for Hex A. One of these alleles (A'), which in the homozygote results in Tay-Sachs disease, is responsible for the formation of Hex A, which lacks activity against the natural (Gm_2 ganglioside) and the synthetic substrate. The other mutant allele (A'') results in the formation of Hex A, which is inactive against the artificial 4-MU substrate, but retains activity against Gm_2 ganglioside. Thus, the obligate heterozygote, who is a genetic compound (A'A''), may appear to lack Hex A activity when tested with synthetic substrate, but has sufficient activity against the natural substrate *in vivo* to prevent abnormal sphingolipid accumulation. Support for this hypothesis has been adduced by the demonstration of normal Gm_2 catabolism *in vitro* by the use of leukocyte extracts from healthy adults apparently deficient in Hex A activity.[33]

In some inherited biochemical abnormalities, properties of the enzyme other than activity may be utilized as a means of carrier identification. These include alterations of affinity for substrate[34] or coenzyme,[26] differential response to inhibitors,[35] or the demonstration of a disparity between the amount of a protein measured by its functional and immunologically reactive properties.[36, 37] Indeed, qualitative or quantitative variations obtained in the application of these methods suggest the presence of genetic heterogeneity in several disease states.[38, 39]

In some disorders, difficulties in obtaining tissue make it impossible to screen for potential carriers by quantitative enzymology. Measurement of circulating or excreted metabolites may offer a useful alternative, although from the results obtained it may be difficult to distinguish between a heterozygote and a normal homozygote. In patients with phenylketonuria, for example, there is no measurable phenylalanine hydroxylase activity in liver, to which the enzyme normally is restricted.[40] However, there is sufficient enzyme activity in the heterozygous carrier to maintain normal fasting blood levels of phenylalanine. When the subject is stressed with an oral load of this amino acid (100 mg per kg), an abnormal response in blood phenylalanine, as well as in tyrosine levels, may permit identification of the heterozygous carrier.[41] Similarly, in cystathioninuria, a methionine load may be used to monitor urinary cystathionine excretion; heterozygotes have excretion rates well above those of normal people.[42] Because orotic aciduria, accompanied by hyperammonemia, was found in a patient with ornithine transcarbamylase deficiency, increased excretion of orotic acid after a protein load was employed as one measure of partial enzyme deficiency in an obligate carrier and other presumed heterozygotes for this X-linked disorder in a family.[43] Similar loading tests have been useful for carrier detection in other inborn metabolic disorders.[5]

Other metabolic parameters may be useful, such as measurement of the concentration of a metabolite derived from an overall reaction sequence in which there is a decrease in enzyme activity[44, 45] or the excessive accumulation of a normal or abnormal constituent within cells similarly reflects a decrease in enzyme activity.[46, 47] An example of this approach is provided by studies of the activity of cell cultures derived from patients with galactosemia, from heterozygous carriers, and from normals in producing radioactively labeled CO_2 from ^{14}C-labeled galactose.[48] Since this overall reaction depends upon the presence of an active galactose-1-phosphate uridyl transferase, and since this step appears to be among the rate-limiting reactions in the overall metabolic sequence, this procedure can be used to distinguish the heterozygote from normal and homozygous-deficient subjects. However, unlike the findings with direct assay of enzyme activity, no clear-cut gene-dose relationship has been observed between the amount of labeled CO_2 produced and the genotype of the subject. Similar disparities have been found in metabolic or biochemical tests for the carrier state which do not involve a relatively direct measurement of the potentially abnormal primary gene product.

Even when the primary lesion in a genetic disorder is unknown, a suitable enzymatic or chemical marker may be found. The metachromatic staining of cultured fibroblasts by toluidine blue has been widely used to detect lipid, mucopolysaccharide, and other storage disorders.[49] Although it has been claimed that this technique will detect heterozygotes in a number of inborn errors of metabolism,[50] metachromasia has been criticized because of a lack of specificity and reproducibility, as well as for a lack of sensitivity as a discriminant between affected homozygotes and heterozygotes.[51] Attempts to circumvent this problem have been successful in some of the lysosomal storage disorders. For example, studies of the kinetics of incorporation of radioactively labeled sulfate into cellular mucopolysaccharide have been useful in carrier identification in the mucopolysaccharidoses.[52] When radioactive sulfate is added to medium overlying fibroblast cultures from normals, the incorporation reaches a plateau within 24 hours as synthesis, excretion, and breakdown of mucopolysaccharides reach equilibrium. In contrast, with cell cultures from patients with Hurler's syndrome, catabolism is delayed and radioactivity continues to increase for many days. Although cells from the heterozygote have been reported to show metachromasia, the pattern of incorporation of the radioactive label resembles that of normal cells.

In this regard, it should be pointed out that environmental alterations may produce phenocopies *in vitro*. Thus, when the medium pH is maintained at about 8.0, Hurler-like radioactively labeled sulfate uptake patterns can be induced in fibroblast cultures from normal individuals.[53]

Another approach to carrier detection involves the use of genetic

TABLE 5–3. CARRIER DETECTION IN VARIOUS METABOLIC DISORDERS ASSOCIATED WITH MENTAL RETARDATION

Disorder	Abnormality in Carrier	Reference
DISORDERS OF AMINO ACID METABOLISM OR TRANSPORT		
Urea Cycle Disorders		
Argininemia	Elevated plasma arginine; variable reduction of red cell arginase	57
Argininosuccinic acidemia	Reduced red cell argininosuccinase; small elevation of urinary argininosuccinate excretion	58,59,60
Citrullinemia	Elevated plasma citrulline; reduced argininosuccinate synthetase activity in cell cultures	61,62
Ornithine transcarbamylase (OTC) deficiency*	Reduced liver or intestinal OTC activity; increased urine orotic acid	43,63
Carbamyl phosphate synthetase deficiency	None detected as yet	64
Disorders of Branched Chain Amino Acids		
Maple syrup urine disease	Disappearance of leucine or keto acid from plasma; oxidation of labeled α-keto-isocaproate by leukocytes, cell cultures	65,66
Variants of maple syrup urine disease	Decreased activity of branched chain decarboxylase	67
Propionic acidemia and the ketotic hyperglycinemia syndrome	Decreased propionyl-CoA carboxylase activity in cultured cells	68
Methyl malonic acidemia	None detected as yet	69,70
Isovaleric acidemia	None detected as yet	71
α-Methyl-β-hydroxybutyric acidemia	Increase in urine α-methyl-β-hydroxybutyric acid after leucine load	72
Hypervalinemia	None detected as yet	73
Disorders of Aromatic Amino Acids		
Phenylketonuria	Disappearance of phenylalanine from plasma and appearance of tyrosine after phenylalanine load	41,74,75
Tyrosinemia	None detected as yet	76
Disorders of Histidine Metabolism		
Histidinemia	Abnormal excretion of histidine after load; decreased histidase activity of skin	77,78
Carnosinemia	Reduced levels of serum carnosinase activity	79
Disorders of Glycine, Proline Metabolism		
Nonketotic hyperglycinemia	None detected as yet	80
Sarcosinemia	Increased sarcosine plasma levels after oral load (variable); increased urinary sarcosine excretion	81,82,83
Hydroxyprolinemia	None detected as yet	84
Hyperprolinemia	None detected as yet	84
Sulfur-Containing Amino Acids		
Homocystinuria	Decreased cystathionine synthase in lymphocytes and cell cultures	19,22,85
Cystathioninuria	Increase urinary cystathionine	42,86,87
β-Mercaptolactate-cysteine disulfiduria	None detected as yet	88
Sulfite oxidase deficiency	None detected as yet	89
Methionine malabsorption and hypermethioninemia	None detected as yet	90,91
Other Disorders of Amino Acid Metabolism		
Pyroglutamic acidemia	None detected as yet	92
Disorders of Amino Acid Transport		
Hartnup's disease	Increased urinary excretion of neutral amino acids	93
Hyper dibasic aminoaciduria	Increased urinary excretion of dibasic amino acids	94
Oculocerebrorenal syndrome*	None detected as yet	95
DISORDERS OF MUCOPOLYSACCHARIDE METABOLISM		
Hurler syndrome	Decreased activity of α-L-iduronidase in cell extracts	96
Hunter's syndrome*	Decreased activity of sulfoiduronate sulfatase in clonal cell extracts	97
Sanfilippo A syndrome	Decreased activity of heparin-N-sulfatase in cell extracts	98

Disorder	Abnormality in Carrier	Reference
Sanfilippo B syndrome	Decreased activity of α-N-acetyl-glucosaminidase in cell extracts	99
Morquio's syndrome	None detected as yet	100
Scheie syndrome	Decreased activity of α-L-iduronidase in cell extracts	96
Maroteaux-Lamy syndrome	Decreased activity of arylsulfatase B in cell extracts	101
β-Glucuronidase deficiency	Decreased activity of β-glucuronidase in cell extracts	102
DISORDERS OF LIPID METABOLISM		
Sphingolipid Storage Diseases		
Generalized gangliosidosis (GM$_1$ gangliosidosis, type 1)	Decreased β-galactosidase activity of tissue and cell extracts	103,104
Juvenile (GM$_1$ gangliosidosis, type 2)	Decreased β-galactosidase activity of tissue and cell extracts	105
Tay-Sachs disease (GM$_2$ gangliosidosis, type 1)	Decreased activity of N-acetyl-β-D-hexosaminidase A in serum, urine, tears, tissue and cell extracts	28,106–110
Sandhoff's disease (GM$_2$ gangliosidosis, type 2)	Decreased activity of N-acetyl-β-D-hexosaminidase A and B in tissue and cell extracts	111,112
Juvenile (GM$_2$ gangliosidosis, type 3)	Decreased activity of N-acetyl-β-D-hexosaminidase A in cell extracts	112
Niemann-Pick disease	Decreased activity of sphingomyelinase in leukocyte extracts (inconsistent results with cell extracts)	114,115
Gaucher's disease (infantile and late onset)	Decreased activity of β-glucosidase in tissue and cell extracts	116,117
Metachromatic leukodystrophy (infantile and late onset)	Decreased activity of arylsulfatase A in tissue and cell extracts	118,119
Krabbe's disease	Decreased activity of galactocerebroside β-galactosidase in serum and cell extracts	120
Ceramide lactoside lipidosis	None detected as yet	121
GM$_3$ (hematoside) sphingolipodystrophy	None detected as yet	122
Other Disorders of Lipid Metabolism		
Refsum's disease	Decreased activity of phytanic acid α-hydroxylase in cell extracts	123
Wolman's disease	Decreased activity of acid esterase in cell extracts	124
Mucolipidosis II (I-cell disease)	Increase in serum activity of several lysosomal hydrolases (variable findings)	125,126
DISORDERS OF CARBOHYDRATE METABOLISM		
Galactosemia	Decreased activity of galactose-1-phosphate uridyl transferase in red cells, leukocytes, tissue and cell extracts	127,128
Hereditary fructose intolerance	None detected as yet	129
Fucosidosis	Decreased activity of α-fucosidase in cell extracts	130
Mannosidosis	None detected as yet	131
Pyruvate decarboxylase	Decreased activity of pyruvate decarboxylase in cell extracts	132
Pyruvate carboxylase deficiency	None detected as yet	133,134
OTHER METABOLIC DISORDERS		
Defects in Purine, Pyrimidine Metabolism		
Lesch-Nyhan syndrome*	Decreased activity of hypoxanthine-guanine phosphoribosyl transferase activity in clonal cells or cell extracts	135,136
Hereditary orotic aciduria	Decreased activity of red cell orotidylic decarboxylase	137
Lysosomal acid phosphatase deficiency	Decreased acid phosphatase activity in cell extracts	18
Aspartylglucosaminuria	Decreased activity of N-aspartyl-β-glucosaminidase in cell extracts	138

*Transmitted as X-linked abnormality

linkage, such as the method employed in detection of female carriers for hemophilia A.[54] The genetic locus for factor VIII deficiency is closely linked on the X chromosome to that for G6PD. Since approximately 30 per cent of American Negro females are heterozygous for the G6PD electrophoretic variants (B+ normal, and A+ or A−), heterogeneity at the G6PD locus and genetic linkage data may be useful in determining, with a high degree of probability, whether a close female relative of a hemophilic male carries the mutant gene. In principle, this method should be applicable, under favorable circumstances, to identification of carriers of autosomal mutant genes. Moreover, this method can be used for prenatal prediction of fetal disease. For example, prenatal diagnosis of myotonic dystrophy, an autosomal dominant disorder, is feasible because the gene for this disorder is closely linked to the secretor (Se) locus. The Se status of the fetus may be determined by examination of the amniotic fluid. In a recent report in which the genotypes of the parents represented a favorable situation for prediction with a high degree of confidence of the risk of the fetus for myotonic dystrophy, a decision could be made not to terminate pregnancy.[55] With the recent rapid developments in knowledge of the human chromosome map, genetic linkage on the basis of somatic cell hybridization studies, and pedigree analysis,[56] there should be an increasing number of occasions for heterozygote detection, as well as for prenatal diagnosis of genetic abnormalities.

AUTOSOMAL RECESSIVE DISORDERS FOR WHICH CARRIER DETECTION IS POSSIBLE

With the expansion in knowledge of biochemistry and metabolic diseases, the list of autosomal recessive disorders for which carrier identification is possible is increasing. Table 5–3 summarizes the currently available information concerning the feasibility and approach to identification of heterozygotes in these disorders associated with mental retardation.

CONCLUSION

In recent years there has been dramatic progress in the development of biochemical and other techniques for the detection of carriers of a single dose of a mutant gene. Clearly, there are important applications of heterozygote detection to genetic counseling. Although there seems little justification or need for the widespread application of the methods for carrier detection in the general population, the techniques may be of value in high-risk sub-populations. In situations where antenatal diagnosis is possible for the identification of a lethal or seriously incapacitating genetic disorder, screening to identify couples at-risk offers even more cogent reasons for mass screening. This presentation summarizes the purpose, general approaches, and some possible limitations in screening methods for identification of heterozygotes for a variety of recessively inherited inborn errors of metabolism associated with mental retardation.

References

1. Scriver, C. R.: Screening and treatment of hereditary metabolic disease. Proceedings of the Fourth International Congress of Human Genetics, Paris, 1971. Excerpta Medica Foundation, 1972.
2. Heber, R.: Epidemiology of Mental Retardation. Springfield, Ill., Charles C Thomas, Publisher, 1970.
3. Goodman, S. I.: Some advances in prevention of mental retardation. *In* Schulman, I. (ed.): Advances in Pediatrics. Vol. 19. Chicago, Yearbook Medical Publishers, 1972, p. 257.
4. Milunsky, A.: The Prenatal Diagnosis of Hereditary Disorders. Springfield, Ill., Charles C Thomas, Publisher, 1973.
5. Hsia, D. Y.: The detection of heterozygous carriers. Med. Clin. North Am., 53:857, 1969.
6. Nitowsky, H. M., and Legum, C.: Genetic counseling: General principles and clinical applications. *In* Schulman, I. (ed.): Advances in Pediatrics. Vol. 18. Chicago, Yearbook Medical Publishers, 1971, p. 13.
7. Nitowsky, H. M.: Prescriptive screening for inborn errors of metabolism. A critique. Am. J. Ment. Defic., 77:538, 1973.
8. Rosenberg, L. E.: Genetic heterogeneity in cystinuria. *In* Nyhan, W. L. (ed.): Amino Acid Metabolism and Genetic Variation. New York, McGraw-Hill Book Co., 1967, p. 341.
9. Nadler, H. L., Inouye, T., and Hsia, D. Y. Y.: Clinical galactosemia: A study of fifty-five cases. *In* Hsia, D. Y. Y. (ed.): Galactosemia. Springfield, Ill., Charles C Thomas, Publisher, 1969, p. 127.
10. Donnell, G. M., Bergren, W. R., Perry, G., and Koch, R.: Galactose-1-phosphate in galactosemia. Pediatrics, 31:802, 1963.
11. Kirkman, H. M., and Bynum, E.: Enzymic evidence of a galactosemic trait in parents of galactosemic children. Ann. Hum. Genet., 23:117, 1959.
12. Singer, J. D., and Brock, D. J. H.: Half-normal adenylate kinase activity in three generations. Ann. Hum. Genet., 35:109, 1971.
13. Hopkinson, D. A., Spencer, N., and Harris, H.: Genetical studies on human red cell acid phosphatase. Am. J. Hum. Genet., 16:141, 1964.
14. Mellman, W. J., Tedesco, T. A., and Feigl, P.: Estimation of the gene frequency of the Duarte variant of galactose-1-phosphate uridyl transferase. Ann. Hum. Genet., 32:1, 1968.
15. Motulsky, A.: Hemolysis in glucose-6-phosphate dehydrogenase deficiency. Fed. Proc., 31:1286, 1972.
16. Hirschhorn, K., Nadler, H. L., Waithe, W. I., Brown, B. I., and Hirschhorn, R.: Pompe's disease: Detection of heterozygotes by lymphocyte stimulation. Science, 166:1632, 1969.
17. Hirschhorn, K., and Hirschhorn, R.: Role of lysosomes in the lymphocyte response. Lancet, 1:1046, 1965.
18. Nadler, H. L., and Egan, T. J.: Deficiency of lysosomal acid phosphatase: A new familial metabolic disorder. N. Engl. J. Med., 282:302, 1970.
19. Goldstein, J. L., Campbell, B. K., and Gartler, S. M.: Homocystinuria: Heterozygote detection using phytohemagglutinin-stimulated lymphocytes. J. Clin. Invest., 52:218, 1973.

20. Kashiwamata, S., Kotake, Y., and Greenberg, D. M: Studies of cystathionine synthase of rat liver: Dissociation into two components by sodium dodecyl sulfate disc electrophoresis. Biochim. Biophys. Acta, 212:501, 1970.
21. Uhlendorf, B. W., Conerly, E. B., and Mudd, S. H.: Homocystinuria. Studies in tissue culture. Pediatr. Res., 7:645, 1973.
22. Fleisher, L. D., Beratis, N. G., Tallon, H. H., Hirschhorn, K., and Gaull, G. E.: Homocystinuria due to cystathionine synthase deficiency: Detection of heterozygotes and homozygotes using cultured skin fibroblasts. Pediatr. Res., 7:158, 1973.
23. Frieden, C.: Protein–protein interaction and enzymatic activity. Ann. Rev. Biochem., 40:653, 1971.
24. Harris, H.: The Principles of Human Biochemical Genetics. Amsterdam, North-Holland Publishing Co., 1970.
25. Gaul, G. E., and Sturman, J. A.: Vitamin B_6 dependency in homocystinuria. Br. Med. J., 3:532, 1971.
26. Seashore, M. R., Durant, J. L., and Rosenberg, L. E.: Studies of the mechanism of pyridoxine-responsive homocystinuria. Pediatr. Res., 6:187, 1972.
27. O'Brien, J. S.: Tay-Sachs disease: From enzyme to prevention. Fed. Proc., 32:191, 1973.
28. Okada, S., and O'Brien, J. S.: Tay-Sachs disease: Generalized absence of a beta-D-N-acetyl-hexosaminidase component. Science, 165:698, 1969.
29. Robinson, D., and Stirling, J. L.: N-acetyl-β-glucosaminidases in human spleen. Biochem. J., 107:321, 1968.
30. Kaback, M. M., Zeiger, R. S., Reynolds, L. W., and Sonnenborn, M.: Tay-Sachs disease: A model for control of recessive genetic disorders. *In* Motulsky, A. G., and Lenz, W. (eds.): Birth Defects. Amsterdam, Excerpta Medica, 1974, p. 248.
31. Navon, R., Padeh, B., and Adam, A.: Apparent deficiency of hexosaminidase A in healthy members of a family with Tay-Sachs disease. Am. J. Hum. Genet., 25:287, 1973.
32. Vidgoff, J., Buist, N. R. M., and O'Brien, J. S.: Absence of β-N-acetyl-D-hexosaminidase A activity in a healthy woman. Am. J. Hum. Genet., 25:372, 1973.
33. Tallman, J. F., Brady, R. O., Navon, R., and Padeh, B.: Ganglioside catabolism in hexosaminidase A-deficient adults. Nature, 252:254, 1974.
34. Tedesco, T. A., and Mellman, W. J.: Argininosuccinate synthetase activity and citrulline metabolism in cells cultured from a citrullinemic subject. Proc. Natl. Acad. Sci. U.S.A., 57:829, 1967.
35. Kalow, W., and Genest, K.: A method for the detection of atypical forms of human cholinesterase: Determination of dibucaine numbers. Can. J. Biochem. Physiol., 35:339, 1957.
36. Zimmerman, T. S., Ratnoff, O. D., and Littell, A. S.: Detection of classic hemophilia using an immunologic assay for antihemophilic factor (factor VIII). J. Clin. Invest., 50:255, 1971.
37. Bennett, B., Ratnoff, O. D., and Levin, J.: Immunologic studies on Von Willebrand's disease. J. Clin. Invest., 51:2597, 1972.
38. Bennett, E., and Huehns, E. R.: Immunological differentiation of three types of haemophilia and identification of some female carriers. Lancet, 2:956, 1970.
39. Holmberg, L., and Nilsson, I. M.: Two genetic variants of Von Willebrand's disease. N. Engl. J. Med., 288:595, 1973.
40. Friedman, P. A., and Kaufman, S.: A study of the development of phenylalanine hydroxylase in fetuses of several mammalian species. Arch. Biochem. Biophys., 146:321, 1971.
41. Gold, R. J. M., Maag, U. R., Neal, J. L., and Scriver, C. R.: The use of biochemical data in screening for mutant alleles and in genetic counseling. Ann. Hum. Genet., 37:315, 1974.
42. Mongeau, J. G., Hilgartner, M., Worthen, H. G., and Frimpter, G. W.: Cystathioninuria: A study of an infant with normal mentality, thrombocytopenia, and renal calculi. J. Pediatr., 69:1113, 1966.
43. Goldstein, A. S., Hoogenraad, N. J., Johnson, J. D., Fukanaga, K., Swierczewski, E., Cann, H. M., and Sunshine, P.: Metabolic and genetic studies of a family with ornithine transcarbamylase deficiency. Pediatr. Res., 8:5, 1974.

44. Hommes, F. A., Kuipers, J. R. G., Elema, J. D., Jansen, J. F., and Jonxis, J. H.: Propionic acidemia: A new inborn error of metabolism. Pediatr. Res., 2:519, 1968.
45. Rosenberg, L. E., Lilljegrist, A. C., and Hsia, Y. E.: Methylmalonic aciduria: An inborn error leading to metabolic acidosis, long chain ketonuria and intermittent hyperglycinemia. N. Engl. J. Med., 278:1319, 1968.
46. Howell, R. R.: The glycogen storage diseases. *In* Stanbury, J. B., Wyngaarden, J. B., and Frederickson, D. S. (eds.): The Metabolic Basis of Inherited Diseases. 3rd. ed. New York, McGraw-Hill Book Company, 1972, p. 149.
47. Schneider, J. A., Wong, V., Bradley, K. H., and Seegmiller, J. E.: Biochemical comparisons of the adult and childhood forms of cystinosis. N. Engl. J. Med., 279:1253, 1968.
48. Krooth, R. S., and Weinberg, A. N.: Studies on cell lines developed from the tissues of patients with galactosemia. J. Exp. Med., 113:1155, 1961.
49. Danes, B. S., and Bearn, A. G.: Hurler's syndrome. A genetic study in cell culture. J. Exp. Med., 123:1, 1967.
50. Danes, B. S., and Bearn, A. G.: Cystic fibrosis of the pancreas. A study in cell culture. J. Exp. Med., 129:775, 1969.
51. Taysi, K., Kistenmacher, M. L., Punnett, H. H., and Mellman, W. J.: Limitations of metachromasia as a diagnostic aid in pediatrics. N. Engl. J. Med., 281:1108, 1969.
52. Fratantoni, J. C., Hall, C. W., and Neufeld, E. F.: The defect in Hurler's and Hunter's syndromes: Faulty degradation of mucopolysaccharide. Proc. Natl. Acad. Sci. U.S.A., 60:699, 1968.
53. Lie, S. O., McKusick, V. A., and Neufeld, E. F.: Simulation of genetic mucopolysaccharidoses in normal human fibroblasts by alteration of pH of the medium. Proc. Natl. Acad. Sci. U.S.A., 69:2361, 1972.
54. McCurdy, P. R.: Use of genetic linkage for the detection of female carriers of hemophilia. N. Engl. J. Med., 285:218, 1971.
55. Schrott, H. G., Karp, L., and Omenn, G. S.: Prenatal prediction in myotonic dystrophy: Guideline for genetic counseling. Clin. Genet., 4:38, 1973.
56. New Haven Conference (1973): First International Workshop on Human Gene Mapping. Birth Defects: Original Article Series Vol. 10, No. 3. New York, The National Foundation, 1974.
57. Terheggen, H. G., Lavinha, F., Colombo, J. P., Van Sande, M., and Lowenthal, A.: Familial hyperargininemia. J. Génét. Hum., 20:69, 1972.
58. Tomlinson, S., and Westall, R. G.: Argininosuccinic aciduria. Argininosuccinase and arginase in human blood cells. Clin. Sci., 26:261, 1964.
59. Coryell, M. E., Hall, W. K., Thevaos, T. G., Welter, D. A., Gatz, A. J., Horton, B. F., Sisson, B. D., Looper, J. W., and Farrow, R. T.: A familial study of a human enzyme defect, argininosuccinic aciduria. Biochem. Biophys. Res. Commun., 14:377, 1964.
60. Shih, V. E., Littlefield, J. W., and Moser, H. W.: Argininosuccinase deficiency in fibroblasts cultured from patients with argininosuccinic aciduria. Biochem. Genet., 3:81, 1969.
61. Wick, H., Brechbühler, T., and Girard, J.: Citrullinemia: Elevated serum citrulline levels in healthy siblings. Experientia, 26:823, 1970.
62. Wick, H., Bachmann, C., Baumgartner, R., Brechbühler, T., Colombo, J. P., Wiesmann, V., Mihatsch, M. J., and Ohnacker, H.: Variants of citrullinemia. Arch. Dis. Child, 48:636, 1973.
63. Short, E. M., Conn, H. O., Snodgrass, P. J., Campbell, A. G. M., and Rosenberg, L. E.: Evidence for X-linked dominant inheritance of ornithine transcarbamylase deficiency. N. Engl. J. Med., 288:7, 1973.
64. Gelehrter, T. D., and Snodgrass, P. J.: Lethal neonatal deficiency of carbamyl phosphate synthetase. N. Engl. J. Med., 290:430, 1974.
65. Dancis, J., Hutzler, J., and Levitz, M.: Detection of the heterozygote in maple syrup urine disease. J. Pediatr., 66:595, 1965.
66. Langenbeck, U., Rüdiger, H. W., Schulze-Schencking, M., Keller, W., Brackertz, D., and Goedde, H. W.: Evaluation of a heterozygote test for maple syrup urine disease in leucocytes and cultured fibroblasts. Humangenetik, 11:304, 1971.

67. Dancis, J., Hutzler, J., Snyderman, S. E., and Cox, R. P.: Enzyme activity in classical and variant forms of maple syrup urine disease. J. Pediatr., 81:312, 1972.

68. Hsia, Y. E., Scully, K. J., and Rosenberg, L. E.: Inherited propionyl-CoA carboxylase deficiency in "ketotic hyperglycinemia." J. Clin. Invest., 50:127, 1971.

69. Morrow, G., Barness, L. A., Auerbach, V. H., DiGeorge, A. M., Ando, T., and Nyhan, W. L.: Observations on the coexistence of methylmalonic acidemia and glycinemia. J. Pediatr., 174:680, 1969.

70. Morrow, G.: Methylmalonic acidemia. In Nyhan, W. L. (ed.): Heritable Disorders of Amino Acid Metabolism. New York, John Wiley & Sons, Inc., 1974, p. 61.

71. Levy, H. L., and Erickson, A. M.: Isovaleric acidemia. In Nyhan, W. L. (ed.): Heritable Disorders of Amino Acid Metabolism. New York, John Wiley & Sons, Inc., 1974, p. 81.

72. Daum, R. S., Scriver, C. R., Mamer, O. A., Delvin, E., Lamm, P., and Goldman, H.: An inherited disorder of isoleucine catabolism causing accumulation of α-methyl-β-hydroxy-butyrate, and intermittent metabolic acidosis. Pediatr. Res., 7:149, 1973.

73. Dancis, J., Hutzler, J., Tada, K., Wada, Y., Morikawa, T., and Arakawa, T.: Hypervalinemia: A defect in valine transamination. Pediatrics, 39:813, 1967.

74. Woolf, L. I., Cranston, W. I., and Goodwin, B. L.: Genetics of phenylketonuria. Heterozygosity for phenylketonuria. Nature (London) 213:882, 1967.

75. Rampini, S., Anders, P. W., Curtius, H. C., and Marthaler, T.: Detection of heterozygotes for phenylketonuria by column chromatography and discriminatory analysis. Pediatr. Res., 3:287, 1969.

76. Buist, N. R. M., Kennaway, N. G., and Fellman, J. H.: Disorders of tyrosine metabolism. In Nyhan, W. L. (ed.): Heritable Disorders of Amino Acid Metabolism. New York, John Wiley & Sons, Inc., 1974, p. 60.

77. Ghadimi, H., Partington, M. W., and Hunter, A.: Inborn error of histidine metabolism. Pediatrics, 29:714, 1962.

78. La Du, B. N.: Histidinemia. In Stanbury, J. B., Wyngaarden, J. B., and Fredrickson, D. S. (eds.): Metabolic Basis of Inherited Disease. New York, McGraw-Hill Book Company, 1972, p. 338.

79. Perry, T. L., Hansen, S., and Love, D. L.: Serum-carnosinase deficiency in carnosinemia. Lancet, 1:1229, 1968.

80. Nyhan, W. L.: Nonketotic hyperglycinemia. In Nyhan, W. L. (ed.): Heritable Disorders of Amino Acid Metabolism. New York, John Wiley & Sons, 1974, p. 309.

81. Gerritsen, T., and Waisman, H. A.: Hypersarcosinemia: An inborn error of metabolism. N. Engl. J. Med., 275:66, 1966.

82. Hagge, W., Brodehl, J., and Gellissen, K.: Hypersarcosinemia. Pediatr. Res., 1:409, 1967.

83. Willems, C., Heusden, A., Hainout, A., and Chapelle, P.: Hypersarcosinéme avec sarcosinurie. Etude d'une nouvelle famille. J. Génet. Hum., 19:110, 1971.

84. Scriver, C. R., and Efron, M. L.: Disorders of proline and hydroxyproline metabolism. In Stanbury, J. B., Wyngaarden, J. B., and Fredrickson, D. S. (eds.): Metabolic Basis of Inherited Disease. New York, McGraw-Hill Book Company, 1972, p. 351.

85. Fleisher, L. D., Tallan, H. H., Beratis, N. G., Hirschhorn, K., and Gaull, G. E.: Cystathionine synthase deficiency: Heterozygote detection using cultured skin fibroblasts. Biochem. Biophys. Res. Commun., 55:38, 1973.

86. Frimpter, G. W.: Cystathioninuria in a patient with cystinuria. Am. J. Med., 46:832, 1969.

87. Whelan, D. T., and Scriver, C. R.: Cystathioninuria and renal iminoglycinuria in a pedigree. N. Engl. J. Med., 278:925, 1968.

88. Crawhall, J. C., Bu, K., Purkiss, P., and Stanbury, J. B.: Sulfur amino acids as precursors of β-mercaptolactate-cysteine disulfide in human subjects. Biochem. Med., 5:109, 1971.

89. Irreverre, F., Mudd, S. H., Heizer, W. D., and Laster, L.: Sulfite oxidase deficiency: Studies of a patient with mental retardation, dislocated ocular lenses, and abnormal urinary excretion of S-sulfo-L-cysteine, sulfite, and thiosulfate. Biochem. Med., 1:187, 1967.

90. Hooft, C., Tinnermous, J., Snoeck, J., Antever, I., Oyaert, W., and Van Den Hende, C.: Methionine malabsorption syndrome. Ann. Paediatr., 205:73, 1965.

91. Raine, D. N.: Methioninaemia in infancy. *In* Carson, N. A. J., and Raine, D. N. (eds.): Inherited Disorders of Sulphur Metabolism. Proceedings of the Eighth Symposium Society for the Study of Inborn Errors of Metabolism. Edinburgh. Churchill Livingstone, 1971, p. 40.

92. Eldjarn, L., Jellum, E., and Stokke, O.: Pyroglutamic acidemia. *In* Nyhan, W. L. (ed.): Inherited Disorders of Amino Acid Metabolism. New York, John Wiley & Sons, Inc., 1974, p. 479.

93. Wong, P. W. K., and Pillai, P. M.: Clinical and biochemical observations in two patients with Hartnup disease. Arch. Dis. Child., 41:383, 1966.

94. Oyanagi, K., Miura, R., and Yamanouchi, T. J.: Congenital lysinuria: A new inherited transport disorder of dibasic amino acids. J. Pediatr., 77:259, 1970.

95. Holmes, L. B., McGowan, B. L., and Efron, M. L.: Lowe's syndrome: A search for the carrier state. Pediatrics, 44:358, 1969.

96. Bach, G., Friedman, R., Weismann, B., and Neufeld, E. F.: The defect in the Hurler and Scheie syndromes: Deficiency of an α-L-iduronidase. Proc. Natl. Acad. Sci. U.S.A., 69:2048, 1972.

97. Bach, G., Eisenberg, F. Jr., Cantz, M., and Neufeld, E. F.: The defect in the Hunter syndrome: Deficiency of sulfoiduronate sulfatase. Proc. Natl. Acad. Sci. U.S.A., 70:2134, 1973.

98. Kresse, H., and Neufeld, E. F.: The Sanfilippo A corrective factor. Purification and mode of action. J. Biol. Chem., 247:2164, 1972.

99. O'Brien, J. S.: Sanfilippo syndrome: Profound deficiency of alpha-acetyl glucosaminidase activity in organs and skin fibroblasts from type-B patients. Proc. Natl. Acad. Sci. U.S.A., 69:1720, 1972.

100. Bach, G., Cantz, M., Hall, C. W. and Neufeld, E. F.: Genetic errors of mucopolysaccharide degradation. Biochem. Soc. Trans., 1:231, 1973.

101. Stumph, D. A., Austin, L. H., Crocker, A. C., and LaFrance, M.: Mucopolysaccharidosis type VI (Maroteaux-Lamy syndrome). I. Sulfatase B deficiency in tissues. Am. J. Dis. Child., 126:747, 1973.

102. Hall, C. W., Cantz, M., and Neufeld, E. H.: A β-glucuronidase deficiency mucopolysaccharidosis: Studies in cultured fibroblasts. Arch. Biochem. Biophys., 155:32, 1973.

103. O'Brien, J. S.: Generalized gangliosidosis. J. Pediatr., 75:167, 1969.

104. Singer, H. S., and Schaffer, I. A.: White cell β-galactosidase activity. N. Engl. J. Med., 282:571, 1970.

105. O'Brien, J. S., Ho, M. W., Veath, M. L., Wilson, J. F., Myers, G., Opitz, J. M., Zurhein, G. M., Spranger, J. W., Hartmann, H. A., Haneberg, B., and Grosse, F. R.: Juvenile GM_1 gangliosidosis: Clinical, pathological, chemical, and enzymatic studies. Clin. Genet., 3:411, 1972.

106. Sandhoff, K.: Variation of β-N-acetyl hexosaminidase pattern in Tay-Sachs disease. F.E.B.S. Letters, 4:351, 1969.

107. O'Brien, J. S., Okada, S., Chen, A., and Fillerup, D. L.: Tay-Sachs disease: Detection of heterozygotes and homozygotes by serum hexosaminidase assay. N. Engl. J. Med., 283:15, 1970.

108. Suzuki, T., Berman, P. H., and Suzuki, K.: Detection of Tay-Sachs disease heterozygote by assay of hexosaminidase A in serum and leukocytes. J. Pediatr., 78:643, 1971.

109. Navon, R., and Padeh, B.: Urinary test for identification of Tay-Sachs genotypes. J. Pediatr., 80:1026, 1972.

110. Carmody, P., Rattazzi, M. C., and Davidson, R. G.: Tay-Sachs disease—the use of tears for the detection of heterozygotes. N. Engl. J. Med., 289:1072, 1973.

111. Sandhoff, K., Harzer, K., Wässle, W., and Jatkewitz, H.: Enzyme alterations and lipid storage in three variants of Tay-Sachs disease. J. Neurochem., 18:2469, 1971.

112. Suzuki, Y., Jacob, J. C., Suzuki, K., Kutty, K. M., and Suzuki, K.: GM_2 gangliosidosis with total hexosaminidase deficiency. Neurology (Minneap.), 20:848, 1970.

113. O'Brien, J. S., Okado, S., Ho, M. W., Fillerup, D. L., Veath, M. L., and Adams, K.: Ganglioside-storage diseases. Fed. Proc., 30:956, 1971.

114. Kampine, J. P., Brady, R. O., Kanfer, J. N., Feld, M., and Shapiro, D.: Diagnosis of Gaucher's disease and Niemann-Pick disease with small samples of venous blood. Science, 155:86, 1967.

115. Fredrickson, D. S., and Sloan, H. R.: Sphingomyelin lipidoses: Niemann-Pick disease. *In* Stanbury, J. B., Wyngaarden, J. B., and Fredrickson, D. S. (eds.): The Metabolic Basis of Inherited Disease. New York, McGraw-Hill Book Company, 1972, p. 783.

116. Brady, R. O., Jolinson, W. G., and Uhlendorf, B. W.: Identification of heterozygous carriers of lipid storage diseases. Current status and clinical applications. Am. J. Med., 51:423, 1971.

117. Beutler, E., Kuhl, W., Trindad, R., Teplitz, R., and Nader, H.: β-glucosidase activity in fibroblasts from homozygotes and heterozygotes for Gaucher's disease. Am. J. Hum. Gen., 23:62, 1971.

118. Kaback, M. M., and Howell, R. R.: Infantile metachromatic leukodystrophy: Heterozygote detection in skin fibroblasts and possible applications to intrauterine diagnosis. N. Engl. J. Med., 282:1336, 1970.

119. Percy, A. K., and Kaback, M. M.: Infantile and adult-onset metachromatic leukodystrophy. Biochemical comparison and predictive diagnosis. N. Engl. J. Med., 285:785, 1971.

120. Suzuki, Y., and Suzuki, K.: Krabbe's globoid cell leukodystrophy: Deficiency of galactocerebrosidase in serum, leukocytes and fibroblasts. Science, 171:73, 1971.

121. Dawson, G., and Stern, A. O.: Lactosyl ceramidosis: Catabolic enzyme defect of glycosphingolipid metabolism. Science, 170:556, 1970.

122. Max, S. R., Maclaren, N. K., Brady, R. O., Bradley, R. M., Rennels, M. D., Tanaka, J., Garciá, J. H., and Cornblath, M.: GM_3 (hematoside) sphingolipodystrophy. N. Engl. J. Med., 291:929, 1974.

123. Herndon, J. H., Jr., Steinberg, D., and Uhlendorf, B. W.: Refsum's disease: Defective oxidation of phytanic acid in tissue cultures derived from homozygotes and heterozygotes. N. Engl. J. Med., 281:1034, 1969.

124. Young, E. P., and Patrick, A. D.: Deficiency of acid esterase activity in Wolman's disease. Arch. Dis. Child., 45:664, 1970.

125. Lie, K. K., Thomas, G. H., Taylor, H. A., and Sensenbrenner, J. A.: Analysis of N-acetyl-β-D-glucosaminidase in mucolipidosis II (I cell disease). Clin. Chim. Acta, 45:243, 1973.

126. Leroy, J. G., and Van Elsen, A. F.: I-cell disease (mucolipidosis II). Serum hydrolases in obligate heterozygotes. Humangenetik, 20:119, 1973.

127. Beutler, E., Boluda, M. C., and Day, R. W.: The genetics of galactose-1-phosphate uridyl transferase deficiency. J. Lab. Clin. Med., 68:646, 1966.

128. Siegel, S.: Disorders of galactose metabolism. *In* Stanbury, J. B., Wyngaarden, J. B., and Fredrickson, D. S. (eds.): The Metabolic Basis of Inherited Disease. New York, McGraw-Hill Book Company, 1972, p. 174.

129. Raivio, K., Perheentupa, J., and Nikkilä, E. A.: Aldolase activities in the liver in parents of patients with hereditary fructose intolerance. Clin. Chim. Acta, 17:275, 1967.

130. Patel, V., Watanabe, I., and Zeman, W.: Deficiency of α-L-fucosidase. Science, 176:426, 1972.

131. Autio, S., Norden, N. E., Öckerman, P., Riekkinen, P., Rapola, J., and Louhimo, T.: Mannosidosis: Clinical, fine structural and biochemical findings in three cases. Acta Paediatr. Scand., 62:555, 1973.

132. Blass, J. P., Avigon, J., and Uhlendorf, B. W.: A defect in pyruvate decarboxylase in a child with an intermittent movement disorder. J. Clin. Invest., 49:423, 1970.

133. Brunette, M. G., Delvin, E., Hazel, B., and Scriver, C. R.: Thiamin-responsive lactic acidosis in a patient with deficient low-Km pyruvate carboxylase activity in liver. Pediatrics, 50:702, 1972.

134. Grover, W. D., Auerbach, V. H., and Patel, M. S.: Biochemical studies and therapy

in subacute necrotizing encephalomyelopathy (Leigh's syndrome). J. Pediatr., 81:39, 1972.

135. Migeon, B. R., Der Kaloustian, V. M., Nyhan, W. L., Young, W. J., and Childs, B.: X-linked hypoxanthine-guanine phosphoribosyl transferase deficiency: Heterozygote has two clonal populations. Science, 160:425, 1968.

136. Salzmann, J., De Mars, R., and Benke, P.: Single allele expression at an X-linked hyperuricemia locus in heterozygous human cells. Proc. Natl. Acad. Sci. U.S.A., 60:545, 1968.

137. Fallon, H. J., Smith, L. H., Jr., Graham, J. B., and Burnett, C. N.: A genetic study of hereditary orotic aciduria. N. Engl. J. Med., 270:878, 1964.

138. Aula, P., Näntö, V., Laipio, M. L., and Autio, S.: Aspartylglucosaminuria: Deficiency of aspartylglucosaminidase in culture of fibroblasts of patients and their heterozygous parents. Clin. Genet., 4:297, 1973.

CHAPTER 6

CARRIER DETECTION IN X-LINKED DISEASE

Leonard Pinsky, M.D.*

Mass screening programs to identify carriers of the autosomal recessive Tay-Sachs gene exemplify perfectly the exploitation of human genetic knowledge for improving public health on the basis of a profitable cost:benefit ratio. One element responsible for the profitability of the ratio in these programs is the high frequency (1/30) of carriers of the Tay-Sachs gene in populations of Ashkenazic Jews. In this sense, *each human family segregating for a deleterious X-linked gene mutation should be considered a micropopulation within which the frequency of females carrying the gene may be very high.* The rules of Mendelian inheritance specify that the finding of one Tay-Sachs gene carrier will lead to the birth of a homozygous child with Tay-Sachs disease at a rate of 1/120. Indeed, it will reach this probability only if a carrier chooses another Ashkenazic Jew for a mate. This follows from the fact that after the establishment of conjugal heterozygosity at the Tay-Sachs gene locus (1/30), the likelihood of having Tay-Sachs disease for each child of such a mating is 1/4. In contrast, the same rules of Mendelian inheritance dictate that each female carrying a deleterious X-linked gene can produce affected offspring without regard for the genotype of her mate. For X-linked genes which are recessive, the risk of an affected child is 1/4; of an affected male, 1/2. For X-linked genes that are dominant, the risk of having an affected child is 2/4, since female offspring may express the disease also, although to a lesser extent, on the average. It follows from the foregoing that, all other factors being equal, a population with a frequency of female X-linked gene carriers much less than one thirtieth may merit consideration for mass screening. The populations in Lunenberg and Colchester counties of the Province of Nova Scotia, Canada, for example, may qualify for such screening in regard to Fabry's disease and nephrogenic diabetes insipidus, respectively.

*Supported by a grant (MT-2830) from the Medical Research Council of Canada.

Putting all ethical, social, cultural, and religious factors aside, other aspects of the relative prophylactic efficacy of carrier detection in X-linked genetic disease merit discussion. Prenatal sex determination warrants consideration at this point. For example, a woman who is a carrier of a deleterious X-linked recessive gene can choose to abort all of her male fetuses and can expect to have only healthy daughters. In so doing she will, of course, dispose of healthy male fetuses half the time, and she will, on the average, contribute to the population more than her share of daughters who are healthy carriers of the gene. These female offspring will thus have the burden of transmitting it to their children.

The societal burden resulting from such a program of selective abortion will be reflected in an increased number and cost of these abortions. Where mutant hemizygote detection in utero is possible, a vastly increased number of amniocenteses followed by diagnostic amniotic cell culture will be required. The relative eugenic and dysgenic value of blind male abortion, selective male abortion, and reproductive compensation leading to excess births of heterozygous females has been discussed elsewhere.[1,2] Table 6-1 presents a sample of the conclusions.

On the other hand, by choosing to conceive no children, a female X-linked gene carrier will remove her gene from the germ line of her family and the gene pool of her population. The need for profound, personal de-

TABLE 6–1. EUGENIC AND DYSGENIC EFFECTS OF VARIOUS SELECTIVE ABORTION PLANS FOR A LETHAL† X-LINKED RECESSIVE DISEASE WHEN THE GENE FREQUENCIES HAVE REACHED A NEW EQUILIBRIUM*

	Incidence of Affected Males			Prevalence of Heterozygous Females		
	A	B	C	A	B	C
	(x = initial incidence)			(y = initial prevalence)		
1	0.33x	x	1.67x	$y + \dfrac{ny\ddagger}{2}$	1.14y	2.29y
2	0.33x	0.94x	1.21x	1.5y	1.04y	1.5y
3	0.33x	0.85x	0.85x	0.5y	0.89y	0.89y

Abortion Plan	Diagnosis of Maternal Heterozygosity	and	Reproduction Ceases After
1. Abortion of all males	A. Prospective		Two normal
2. Abortion of affected male	B. Retrospective		Any two
3. Abortion of affected male and heterozygous female fetuses	C. Retrospective		Two normal

*From Motulsky, A. G., Fraser, G. R., and Felsenstein, J.: Public health and long-term genetic implications of intrauterine diagnoses and selective abortions. *In* Bergsma, D. (ed.): Birth Defects: Orig. Art. Ser., Intrauterine Diagnosis. Vol. 7, No. 5. The National Foundation—March of Dimes, White Plains, N.Y. 1971, pp. 22–32.
†Indicates lethal before reproduction.
‡"n" = number of generations after introduction of plan 1A.

cisions, such as voluntary infertility and selective male abortion presupposes the absence of a viable alternative, such as successful therapy of the disease in question, or segregation of X- from Y-bearing sperm, to avoid the birth of males.

There are two benefits of carrier detection in X-linked genetic disease which accrue directly to those at risk. First, for those proved to be carriers, it may be possible to prevent, minimize, or at least anticipate, the appearance of certain expressions of the full disease as seen in affected hemizygotes. Second, those proved not to be carriers can shed their accumulated doubts and fears regarding their genetic prognosis. The prevalence of misinformation and misinterpretation regarding the facts of human genetics often generates a great number of such emotional burdens.

TYPES OF X-LINKED GENES

The intergenerational pattern of transmission of X-linked recessive genes can be mimicked by autosomal dominant genes whose full expression is limited to males. How does one recognize X-linked recessive genes? There are two absolute ways: (1) The gene responsible for a suspect disorder is demonstrably linked to another gene known to be on the X-chromosome. (The absence of demonstrable linkage, however, may simply mean that the gene in question is indeed on the X-chromosome, but is too far away from a known X-linked marker gene; independent assortment, therefore, appears to occur through the process of recombination resulting from crossing-over.) (2) The gene responsible for the suspect disorder can be shown to be expressed only in some cells of female carriers as a result of random X-chromosome inactivation (the Lyon Rule).

There are several situations which point to an X-linked recessive gene: (1) The full disorder occurs in males almost exclusively, they seldom live long enough, or are healthy enough, to reproduce (X-linked recessive lethal); and one-third of them lack a positive family history. (That is, they appear to result from *de novo* mutation.) (2) The disorder occurs in females with a deficiency of X-chromosome material, such as those with the Turner syndrome. (3) Mothers, but not fathers, of affected boys sometimes express the disorder, usually to an intermediate degree. (4) An apparently homologous disorder, known to be determined by a gene on the X-chromosome, occurs in a mammal. (This occurrence is based on the evolutionary stability of the X-chromosome.[169])

It is interesting to review the adjunctive situations that indicate X-linked recessive inheritance from the viewpoint of those disorders in which the communal phenotype is male pseudohermaphroditism of one degree or another. Many of the disorders appear to be transmitted by carrier females to half their sons. Thus, one cannot expect expression of these disorders in X-monosomy because such human beings differentiate as females. Nor can one expect fathers of affected males to express intermediate degrees of male sexual maldevelopment because, if they did, they would likely be infertile. On the other hand, testicular feminization, a classic form of human gene-determined male pseudohermaphroditism does have a homologue in the mouse, in which it is known to be X-linked.[3]

Two ways to distinguish females who carry autosomal dominant genes whose expression is limited to males are: (1) evoke expression of the gene by providing a male hormonal milieu to the female, *in vivo*, or to her cells, *in vitro* (the former procedure has been done for the sex-limited serum protein in the mouse);[4] (2) remove cells of the female to an *in vitro* (cell culture) situation and determine whether the gene is expressed in a sexually neutral environment provided by the culture medium (it is possible to interpret in this way the early results on skin fibroblast cultures from female carriers of the gene responsible for testicular feminization).[5] This last procedure is discussed further on p. 146.

Female carriers of X-linked dominant genes yield segregation patterns superficially equal to those who carry autosomal dominant genes; that is, half their sons and half their daughters are affected. However, all the daughters of affected males are affected, and all their sons are normal—the latter also being true of males affected by X-linked recessive genes. Furthermore, as can be predicted from random X chromosome inactivation, females affected by an X-linked dominant disorder, compared to their male siblings, are, on the average, less severely affected individually, and more variably affected as a group. X-linked dominant disorders, all of which are rare, have the additional property of being twice as frequent in females as in males. (If p is the frequency of the mutant allele, the F/M ratio, $\dfrac{p^2 + 2pq}{p}$, reduces to $\dfrac{2}{1}$, because p is small, p^2 is even smaller, and q (which is $1 - p$) is essentially 1.)

X-linked genes that are lethal for hemizygous males before birth produce a distinctive pattern of inheritance which features mother-to-daughter transmission, an increase in abortions in affected women, and a concomitant deficiency of sons of affected women. If a disorder postulated to be due to an X-linked dominant with intrauterine male lethality is found in a living male with a multiple-X,Y syndrome, the postulate is strengthened.

EXPRESSION OF X-LINKED RECESSIVE GENES IN FEMALES

Females may express X-linked recessive disorders fully under exceptional circumstances.[64] These include: (a) females with X-monosomy in pure or mosaic form and those with other kinds of deficient X-chromosome material; (b) random X-chromosome inactivation (Lyonization) resulting in a great majority of cells whose "active" X-chromosome bears the mutant gene; (c) female offspring resulting from the mating of an affected male with an asymptomatic carrier female; (d) various non-disjunctional events leading to disomy for an X-chromosome bearing the mutation; and (e) the association in a female of a classic X-linked recessive mutation with a "normal" allele whose function is significantly inferior to that of the common "wild-type" allele at that locus. One should add to this list "females" who are really males with extreme pseudohermaphroditism (or those who are true hermaphrodites), and those who have an autosomal disorder which is not easily distinguishable from the X-linked one in question.

In accord with accepted practice, the recessiveness or dominance of the various X-linked genes referred to so far merely reflects their respective degrees of spontaneous clinical expression. It will be appreciated that when measurement of gene action approaches the primary gene product closely, more genes that are clinically "recessive" can be classified as biochemically "dominant."

X-LINKED GENE MARKERS

Certain families have normal or relatively benign X-linked traits which can be exploited as markers for predicting the carrier status of those at risk of inheriting a deleterious X-linked trait that segregates concurrently. Some of these markers are sufficiently common to merit the label of polymorphisms. They include the recessive genes for the two types of colorblindness, and the numerous types of glucose-6-phosphate dehydrogenase (G6PD) deficiency and isozyme variants, as well as the dominant genes responsible for the Xg^a erythrocyte type and the X_m serum protein. Deficiency and excess, respectively, of the thyroxine-binding globulin can also function as X-chromosome markers,[6, 7] but the responsible genes are not as common as those previously mentioned.

The ability of human beings to visualize spatial relations and to manipulate visual images mentally appears to be influenced to a great extent by an X-linked gene.[9] The precision of techniques for measuring this ability, however, is not yet great enough to handle this gene as a marker.

GENETIC HETEROGENEITY

It is well known that clinical homogeneity may conceal genetic heterogeneity. This problem impinges on the provision of many types of clinical genetic services; detection of X-linked carriers is no exception. Deleterious X-linked genes are represented in almost every class of human inherited disease, as shown in the clinical catalogue at the end of this chapter. The application of rigid differential diagnostic criteria is a prerequisite of valid genetic counseling information for any X-linked disorder in each of these classes. Recognition of the X-linked variety of a clinical phenotype may be possible by more or less subtle clinical distinction (the Hunter syndrome), by histopathological criteria (X-linked congenital adrenal hypoplasia; X-linked ichthyosis), by radiography (hydrocephalus due to X-linked aqueductal stenosis), or by discriminating biochemical procedures that challenge the patient, or his cultured cells, to reveal their precise genetic identity. On occasion, the ability to recognize female carriers of an X-linked gene is the pivotal criterion for discriminating the X-linked form of a phenotype from one (or more) of its clinical relatives that is determined by another type of mutant gene. The latter is true, for instance, in the distinction of the X-linked from the autosomal forms of chronic granulomatous disease,[8] and anhidrotic ectodermal dysplasia.[34]

Distinctive qualitative features of different carriers may help to distinguish among different types of X-linked genes—all of which yield clinically indistinguishable, fully affected hemizygotes. It has been suggested that this may be true for different X-linked forms of retinitis pigmentosa.[10] For didactic purposes only, one might also include here the use of distinctive carrier characteristics to distinguish between the two X-linked types of nonspherocytic hemolytic anemia; namely, those resulting from phosphoglycerate kinase and G6PD enzyme deficiencies in erythrocytes. Thus, if a male suspected of having one of these anemias were unavailable for study, a retrospective distinction between the two possible diagnoses could be easily made by demonstrating a subpopulation of erythrocytes with one or the other enzyme deficiency in a female relative who was a carrier.

APPROACHES TO CARRIER DETECTION IN X-LINKED DISEASE

There are two general approaches:

(1) The *direct approach* depends upon (a) the measurement of a biochemical parameter closely related to the mutant gene's primary action, or (b) the recognition of one or more clinical or laboratory features that are sufficiently distinctive to betray heterozygosity at the X-linked locus in question.

(2) The *indirect approach,* which may also be called the "pedigree approach," depends upon the use of posterior and collateral genetic information available from the pedigree. Such information may include the presence of X-linked marker genes that are segregating in the family, along with the mutant gene in question.

By either approach alone, diagnosis of the X-linked gene carrier state may be less than 100 per cent certain. Occasionally, one can combine information obtained from both approaches to yield the best estimate of a consultand's carrier status. Various specific examples of both approaches are given later in this chapter.

X-CHROMOSOME INACTIVATION (THE LYON RULE)

Inactivation of one X-chromosome in each somatic cell of a diploid female is central to the theory of carrier detection in X-linked disease (Fig. 6–1). The rule has important practical corollaries: (1) inactivation is random in the tissues of all organs; (2) inactivation is stable, that is, somatically heritable; and (3) the two cell types resulting from random inactivation are selectively neutral so that their ratio at the time of inactivation in the embryonic anlagen of each organ is reflected in the cells of each organ at maturity.

Several additional corollaries have to be appended to this list in order to exploit cell culture technology for recognition of X-linked gene carriers. These include: (1) a reasonably homogeneous distribution of the two cell types within an organ so that a relatively small biopsy is likely to sample both cell types; (2) preservation of selective neutrality of each cell type during serial subcultivation; (3) equal cloning efficiency and phenotypic expression of each cell type when their vitality is challenged by the rigors of single-cell isolation; (4) lack of interaction between the cell types in uncloned cultures; and (5) expression of each cell type in the sexually neutral environment provided by the culture medium; that is, absence of a sex-limited influence on the expression of each cellular phenotype.

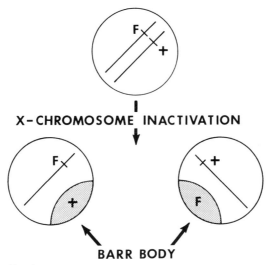

Figure 6–1. Random X-chromosome inactivation. F denotes the mutant allele.

MEASUREMENTS WHICH REFLECT BOTH OF THE
LYONIZED CELL TYPES

As one would predict, measurements of mutant gene action which encompass the expression of both cellular phenotypes resulting from random single X-inactivation yield values for obligate heterozygotes which span the spectrum from clearly normal to clearly abnormal. This is true, for example, of heterozygotes for the relatively common mutant gene responsible for erythrocyte G6PD deficiency in the American Negro. Measurements less closely related to primary mutant gene action are likely to be even less reliable as an index of carrier detection. For instance, the value of creatine phosphokinase (CPK) levels in the detection of carriers bearing the gene for X-linked Duchenne muscular dystrophy[11] and of serum phosphorus levels in the detection of carriers of X-linked hypophosphatemic rickets[12] are well known to depend upon multiple determinations; nevertheless, obligate carriers of each gene may have persistently normal values of each measurement.

Occasionally, a female heterozygous for an X-linked enzyme deficiency may yield a value for total enzyme activity in the normal range when a simultaneous test to reveal cell mosaicism is affirmative.

Repeated determinations of erythrocyte G6PD activity in any one potential carrier will also vary as a function of the proportion of young red cells in the circulation at any given time, young erythrocytes being relatively G6PD-rich. This exemplifies a form of physiological mosaicism superimposed upon that decreed by random X-chromosome inactivation. It is self-evident that the reliability of detecting carriers of X-linked genes afforded by a measurement of total enzyme activity is influenced greatly by the range of biological variation in the measurement for a sample of normal women. Thus, the selection of a female control sample should reflect rigidly the sources of variation to be expected in a sample of potential carriers. This means that variation arising from endocrine and metabolic changes attendant on the menarche, menstrual cycle, pregnancy, oral contraceptives, and the menopause must be considered. For instance, antihemophilic factor (AHF) levels are increased,[13] but creatine phosphokinase levels are decreased[14] in pregnancy. There is no need to mention that the measurement of a Gaussian parameter to distinguish between normal females and those heterozygous for a mutant X-linked gene demands the highest standards of personnel and material quality control. Only in this way can one minimize the frequency of uninformative values which fall in the zone of overlap between the distribution of normal and heterozygous values.

One way of minimizing the extent to which biological (and occasionally technical) variation interferes with the ability to discriminate heterozygous from normal, using a measurement such as total G6PD activity, is to normalize the results against the activity of a control enzyme—ideally one which is physiologically correlated with the mutant enzyme in question. This principle has been exploited to maximize the efficiency of carrier detection for carriers of both autosomal and X-linked genes. Examples in the latter category are the use of the hypoxanthine-guanine phosphoribosyl transferase (HGPRT) to adenine PRT ratio in single hair follicles[30] and in uncloned skin fibroblast cultures of potential Lesch-Nyhan heterozygotes.[16] In the case of the hair follicles, the value of APRT activity controls for the size and health of the hair follicle cell sheath. In the case of the monolayer fibroblast cultures, the APRT activity controls for the mitotic index of the cells at the time they are harvested for assay, as well as for other sources of natural variation.

EXPRESSION OF MUTANT X-LINKED GENES RESTRICTED TO CERTAIN ORGANS

It is self-evident that organs (or tissues) which do not express the enzyme activity (or another gene product) determined by the normal allele at an X-linked locus will not be useful for heterozygote detection. For instance, the X-linked gene responsible for congenital hyperammonemia due to ornithine transcarbamylase (OTC) deficiency is not expressed in hematocytes or skin; direct evidence of an intermediate enzyme deficiency in its carrier state must be obtained from the liver.[17] A noninvasive approach to recognition of an X-linked gene carrier when the mutant's expression is limited to an inaccessible organ is to challenge the function of the organ from without in the form of a "tolerance test." For OTC deficiency, this can be done with a high protein[18] or ammonia[19] load; for hypophosphatemic rickets, with a phosphate load;[12] and for congenital vasopressin-resistant diabetes insipidus, with a water deprivation test.[21]

Consistent expression of an X-linked mutant gene in an organ-restricted fashion may result from preferential survival of the cell with the normal phenotype or from nonrandom inactivation of the X-chromosome bearing the mutant allele in certain organs. For instance, the total erythrocyte HGPRT activity of most obligate carriers of the X-linked Lesch-Nyhan mutation is normal,[22] and their peripheral lymphocytes do not manifest cellular mosaicism, either.[23] It has now been shown that the erythrocytes of two obligate L-N heterozygotes consist only of cells in which the X-chromosome bearing the normal allele is active. This has been done by exploiting an X-linked G6PD marker in one case,[24] and a kinetically distinctive L-N mutation, in the other.[25] Teleologically, it is easy to believe that paralysis of the purine salvage pathway resulting from HGPRT-deficiency would not favor the survival of rapidly differentiating hematopoietic cells. In addition, there is independent evidence[26] that bone marrow cells utilize preformed purine precursors preferentially in the conduct of their nucleotide metabolism. Furthermore, female carriers of genes which confer partial, rather than complete, HGPRT deficiency on their hemizygous sons, do display low levels of total erythrocyte HGPRT activity.[27] This suggests that the amount of HGPRT activity in a hematopoietic stem cell must exceed a threshold to survive.

Occasionally, expression of an X-linked mutant gene in a carrier female may be restricted to certain tissues or organs, purely as a result of the randomness of X-chromosome inactivation. Therefore, one may have to sample two or more tissues to be certain of a potential carrier's genotype.

One can imagine another hindrance to detection of X-linked gene carriers resulting from organ-restricted expression of the mutant gene. Imagine, for example, that the primary site of mutant gene action occurs in an organ which is known, but inaccessible (or is entirely unknown), and that cells remote from it effect clinical expression of the mutant gene's primary action secondarily. For instance, an abnormal metabolite produced by Lyonized mutant cells in liver may not be toxic for liver and yet may have distant effects—for example, dermatopathology. Hence, one might indeed see intermediate expression of the disorder in the skin of females carrying the mutant X-linked gene, but this would not be the result of cellular mosaicism due to random X-chromosome inactivation in embryonic ectoderm. A search for such mosaicism would necessarily fail.

Another equally reasonable, but still hypothetical, situation with the same consequences is one in which a substance responsible for "organizing" or "inducing" normal function in "target" cells was normally produced in a remote cell type under the control of an X-linked gene. (This was the first explanation—now disproved[20]—for failure to observe erythrocyte mosaicism in females heterozygous at the Xg locus.) If androgen responsiveness in at least some mature target organs of human beings is dependent on an inductive process during a "critical period" in early prenatal development (as is true in the rat), then some forms of androgen-refractory male pseudohermaphroditism in man which appear to be due to X-linked gene mutations may not display target cell mosaicism. In contrast, those forms of presumably X-linked androgen-refractory male pseudohermaphroditism which derive from mutant genes whose primary action does occur in or on target cells would, of course, be expected to manifest target cell mosaicism in heterozygotes.

Organ-restricted expression of X-linked mutant genes presents a special problem for carrier detection when the full phenotype is expressed only in male sexual organs. Thus, for several forms of hereditary male hypogonadism whose pattern of inheritance is compatible with that of an X-linked gene, minimal clinical or laboratory expression in female carriers is impossible. On the other hand, in other forms, the phenotype involves organs which are not sex-specific. For instance, the anosmia (or hyposmia) which accompanies one type of hypogonadotropic male hypogonadism may be recognized in female carriers.[28] The same could apply to the ichthyosis which occurs in a second form of hypogonadotropic male hypogonadism.[29]

RECOGNITION OF CELLULAR (OR CLONAL) MOSAICISM IN X-LINKED GENE CARRIERS WITHOUT CELL CULTURE

Hair Follicles

Because "... cells of the scalp exhibit a certain degree of clonal growth" and "... because the hair follicle starts from a small number of cells,"[30] it has been possible to diagnose Lesch-Nyhan heterozygosity by direct assay of HGPRT activity in single hair follicles. This diagnosis has been accomplished by test tube quantitative assays[30, 31] and by electrophoretic separation of HGPRT activity with subsequent measurement of the enzyme zone on the gel.[32] As discussed earlier, the use of APRT activity as a control enzyme activity in each of these procedures minimizes the scatter of, and therefore the overlap between, the data for normal hair follicles and those of heterozygotes. As expected, proved heterozygotes have three kinds of follicles—the polar clonal types and those with an intermediate HGPRT/APRT ratio, representing hair follicles of mixed cell origin.

The clonal nature of many hair follicles is of profound significance. Any X-linked mutation affecting an enzyme which has a widespread tissue distribution because it controls a very basic step in cell metabolism is likely to reveal its female carrier state in hair follicles.

Sweat Glands

Female carriers of the X-linked gene responsible for anhidrotic ectodermal dysplasia may not show focal expressions of the disorder such as patchy alopecia or partial anodontia. A patchy distribution of sweating[149] is not a reliable indicator of heterozygosity. A reduced concentration of sweat gland pores on the epidermal ridges was found in five out of six obligate carriers, three of the five not having any other clinical expression of the disease.[33] Nevertheless, one carrier had dental defects with a normal sweat pore count. It has been suggested recently that direct observation of sweat gland pores through a stereomicroscope at 16-fold magnification can reveal not only a reduced concentration of pores per unit area but also a patchy distribution of them on the skin of heterozygous carriers.[34] The size of the patches (clonal mosaicism) is consistent with the tendency of primitive cells of epidermal origin to grow without much mingling. The greater tendency of cells of mesodermal origin to mingle, giving rise to fine-grained mosaicism reflected in much smaller patch sizes, will be referred to later when the efficacy of skin fibroblast cultures for X-linked carrier detection is discussed.

Teeth

The teeth of females carrying the X-linked gene for amelogenesis imperfecta have vertical bands of opaque enamel alternating with bands of normal enamel.[35] It is noteworthy that under scanning electron microscopy clearly demarcated zones of mutant enamel in heterozygous teeth are not as abnormal as in hemizygous teeth. This reflects either mingling of cells with normal phenotype in zones made up predominantly of cells with the mutant phenotype, or some normalizing interaction between normal ameloblasts in one clonal zone and mutant ameloblasts in an adjacent clone.

Eyes

The patchy, mosaic distribution of lesions in females heterozygous for the various X-linked fundus dystrophies is well known. Warburg[36] has recently called attention to an observation of Krill[37] that while intrafamilial variation among carriers for X-linked retinitis pigmentosa is very variable, interocular variation within a carrier is minimal. One way of explaining this situation is to postulate that random X-chromosome inactivation and thorough mingling of primordial retinal cells occur before cephalic lateralization of the embryo. Another type of explanation has been mentioned previously: that X-linked retinitis pigmentosa is primarily an extraocular disorder with both eyes equally susceptible to a remote effect.

Erythrocytes

Two erythrocyte phenotypes resulting from differential X-chromosome inactivation may be recognized in smears by histochemical procedures (G6PD deficiency);[38] by differential centrifugation of young and old erythrocytes to maximize the efficiency of carrier detection in relation to those mutant enzymes whose instability results in a greater deficit of enzyme activity within old than within young cells; by exposure of a mixed population of the erythrocyte types to an agent which selectively hemolyzes one of the types (primaquine destroys G6PD-deficient erythrocytes);[39] or by measurement of the mixed cell population with a technique sufficiently sensitive to distinguish the *qualitative* contribution of the cell with the mutant phenotype to the total activity of the mixture (electrophoretic separation of a mutant G6PD isozyme).

Neutrophils

Although the basic metabolic defect in X-linked chronic granulomatous disease is unknown, it is clearly intrinsic to neutrophils and interferes with their ability to kill a variety of catalase-positive organisms following normal phagocytic ingestion.[65] Functional cellular mosaicism in females carrying the gene can be evoked by incubating their neutrophils with latex particles and the pale yellow, soluble dye, nitroblue tetrazolium (NBT). Neutrophils with the mutant phenotype cannot reduce the dye to its blue-black, insoluble product, formazan.[8]

Serum Protein

It is now clear that males affected with classical hemophilia (type A) produce a functionally deficient antihemophilic factor which is antigenically intact. They exemplify a human CRM^+ mutation. This fact has been exploited to bypass the notorious unreliability of carrier female detection on the basis of quantitative AHF-factor procoagulant measurements. It has been reported that at least 90 per cent of females who are obligate carriers of the gene for classical hemophilia can be recognized by their possession of AHF factor antigenic activity in excess of AHF factor procoagulant activity.[40] Indeed, the disproportionate deficit of AHF factor procoagulant activity can also serve to distinguish AHF procoagulant deficiency due to hemophilia from AHF procoagulant deficiency due to Von Willebrand's disease. In the latter, the deficits of AHF procoagulant and antigenic activity are coordinate. The principle embodied in this new test for detecting female carriers of the gene for classical hemophilia A should be applicable whenever X-linked structural gene mutations yield a product whose physiologic function is much below normal, but which is immunologically competent to react with antibody made against the product of the normal allele.

USE OF SKIN FIBROBLAST CELL CULTURE TO RECOGNIZE THE FEMALE CARRIER STATE OF X-LINKED DISORDERS

The corollaries of the Lyon Rule which have practical significance for recognition of differential X-chromosome inactivation in cell culture have been listed earlier in this chapter. The experience obtained with several X-linked disorders in relation to these corollaries will now be analyzed critically.

Concealment of Cellular Heterogeneity by Cellular Interaction Requiring Cell Contact

Shortly after the basic enzyme defect in the Lesch-Nyhan syndrome had been defined, Rosenbloom et al.[42] were able to demonstrate the carrier state in uncloned fibroblast cultures derived from the skin of the mother. They used ^{3}H-hypoxanthine radioautography (^{3}H-adenine serving as a control) to show that the mother's cultures contained some cells which could, and others which could not, incorporate hypoxanthine into insoluble polynucleotides. Others found insufficient numbers of radioautographically-negative[43] or too many intermediately labeled cells[44] in cultures of obligate L-N heterozygotes to be confident of this approach to heterozygote detection. Therefore, other techniques for heterozygote detection were evolved. One of these was to isolate single cell clones from heterozygous cultures and then to seek phenotypic mosaicism. This was accomplished readily,[45] but did not seem to be a facile technique because of the labor and time involved. For this reason, another system was developed: one based on the selective advantage of HGPRT-negative cells in the presence of a medium containing toxic (8-aza- or 6-thio-) analogues of guanine. Three laboratories published their results with the technique at about the same time.[43, 47, 48] One of the published reports indicated that the problem of "intermediate" phenotypes was overcome by use of the selective medium if 10^4 or 10^5 cells were planted per 60 mm. petri dish;[43] a second report showed that initial cell density should not exceed 10^4 cells per 60 mm. dish in order to avoid reduced recoverability of HGPRT-negative cells because of "cellular interaction."[48] The third laboratory reported, in a separate publication[16] that artificial and natural mixtures of HGPRT-positive and HGPRT-negative cells progressively increased their total incorporation of hypoxanthine relative to adenine as culture cell density increased from 2 to 8×10^4 cells per 60 mm. dish. All three laboratory reports were referring to the intercellular phenomenon which has been labeled "metabolic cooperation." The term was coined by Subak-Sharpe[49] to describe the ability of HGPRT-positive cells in contact with HGPRT-negative cells to confer upon the latter the ability to incorporate hypoxanthine. The greater the cell density of a culture, the more likely is "metabolic cooperation" to interfere with recognition of HGPRT-negative cells in a culture derived from a potential heterozygote.

The time necessary for HGPRT-pseudopositive cells to revert to their HGPRT-negative phenotype after loss of contact with HGPRT-positive cells has been observed to occur promptly by one group[50] and after two or three days by another.[51] Accordingly, the first group postulated that metabolic cooperation involves the transfer of an HGPRT-dependent product from one cell to the other, whereas the second group visualized the transfer of HGPRT itself or of some informational macromolecule allowing the HGPRT-negative cell to make its own enzyme. The resolution of this disagreement is of less importance to the reader than the fundamental observation that the two cellular phenotypes resulting from X-chromosome inactivation can, under certain conditions, interact in cell cultures of X-linked heterozygotes to mask the presence of cellular mosaicism and, therefore, conceal the heterozygous state.

Concealment of Cellular Heterogeneity by Cellular Interaction Not Requiring Cell Contact

Shortly after discovery of the phenomenon of "reciprocal correction" in cocultures (that is, mixed cell cultures) representing different mucopolysaccharide storage diseases,[52] its significance for recognizing the heterozygous state of the X-linked Hunter syndrome was discussed. The cellular interaction underlying "correction" does not require cell contact: evidently the agent of correction (in this case, the enzyme itself) can travel through the medium. Initial observations relating to the question of correction between the two cell types in cultures of Hunter heterozygotes were based on the property of stored mucopolysaccharides to stain metachromatically with dyes, such as toluidine blue. The limitations of this tinctorial technique have been well publicized,[53] and the consensus is that metachromasia is much too nonspecific to serve as the basis for distinguishing cultured cell heterogeneity in heterozygous carriers of the Hunter gene mutation.

Recently, Neufeld and Migeon[88] have confirmed that uncloned fibroblast cultures derived from Hunter heterozygotes have normal $^{35}SO_4$ accumulation into mucopolysaccharides, as would be predicted by "correction."

Although the isolation of single cell clones can prove heterozygosity at the X-linked Hunter locus,[54] there is some suggestion that preferential proliferation of cells with the mutant phenotype can occur.[55] If this occurred in the opposite direction, there would be a risk of misdiagnosing a heterozygote as normal. For this reason, Neufeld[88] recommends that women with a high pedigree-probability of being Hunter heterozygotes undergo prenatal diagnosis.

Concealment of Cellular Heterogeneity by Wild-Type Allele with Hypernormal Activity or by Nonallelic Modifier Gene(s) with This Effect

The situation converse to that presented in the title immediately preceding was mentioned as one basis for the expression of an X-linked recessive disorder in a female. Romeo and Migeon[56] found that in a family segregating for Fabry's disease, X-linked α-galactosidase deficiency, one heterozygote had normal levels of α-galactosidase activity in her uncloned skin fibroblast cultures, in part because her normal cell clones had significantly more activity than the normal cell clones of her daughter or her father. It could not be determined whether a gene allelic or nonallelic with the α-galactosidase locus was responsible.

Experience with Other Possible Complications of Single X-inactivation for Detection of X-linked Gene Carriers in Cell Culture

Considerations related to the randomness, the timing (the number of cells in a primordial organ or tissue at the time of inactivation), and the migratory behavior (degree of mingling) of the two post-inactivation cell types had generated much fear that the small skin biopsies (ca. 3–4 mm.) used routinely for originating fibroblast cell strains, frequently might fail to sample the mutant cell type in X-linked gene carriers. These fears have not been realized! Although the proportion of mutant clones isolable from strains derived from different skin biopsies of a single heterozygote has varied significantly, there are no published reports of obligate heterozygotes who have not displayed their mosaicism in cell culture. Indeed, a recent report[161] on X-linked glycogen storage disease has described the recovery of the mutant clonal fibroblast type from an obligate heterozygote whose erythrocytes, leukocytes, and uncloned fibroblast cultures had phosphorylase kinase activity in the normal range. In the case of Lesch-Nyhan heterozygosity, the conduct and design of the cell culture techniques in routine use are such that the chance of missing the diagnosis would be 10^{-2} or less, even if only 1 per cent of the cells sampled had the mutant phenotype.[15, 48] This calculation is based on the fact that the two cell types in L-N heterozygote cell cultures appear to grow equally, despite prolonged periods of serial subculture.[15] This is true, however, only if adequate levels of folic acid are in the growth medium to fulfill the increased demand of the mutant cell type for this vitamin, which is consumed in the pathway of *de novo* purine biosynthesis.[57] Eagle's minimal essential medium, supplemented with 10 per cent fetal bovine serum, happens to contain enough folate to satisfy the excessive requirement of L-N hemizygous cells. If it did not contain enough, then in many laboratories using Eagle's MEM, preferential survival of cells with the normal phenotype might have occurred in potential L-N heterozygotes with resultant errors in diagnosis of the carrier state.

A theoretical threat to the use of cultured cell clones for recognizing differential X-chromosome inactivation is that the mutant cell phenotype might be critically dependent on the *in vitro* age of the cells. In this event, even presenescent clones might then present a normal phenotype, despite the fact that their active X-chromosome bore the mutant allele at the locus in question. This would, of course, minimize the chance of detecting the mutant cellular phenotype in potential X-linked gene heterozygotes. There is no precedent for this, but it is noteworthy, nevertheless. The converse situation is equally probable, on theoretical grounds — that a normal clone might present a mutant phenotype if much of its lifespan were consumed by the time it was tested. This would increase the apparent frequency of cells with the mutant phenotype in obligate heterozygotes, but more importantly, it would bring the hazard of false-positive diagnoses of heterozygosity in facultative heterozygotes.

THE INDIRECT APPROACH TO CARRIER DETECTION IN X-LINKED GENETIC DISEASE

The object of this approach is to arrive at the best estimate of the probability that a given female is a carrier when proof of the carrier state cannot be obtained from her directly. This situation can arise when tests to discriminate carriers for the disorder in question are not available, or are not completely reliable because of their empiric origin. The indirect approach must, of course, be used if the carrier is no longer alive or, for some other reason, is not available for testing.

Use of Genetic Information from the Pedigree

Some very simple examples are presented below. For any X-linked recessive disorder:

1. The daughter of an affected male may be considered a carrier (posterior information).

2. A woman who has an affected uncle or an affected nephew (collateral information), as well as an affected son (anterior information), may be considered a carrier.

3. The mother of two affected sons must be considered a carrier.

4. The mother of a sporadic case of Duchenne muscular dystrophy has repeatedly normal creatine phosphokinase (CPK) levels in her blood. If her sister has elevated levels typical of heterozygotes, it is likely that the mother is a silent carrier.

If a mother, about whom there is neither posterior nor collateral information, has had only one son, and he is affected with a lethal, X-linked recessive disorder, one can calculate that the risk to her next son is 1/3.[58] The difference between 1/2 (the risk to her next son if she is certain to be a carrier) and 1/3 represents the likelihood that her first son was the product of a fresh mutation.

When there is posterior or collateral genetic information about the mother, one can use it to improve the estimate that she is a carrier. For instance, the greater the number of unaffected sons she has had, and the greater the number of her unaffected brothers, the less is the likelihood that she is a carrier. This information is provided in Table 6-2, published by Murphy.[59] Similarly, a woman with an antecedent family history of an X-linked recessive disorder, who has not had an affected son, will have a likelihood of being a carrier which is lowered by (a) the number of generations which intervene between her and the last certain

TABLE 6–2.   PROBABILITY THAT THE MOTHER OF A SPORADIC CASE OF AN X-LINKED LETHAL TRAIT IS HETEROZYGOUS*

No. of Unaffected Sons (m)	Number of Unaffected Brothers (u)										
	0	1	2	3	4	5	6	7	8	9	10
0	0.66667	0.60000	0.55556	0.52941	0.51515	0.50769	0.50388	0.50195	0.50097	0.50049	0.50024
1	0.50000	0.42857	0.38462	0.36000	0.34694	0.34021	0.33679	0.33506	0.33420	0.33377	0.33355
2	0.33333	0.27273	0.23810	0.21951	0.20988	0.20497	0.20249	0.20125	0.20062	0.20031	0.20016
3	0.20000	0.15789	0.13514	0.12329	0.11724	0.11419	0.11265	0.11188	0.11150	0.11130	0.11121
4	0.11111	0.08571	0.07246	0.06569	0.06227	0.06055	0.05969	0.05926	0.05904	0.05893	0.05888
5	0.05882	0.04478	0.03759	0.03396	0.03214	0.03122	0.03076	0.03053	0.03042	0.03036	0.03033
6	0.03030	0.02290	0.01916	0.01727	0.01633	0.01586	0.01562	0.01550	0.01544	0.01541	0.01540
7	0.01538	0.01158	0.00967	0.00871	0.00823	0.00799	0.00787	0.00781	0.00778	0.00777	0.00776
8	0.00775	0.00583	0.00486	0.00438	0.00413	0.00401	0.00395	0.00392	0.00391	0.00390	0.00389
9	0.00389	0.00292	0.00244	0.00219	0.00207	0.00201	0.00198	0.00196	0.00196	0.00195	0.00195
10	0.00195	0.00146	0.00122	0.00110	0.00104	0.00101	0.00099	0.00098	0.00098	0.00098	0.00098

*From Murphy, E. A.: The rationale of genetic counseling. J. Pediatr., 72:121, 1968. Reprinted with permission.

carrier, and (b) the number of normal sons she has had. This information is provided in Table 6–3, taken from Murphy.[59]

To illustrate the background of these tables a simple pedigree may be examined. Figure 6–2 illustrates a hypothetical family in which a consultand (c) has had two maternal uncles affected with Duchenne muscular dystrophy and she also has two normal brothers. The prior probability that the consultand's mother is a carrier is 1/2, but since she is in the "condition" of having had two normal sons, the likelihood that the consultand's mother is a carrier is reduced by the "conditional" probability of 1/4 ($1/2 \times 1/2$). The product of her prior and conditional probabilities for being a carrier is $1/2 \times 1/4$ or 1/8. On the other hand, the consultand's mother had a prior probability of 1/2 that she was not a carrier. In this event, the likelihood of her having normal sons was 1. Therefore, the product of her prior conditional probabilities for not being a carrier is 1/2. It is four times more likely that she is not a carrier than that she is a carrier. Therefore, she has a 1/5 probability of being a carrier and a 4/5 probability of not being a carrier. This means that our consultand has a 1/10 chance of being a carrier, a 1/20 chance of bearing an affected son, and a 1/40 chance of bearing an affected child.

In addition to the use of posterior and collateral information from the pedigree, one may also be able to derive information from the consultand herself in the form of empiric laboratory data which have predictive value. For instance, three fourths of obligate heterozygotes for the gene causing Duchenne muscular dystrophy have CPK values which exceed the level in 95 per cent of normal females. Therefore, if in the example given above, the consultand's mother had a normal CPK value on at least four determinations,[11] the probability of this result if she were a carrier would be 1/4. Conversely, the probability of this result, if she were not a carrier, would be 19/20. By employing these values as additional conditional probabilities one would arrive at an overall risk that the consultand's mother was a carrier which was lower than the 1/5 predicted without the empiric biochemical information. Of course, one should determine the consultand's CPK level, as well.

Use of X-linked Gene Markers for Carrier Detection

If a particular X-linked marker is segregating in a family, if the heterozygous phenotype of the marker can be recognized confidently, and if the marker locus is closely linked to the locus at which the mutant gene in question resides, then linkage analysis may be used to identify female heterozygotes. For instance, a boy with hemophilia who has electrophoretic type A G6PD is born to a mother who is heterozygous for types A and B G6PD. Any daughter born to such a mother who does not have type A G6PD is very unlikely to have inherited the gene for hemophilia. Since there is about 4 per cent recombination between these

TABLE 6–3. THE PROBABILITY THAT THE CONSULTAND IS A CARRIER OF AN X-LINKED RECESSIVE DISORDER IF SHE HAS NO AFFECTED CHILDREN, m NORMAL SONS, AND n GENERATIONS HAVE ELAPSED SINCE THE LAST CERTAIN CARRIER*

Generations Since the Last Certain Carrier (n)	Number of Normal Sons of the Consultand (m)										
	0	1	2	3	4	5	6	7	8	9	10
1	0.5000	0.3333	0.2000	0.1111	0.0588	0.0303	0.0154	0.0078	0.0039	0.0019	0.0010
2	0.2500	0.1429	0.0769	0.0400	0.0204	0.0103	0.0052	0.0026	0.0013	0.0006	0.0003
3	0.1250	0.0667	0.0345	0.0175	0.0088	0.0044	0.0022	0.0011	0.0006	0.0003	0.0001
4	0.0625	0.0323	0.0164	0.0083	0.0041	0.0021	0.0010	0.0005	0.0003	0.0001	0.0001

*From Murphy, E. A.: The rationale of genetic counseling. J. Pediatr., 72:121, 1968. Reprinted with permission.

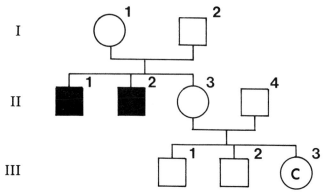

Figure 6-2. A simple pedigree of a hypothetical family. Individual III-3 is the consultand. The possibility that she is a carrier is influenced by the fact that she has two affected maternal uncles and two normal brothers. Circles represent females; squares represent males.

Tentative map* of Xq (long arm of X chromosome):
(centromere) 0____?____PGK____?____HGPRT____?____heA-G6PD-mdc-chD-chP

Tentative map* of ?Xp (short arm of X chromosome):
rs-oa-mr-Xg-ich-Fa
-17-17-11   11 24 (numbers = interval from Xg in cM)

*polarity arbitrary.

Figure 6-3. Tentative gene map of the X-chromosome, as of February, 1974. (Numbers in parentheses refer to the catalogue number of genetic disorders listed in McKusick, V. A.: Mendelian Inheritance in Man. 3rd ed. Baltimore, The Johns Hopkins Press, 1971.)

rs,	retinoschisis (31270)	heA,	hemophilia A (30670)
oa,	ocular albinism (30050)	G6PD,	glucose-6-phosphate
Xg,	Xg blood group (31470)		dehydrogenase (30590)
mr,	mental retardation ±	cbD,	deutan colorblindness (30380)
	hydrocephalus (30950)	cbP,	protan colorblindness (30390)
ich,	ichthyosis (30810)	mdc,	muscular dystrophy with con-
Fa,	Fabry's disease or α-galactosidase		tractures (31030)
	deficiency (30150)	PGK,	phosphoglycerate kinase
HGPRT,	hypoxanthine guanine		(31180)
	phosphoribosyl transferase		
	(30800)		

TABLE 6–4. PAIRS OF LOCI SHOWN TO BE GENETICALLY LINKED
BY FAMILY STUDIES *

Colorblindness loci (30380, 30390) and G6PD locus (30590)
Colorblindness loci and hemophilia A (30690)
G6PD locus and hemophilia A locus
Xg blood group locus (31470) and locus for X-linked ichthyosis (30810)
Xg blood group locus and ocular albinism locus (30050)
Xg blood group locus and Fabry locus (30150)
Xm serum protein locus (31490) and the locus for Hunter syndrome (30990)
Xm and colorblindness loci
Deutan and protan colorblindness loci
Xg blood group locus and locus for retinoschisis (31270)
Xg and mental retardation with or without hydrocephalus (30950)
Colorblindness loci and muscular dystrophy with contractures (31030)

*From Gerald, P. S., and Brown, J. A.: Report of the committee on the genetic constitution of the X chromosome. *In* Bergsma, D. (ed.): Birth Defects: Orig. Art. Ser., Human Gene Mapping: New Haven Conference (1973). Vol. 10, No. 3. Published by Symposia Specialists, Miami, for the National Foundation—March of Dimes, White Plains, N.Y., 1974, p. 29.

loci, the probability that such a daughter is not a carrier of hemophilia A exceeds 90 per cent.[13] This approach is rarely used because there are relatively few X-linked markers known to be distributed along the X-chromosome, and because there are few mutations with which they are in close linkage. Indeed, the newest serologic test for detecting hemophilia A carriers has the same efficiency as that described in the example above.

The tentative gene map of the X-chromosome as of February, 1974, is shown in Figure 6–3. A list of linked loci is given in Table 6–4. A list of loci proved not to be closely linked is given in Table 6–5. The linkages known include some for which, at the present time, the carrier state is not easily ascertained directly. These include: retinoschisis, mental retardation with or without hydrocephalus, one form of ocular albinism, and the form of muscular dystropy (Dreifuss) associated with contractures.

Further definition of the X-chromosome gene map can be anticipated from the application of new chromosome staining methods to various chromosomal rearrangements involving the X-chromosome. The segregational behavior of normal, as well as of such rearranged X-chromosomes in somatic cell hybrids, should be particularly useful for this purpose.[61]

A CLINICAL CATALOGUE OF X-LINKED, SEX-LIMITED, AND SEX-MODIFIED GENETIC DISEASE WITH NOTES ON CARRIER DETECTION AND DISCRIMINATION OF SIMILAR PHENOTYPES

The preparation of this catalogue was motivated by the belief that the critical step in recognizing carriers of x-linked and sex-limited ge-

TABLE 6-5. PAIRS OF X-LINKED LOCI PROVEN
TO BE NOT CLOSELY LINKED *

Colorblindness loci and retinitis pigmentosa (31260)
Colorblindness loci and thyroxine-binding globulin (31420)
G6PD and thyroxine-binding globulin (31420)
Hemophilia A (30670)and hypophosphatemia (30780)
HGPRT (30800) and G6PD (30590)
HGPRT and cbD
Ichthyosis (30810) and colorblindness loci
Ichthyosis and G6PD
Xg and Addison disease with cerebral sclerosis (30010)
Xg and agammaglobulinemia (30030)
Xg and anemia, hypochromic (30130)
Xg and choroideremia (30310)
Xg and colorblindness loci
Xg and Duchenne muscular dystrophy (31020)
Xg and ectodermal dysplasia, anhidrotic (30510)
Xg and G6PD
Xg and hemophilia A
Xg and hemophilia B (30690)
Xg and HGPRT (30800)
Xg and keratosis follicularis (30880)
Xg and mucopolysaccharidosis II (30990)
Xg and retinitis pigmentosa (31260)
Xg and testicular feminization (31370)
Xg and thrombocytopenia (31390)
Xg and thyroxine-binding globulin (31420)
Xg and Xm (31490)

---

*From Gerald, P. S., and Brown, J. A.: Report of the committee on the genetic constitution of the X chromosome. *In* Bergsma, D. (ed.): Birth Defects: Orig. Art. Ser., Human Gene Mapping: New Haven Conference (1973). Vol. 10, No. 3. Published by Symposia Specialists, Miami, for the National Foundation—March of Dimes, White Plains, N.Y., 1974, p. 30.

netic disease is, simply, recognition of those diseases. To facilitate the latter, the diseases have been arranged in clinical classes. A specific entity may appear in more than one class because of its pleiotropism. I have not made an exhaustive attempt to list every known disorder which might be X-linked or whose expression is sex-influenced for other reasons. Nevertheless, all those conditions that are reasonably well known are included. To reduce the number of bibliographic citations necessary, and for purposes of cross reference, the classified disorders are identified by McKusick's[62,]* catalogue number, when it is available. Two- and three-digit numbers refer to the present bibliography.

Adrenal Gland

1. Congenital adrenal hypoplasia (30020)
2. Hereditary unresponsiveness to ACTH (20220)
3. Sudanophilic leukodystrophy with adrenocortical atrophy (Schilder's disease; 30010; see CNS in catalogue)

---

*The reader is advised to consult the fourth edition (1974) of McKusick's *Mendelian Inheritance in Man,* which was not available when this chapter was prepared.

The X-linked disorders which may present with adrenal insufficiency (in the absence of anterior pituitary disease) are listed above. Congenital adrenal hypoplasia of the X-linked type is distinguishable from that of autosomal recessive origin.[63] In the former, fetal cortical zones are present, but poorly differentiated, and there are large, scattered areas of eosinophilia. In the latter, the fetal cortex is absent or markedly diminished, whereas the adult zones are well differentiated, if miniature. Although most patients with congenital adrenal hypoplasia of either genetic type are ill within the first few days and weeks of life, 15 per cent are apparently well until one year of age or more. This delayed presentation overlaps with a second category, known as hereditary unresponsiveness to ACTH.

Patients with hereditary unresponsiveness to ACTH present in childhood with glucocorticoid insufficiency; their mineralocorticoid function, however, is normal. Evidence for heterogeneity of this phenotype comes from several sources:[60] (1) in some families, parental consanguinity and affected offspring of both sexes indicate autosomal recessive inheritance; (2) in other families, only males are affected in the absence of parental consanguinity; (3) one sibship which included affected females yielded clear evidence of delayed onset and progressive severity, suggesting a degenerative process;[66] in others, hypermelanosis at birth indicated developmental failure in utero of the ACTH-dependent zones of the adrenal cortex; and (4) the histologic appearance of the adrenals differs between families.[67]

Anemias

1. Glucose-6-phosphate dehydrogenase deficiency (30590)
2. Phosphoglycerate kinase deficiency (31180)
3. Familial sideroblastic anemia (30130)
4. Zinsser-Cole-Engman syndrome (30500)

Most female carriers of the genes commonly responsible for severe G6PD in the American Negro, the Mediterranean countries, and the Orient can be detected by their intermediate erythrocyte enzyme values. Those carriers with a small proportion of G6PD-negative cells may be recognized more reliably by techniques which assess the phenotype of individual erythrocytes. The most dependable of these are the methemoglobin elution[68] and G6PD-tetrazolium[69] methods. These techniques must be employed under standard conditions with appropriate controls in order to avoid the complication of pseudomosaicism.

Phosphoglycerate kinase deficiency can be recognized in carrier females by a mild hemolytic anemia,[70] by an intermediate degree of erythrocyte and leukocyte enzyme deficiency, and by elevated levels of dihydroxyacetone phosphate and 2,3-diphosphoglycerate, two precursors of phosphoglycerate in the glycolytic pathway.[71]

Carriers of familial sideroblastic anemia may have splenomegaly and a subpopulation of erythrocytes with morphological anomalies as an expression of cell mosaicism.

Bleeding Disorders

1. Hemophilia A (30670)
2. Hemophilia B (30690)
3. Fibrin-stabilizing factor (factor XIII) deficiency (30550)
4. Thrombocytopenia
   a. isolated (31390)
   b. part of Aldrich syndrome with increased infection (30100)
   c. with increased serum IgA and renal disease (31400)
   d. part of Zinsser-Cole-Engman syndrome (30500)

Asymptomatic carrier detection for hemophilia A has been discussed previously in relation to use of the G6PD locus as an X-linked marker[13] and use of a serologic test which differentiates between the procoagulant and antigenic levels of antihemophilic factor activity.[40]

Deficient AHF activity occurs in Von Willebrand's disease and in combination with factor V deficiency (22730) — both autosomal diseases. In rare families, AHF deficiency is accompanied by an increased bleeding time with an X-linked recessive pattern of inheritance.[72]

Asymptomatic carrier detection for hemophilia B based upon levels of Christmas factor procoagulant activity appears to be as unreliable as the similar test for hemophilia A. It is noteworthy that the occasional female carrier of hemophilia B may be asymptomatic, despite having levels of Christmas factor typical of severely affected hemizygotes.

Deficiency of factor XIII can result from autosomal or X-linked recessive mutations. Some patients synthesize an immunologically competent, but functionally inactive, factor.[73] This may be exploitable as a test for carriers.

Cell Membrane Transport Disorders and Diseases which Manifest as Such

1. Diabetes insipidus
   a. vasopressin-sensitive (neurohypophyseal; 30490)
   b. vasopressin-resistant (nephrogenic; 30480)
2. Hypophosphatemic (vitamin D–resistant) rickets (30780)
3. Oculocerebrorenal syndrome (30900)
4. Pseudohypoparathyroidism (30080)
5. Primary, infantile-onset, hypoparathyroidism (30770)

In aggregate, the familial forms of diabetes insipidus constitute an excellent example of genetic heterogeneity underlying a single phenotype. The autosomal dominant form of neurohypophyseal diabetes insipidus is more common than the X-linked form.[74] Females carrying the gene for the latter may be minimally affected.

Females who are heterozygous for X-linked nephrogenic diabetes insipidus may have overt disease or may be asymptomatic. Those who are asymptomatic may reveal their carrier status when challenged by a water-deprivation test,[21] but many obligate carriers do not do so.[46, 75] The demonstration that vasopressin-resistance is reflected in decreased nephrogenous cAMP[144] may lead to an improved test for carrier detection. X-linked nephrogenic diabetes insipidus must be distinguished from still other monogenic causes of this phenotype: juvenile nephronophthisis (25610), renal-retinal dysplasia (26690), and medullary cystic disease (17400).

Females heterozygous for hypophosphatemic rickets are often symptomatic; hence the designation of the responsible gene as X-linked dominant. Most subclinical carriers can be identified by low serum phosphorus levels, *if repeated estimations are performed.* Occasionally, the carrier status of a female can be revealed only by estimating the Tm of renal tubular phosphate reabsorption.

It is well known that various monogenic forms of the Fanconi syndrome of renal tubular malabsorption can mimic hypophosphatemic rickets. These include, primarily, cystinosis, tyrosinemia, the oculocerebrorenal syndrome, and cases in which renal malabsorption appears to be unassociated with extrarenal disease. In addition, an autosomal recessive disorder of vitamin D metabolism,[76] known as vitamin D-dependent rickets, resembles, to some extent, hypophosphatemic rickets.

There is no reliable method for detecting the carrier state of oculocerebrorenal syndrome of Lowe, because the primary metabolic error in this disorder remains unknown more than 20 years after it was first described. It has been suggested that punctate lenticular opacities in carrier females are a minimal expression of the cataracts seen in affected boys. These are sufficiently common in the general population (1–2 per cent) and sufficiently uncommon among obligate carriers to make the sign unreliable. The ornithine tolerance test has not been proven to be reliable, either.[77]

The precise genetic basis of pseudohypoparathyroidism has not been established. In most families, the pattern of inheritance is compatible with that of an X-linked dominant gene, because the pedigrees lack examples of male-to-male transmission and their ratio of affected males to females is 1:2. However, females do not have milder expression than males—a finding contrary to that expected from X-linked genes. It is possible that the gene is X-linked, and that its expression is sex-influenced. It is less likely to be autosomal and sex-influenced, because it is difficult, on this basis, to account for a 2:1 preponderance of females without more severe expression in females.

In some families, male-to-male transmission has occurred, suggesting an autosomal dominant form of the phenotype.[135] In others, an autosomal recessive mode of inheritance is suggested.[78]

Whatever the mode of inheritance, the value of measuring nephrogenous cAMP after parathormone stimulation as a test for carriers is uncertain—largely because of the inconstancy of the measurement among affected individuals within one family.[79] The inconstancy may originate, in part, with the unknown basis upon which hypocalcemic patients become normocalcemic (pseudo-pseudohypoparathyroidism).[136, 137] The situation is complicated further by recognition of so-called type II pseudohypoparathyroidism[138] in which parathormone generates a normal nephrogenic cAMP response that does not culminate in phosphaturia.

Central Nervous System

1. Hydrocephalus (aqueductal stenosis) and mental retardation ± spastic paraplegia (30700)
2. Mental retardation ± hydrocephalus[80]
3. Lesch-Nyhan syndrome (choreoathetotic cerebral palsy, mental retardation, self-mutilation; 30800)
4. Spastic paraplegia and optic atrophy (31110)
5. Spastic paraplegia (31290)
6. Pelizaeus-Merzbacher disease (31160)
7. Schilder's adrenoleukodystrophy (diffuse cerebral sclerosis with adrenal atrophy; 30010)[81]
8. Menkes' syndrome (30940)
9. Laryngeal abductor paralysis with psychomotor retardation (24570)[82]
10. Cerebellar ataxia (30250)
11. Cerebellar ataxia with extrapyramidal involvement (30260)
12. Spinocerebellar disease with gout (partial HGPRT deficiency; 30620)
13. Spinal ataxia (31330)
14. Spinal and bulbar muscular atrophy (31320)
15. Peroneal muscular atrophy (30280)
16. Agenesis (partial) of the corpus callosum with mental retardation (30410)
17. Borjeson syndrome (severe mental defect, convulsions, polyendocrinopathy; 30190)
18. Familial convulsive disorder and mental retardation limited to females[151]

It would seem that hair follicles provide the simplest route to heterozygote detection for the Lesch-Nyhan syndrome.[30-32]

Unlike female carriers of complete HGPRT deficiency (Lesch-Nyhan), those who carry a gene responsible for partial HGPRT deficiency are likely to have intermediate values of the enzyme activity in their peripheral red and white blood cells.[27]

In one family, an X-linked gene causing mental retardation with or without hydrocephalus may be closely linked (recombination fraction: 0.11) to the Xg locus.[83] Classical X-linked hydrocephalus due to aqueductal stenosis appears to represent a different gene.

Pelizaeus-Merzbacher disease is an X-linked type of diffuse cerebral sclerosis. Its onset is in early infancy and it is characterized by nystagmus. Female carriers may express the disorder. Histopathologically, there is irregular demyelination. The basic defect is unknown.

Adrenocortical atrophy of varying degrees occurs in most boys with Schilder's disease, a second X-linked type of cerebral sclerosis.[81] Recent data from cultured skin fibroblasts indicate that the disorder is associated with a widespread defect in cholesterol metabolism.[84] Hopefully, this will lead to a method for detecting the female carrier state of this fatal X-linked disorder.

In Menkes' syndrome, there is onset of neurologic impairment in early infancy. Head hair is morphologically abnormal and fragile. The basic defect appears to involve intestinal copper absorption.[85] Recent recognition of a mutant phenotype in cultured skin fibroblasts[86] may lead to an accurate test for female carriers.

Charcot-Marie-Tooth disease (peroneal muscular atrophy) may be the result of autosomal, as well as X-linked, genes. Both X-linked dominant and recessive types have been described. Especially in the former, carrier females may have subtle signs which betray their genotype.

Deafness

1. Conductive, with stapes fixation (30440)
2. Neural, congenital (30450)
3. Neural, early childhood onset (30480)
4. Albinism–neural deafness syndrome (30070)
5. Norrie's disease (31060; see "ocular disease" and "malformation syndromes" in this catalogue)
6. Oto-palato-digital syndrome (31130; see "malformation syndromes")
7. Hunter syndrome (30990; see "storage disorders")

In the first four X-linked disorders listed, deafness is a major manifestation. Each can be mimicked by mutant genes not on the X-chromosome. The albinism-deafness syndrome is particularly interesting in this regard, because it belongs to a community of disorders that feature deafness in association with pigmentary disturbance.[87] Information on carrier detection is essentially nonexistent. In one family with the albinism-deafness syndrome, hearing impairment in heterozygotes was reported to be bilaterally asymmetric.

In the oto-palato-digital syndrome, conductive deafness may be secondary to cleft palate. Malformations of the ossicles were observed in two patients.

Epiphyseal Dysplasia

1. Spondylo-epiphyseal dysplasia tarda (31340)
The radiologic onset of this disorder is between 10 and 15 years of age. Carriers cannot be detected. Linkage with the Xg locus is loose.[139]

Heart

1. Primary endocardial fibroelastosis (30530)
2. Congenital valvular disease (31440)
3. Part of Lenz syndrome (30980)
4. Dextrocardia with other cardiac defects[163]

Endocardial fibroelastosis may be secondary to cardiovascular anomalies and inherited cardiomyopathies, or it may be primary. Primary endocardial fibroelastosis is usually sporadic. Multiple affected siblings occur often enough in the absence of consanguinity to indicate multifactorial inheritance. There may also be an autosomal recessive form (22660). Familial, primary, contracted, endocardial fibroelastosis with an X-linked recessive pattern of inheritance has been described once. A second family with the primary, dilated type[140] may also exemplify this situation. An affected mother and son have been described once.[141]

Neonatal Hyperammonemia Due to Ornithine Transcarbamylase (OTC) Deficiency

This disorder is determined by an X-linked dominant gene.[19] The gene is lethal for males in the neonatal period, despite all measures known to reduce ammonia build-up.[17, 18] Females express the disorder to various degrees.

Two noninvasive tests to detect female carriers have been used. The first involves an oral ammonium chloride load with sequential determinations of serum ammonia. This test was abnormal in one carrier whose hepatic levels of OTC were normal.[19] The second involves the measurement of urinary orotic acid (an intermediate in pyrimidine biosynthesis) after a high protein meal.[18] The test is based on the fact that carbamyl phosphate, which accumulates because of OTC deficiency, serves as a substrate for aspartate transcarbamylase, and thus drives the pathway of *de novo* pyrimidine biosynthesis. Excessive orotic aciduria after a protein load has helped in the recognition of two carriers in one family whose ammonium tolerances were normal. A third, related, carrier excreted a large amount of orotic acid in her urine without ammonium or protein loading.

Hypomagnesemia-Hypocalcemia

1. X-linked (?) hypomagnesemic tetany (30760)

Several families with apparently X-linked hypomagnesemia have been reported. An autosomal recessive variety is possible (24130) in one family with parental consanguinity. The hypomagnesemia is associated with hypocalcemia.

Familial, neonatal-onset, hypoparathyroidism (30770) may also be an X-linked cause of hypocalcemia in early life.

Increased Susceptibility to Infection

1. Infantile (Bruton-type) agammaglobulinemia (30030)
2. X-linked immunodeficiency with hyper-IgM[115]
3. Immunodeficiency with thrombocytopenia and eczema (Aldrich syndrome; 30100)
4. X-linked severe combined immunodeficiency (30040)
5. Chronic granulomatous disease (30640)

Among the five X-linked disorders in this group, there is a reliable test for recognizing heterozygotes only for one of them — chronic granulomatous disease. Indeed, a positive nitroblue tetrazolium reduction test in the mother of a sporadic male patient with chronic granulomatous disease serves to distinguish the X-linked from the otherwise indistinguishable autosomal recessive form of the disease. Unfortunately, a test is not available for identifying female carriers of severe combined immunodeficiency. Such a test would serve, likewise, to distinguish the X-linked and autosomal recessive (Swiss type) forms of this phenotype. Clinico-pathologic evidence of two types favors the X-linked form of severe combined immunodeficiency: Circulating lymphopenia tends to have a later onset and is of less severe degree than in the autosomal recessive form. Likewise, histologic examination reveals less severe depletion of lymphoid elements in the X-linked form.[116] Severe combined immunodeficiency associated with adenosine deaminase deficiency is caused by homozygosity for an autosomal recessive gene.[146] Such patients have, in addition, distinctive abnormalities of metaphyseal ossification. Most patients with severe combined immunodeficiency resulting from thymus and parathyroid hypoplasia are sporadic. A family has been described recently.[147] Typical patients have characteristic facial features. Combined immunodeficiency also occurs with short-limbed dwarfism on an autosomal recessive basis.[152]

It has recently been appreciated that erythrocytes and lymphocytes, as well as thrombocytes, are abnormal in the Aldrich syndrome.[148] This suggests that the basic defect is expressed in primitive cells of the marrow and may lead to a worthwhile test for carrier detection.

The delineation of various primary "agammaglobulinemias" in terms of their underlying B cell defects has begun.[165] Hopefully, this will lead to the appearance of tests designed to detect female carriers of the first two disorders listed in this section.

Integumentary System

Males only, or predominantly:
1. Anhidrotic ectodermal dysplasia (30510)
2. Bullous dystrophy (30200)[122]*
3. Cantu syndrome[123]
4. Dyskeratosis congenita (30500)*
5. Ehlers-Danlos (one type)
6. Fabry's disease (30150)
7. Gerodermia osteodysplastica (30580)
8. Goeminne syndrome (31430)
9. Ichthyosis (30810)
10. Ichthyosis with hypogonadotropic male hypogonadism (30820)[29]
11. Keratosis follicularis (30880)
12. Menkes' kinky hair syndrome (30940)
13. Van den Bosch syndrome (31450)

Females only, or predominantly:
1. Focal dermal hypoplasia (30560)
2. Incontinentia pigmenti (30830)
3. Netherton syndrome[125]

Anhidrotic ectodermal dysplasia is genetically heterogeneous. Female carrier detection of the X-linked form has been discussed on p. 144.

Female carriers of dyskeratosis congenita (Zinsser-Cole-Engman syndrome) may have poikiloderma and oral leukoplakia. A fully affected female has been described.[127]

Goeminne syndrome has been recognized in one family. Some female carriers had torticollis and cutaneous nevi. The full syndrome includes multiple spontaneous keloids, cryptorchidism, and renal dysplasia.

Ichthyosis, which is X-linked, can be distinguished by a combination of histological and clinical criteria from the autosomal dominant (vulgaris) and recessive forms (lamellar).[128, 129] In adult female heterozygotes, scaling of the legs is the most common manifestation. Females also exhibit characteristic corneal opacifications. In one study, of eight carriers, four had this condition. In ichthyosis vulgaris, the granular layer is hypoplastic. In X-linked ichthyosis, it is normal. In one study,[129] of nine obligate X-linked carriers, two had normal histology, but one half of the sisters of affected boys were abnormal, histologically. Several types of X-linked ichthyosis may exist.[155, 156]

Keratosis follicularis can be expressed in females to various degrees.

*May be the same syndrome[126]

The skin lesions of Fabry's disease, angiokeratoma corporis diffusum, may also occur in a variant of fucosidosis.[157]

Kidney

1. Alport syndrome (10420)
2. Fabry's disease* (30150)
3. Goeminne syndrome (31430)
4. Lenz syndrome (30980)
5. HGPRT deficiency (nephrolithiasis; 30620; 30800)
6. Hypophosphatemic rickets (30780)
7. Nephrogenic diabetes insipidus (30480)
8. Oculocerebrorenal syndrome (30900)
9. Pseudohypoparathyroidism (30080)

The mode of inheritance of the Alport syndrome remains in doubt.[133] Genetic heterogeneity may be responsible, in part, for this doubt. In any event, females are typically less severely affected than males. Their carrier status may have to be revealed by systematic investigation for proteinuria, hematuria, deafness, or cataract.

Nephrolithiasis can be associated with complete or partial HGPRT deficiency.

Disorders 6–9 in the list above are discussed in this catalogue as "transport disorders" (see p. 158).

Male Sexual Maldevelopment

1. Familial pure XY gonadal dysgenesis (30610)
2. Familial testicular dysgenesis[89, 90]
3. Familial male pseudohermaphroditism (30730; 31210; 31230)
4. Testicular feminization (31370; 31380)
5. Familial gynecomastia (30570; 30650)
6. Familial hypogonadotropic hypogonadism
   a. with anosmia (30870)
   b. with ichthyosis (30820)
   c. with ataxia (30740)
7. Familial panhypopituitarism (see "pituitary gland" in this catalogue)
8. Familial male isosexual precocious puberty
9. Aarskog syndrome (30540; also 23970)
10. Hypospadias-dysphagia syndrome (30710)
11. Hypospadias with telecanthus, cleft lip and palate, and mental retardation (30790)

None of the monogenic disorders of male sexual development listed above has been proven to be the result of an X-linked recessive gene

---

*See Chapter 7.

mutation. For each, an autosomal dominant, whose full expression is limited to, or more prevalent in, males, remains possible. However, there is substantial evidence that loci on the X chromosome participate in normal male sexual development.[91] Furthermore, testicular feminization (Tfm; universal refractoriness to androgen) is X-linked in the mouse,[3] and the occasional fertility of males with the Aarskog syndrome has resulted in seven sons, none affected.[92, 124] In either event, one can discuss detection of carrier females for each of these disorders without certainty of their precise mode of inheritance.

Familial pure XY gonadal dysgenesis, familial testicular dysgenesis, and some types of familial male pseudohermaphroditism result from defects in the determination, differentiation, and function of the testis, so that expression in female carriers would appear to be ruled out. Other types of familial male pseudohermaphroditism may result from functional defects of the testes and adrenals, so that expression in carrier females might result from adrenal involvement.

Another category of familial male pseudohermaphroditism involves target organ unresponsiveness to androgen. Since nonreproductive structures are targets for androgen, expression in carrier females might be expected. Uncommonly, female carriers of the gene for complete testicular feminization (Tfm) have delayed menarche and/or reduced sexual hair. One such woman had bilateral asymmetry of the breasts and pubic hair, a distribution compatible with an unusual expression of random X-chromosome inactivation.[93] In mouse Tfm, carrier females may have a blunted or absent response to androgen of kidney β-glucuronidase[94] and of submandibular gland production of nerve growth factor[95] and protease "D."[96] Shanies et al.[5] found that early generation fibroblasts cultured from the forearm skin of affected Tfm males metabolize testosterone predominantly to products which retain the 17C=O group. Two carriers yielded skin fibroblast cultures whose testosterone metabolism was indistinguishable from that of their sons. This result must be confirmed, particularly with cultured skin fibroblasts that have undergone prolonged serial subcultivation. The published result is compatible with the idea that the expression of the X-linked gene is sex-limited *in vivo*, but not *in vitro*. It is also compatible with an autosomal dominant gene whose expression is sex-limited *in vivo*, but not *in vitro*. In addition, preferential survival *in vitro* of cells with the mutant phenotype could account for the results, but the early subculture age at which the strains were studied weakens this idea. Successful cloning of carrier skin fibroblasts to generate the two cell types expected from an X-linked gene mutation would resolve these questions. Individual hair follicles could serve the same purpose. Very recently, it has been reported that cultured skin fibroblasts from patients with complete Tfm lack receptors for 5α-dihydrotestosterone.[145] Once confirmed, this may provide a reliable microassay for testing the Lyon Rule on cultured cell clones.

One class of familial male pseudohermaphroditism may result from target organ unresponsiveness to androgen, which is universal in distribution, but incomplete in degree (incomplete Tf). Another class may result from androgen unresponsiveness that is restricted to certain target organs. The last category may include an autosomal recessive disorder known as pseudovaginal perineoscrotal hypospadias (PPSH).[97] Familial gynecomastia can also be viewed as the result of a focal defect in androgen unresponsiveness during a critical period in development of the breast primordia.[98]

Familial hypogonadotropic male hypogonadism (FHMH) is heterogeneous, clinically and genetically (24420; 24430; Kraus-Ruppert syndrome, cited by Rimoin and Schimke, 1971, p. 321).[102] In familial hypogonadotropic hypogonadism with anosmia (Kallmann syndrome), males are affected more often than females, but male-to-male transmission has been observed in some kindreds which, therefore, must be segregating for an autosomal gene.[28] Females carrying the gene may have minimal expressions, such as delayed menarche, and menstrual irregularity, as well as full expression. Anosmia must be identified by specific testing; it can occur without hypogonadism in affected males or females of these families.[28]

FHMH with ataxia has been described in two families. Females were unaffected. Two sisters with hypogonadotropic hypogonadism and ataxia also had choroidoretinal degeneration, suggesting an autosomal recessive mutation.[100]

Several families with isosexual precocious puberty restricted to males have been described. In some of these, transmission by males or females has been described; an autosmal dominant with male-limited expression, therefore, is probable. In others, transmission by females has been the rule.[101] In one such family, the carrier females had signs of virilization, such as male muscularity, frontal baldness, and hirsutism.[102]

Malformation (Syndromes)

Males fully affected:
1. Aarskog syndrome (23970; 30540)
2. Borjeson syndrome (30190)
3. Cuendet syndrome[111]
4. Cantu syndrome[123]
5. Goeminne syndrome (31430; see Integumentary System)
6. Hypospadias-dysphagia-stridor syndrome (30710)
7. Lenz syndrome (30980)
8. Norrie syndrome (31060)
9. Oto-palato-digital syndrome (31130)
10. "Prune belly" syndrome (10010)
11. Telecanthus-hypospadias syndrome (31360)
12. Van den Bosch syndrome (31450)

TABLE 6-6. MALFORMATION SYNDROMES (? X-LINKED) AND THEIR EXPRESSION IN OBLIGATE FEMALE CARRIERS

Syndrome	Possible Expression in Female Carriers
Aarskog	Short stature, widow's peak, telecanthus, megalocornea, pouty lower lip, pectus excavatum, single crease fifth digit plus camptodactyly, hyperextensible interphalangeal joints, wide feet[92]
Borjeson	Mental retardation
Hypospadias-dysphagia-stridor	Dysphagia, stridor, short trachea, pulmonary hypoplasia[130]
Lenz	? Minor skeletal and ocular defects[131]
Norrie	? Short thumbs[114]
Oto-palato-digital	Short, vertical nasal bridge, telecanthus, supraorbital fullness, down-tilted lateral portion of upper eyelid, carpal, tarsal, metatarsal fusions, abnormal angles and ratios of hand bones[132]
Telecanthus-hypospadias	Telecanthus

Females fully affected (? males die in utero):

1. Focal dermal hypoplasia syndrome (30560)
2. Incontinentia pigmenti (30830)
3. Oro-facio-digital syndrome I (31120)[112]
4. Wildervanck syndrome (31460)

Several of the malformation syndromes in which only males are fully affected appear to find minimal expression in obligate female carriers or close female relatives. These syndromes and their expression are listed in Table 6-6.

Any of the syndromes which affect females, presumably because of their lethality for males *in utero,* may express themselves to an intermediate degree if the responsible genes are X-linked and are subject to random X-chromosome inactivation.

Malformation (Localized)

1. Imperforate anus (31080)
2. Metacarpal 4-5 fusion[99]
3. Cleft palate[162]
4. Dextrocardia with other cardiac defects[163]
5. Aqueductal stenosis (30700)
6. Agenesis of the corpus callosum, partial (30410)
7. Fixation of the stapedial foot plate (30440)

Manic-Depressive Illness

Manic-depressive illness of one type may depend, in part, on an X-linked dominant gene.[142] The gene appears to be detectably linked with the color-blindness loci.[105] For the protan locus, the recombination fraction estimate is 0.11, with a standard error of 0.08 and a 95 per cent confidence interval of ≤ 0.24. These results await confirmation.[167]

Mental Retardation without Physical Abnormality

There appears to be a form of X-linked recessive mental retardation whose distinguishing features are its nonprogressive course and its freedom from physical or structural abnormalities (30950). Under these circumstances, the location of the mutant gene on the X-chromosome could be very useful for female carrier detection, but it is not known at this time.

Myopathies

1. Muscular dystrophy
   a. Duchenne (31020)
   b. Becker (31010)
   c. Dreifuss (31010)
   d. Mabry (31000)
2. Centronuclear myopathy (31040)
3. Hypokalemic periodic paralysis (31170)

About 10 per cent of women who carry the gene for Duchenne muscular dystrophy (md) have mild signs of the disease; for example, hypertrophy of the calf muscles. About 10 per cent of these carriers have R waves of increased amplitude on the right precordial lead of the electrocardiogram.[103] About 75 per cent of them have elevated levels of creatine phosphokinase (CPK) activity in their serum.[104] It is essential to be aware of the inconstancy of the elevation. In one study[11] each of ten carriers had at least one normal CPK value. The inconstancy is, apparently, not the result of physical exertion in the few days preceding the determination. Serum CPK levels tend to be higher in nonpregnant women than in women between 8 and 20 weeks pregnant.[14, 106] Serum CPK values in normal females increase with age, but the increment is small, compared to variation between subjects grouped by decade.[11] Age may have a significant effect on serum CPK values in carriers.[104, 166] Neither the menstrual cycle nor "the pill" is an important source of variation. Females who apparently are not carriers may have abnormally high CPK values on occasion,[11] thus inviting the risk of being diagnosed as carriers erroneously.

Serum aldolase values, the determination of the lactic dehydrogenase isozymes, LDH_4 and LDH_5, and estimation of total body potassium concentration[107] are unlikely to increase the efficiency of carrier detection appreciably. The value of muscle histology is debatable.[104, 108] Ionasescu[109] has found increased incorporation of amino acid by muscle polyribosomes from carriers of Duchenne md. In the young carriers, much of the increased incorporation was directed to collagen synthesis. Five carriers with normal serum CPK levels and eight with normal muscle histology revealed their genotype on assay of their polyribosomes. It has been suggested that the polyribosome test, coupled with CPK determinations, can identify more than 90 per cent of Duchenne carriers. This experience awaits confirmation in other laboratories. With the scanning electron microscope, dramatic surface deformation of erythrocytes has been observed in four patients with Duchenne muscular dystrophy. Three female carriers appeared to have a smaller proportion of deformed cells.[150] Once confirmed, this approach to carrier detection may be very fruitful. Crystalline perinuclear inclusions have been observed in myoblasts cultured from nine patients with Duchenne md.[153] The validity of this observation is strengthened by the finding of similar inclusion bodies in myoblasts cultured from mothers of affected boys. This preliminary report awaits confirmation.

The last three methods reflect dramatically new approaches to the challenge of female carrier detection in Duchenne md. The prospect of accurate detection of carriers of this tragic X-linked recessive disorder now seems imminent.

Three other X-linked types of muscular dystrophy have been defined—in varying degrees. Each differs from the Duchenne type in one or more of the following ways: age of onset, rate of progression, precise pattern of muscle involvement, association with mental retardation, and possibly, histology. About 50 per cent of females carrying the Becker md gene have levels of CPK which exceed 95 per cent of control values.[41] The Becker gene is loosely linked to the color-blindness loci.

In rare families, a disorder clinically equivalent to Duchenne md affects boys and girls of consanguineous parents.[110] It is not known whether such parents have serum CPK elevations or any other expression of heterozygosity. Such information could be crucial for a genetic counselor to distinguish the X-linked from the autosomal recessive genocopy whenever the information available from the pedigree is inconclusive.

Hypokalemic periodic paralysis is inherited autosomally, but penetrance and expressivity in females are reduced.

Ocular Disease

1. a. Cataract (30220; 30230)
   b. Cataract as part of the oculocerebrorenal syndrome (30900)
2. Hypoplasia of the iris with glaucoma (30850)

3. Fundus dystrophies
   a. ocular albinism isolated (30050)
   b. ocular albinism with other ocular defects (30060)
   c. pigmentary retinopathies (30320; 30330; 31260)
   d. retinoschisis (vitreous veils; 31270)
   e. choroideremia (30310)
   f. choroideremia as part of van den Bosch syndrome (31450)
   g. macular degeneration (30910)
   h. night blindness (non-progressive) with myopia (31050)
4. Microphthalmia
   a. associated with ocular anomalies (30970)
   b. associated with nonocular anomalies (Lenz syndrome, 30980; Cuendet syndrome)[111]
5. Nystagmus (31070; 31080)
6. Norrie syndrome (hereditary pseudoglioma; 31060)

In some families, congenital total cataract is inherited as an X-linked disorder and carrier females have posterior sutural opacities. In one such family, the latter condition was an early expression of the disorder in affected hemizygotes. In other families, cataract is inherited as an X-linked disorder in association with microcornea or microphthalmia.

The fundus dystrophies have received much attention as vehicles for clinical expression of random X-chromosome inactivation. In females carrying the gene for uncomplicated ocular albinism, "dust-like" pigmentation is common in the posterior pole, and the peripheral fundus may show patches of pigmentation. In complicated ocular albinism, female carriers have slight protanomaly and minimal nystagmus, but no evidence of pigmentary disturbance.

The pigmentary retinopathies include disorders with varying degrees of choroidal involvement. There appear to be several X-linked types in this group, based largely on distinctions among their respective carrier states.[10] In one type (30320), female carriers have a pronounced "golden tapetal reflex" and "glistening golden bodies" at the level of Bruch's membrane, even without visual deficit. In a second type, female carriers have "white deposits" in the mid-peripheral fundus, at the level of Bruch's membrane, but they also have variable degrees of pigmentary dystrophy and visual deficit. In a third type, female carriers have pigmentary dystrophy in the lower halves of the eyes, but no white deposits. It has been suggested that fluorescein angiography may reveal in carriers subtle changes that are not apparent on routine examination.[113] Note that autosomal recessive and autosomal dominant forms of pigmentary retinopathy are more common than those determined by X-linked genes. Furthermore, pigmentary retinopathy is a component of several autosomal monogenic syndromes which include nonocular pathology.

In retinoschisis, there is intraretinal degenerative splitting associated with pigmentation, but macular changes may occur as an isolated expression of the disorder. Expression in carrier females is not well defined: it may take the form of macular pathology or of minimal retinoschisis.[154]

Males affected with choroideremia have primary choroidal atrophy associated with pigmentary abnormalities. Most female carriers have no visual deficit despite the presence of "brown deposits" in the fundus.

Norrie syndrome involves retinal malformation (with secondary atrophy of the globe), which may or may not be accompanied by mental retardation, deafness, and short thumbs in individual males. It is possible that females may express their carrier status by short thumbs.[114]

Anterior Pituitary

1. X-linked panhypopituitarism (31200)

This is a rare form of familial functional panhypopituitarism. The autosomal recessive variety is more common (26260). Autosomal forms of dysmorphogenic panhypopituitarism also occur.[134] There is no information on carrier detection of the X-linked disorder.

Reticuloendotheliosis (Lymphohistiocytosis)

Two families with apparently X-linked lymphohistiocytosis have been recorded (31250).[143, 164] Affected boys die of a fulminating mononucleosis-like syndrome characterized by various combinations of fever, sore throat, hepatosplenomegaly, lymphadenopathy, maculopapular rash, purpura, anemia, peripheral atypical lymphocytosis, and hyperimmunoglobulinemia.

Storage Disorders

1. Glycogen storage disease, type VIII (phosphorylase b kinase deficiency; 30600)
2. Mucopolysaccharidosis, type II (Hunter syndrome; sulfoiduronate sulfatase deficiency; 30990)
3. Ceramide trihexoside lipidosis (Fabry's disease; ceramide trihexosidase deficiency; 30150)

Glycogen storage disease due to phosphorylase b kinase deficiency has been extracted from the heterogeneous category of liver "phosphorylase" deficiency. Female carriers of the disorder may have hepatomegaly, and the activity of the enzyme in their leukocytes may be intermediate. Fibroblast mosaicism has been found.[161]

The Hunter syndrome is the only mucopolysaccharide (MPS) storage disease which is X-linked. The enzyme deficiency has recently been identified.[117] It runs a less severe course than its closest relative, the autosomal recessive Hurler syndrome. Corneal opacity does not occur in the Hunter syndrome, but deafness is more common than in the Hurler syndrome. Female carriers of the Hunter gene are asymptomatic. Under rigidly standardized conditions, one can exploit the metachromatic staining reaction of stored mucopolysaccharides with toluidine blue to distinguish the two expected cellular phenotypes in uncloned lymphocyte[118] and skin fibroblast[119] cultures of carrier females. Most laboratories charged with the responsibility of carrier diagnosis do not depend upon this technique for two reasons: (1) it is difficult to define and maintain the standard conditions; and (2) the skin fibroblast cultures of 7 to 27 per cent[120, 121] of control individuals stain metachromatically. Furthermore, the ability of cells with the normal phenotype to "correct" those with the mutant phenotype may serve to mask cellular mosaicism. The reality of this threat is underscored by the normal $^{35}SO_4$ incorporation into the MPS fraction of uncloned heterozygous skin fibroblast cultures which yield the two types of clones.[88] It is preferable, therefore, to isolate single cell clones of fibroblasts from potential Hunter heterozygotes and to assess their individual phenotypes by estimating their intracellular MPS concentration or their rate of $^{35}SO_4$ incorporation into MPS.

It has been suggested that cells with the mutant phenotype may survive preferentially during serial subculture of fibroblast strains developed from the skin of Hunter heterozygotes.[55] This tendency, if prevalent, would minimize the risk of failing to diagnose the carrier state.

Approaches to the recognition of female carriers of the Fabry gene are given in Chapter 7.

CONCLUDING COMMENTS

It is appropriate to conclude this chapter with some comments on the load of X-linked genetic disease in man and on some special problems which can be anticipated as a consequence of screening programs for carriers of X-linked genetic disease, whether the latter are directed at particular kindreds or at larger populations whose genetic relatedness is based on the religious and/or geographic community.

One estimate, based on McKusick's catalogue of Mendelian disease, is that known X-linked disorders (approximately 130) are responsible for roughly one sixth of all Mendelian genetic disease.[158] Another estimate of the health burden attributable to X-linked genetic disease can be generated by scrutinizing the causes of admission to a major children's hospital during a period of one year. One such survey[159] revealed that X-linked diseases were responsible for 40 per cent of all admissions caused by mutant Mendelian genes. These statements would not apply to certain subpopulations which, for any reason, have high frequencies of specific X-linked genes.

Advocacy of, and compliance with, any program to screen for carriers of X-linked genetic disease are most likely to be high in two situations: (1) where a technique for prenatal diagnosis of affected male fetuses is available; and/or (2) where effective therapy is available. At this time (May, 1975), prenatal diagnosis can be made in the Lesch-Nyhan and Hunter syndromes, and in Fabry's disease. Treatment of the former two disorders is not adequate, whereas encouraging progress has been made for the latter (see Chapter 7).

Because stigmatization of the X-linked carrier state falls only on females, two major ethical-moral "issues"[160] in screening for X-linked genetic disease stand out: the age at which screening is done, and the confidentiality (or ownership) of the results. If screening of females is done in the newborn period, who "owns" the positive result — the infant or her parents? At whatever premarital age it is done, does the prospective husband of a known carrier have a "right" to this knowledge? These questions are more challenging when one mate is the "culprit," than when both may be involved — as is true in the Tay-Sachs screening program. If screening is done in the newborn period, when should one inform the carrier of her status? If screening is done on girls in high school (for instance, because compliance is deemed likely to be high at this time), what is the emotional cost of burdening a teenager with the knowledge that she is a carrier of a harmful X-linked gene? Would compliance be as high among females in high school if the result of their screening test were withheld until some later date?[168]

Answers to these questions must await the appearance of carefully structured and carefully evaluated experimental screening programs. It is reasonable to imagine that well designed studies on single kindreds could serve as experimental models of screening programs destined for larger populations.

References

1. Fraser, G. R.: Selective abortion, gametic selection, and the X chromosome. Am. J. Hum. Genet., 24:359, 1972.
2. Motulsky, A. G., Fraser, G. R., and Felsentein, J.: Public health and long-term genetic implications of intrauterine diagnosis and selective abortions. Birth Defects, 7:22, 1971.
3. Lyon, M. F., and Hawkes, S. G.: X-linked gene for testicular feminization in the mouse. Nature (London), 225:1217, 1970.
4. Passmore, H., and Shreffler, D. C.: A sex-limited serum protein variant in the mouse: Hormonal control of phenotypic expression. Biochem. Genet., 5:201, 1971.
5. Shanies, D. D., Hirschhorn, K., and New, M. I.: Metabolism of testosterone-^{14}C by cultured human cells. J. Clin. Invest., 51:1459, 1972.
6. Köbberling, J., and Emrich, D.: The genetic polymorphism of the thyroxine-binding globulin (TBG). Humangenetik, 14:85, 1972.
7. Refetoff, S., Robin, N. I., and Alper, C. A.: Study of four new kindreds with inherited thyroxine-binding globulin abnormalities. Possible mutations of a single gene locus. J. Clin. Invest., 51:848, 1972.
8. Windhorst, D., Holmes, B., and Good, R. A.: A newly defined X-linked trait in man with demonstration of the Lyon effect in carrier females. Lancet, 1:737, 1967.

9. Bock, R. D., and Kolakowski, D.: Further evidence of sex-linked major-gene influence on human spatial visualizing ability. Am. J. Hum. Genet., 25:1, 1973.

10. Bird, A. C., and Blach, R. K.: X-linked recessive fundus dystrophies and their carrier states. Trans. Ophthalmol. Soc. U.K., 90:127, 1970.

11. Perry, T. B., and Fraser, F. C.: Variability of serum creatine phosphokinase activity in normal women and carriers of the gene for muscular dystrophy. Neurology, 23:1316, 1973.

12. Williams, T. F., and Winters, R. W.: Familial (hereditary) vitamin D-resistant rickets with hypophosphatemia. *In* Stanbury, J. B., Wyngaarden, J. B., and Fredrickson, D. S. (eds.): Metabolic Basis of Inherited Disease. 3rd ed. McGraw-Hill Book Company, New York, 1972, Chap. 60.

13. McCurdy, P. R.: Use of genetic linkage for the detection of female carriers of hemophilia. New Engl. J. Med., 285:218, 1971.

14. Emery, A. E. H., and King, B.: Pregnancy and serum creatine kinase levels in potential carriers of Duchenne X-linked muscular dystrophy. Lancet, 1:1013, 1971.

15. Migeon, B. R.: Studies of skin fibroblasts from 10 families with HGPRT deficiency, with reference in X-chromosomal inactivation. Am. J. Hum. Genet., 23:199, 1971.

16. Wood, S., and Pinsky, L.: Lesch-Nyhan mutation: The influence of population density on purine phosphoribosyltransferase activities and exogenous purine utilization in monolayer cultures of skin fibroblasts. J. Cell. Physiol., 80:33, 1972.

17. Campbell, A. G. M., Rosenberg, L. E., Snodgrass, P. J., and Nuzum, C. T.: Ornithine transcarbamylase deficiency: neonatal hyperammonemia in males. N. Engl. J. Med., 288:1, 1973.

18. Goldstein, A. S., Hoogenraad, N. J., Johnson, J. D., et al.: Metabolic and genetic studies of a family with ornithine transcarbamylase deficiency. Pediatr. Res., 8:5, 1974.

19. Short, E. M., Conn, H. O., Snodgrass, P. J., Campbell, A. G., and Rosenberg, L. E.: Evidence for X-linked dominant inheritance of ornithine transcarbamylase deficiency. N. Engl. J. Med., 288:7, 1973.

20. Ducos, J., Marty, Y., Sanger, R., and Race, R. R.: Xg and X chromosome inactivation. Lancet, 2:219, 1971.

21. Carter, C., and Simpkiss, M.: The "carrier" state in nephrogenic diabetes insipidus. Lancet, 2:1069, 1956.

22. Kelley, W. N., Greene, M. L., Rosenbloom, F. M., et al.: Hypoxanthine-guanine phosphoribosyltransferase deficiency in gout. Ann. Intern. Med., 70:155, 1969.

23. Dancis, J., Berman, P., Jansen, V., and Balis, M. E.: Absence of mosaicism in the lymphocyte in X-linked congenital hyperuricosuria. Life Sci., 7:587, 1968.

24. Nyhan, W. L., Bakay, B., Connor, J. D., Marks, J. F., and Keele, D. K.: Hemizygous expression of glucose-6-phosphate dehydrogenase in erythrocytes of heterozygotes for the Lesch-Nyhan syndrome. Proc. Natl. Acad. Sci. U.S.A., 65:214, 1970.

25. McDonald, J. A., and Kelley, W. N.: Lesch-Nyhan syndrome: Absence of the mutant enzyme in erythrocytes of a heterozygote for both normal and mutant hypoxanthine-guanine phosphoribosyl transferase. Biochem. Genet., 6:21, 1972.

26. Lajtha, L. A., and Vane, J. R.: Dependence of bone marrow cells on the liver for purine supply. Nature (London), 182:191, 1958.

27. Emmerson, B. T., Thompson, C. J., and Wallace, D. C.: Partial deficiency of hypoxanthine-guanine phosphoribosyltransferase: Intermediate enzyme deficiency in heterozygote red cells. Ann. Intern. Med., 76:285, 1972.

28. Santen, R. J., and Paulsen, C. A.: Hypogonadotropic eunuchoidism. I. Clinical study of the mode of inheritance. J. Clin. Endocrinol. Metab., 36:47, 1973.

29. Maurer, W. F., and Sotos, J. F.: Sex-linked familial hypogonadism and ichthyosis. Soc. Pediat. Res., 39th Ann. Meeting, 1969, p. 181.

30. Gartler, S. M., Scott, R. C., Goldstein, J. L., Campbell, B., and Sparkes, R.: Lesch-Nyhan syndrome: Rapid detection of heterozygotes by use of hair follicles. Science, 172:572, 1971.

31. Silvers, D. N., Cox, R. P., Balis, M. E., and Dancis, J.: Detection of the heterozygote in Lesch-Nyhan disease by hair-root analysis. N. Engl. J. Med., 286:390, 1972.

32. Francke, U., Bakay, B., and Nyhan, W. L.: Detection of heterozygous carriers of the

Lesch-Nyhan syndrome by electrophoresis of hair root lysates. J. Pediatr., 82:472, 1973.

33. Frias, J. L., and Smith, D. W.: Diminished sweat pores in hypohidrotic ectodermal dysplasia: A new method for assessment. J. Pediatr., 72:606, 1968.

34. Passarge, E., and Fries, E.: X chromosome inactivation in X-linked hypohidrotic ectodermal dysplasia. Nature [New Biol.], 245:58, 1973.

35. Sauk, J. J., Jr., Lyon, H. W., and Witkop, C. J., Jr.: Electron optic microanalysis of two gene products in enamel of females heterozygous for X-linked hypomaturation amelogenesis imperfecta. Am. J. Hum. Genet., 24:267, 1972.

36. Warburg, M.: Random inactivities of the X chromosome in intermediate X-linked retinitis pigmentosa. Two hypotheses. Trans. Ophthalmol. Soc. U.K., 91:553, 1971.

37. Krill, A. E.: Observations of carriers of X-chromosomal linked chorioretinal degenerations. Am. J. Ophthalmol., 64:1029, 1967.

38. Fairbanks, V. F., and Fernandez, M. N.: The identification of metabolic errors associated with hemolytic anemia. J.A.M.A., 208:316, 1969.

39. Panizon, F., Zacchello, F., Sartori, E., and Addis, S.: The ratio between normal and sensitive erythrocytes in heterozygous glucose-6-phosphate dehydrogenase deficient women. Acta Haematol. (Basel), 43:291, 1970.

40. Bennett, B., and Ratnoff, O. D.: Detection of the carrier state for classic hemophilia. N. Engl. J. Med., 288:342, 1973.

41. Emery, A. E. H., Clack, E. R., Simon, S., and Taylor, J. L.: Detection of carriers of benign X-linked muscular dystrophy. Br. Med. J., 4:522, 1967.

42. Rosenbloom, F. M., Kelley, W. N., Henderson, J. F., and Seegmiller, J. E.: Lyon hypothesis and X-linked disease. Lancet, 2:305, 1967.

43. Migeon, B. R.: X-linked hypoxanthine-guanine phosphoribosyl transferase deficiency: Detection of heterozygotes by selective medium. Biochem. Genet., 4:377, 1970.

44. Salzmann, J., De Mars, R., and Benke, P.: Single-allele expression at an X-linked hyperuricemia locus in heterozygous human cells. Proc. Nat. Acad. Sci. U.S.A., 60:545, 1968.

45. Migeon, B. R., Der Kaloustian, V. M., Nyhan, W. L., Young, W. L., and Childs, B.: X-linked hypoxanthine-guanine phosphoribosyl transferase deficiency: Heterozygote has two clonal populations. Science, 160:425, 1968.

46. Uttley, W. S., and Thistlethwaite, D.: Failure to detect the carrier in congenital nephrogenic diabetes insipidus. Arch. Dis. Child., 47:137, 1972.

47. Wood, S., and Pinsky, L.: Lesch-Nyhan syndrome: Rapid detection of heterozygotes. Clin. Genet., 1:216, 1970.

48. Felix, J. S., and DeMars, R.: Detection of females heterozygous for the Lesch-Nyhan mutation by 8-azaguanine-resistant growth of cultured fibroblasts. J. Lab. Clin. Med., 77:596, 1971.

49. Subak-Sharpe, H., Burk, R. R., and Pitts, J. D.: Metabolic cooperation between biochemically marked mammalian cells in tissue culture. J. Cell Sci., 4:353, 1969.

50. Cox, R. P., Krauss, M., Balis, M. E., and Dancis, J.: Evidence for transfer of enzyme product as the basis of metabolic cooperation between tissue culture fibroblasts of Lesch-Nyhan disease and normal cells. Proc. Natl. Acad. Sci. U.S.A., 67:1573, 1970.

51. Fujimoto, W. Y., and Seegmiller, J. E.: Hypoxanthine-guanine phosphoribosyltransferase deficiency: Activity in normal, mutant, and heterozygote cultured human skin fibroblasts. Proc. Natl. Acad. Sci., U.S.A., 65:577, 1970.

52. Fratantoni, J. C., Hall, C. W., and Neufeld, E. F.: Hurler and Hunter syndromes. I. Mutual correction of the defect in cultured fibroblasts. Science, 162:570, 1968.

53. Milunsky, A., and Littlefield, J. W.: Diagnostic limitations of metachromasia. N. Engl. J. Med., 281:1128, 1969.

54. Danes, B. S., and Bearn, A. G.: Hurler's syndrome (X-linked): A genetic study of clones in cell culture with particular reference to the Lyon hypothesis. J. Exp. Med., 126:509, 1967.

55. Booth, C. W., and Nadler, H. L.: In vitro selection for the Hunter gene. N. Engl. J. Med., 288:636, 1973.

56. Romeo, G., and Migeon, B. R.: Genetic inactivation of the alpha-galactosidase locus in carriers of Fabry's disease. Science, 170:180, 1970.

57. Felix, J. S., DeMars, R.: Purine requirement of cells cultured from humans affected with Lesch-Nyhan syndrome (hypoxanthine-guanine phosphoribosyltransferase deficiency). Proc. Nat. Acad. Sci. U.S.A., 62:536, 1969.

58. Murphy, E. A.: Probabilities in genetic counseling. Birth Defects, 9:19, 1973.

59. Murphy, E. A.: The rationale of genetic counseling. J. Pediatr., 72:121, 1968.

60. Franks, R. C., and Nance, W. E.: Hereditary adrenocortical unresponsiveness to ACTH. Pediatrics, 45:43, 1970.

61. Croce, C. M., Litwack, G., and Koprowski, H.: Human regulatory gene for inducible tyrosine aminotransferase in rat-human hybrids. Proc. Natl. Acad. Sci. U.S.A., 70:1268, 1973.

62. McKusick, V. A.: Mendelian Inheritance in Man. 3rd ed. Baltimore, Johns Hopkins Press, 1971. (Fourth edition now available.)

63. Pakravan, P., Kenny, F. M., Depp, R., and Allen, A. C.: Familial congenital absence of adrenal glands: Evaluation of glucocorticoid, mineralocorticoid, and estrogen metabolism in the perinatal period. J. Pediatr., 84:74, 1974.

64. Czapek, E. E., Hoyer, L. W., and Schwartz, A. D.: Hemophilia in a female: Use of factor VIII antigen levels as a diagnostic aid. J. Pediatr., 84:485, 1974.

65. Baehner, R. L.: Molecular basis for functional disorders of phagocytes. J. Pediatr., 84:317, 1974.

66. Moshang, T., Jr., Rosenfeld, R. L., Bongiovanni, A. M., Parks, J. S., and Amrhein, J. A.: Familial glucocorticoid insufficiency. J. Pediatr., 82:821, 1973.

67. Kelch, R. P., Kaplan, S. L., Biglieri, E. G., et al.: Hereditary adrenocortical unresponsiveness to adrenocorticotropic hormone. J. Pediatr., 81:726, 1972.

68. Gall, J. C., Jr., Brewer, G. J., and Dern, D. J.: Studies of glucose-6-phosphate dehydrogenase activity of individual erythrocytes: The methemoglobin-elution test for identification of females heterozygous for G6PD deficiency. Am. J. Hum. Genet., 17:359, 1965.

69. Fairbanks, V. F., and Lampe, L. T.: A tetrazolium-linked cytochemical method for estimation of glucose-6-phosphate dehydrogenase activity in individual erythrocytes: Application in the study of heterozygotes for glucose-6-phosphate dehydrogenase deficiency. Blood, 31:589, 1968.

70. Kraus, A. P., Langston, M. F., Jr., and Lynch, B. L.: Red cell phosphoglycerate kinase deficiency: A new cause of non-spherocytic hemolytic anemia. Biochem. Biophys. Res. Commun., 30:173, 1968.

71. Konrad, P. N., McCarthy, D. J., Mauer, A. M., Valentine, W. N., and Paglia, D. E.: Erythrocyte and leukocyte phosphoglycerate kinase deficiency with neurologic disease. J. Pediatr., 82:456, 1973.

72. Holmberg, L., and Nilsson, I. M.: Two genetic variants of Von Willebrand's disease. N. Engl. J. Med., 288:595, 1973.

73. Duckert, F.: Le facteur XIII et la proteine XIII. Nouv. Rev. Fr. Hematol., 10:685, 1970.

74. Braverman, L. E., Mancini, J. P., and McGoldrick, D. M.: Hereditary idiopathic diabetes insipidus. Ann. Intern. Med., 63:503, 1965.

75. Schoen, E. J.: Renal diabetes insipidus. Pediatrics, 26:808, 1960.

76. Fraser, D., Kooh, S. W., Kind, P. H., et al.: Pathogenesis of hereditary vitamin-D-dependent rickets. N. Engl. J. Med., 289:817, 1973.

77. Holmes, L. B., McGowan, B. L., and Efron, M. L.: Lowe's syndrome: A search for the carrier state. Pediatrics, 44:358, 1969.

78. Cederbaum, S., and Lippe, B. M.: Possible autosomal recessive inheritance in a family with Albright's hereditary osteodystrophy and an evaluation of the genetics of the disorder. Am. J. Hum. Genet., 25:638, 1973.

79. Marcus, R., Wilber, J. F., and Aurbach, G. D.: Parathyroid-hormone sensitive adenyl cyclase from the renal cortex of a patient with pseudohypoparathyroidism. J. Clin. Endocrinol. Metab., 33:537, 1971.

80. Fried, K.: X-linked mental retardation and/or hydrocephalus. Clin. Genet., 3:258, 1972.

81. Schaumberg, H. H., Richardson, E. P., Johnson, P. C., et al.: Schilder's disease: Sex-linked recessive transmission with specific adrenal changes. Arch. Neurol., 27:458, 1972.

82. Watters, G. V., and Fitch, N.: Familial laryngeal abductor paralysis and psychomotor retardation. Clin. Genet., 4:429, 1973.

83. Fried, K., and Sanger, R.: Possible linkage between Xg and the locus for a gene causing mental retardation with or without hydrocephalus. J. Med. Genet., 10:17, 1973.

84. Burton, B. K., and Nadler, H. L.: Schilder's disease: Abnormal cholesterol retention and accumulation in cultivated fibroblasts. Pediatr. Res., 8:170, 1974.

85. Danks, D. M., Campbell, P. E., Stevens, B. J., Mayne, V., and Cartwright, E.: Menkes' kinky hair syndrome. An inherited defect in copper absorption with widespread effects. Pediatrics, 50:188, 1972.

86. Danks, D. M., Cartwright, E., Stevens, B. J., and Townley, R. R. W.: Menkes' kinky hair disease: Further definition of the defect in copper transport. Science, 179:1140, 1973.

87. Konigsmark, B. W.: Hereditary congenital severe deafness syndromes. Ann. Otol. Rhinol. Laryngol., 80:269, 1971.

88. Neufeld, E. F.: Personal communication. April, 1974.

89. Jirasek, J. E.: Testicular dysgenesis syndromes. Birth defects, 7:159, 1972.

90. Bartlett, D. J., Grant, J. K., Pugh, M. A., and Aherne, W.: A familial feminizing syndrome. J. Obstet. Gynaecol. Br. Commonw., 75:199, 1968.

91. Pinsky, L.: Human male sexual maldevelopment: A teratogenetic classification of the monogenic forms. Teratology, 10:193, 1974.

92. Berman, P., Desjardins, C., and Fraser, F. C.: The inheritance of the Aarskog syndrome. Birth Defects, 10:151, 1974.

93. Gayral, L., Barraud, M., Carrie, J., and Candebat, L.: Pseudohermaphrodisme à type de testicule féminisant. Étude hormonale et étude psychologique. Toulouse Med., 61:637, 1960.

94. Tettenborn, V., Dofaku, R., and Ohno, S.: Non-inducible phenotype exhibited by a proportion of female mice heterozygous for the X-linked testicular feminization mutation. Nature [New Biol.], 234:37, 1971.

95. Lyon, M. F., Hendry, I., and Short, R. V.: The submaxillary salivary glands as test organs for response to androgen in mice with testicular feminization. J. Endocrinol., 58:357, 1973.

96. Schenkein, I., Levy, M., Bueker, E. D., and Wilson, J. D.: Immunological and enzymatic evidence for the absence of an esteroproteolytic enzyme (Protease "D") in the submandibular gland of the Tfm mouse. Endocrinology, 94:840, 1974.

97. Opitz, J. M., Simpson, J. L., Sarto, G. E., Summitt, R. L., New, M. I., and German, J.: Pseudovaginal perineoscrotal hypospadias. Clin. Genet., 3:1, 1972.

98. Federman, D. D.: Abnormal Sexual Development. Philadelphia, W. B. Saunders Company, 1967, p. 119.

99. Holmes, L. B., Wolf, E., and Miettinen, O. S.: Metacarpal 4-5 fusion with X-linked recessive inheritance. Am. J. Hum. Genet., 24:562, 1972.

100. Boucher, B. J., and Gibberd, F. B.: Familial ataxia, hypogonadism, and retinal degeneration. Acta Neurol. Scand., 45:507, 1969.

101. Jungck, E. C., Brown, N. H., and Carmona, N.: Constitutional precocious puberty in the male. Amer. J. Dis. Child., 91:138, 1956.

102. Rimoin, D. L., and Schimke, R. N.: Genetic Disorders of the Endocrine Glands. St. Louis, C. V. Mosby Company, 1971, p. 340.

103. Emery, A. E. H.: Abnormalities of the electrocardiogram in hereditary myopathies. J. Med. Genet., 9:8, 1972.

104. Gardner-Medwin, D., Pennington, R. J., and Walton, J. N.: The detection of carriers of X-linked muscular dystrophy. J. Neurol. Sci., 13:459, 1971.

105. Mendlewicz, J., Fleiss, J. L., and Fieve, R. R.: Evidence for X-linkage in the transmission of manic-depressive illness. J.A.M.A., 222:1624, 1972.

106. Blyth, H., and Hughes, G. P.: Pregnancy and serum CPK levels in potential carriers of severe X-linked muscular dystrophy. Lancet, 1:855, 1971.

107. Bland, W. H., Lederer, M., and Cassen, B.: The significance of decreased body potassium concentrations in patients with muscular dystrophy and nondystrophic relatives. N. Engl. J. Med., 276:1349, 1967.

108. Roy, S., and Dubowitz, V.: Carrier detection in Duchenne muscular dystrophy. A comparative study of electron microscopy, light microscopy, and serum enzymes. J. Neurol. Sci., 11:65, 1970.

109. Ionasescu, V., Zellweger, H., Shirk, P., and Conway, T. W.: Identification of carriers of Duchenne muscular dystrophy by muscle protein synthesis. Neurology, 23:497, 1973.
110. Boyer, S. H., and Fainer, D. C.: Genetics and disease of muscle. Am. J. Med., 35:622, 1963.
111. Cuendet, J. F.: La microphthalmie compliquee. Ophthalmologica, 141:380, 1961.
112. Rimoin, D. L., and Edgerton, M. T.: Genetic and clinical heterogeneity in the oral-facial-digital syndromes. J. Pediatr., 71:94, 1967.
113. Weinstein, G. W., Maumence, A. E., and Hyvarinen, L.: On the pathogenesis of retinitis pigmentosa. Ophthalmologica, 162:82, 1971.
114. Holmes, L. B.: Norrie's disease: An X-linked syndrome of retinal malformation, mental retardation, and deafness. J. Pediatr., 79:89, 1971.
115. Rosen, F. S., Craig, J. M., Vawter, G., and Janeway, C.: The dysgammaglobulinemias and X-linked thymic hypoplasia. Birth Defects, 4:67, 1968.
116. Gatti, R. A.: X-linked severe dual system immunodeficiency. In Bergsma, D. (ed.): Birth Defects Atlas and Compendium. Baltimore, The Williams & Wilkins Co., 1973, p. 901.
117. Bach, G., Eisenberg, F., Jr., Cantz, M., and Neufeld, E.: The defect in the Hunter syndrome: Deficiency of sulfoiduronate sulfatase. Proc. Natl. Acad. Sci. U.S.A., 70:2134, 1973.
118. Foley, K. M., Danes, B. S., and Bearn, A. G.: White blood cell cultures in genetic studies on the human mucopolysaccharidoses. Science, 164:424, 1969.
119. Danes, B. S., and Bearn, A. G.: Hurler's syndrome: A genetic study in cell culture. J. Exp. Med., 123:1, 1966.
120. Dorfman, A., and Matalon, R.: The Hurler and Hunter syndromes. Am. J. Med., 47:691, 1969.
121. Taysi, K., Kistenmacher, M. L., Punnett, H. H., and Mellman, W. J.: Limitations of metachromasia as a diagnostic aid in pediatrics. N. Engl. J. Med., 281:1108, 1969.
122. Pegum, J. S., and Ramsay, C. A.: X-linked epidermolysis bullosa (mendes da costa), poikiloderma, retarded growth. Proc. R. Soc. Med., 66:234, 1973.
123. Cantu, J.-M., Hernandez, A., Laracilla, J., et al.: A new X-linked recessive disorder with dwarfism, cerebral atrophy, and generalized keratosis follicularis. J. Pediatr., 84:564, 1974.
124. Furukawa, C. T., Hall, B. D., and Smith, D. W.: The Aarskog syndrome. J. Pediatr., 81:1117, 1972.
125. Netherton, E. W.: A unique case of trichorrhexis nodosa—"Bamboo hairs." Arch. Dermatol., 78:483, 1958.
126. Nowakowski, H., and Lenz, W.: Genetic aspects in male hypogonadism. Recent Progr. Horm. Res., 17:53, 1961.
127. Gorlin, R. J., and Sedano, H.: Dyskeratosis congenita with pigmentation, dystrophia unguium, and leukoplakia oris. Mod. Med., July 28, 1960, p. 160.
128. Kuokkanen, K.: Ichthyosis vulgaris. A clinical and histopathological study of patients and their close relatives in the autosomal dominant and sex-linked forms of the disease. Acta Derm. Venereol. [Suppl.] (Stockh.), 62:1, 1969.
129. Frost, P.: Ichthyosiform dermatoses. J. Invest. Dermatol., 60:541, 1973.
130. Opitz, J. M.: G syndrome. In Bergsma, D. (ed.): Birth Defects Atlas and Compendium. Baltimore, The Williams & Wilkins Co., 1973, p. 420.
131. Goldberg, M. F., and McKusick, V. A.: X-linked colobomatous microphthalmos and other congenital anomalies. A disorder resembling Lenz's dysmorphogenetic syndrome. Am. J. Ophthalmol., 71:1128, 1971.
132. Gall, J. C., Jr., Stern, A. M., Poznanski, A. K., Garn, S. M., Weinstein, E. D., and Hayward, J. R.: Oto-palato-digital syndrome comparison of clinical and radiographic manifestations in males and females. Am. J. Hum. Genet., 24:24, 1972.
133. Pashayan, H., Fraser, F. C., and Goldbloom, R. B.: A family showing hereditary nephropathy. Am. J. Hum. Genet., 23:555, 1971.
134. Sadeghi-Nejad, A., and Senior, B.: A familial syndrome of isolated "aplasia" of the anterior retinitary. J. Pediatr., 84:79, 1974.
135. Weinberg, A. G., and Stone, R. T.: Autosomal dominant inheritance in Albright's hereditary osteodystrophy. J. Pediatr., 79:996, 1971.
136. Mautalen, C. A., Dymling, J-F., and Harwith, M.: Pseudohypoparathyroidism 1942–1966. A negative progress report. Am. J. Med., 42:977, 1967.

137. Chase, L. R., Melson, G. L., and Aurbach, G. D.: Pseudohypoparathyroidism: Defective excretion of 3',5'-AMP in response to parathyroid hormone. J. Clin. Invest., 48:1832, 1969.

138. Drezner, M., Neelon, F. A., and Lebovitz, H. E.: Pseudohypoparathyroidism type II: A possible defect in the reception of the cyclic AMP signal. N. Engl. J. Med., 289:1056, 1973.

139. Bannerman, R. M., Ingall, G. B., and Mohn, J. F.: X-linked spondyloepiphyseal dysplasia tarda. Clinical and linkage data. J. Med. Genet., 8:291, 1971.

140. Lindenbaum, R. H., Andrews, P. S., and Khan, A. S.: Two cases of endocardial fibroelastosis—possible X-linked determination. Br. Heart J., 35:38, 1973.

141. Moller, J. H., Fisch, R. O., Fromm, A. L., and Edwards, J. E.: Endocardial fibroelastosis occurring in a mother and son. Pediatrics, 38:918, 1966.

142. Reich, T., Clayton, P. J., and Winokur, G.: The genetics of mania. Am. J. Psychiatry, 125:(Part 2, Suppl.),1358, 1969.

143. Falletta, J. M., Fernbach, D. J., Singer, D. B., et al.: A fatal X-linked recessive reticuloendothelial syndrome with hyperglobulinemia. J. Pediatr., 83:549, 1973.

144. Bell, N. H., Clark, C. M., Avery, S., Sinha, T., Trygstad, C. W., and Allen, D. O.: Demonstration of a defect in the formation of adenosine 3',5'-monophosphate in vasopressin-resistant diabetes insipidus. Pediatr. Res., 8:223, 1974.

145. Meyer, W. J., Keenan, B. S., Park, J. I., et al.: Dihydrotestosterone (DHT) binding activity of cultured skin fibroblasts from familial male pseudohermaphrodites with virilization at puberty. Pediatr. Res., 8:372, 1974.

146. Yount, J., Nichols, P., Ochs, H. D., et al.: Absence of erythrocyte adenosine deaminase associated with severe combined immunodeficiency. J. Pediatr., 84:173, 1974.

147. Steele, R. W., Limas, C., Thurman, G. B., et al.: Familial thymic aplasia: Attempted reconstitution with fetal thymus in a Millipore diffusion chamber. N. Engl. J. Med., 287:787, 1972.

148. Ochs, H. D., Slichter, S. J., Harker, L. A., et al.: The Wiskott-Aldrich syndrome: A genetic defect functionally expressed in marrow stem cell derivatives. Pediatr. Res., 8:416, 1974.

149. Kerr, C. B., Wells, R. S., and Cooper, K. E.: Gene effect in carriers of anhidrotic ectodermal dysplasia. J. Med. Genet., 3:169, 1966.

150. Matheson, D. W., and Howland, J. L.: Erythrocyte deformation in human muscular dystrophy. Science, 184:165, 1974.

151. Juberg, R. C., and Hellman, C. D.: A new familial form of convulsive disorder and mental retardation limited to females. J. Pediatr., 79:726, 1971.

152. Amman, A. J., Sutliff, W., and Millinchick, E.: Antibody-mediated immunodeficiency in short-limbed dwarfism. J. Pediatr., 84:200, 1974.

153. Morgan, J., and Cohen, L.: Crystalline perinuclear inclusions in myoblasts from human dystrophic muscle. N. Engl. J. Med., 290:863, 1974.

154. Goodman, G., Ripps, H., and Siegel, I. M.: Sex-linked ocular disorders: Trait expressivity in males and carrier females. Arch. Ophthalmol., 73:387, 1965.

155. Passarge, E., Post, B., and Schopt, E.: Possible genetic heterogeneity of X-linked ichthyosis. Birth Defects, 7:46, 1971.

156. Vibrans, V., and Altwein, J.: An X-linked recessive variety of ichthyosis different from the X-linked ichthyosis of Wells and Kerr. Clin. Genet., 1:304, 1970.

157. Borrone, C., Gatti, R., Trias, X., and Durand, P.: Fucosidosis: Clinical, biochemical, immunologic, and genetic studies in two new cases. J. Pediatr., 84:727, 1974.

158. Genetic Disorders: Prevention, treatment, rehabilitation. WHO Tech. Rep. Ser. No. 497, 1972.

159. Clow, C. L., Fraser, F. C., Laberge, C., and Scriver, C. R.: On the application of knowledge to the patient with genetic disease. In Steinberg, A. G., and Bearn, A. G. (eds.): Progress in Medical Genetics, IX. New York, Grune & Stratton, Inc., 1973.

160. Ethical and social issues in screening for genetic disease: A report of the Institute of Society, Ethics and the Life Sciences. N. Engl. J. Med., 286:1129, 1972.

161. Migeon, B. R., and Huijing, F.: Glycogen storage disease associated with phosphorylase kinase deficiency: Evidence for X inactivation. Am. J. Hum. Genet., 26:360, 1974.

162. Lowry, R. B.: Sex-linked cleft palate in a British Columbia Indian family. Pediatrics, 46:123, 1970.
163. Soltan, H. C., and Li, M. D.: Hereditary dextrocardia associated with other congenital heart defects: report of a pedigree. Clin. Genet, 5:51, 1974.
164. Purtillo, D. T., Cassel, C., and Yang, J. P. S.: Fatal infectious mononucleosis in familial lymphohistiocytosis. N. Engl. J. Med., 291:736, 1974.
165. Goldblum, R. M., Lord, R. A., Cooper, M. D., Gathings, W. E., and Goldman, A. S.: X-linked B lymphocyte deficiency. J. Pediatr., 85:188, 1974.
166. Moser, H., and Vogt, J.: Follow-up study of serum-creatinine-kinase in carriers of Duchenne muscular dystrophy. Lancet, 2:661, 1974.
167. Mendlewicz, J., and Rainier, J. D.: Morbidity risk and genetic transmission in manic-depressive illness. Am. J. Hum. Genet., 26:692, 1974.
168. Beck, E., Blaichman, S., Scriver, C. R., and Clow, C. L.: Advocacy and compliance in genetic screening. N. Engl. J. Med., 291:1166, 1974.
169. Ohno, S.: Ancient linkage groups and frozen accidents. Nature [New Biol.], 244:259, 1973.

CHAPTER 7

HETEROZYGOTE DETECTION IN THE LIPIDOSES

Edwin H. Kolodny, M.D.

Mental retardation commonly occurs in the course of many of the genetically determined diseases of metabolism. Early clinical recognition affects the course of only those few conditions which can be treated by special diets or vitamins. In the great majority of cases, little can be done to modify the natural history of the disease. Interest in the prevention of these diseases has, therefore, shifted to the unborn child.

Antenatal diagnosis has now been demonstrated for most of the inherited lipid storage diseases.[1,2] Ordinarily, two spouses do not become aware of their risk for producing affected offspring until the disease is recognized in one of their other children. However, it is theoretically possible for the biochemist to identify couples at risk for having children with any of these diseases without their first having an affected child. This possibility depends upon the availability of relatively simple, yet specific, methods for the detection of heterozygotes for these diseases.

CLINICAL MANIFESTATIONS OF THE LIPIDOSES

The principal clinical manifestations of the lipid storage diseases and their biochemical disturbances are summarized in Table 7–1. Each of these diseases is extremely disabling. Progressive nervous system degeneration occurs in almost all of these conditions. Disturbances in mental, motor, and visual system functioning are especially common. In many of these diseases, the visceral tissues are also involved. The majority of those afflicted are infants and young children. Patients with lipidoses that may become clinically manifest later in childhood, such as

Fabry's disease and the adult form of Gaucher's disease, tend to have less nervous system involvement, and their survival may be measured in years, rather than in months. Genetic transmission of each of the lipidoses is through an autosomal recessive mechanism — with the sole exception of Fabry's disease, an X-linked disorder.

Clinical subtypes have been described in almost all of the inherited lipidoses. These are usually differentiated by the age at which clinical symptoms appear and by the presence or absence of central nervous system and/or visceral involvement. Subtle differences in lysosomal enzyme activity may also serve to distinguish the various subtypes.

The Niemann-Pick group of diseases may be used to illustrate these concepts. Deficiency in sphingomyelinase activity occurs in Types A, B, and C of Niemann-Pick disease. However, the deficiency in Type A (infantile form) is nearly complete, whereas in Type B (chronic form), 15 per cent or more of the sphingomyelinase activity may remain. In Type C (juvenile form), total sphingomyelinase activity is normal, but the activity of the enzyme in the membrane-bound fraction is deficient.

Two other Niemann-Pick disease variants — Types D and E — have also been described. Neither is presented in Table 7–1. Accumulations of sphingomyelin have been demonstrated in visceral tissues from these patients, but their sphingomyelinase activity is apparently normal. Individuals with Type E are all of Nova Scotian ancestry, but they are otherwise clinically similar to Type C (juvenile form) patients. The designation Type E has been used to refer to those adults with a clinical presentation resembling a mild form of Type B (chronic form) Niemann-Pick disease.

The neuronal ceroid-lipofuscinoses (also referred to as Batten's disease or Batten-Spielmeyer-Vogt syndrome) are also characterized by a broad clinical heterogeneity, but qualify under a single diagnostic grouping because of close similarities in their chemical pathology. These conditions are probably less well known than the sphingolipidoses, but occur in the aggregate more commonly than many of the sphingolipidoses. Thus, they deserve special mention.

Seizures beginning between two and five years of age herald the onset of the Jansky-Bielschowsky form of the disease. Ocular signs of this disease include optic atrophy, granular macular degeneration, and eventually, blindness. Myoclonic hyperexcitability and pyramidal tract signs also develop. Death occurs before six years of age.

The Spielmeyer-Sjögren variant is the most common form of neuronal ceroid-lipofuscinosis. The first sign of this disease is usually a loss of central visual acuity at age five or six. Retinal pigmentary degeneration and intellectual regression are also observed early in the course of this disease. By 10 to 12 years of age, grand mal convulsions appear, and a few years later signs of an extrapyramidal movement disorder become obvious. Progression of the disease is slow, with death occurring most often around age 20.

TABLE 7-1. CLINICAL AND BIOCHEMICAL FEATURES

Disease	Subtype	Age at Onset	Age at Death	OCULAR
G_{M2}-gangliosidoses	Tay-Sachs	4–6 mo	2–5 yr	Cherry-red maculae, optic atrophy, blindness
	Sandhoff-Jatzkewitz	4–6 mo	2–5 yr	Same
	Late Infantile	1½–2 yr	2–6 yr	Same & strabismus
	Juvenile	4–9 yr	15 yr	Cherry-red maculae
G_{M1}-gangliosidosis	Infantile	Near birth	1½–2 yr	Cherry-red maculae, optic atrophy, blindness
	Late Infantile	7 mo–7½ yr	3–10 yr	
Gaucher's Disease	Infantile	Near birth	1–3 yr	
	Adult	Late childhood	20's–40's	Pingueculae
Niemann-Pick Disease	Type A (infantile)	Near birth	1–4 yr	Cherry-red maculae
	Type B (chronic form)	1–4 yr	Adulthood	
	Type C (juvenile form)	1–6 yr	5–15 yr	
Fabry's Disease		Childhood	Early to middle adulthood	Telangiectases, corneal opacities
Krabbe's Disease		3–6 mo	2–3 yr	Optic atrophy, blindness
Metachromatic Leukodystrophy	Late Infantile	12–18 mo	2–12 yr	Macula discoloration, optic atrophy, blindness
	Juvenile	3–10 yr	5–20 yr	Disturbance in vision
	Adult	>21 yr	20's–50's	
	Multiple Sulfatase Deficiency Variant	1–3 yr	?	
Farber's Disease		Near birth	2 yr	
Fucosidosis		1–2 yr	3–? yr	Retinal vascular abnormalities

OF THE INHERITED LIPIDOSES

Clinical Manifestations		Biochemistry	
NEUROLOGICAL	VISCERAL	STORED LIPID	ENZYME DEFICIENCY
Increased startle, psychomotor retardation, seizures		G_{M2}-ganglioside	Hexosaminidase A
Same	Foam cells	Same & globoside	Hexosaminidase A & B
Same & ataxia		G_{M2}-ganglioside	Hexosaminidase A
Ataxia, speech deterioration, seizures, dementia, pyramidal tr. signs		G_{M2}-ganglioside	Hexosaminidase A, partial
Psychomotor retardation, seizures	Hurler-like physical features, hepatosplenomegaly, foam cells	G_{M1}-ganglioside	β-galactosidase
Same		G_{M1}-ganglioside	β-galactosidase
Multiple brain stem signs	Hepatosplenomegaly Gaucher cells	Glucocerebroside	β-glucosidase
	Same & anemia, thrombocytopenia, fractures	Glucocerebroside	β-glucosidase
Psychomotor retardation	Hepatosplenomegaly, foam cells	Sphingomyelin	Sphingomyelinase
	Same	Sphingomyelin	Sphingomyelinase
Multiple signs	Same	Sphingomyelin	Membrane-bound sphingomyelinase
Extremity pain	Angiokeratomas, hypertension, vascular occlusions	Ceramide trihexoside	α-galactosidase
Psychomotor retardation, ↑ CSF protein		Galactocerebroside	Galactocerebroside β-galactosidase
Upper motor neuron, cerebellar, & peripheral nerve signs ↑ CSF protein	Metachromatic granules in urine	Sulfatide	Aryl sulfatase A
Mental changes	Same	Sulfatide	Aryl sulfatase A
Dementia, incontinence, seizures		Sulfatide	Aryl sulfatase A
Psychomotor deterioration	Hepatosplenomegaly, skeletal changes	Sulfatide	Aryl sulfatase A, B, C, & steroid sulfatase
Weak hoarse cry, motor & mental retardation	Tender swollen joints, lymphadenopathy, foam cells	Ceramide	Ceramidase
Mental & motor retardation	Hurler-like appearance, hepatomegaly, angiokeratoma, vacuolated lymphs	Fucosphingolipids	α-fucosidase

TABLE 7-1. CLINICAL AND BIOCHEMICAL FEATURES

Disease	Subtype	Age at Onset	Age at Death	OCULAR
Wolman's Disease		Near birth	2–14 mo	
Neuronal Ceroid-Lipofuscinoses	Jansky-Bielschowsky	2–5 yr	6 yr	Optic atrophy, granular macular degeneration
	Spielmeyer-Sjögren	4–8½ yr	15–25 yr	Blindness, retinitis pigmentosa
	Kufs	>20 yr	25–45 yr	

Dementia is the usual presenting sign in the adult form, known as Kufs' disease. Cerebellar or basal ganglia signs also develop. The course of this disease may extend over many years.

Fabry's disease is unique among the inherited lipid storage diseases in its mode of inheritance and the appearance of clinical signs in heterozygotes. The typical patient presenting with Fabry's disease is a young, intellectually normal adult male with whorl-like corneal opacities, clusters of purple punctate angiokeratomas in the body creases, and renal failure.

However, females heterozygous for this X-linked disorder may have one or more of the clinical signs occurring in the hemizygous male—including corneal opacities, angiokeratomas, and painful dysesthesias. Increased urinary excretion of ceramide trihexoside may also occur, with greater amounts of glycolipid appearing in the urine of those heterozygotes who are more clinically affected.[3] The explanation given for this phenotypic variability in the heterozygote is mosaicism for the X-chromosome (Lyon hypothesis) with two cell populations—one in which the normal gene is expressed, the other in which the mutant gene is expressed. A predominance of the latter type of cell would lead to greater clinical expression of the disease in the heterozygous female.

Several of the lipidoses occur much more often among the offspring of Jews of Ashkenazi (Eastern European) descent than in individuals of other ethnic groups. These diseases include the classical form of G_{M_2} gangliosidosis (Tay-Sachs disease), the adult variant of Gaucher's disease, and Type A Niemann-Pick disease. The majority of the thirteen reported cases of fucosidosis are of Italian descent. No ethnic predilection has been determined for the other lipidoses.

OF THE INHERITED LIPIDOSES

Clinical Manifestations		Biochemistry	
NEUROLOGICAL	VISCERAL	STORED LIPID	ENZYME DEFICIENCY
Poor feeding, reduced motor activity	Steatorrhea, hepatosplenomegaly, adrenal gland calcification	Triglycerides, cholesterol esters	Acid esterase
Intellectual loss, pyramidal trait signs, myoclonus		Ceroid-lipofuscin pigment	? Peroxidase
Seizures, pyramidal & extrapyramidal motor system signs		Ceroid-lipofuscin pigment	? Peroxidase
Dementia, cerebellar or extrapyramidal motor system signs		Ceroid-lipofuscin pigment	? Peroxidase

INCIDENCE OF THE LIPIDOSES

The lipidoses are rare diseases, but probably no rarer than most other inborn errors of metabolism. Very few actual data are available on their incidence. Published estimates are based, mainly, on two population groups — Ashkenazic Jews and Swedes.

Frequencies reported for adult Gaucher's disease,[4] Tay-Sachs disease,[5] and Type A Niemann-Pick disease[6] in Ashkenazic Jews are respectively 1 per 2500, 1 per 3600, and 1 per 40,000. Among other population groups, adult Gaucher's disease probably occurs less often than 1 per 40,000,[4] and Tay-Sachs disease, about 1 per 360,000.[5] Studies of the Swedish population, by Hagberg and his colleagues, place the birth incidence of metachromatic leukodystrophy at 1 per 40,000[7] and Krabbe's disease, at 1 per 53,000.[8] Also among Swedes, the juvenile variant (Spielmeyer-Sjögren) of neuronal ceroid-lipofuscinosis has been reported to occur as often as 1 per 25,000.[9] The infantile type (Jansky-Bielschowsky) of this disease probably occurs with the same frequency.[10]

In the author's experience, Fabry's disease occurs as often as any other form of lipidosis, but incidence figures, based upon complete ascertainment in a large population, have not been compiled. G_{M_1} gangliosidosis, Wolman's disease, Farber's disease, and fucosidosis are extremely rare occurrences.

Each year the Shriver Center assists in the diagnosis of approximately 10 new cases of lipidoses. Since there are approximately 180,000 live births each year in the six New England states (the Shriver Center's major area of referral), the incidence for all lipidoses in New England would be, roughly, 1 per 18,000. The ascertainment, however, is not complete; therefore, the real incidence of the inherited lipidoses may be more than twice this estimate.

CLINICAL GENETICS

In all of the inherited lipidoses except Fabry's disease, affected off-spring occur only when both parents are carriers of the same autosomal recessive trait. However, with each pregnancy, such couples have a 75 per cent chance of producing a clinically normal child. Two thirds of such children will, like their parents, also be carriers. In the case of Fabry's disease, an X-linked disorder, there is a 50 per cent chance that each son of a heterozygous mother will have the disease, and the same odds that each of her daughters will be a carrier.

BIOCHEMICAL PATHOLOGY

Nearly all of the inherited lipidoses are sphingolipidoses, i.e., the accumulating substance is a derivative of ceramide, the lipid formed by linkage of the long-chain amino alcohol sphingosine with a fatty acid in an amide bond.

$$
\underbrace{CH_3(CH_2)_{12}CH{=}CH{-}CH{-}CH{-}CH_2OH}_{\text{sphingosine}}
$$

$$
\begin{array}{cc}
 & \overset{|}{OH} \quad \overset{|}{NH} \\
 & \overset{|}{C{=}O} \\
 & \overset{|}{(CH_2)x} \\
 & \overset{|}{CH_3}
\end{array} \left.\vphantom{\begin{array}{c}a\\b\\c\\d\end{array}}\right\} \text{fatty acid}
$$

CERAMIDE

Higher homologues are formed by joining glucose, galactose, or phosphorylcholine to the primary alcohol of ceramide. Abbreviated structures for many of the sphingolipids are shown in Figure 7–1. In the various sphingolipidoses, the accumulating lipid is often found in blood, urine, and cultured skin fibroblasts, as well as in solid tissue.

The acid hydrolases that catalyze the stepwise degradation of the sphingolipids are also shown in Figure 7–1. Deficiency in the activity of any one of these lysosomal enzymes leads to the accumulation of the lipid normally degraded by that particular enzyme. Fucosidosis is not

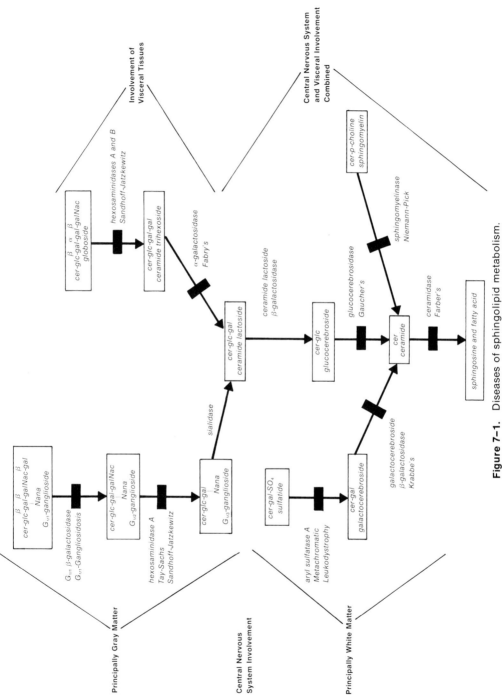

Figure 7–1. Diseases of sphingolipid metabolism.

shown in Figure 7–1, but it could also be considered a sphingolipidosis because fucose-containing sphingolipids, such as the H isoantigen, are among the natural products accumulating in this disease. The other two inherited lipidoses discussed in this chapter—Wolman's disease and the neuronal ceroid-lipofuscinoses—although not sphingolipidoses, are considered lysosomal storage diseases, for in each the accumulating lipid is preferentially stored in lysosomes.

TREATMENT

Once detected, none of these diseases can be cured. Treatment has also been largely unsuccessful, principally because replacement enzyme cannot reach the cells of the central nervous system, where the need for the missing enzyme is greatest. However, even if this difficulty were to be overcome, enzyme replacement therapy would probably have to be instituted early in fetal life. This belief is supported by studies of fetuses afflicted with these diseases that show that beginning very early in fetal life, there are already alterations in morphology and lipid content within the nervous system.[4-8] These limitations do not apply to lipidoses with a later clinical onset that do not affect the central nervous system. Attempts to treat two such diseases—Fabry's disease and Gaucher's disease—with organ transplants have been partly successful in a few cases.[17-20]

Since neither a cure nor effective treatment is available for these diseases, prevention is presently the method of choice for their control. Ideally, if all carriers of each of these disorders were known, then pregnancies at risk for each of these diseases would also be known, and the techniques currently in use for the prenatal diagnosis of many inborn errors of metabolism could be employed. Currently, ascertainment of high risk pregnancies is almost always through the prior birth of an affected sibling. The great majority of carriers for these traits are unaware of their genotype.

The remainder of this chapter is devoted to an appraisal of currently available methods for carrier detection in these diseases and a discussion of probable future trends in this field.

ENZYME ANALYSES IN DETECTION OF HETEROZYGOTES

Of the two biochemical indices useful in the diagnosis of a lipid storage disease, namely, deficiency in the activity of a lysosomal enzyme and storage of a specific lipid or group of structurally related lipids, it is the former which has proved of greatest value in the detection of heterozygotes for these diseases. According to the Gene-Dose Theory of biochemical genetics, the level of enzyme activity in the heterozygote state for each of the lipidoses should be intermediate between that of the homozygous affected individual and the normal. For most of the lipidoses this has proved to be true, in practice. However, reliable heterozygote detection through enzyme assay has not been reproducibly demonstrated for Niemann-Pick disease, Farber's disease, Wolman's disease, or the neuronal ceroid-lipofuscinoses.

Available Sources of Enzymes

Nearly all body tissues and fluids contain lysosomal enzymes. Therefore, a variety of specimens have been employed for demonstrating enzyme deficiency in the lysosomal storage diseases. These include such readily available tissues and fluids as serum, leukocytes, cultured skin fibroblasts, tears, and urine.

The steps employed in the preparation of a leukocyte pellet from whole blood[21, 22] are generally as follows:

1. Sedimentation of red blood cells in a viscous medium, such as dextran

2. Centrifugation of the supernatant to pellet the leukocytes

3. Resuspension of the leukocytes and distilled water lysis of the remaining erythrocytes

4. Centrifugation to repellet the leukocytes

5. Washing of the leukocytes with an isotonic saline solution

Before their actual use as an enzyme source, the leukocytes are disrupted by repeated freezing and thawing or by sonication with ultrasound. Either the entire homogenate is used, or the preparation is first clarified by removal of membranes through high-speed centrifugation prior to use. These procedures require several hours, thereby limiting the number of specimens that can be handled by a single laboratory worker. Nevertheless, their use is generally preferred to that of serum or plasma, because lysosomal enzyme activity is, in most instances, higher in leukocytes, and the results are less ambiguous. An exception is α-fucosidase, which is more active in serum than in leukocytes.[23]

Cultured skin fibroblasts are also used for the diagnosis of the lipidoses and confirmation of the heterozygous state. They are grown under carefully controlled conditions in artificial media to a confluent monolayer—a process which requires from several weeks to two or more months after the biopsy is first planted in tissue culture. The cells are then prepared for use as an enzyme source, by using methods similar to those described for leukocytes. One advantage of using cultured fibroblasts for lysosomal enzyme assays is that a single 3 to 4 mm skin biopsy can yield up to 50 generations of cells. Thus, many millions of cells can be made available for enzyme analyses. Another advantage is that cultured cells can be stored away in liquid nitrogen for future use. The major disadvantage of this approach is the investment in time, equipment, and supplies required for the growth and harvesting of each individual cell line. *In vitro* cell culture remains, however, an essential step in the prenatal diagnosis of the lipidoses.

The use of human tears has been described recently for heterozygote detection in Tay-Sachs disease[24, 25] and Fabry's disease.[26] The activities of the enzymes relevant to these two diseases, β-N-acetylhexosaminidase A and B and α-galactosidase, are higher in tears than in either leukocytes or cultured skin fibroblasts. Another advantage of tears is their ease of collection, transport, and storage through the use of filter paper strips.

The collection of urine is also relatively simple, but enzyme activity in this fluid is lower than in leukocytes, and, except in the case of Tay-Sachs disease,[27] attempts to employ this fluid for heterozygote detection in the lipidoses has not succeeded because of overlap between the values for heterozygotes and for normals.

Artificial Substrates

Specificity of the lysosomal hydrolase is greater for the glycone than for the aglycone portion of the molecule. This property has made possible the development of sensitive assays for these enzymes, with the use of artificial chromogenic and fluorogenic substrates that resemble the natural substrate with respect to the steric configuration of the terminal linkage and the nature of the terminal moiety. The most popular substrates are the derivatives of 4-methylumbelliferone. These conjugates are nonfluorescent, but the free 4-methylumbelliferone released by the hydrolytic action of the relevant hydrolase fluoresces in an alkaline medium. However, not all lysosomal enzymes can be assayed in this way. For example, 4-methylumbelliferyl esters are available but these cannot be used for assaying acid esterase activity for Wolman's disease because these esters are too unstable in the alkaline solution used for developing end-stage fluorescence.

Natural Substrates

Three of the lysosomal enzymes important in sphingolipid metabolism can still be assayed only with their natural substrates. These are ceramidase, for Farber's disease, galactocerebroside β-galactosidase, for Krabbe's disease, and sphingomyelinase, for Niemann-Pick disease. Since these substrates are not hydrolyzed as easily as artificial substrates, it has been the practice to label the substrate with radioactivity to increase the sensitivity of the assay and make it easier to quantitate the amount of reaction product. Unfortunately, these radioactively labeled compounds are not commercially available; therefore, assays for these four enzymes are presently being done only in laboratories which can prepare their own substrates.

Multiple Enzyme Components

Certain of the lysosomal enzymes exist in multiple forms that may differ from each other with respect to charge, neuraminic acid content, pH optima, and thermal stability. These lysosomal enzymes include the β-N-acetylhexosaminidases, β-galactosidases, α-galactosidases, and aryl sulfatases. In diseases associated with a deficiency in the activity of any one of these enzymes, not all enzyme components may be absent. Therefore, when several molecular forms of an enzyme exist, accurate diagnosis and carrier detection of a lysosomal storage disease cannot be accomplished without first carefully adjusting reaction conditions, so that only the activity of the component of interest is examined. This can be done either by physical separation of the components through electrophoresis, isoelectric focusing, ion exchange chromatography, or by inactivation of one or more of the components by treatment with heat or an acid pH environment.

LIPID ANALYSES IN DETECTION OF HETEROZYGOTES

Although lysosomal enzyme activity may be partially deficient in heterozygotes for most of the inherited lipidoses, lipid storage cannot be detected in these individuals. Only in Fabry's disease has increased lipid—in this case, urinary ceramide trihexoside—been demonstrated in the carrier.[3] However, this probably represents a special situation, since the heterozygotes for this X-linked disorder with demonstrable urinary excretion of lipid also have clinical signs of the disease, possibly due to unevenness in the process of X-chromosome inactivation. Therefore, in a predominance of cells, the mutant X-chromosome is retained and allowed to express itself.

Obligate heterozygotes for some of these diseases, however, have manifested certain nonspecific findings. Peripheral lymphocytes from some of the parents of children with Wolman's disease are vacuolated. In the neuronal ceroid-lipofuscinoses, hypergranulation occurs in the circulating polymorphonuclear leukocytes of some of the patients and also in certain of their parents.

PRESENT PROBLEMS OF LARGE-SCALE HETEROZYGOTE DETECTION

Successful large-scale heterozygote detection in the lipidoses depends upon the resolution of several important problems:

Overlap in the Range of Values for Enzyme Activity in Heterozygotes with the Ranges for Normal Controls and Affected Homozygotes. Tests which measure enzyme activity cannot be used for the identification of heterozygotes unless they clearly discriminate heterozygotes from normal controls and patients with the disease. The wider the gap between the ranges for these three genotypes, the more certain one can be of the assigned classification. Ideally, the lowest values for noncarriers should be greater than three standard deviations beyond the mean for obligate heterozygotes.

Overlap in the range of enzyme activity occurs more often between the normal genotype and the carrier group than between carriers and affected homozygotes. Therefore, to improve discrimination of the heterozygote state in each of the lipidoses, further study of the overlap problem is required. Reaction conditions in each laboratory must be carefully investigated so that enzyme activity is optimized.

If more than one component contributes to the enzyme activity in question, refinement in heterozygote detection may require the use of methods in which each of these components can be assayed individually. For example, the enzyme activity might be fractioned by gel electrophoresis, developed with a chromogenic or fluorogenic substrate, and the amount of each component thus separated and exposed can then be quantitated in a scanning device set at the proper wavelengths. Enzyme components with a differential ability to bind to certain ion-exchange resins could be separated by fractional elution with discontinuous salt gradients, followed by assay of the individual fractions. The differential susceptibility of enzyme components to thermal denaturation would be employed for comparisons of total enzyme activity with the activity of the heat-stable components. Each of these methods, as well as differential inactivation by an acid pH environment, has been suggested, or is presently in use, for the large-scale screening of Tay-Sachs heterozygotes.[2]

Many laboratories test the validity of their lysosomal enzyme assays by comparing the values obtained for the enzyme in question with the activity of a second enzyme whose catalytic ability should not be altered. This practice has often allowed the unequivocal classification of heterozygosity where use of the absolute values for the single enzyme in question led to overlap in genotype discrimination. Although the activity of the second enzyme is considered independent of the disease state, this situation may not actually be true. It is not uncommon in the lysosomal storage diseases to observe increases in the activity of certain lysosomal enzymes to compensate for the loss in activity of a particular enzyme. The principle that two tests discriminate between two genotypes more effectively than either test alone has also been demonstrated mathematically.[34]

Factors such as the age of the subject, the presence of chronic illness or pregnancy, and the use of certain drugs may also contribute to the problem of overlap between different genotypes in the lipidoses. Leukocyte acid esterase, for example, is low in infancy and rises slowly through childhood.[35] On the other hand, serum hexosaminidase is high in the first months of life, but reaches adult levels before one year of age.[28] Large-scale screening for Tay-Sachs heterozygotes has revealed alterations in hexosaminidase activity resulting from certain diseases (e.g., diabetes), from drugs (e.g., oral contraceptives), and in pregnancy. The effects of these and other factors on other lysosomal enzymes are unknown, because far fewer subjects have been studied with respect to these other enzymes. Many laboratories attempt to correct for such variables by simultaneously assaying specimens from age-matched controls in similar states of health and on the same medications.

Heterogeneity Among the Lipidoses. Another problem in heterozygote detection of the lipidoses results from clinical heterogeneity among certain of these diseases. At the present time, it is not possible to distinguish carriers of infantile forms of Gaucher's disease, Niemann-Pick disease, or metachromatic leukodystrophy from carriers of the adult forms of these diseases, using *in vitro* assays of enzyme activity. It is not known whether a reduced level of aryl sulfatase A, for example, in an individual not related to a patient with this disease, is indicative of a carrier of the infantile form, or of the adult form of the disease. Additional research efforts directed toward defining optimum conditions of assay and the nature of the molecular defect in each of these diseases may eventually yield tools which will reliably differentiate these genotypes.

Additional Technical Problems. Additional problems must be circumvented if large-scale testing for heterozygotes is ever to become practical. At present, the most conclusive results in carrier detection are obtained with leukocytes and cultured skin fibroblasts. As discussed earlier in this chapter, the preparation of leukocytes is a time-consuming and somewhat tedious process that does not easily lend itself to the management of greatly increased numbers of specimens without scaling up all production factors—including personnel. However, hematologists are especially interested in this problem and are striving for simplification in the preparation of viable leukocytes.

The use of cultured skin fibroblasts requires an even greater investment in time and materials. However, the recent development of a simple and sensitive microfluorimetric technique to assay lysosomal enzyme activity in single cells may circumvent the need for culturing cells.[36] The cells are deposited in microdroplets under oil, ruptured by freezing and thawing, and then allowed to react with a fluorogenic substrate. The fluorescence is developed with an organic base and measured in a fluorescence microscope equipped with a photomultiplier. A theoretical disadvantage of this method, however, is that overlap of the heterozygote range with the range for normals and affected homozygotes is more likely to occur than with assays employing large numbers of cultured cells. This situation arises because of intercell skewing in enzyme activity, which ordinarily is canceled out when a large number of cells is assayed.

Economic Feasibility and Ethics. These two subjects are interrelated. Present methods of lysosomal enzyme assay are expensive; therefore, heterozygote detection in the lipidoses has been confined to high-risk situations — most often the immediate family members of an affected individual. Costs in our laboratory range from $5 for a serum hexosaminidase assay to $200 for a single leukocyte galactocerebroside β-galactosidase assay. Assuming no improvements in testing efficiency, one would find that current resources do not permit wide-scale testing of the general population for heterozygotes of the lipidoses.

Rapid technological progress is now being made in this field. Automation, in particular, offers considerable promise for reducing the cost per test. Therefore, some system of priorities for testing must be established so that the widest use can be made of the available resources.

Here is where matters of ethics must be carefully considered. If the principal purpose of heterozygote detection is to spot those couples at risk for having affected offspring, then one must ask whether society would wish to prevent the birth of all fetuses with lipidoses, regardless of their clinical symptomatology. In particular, should society regard as undesirable the birth of a child with Fabry's disease or the adult form of Gaucher's disease? Both diseases are compatible with useful life, and neither is associated with significant neurological impairment.

Age of the candidates for testing must also be considered. If prevention of lipidoses with mental retardation is the goal, it is almost senseless to test those outside the child-bearing age range. A testing program that concentrates its efforts on couples of reproductive age would cost far less than one that includes all ages; yet, such a program would be just as effective in reducing the incidence of the lipidoses.

A higher proportion of carriers are found among close relatives of patients with lipid storage diseases than in any other single group. For this reason, it is currently our practice to test all close relatives (siblings, uncles, aunts, nieces, nephews, and first cousins) between the ages of 18 and 40, of our patients.

Next in the hierarchy of preferences (after close relatives of patients) would be all those who are married to blood relatives. Since the lipidoses are rare diseases, they are more apt to appear among offspring of a consanguineous marriage. Among marriages between first cousins, there is a 12.5 per cent chance that both spouses have the same autosomal gene trait.

Testing preference should probably be given to couples with a history of infertility or previous family history of genetic defects or mental retardation before other couples of childbearing age are tested. These couples have more at stake with each pregnancy. The infertile couple, that after years of hopeful anticipation gives birth to an infant with a lipidosis, is a particularly poignant tragedy. Similarly, there are cases of several families with two very defective children — one afflicted with a severe neurological impairment of a stable or progressive degenerative type, the other with a lipidosis that proved to be fatal in infancy or childhood. Both situations deserve the attention of the genetic counselor and an offer of heterozygote detection for lipid storage diseases.

The ethnic predilection for certain of the lipidoses must also be considered in the priority scale. The yield of fucosidosis carriers will be higher among Italians than among the general population. Adult Gaucher's disease and infantile type Niemann-Pick disease are, like Tay-Sachs disease, more common in Jews of Ashkenazic descent. Therefore, a higher proportion of carriers of these diseases would be obtained by selective testing of these ethnic groups than by testing the general population.

Finally, if the resources were available, and the technology sufficiently advanced so that reliable determinations could be obtained on a mass scale by using readily available specimens, then all others might be offered testing. Presumably, this would include, first, all other couples in their reproductive years, then all others over age 18, including the curious, who are asserting their right to knowledge of their genotype.

OPTIMUM PRECONDITIONS FOR SCREENING

Thus far, Tay-Sachs disease is the only lipidosis in which large-scale heterozygote screening has been attempted. The factors favoring this development were:

1. The existence of a defined high-risk subgroup in the population
2. The existence of a relatively simple and reliable test for the heterozygote
3. The possibility for a carrier-carrier couple to have normal children through amniocentesis and prenatal diagnosis

A Defined High-Risk Subgroup. A few of the lipidoses, like Tay-Sachs disease, preferentially affect one ethnic group. Testing a relatively small proportion of the general population will, therefore, uncover a major proportion of all carriers for these particular diseases. Most lipidoses, however, do not show an ethnic predilection. Therefore, successful heterozygote screening for these diseases will involve a much larger population base. If one considers also that most of these other lipidoses are less common than Tay-Sachs disease, one can appreciate that with present methodology the expense involved in detecting a single carrier for a lipidosis other than Tay-Sachs disease through mass screening will exceed the cost currently incurred for detection of a Tay-Sachs carrier. Mass screenings for heterozygotes of these other lipidoses will become practical when completely reliable results can be obtained from readily available specimens, and when the test procedures are sufficiently automated to greatly reduce the cost—perhaps to an amount below $1 per test. For the present, the best allocation of available resources would be to conduct heterozygote testing according to priority groups, such as those discussed in the preceding sections. The highest risk category would be individuals in the child-bearing age group who are close relatives of carriers and those who are partners in consanguineous marriages.

A Simple and Reliable Test for the Heterozygote. The principal sources of patient enzyme for heterozygote detection of the lysosomal storage diseases in the future will undoubtedly be serum, tears, and leukocytes — in that order. The filter paper collection of tears will greatly simplify specimen collection. This could be prepared by the test subject at home, placed into a vial containing buffer, and mailed to the testing laboratory. A breakthrough in the methodology for leukocyte isolation from peripheral blood is urgently needed. The present technique, involving resuspension of the leukocyte pellet at two separate points in the procedure, might be modified so that the second resuspension step, following the distilled water lysis of contaminating red blood cells, is omitted.

The problem of overlap in enzyme activities between heterozygotes, normals, and homozygous affected individuals can probably best be managed by making comparisons with the activity of reference enzymes. Variations of this method might include the following:

    a. simultaneous comparison of the activities of two enzymes in each specimen, i.e., the disease-related enzyme and a reference enzyme

    b. simultaneous comparison of the activities of the two forms of the same lysosomal enzyme, for diseases in which only one molecular form of an enzyme is deficient

    c. simultaneous comparison of the activity of the same enzyme in two separate specimens — a control and the specimen in question.

These kinds of comparison can be made by using a dual channel automated apparatus of the type used in several of the Tay-Sachs screening programs for testing serum hexosaminidase activity.[37] The enzyme activity in tears and leukocytes, as well as serum, can be determined with this apparatus.

The final stage of our automated apparatus for lysosomal enzyme determinations consists of a single instrument package containing peak integrator, calculator, and printout capabilities (Fig. 7–2). By sequencing the flow in the two channels so that the peaks of enzyme activity appear alternately, but not simultaneously, in the two detectors (fluorimeters or spectrophotometers), and by alternating the input from the detectors into the integrator, a ratio of the enzyme activity present in the two channels may be obtained. Depending upon the specimens employed and the enzyme activities being measured, the ratio might represent any of the three comparisons referred to above.

The arrangement of instrument components shown in Figure 7–2 has considerable potential for large-scale heterozygote screening for a number of lipidoses. If each well of the timer-activated sampler is filled with serum, tears, or leukocyte enzyme from a different individual, many specimens may be assayed for the carrier state of one disease. Alternatively, if each of the sampler wells contains a different substrate-buffer mixture, then the activities of many different enzymes can be determined in the same specimen. Both sampling techniques have been employed and it was found that the use of the fluorogenic 4-methylumbelliferyl derivatives as substrates and short incubation times allows for the reliable performance of several hundred assays each day, starting with limited amounts of serum, diluted tears, or leukocytes. Enzyme assays for which no fluorogenic substrate is available can be performed using the chromogenic p-nitrophenol derivatives. The apparatus, however, is of no value for the determination of enzyme activities which require the use of radioactive substrates.

Prenatal Diagnosis.   Prenatal diagnosis has been accomplished for almost all of the lipidoses. The current procedure of amniocentesis, amniotic fluid cell culture, and enzyme analysis should also be applicable to those diseases which have not yet been detected *in utero:* Farber's disease, fucosidosis, Wolman's disease, and the neuronal ceroid-lipofuscinoses.

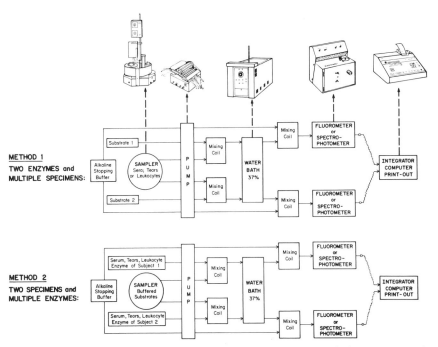

Figure 7–2.   Two methods for determining lysosomal enzyme activity in an automated system.

Prerequisites for Responsible Screening Programs

To be responsible and to gain community acceptance mass screening programs for heterozygote detection in the lipidoses must incorporate within their operational framework the many principles of genetic counseling discussed in Chapter 3. The most important of these are the provisions for:

1. Strict voluntarism
2. Inclusion of an educational program which continuously informs the target population, without creating undue anxiety
3. Preservation of confidentiality
4. Genetic counseling and follow-up testing for relatives of heterozygotes

SUMMARY AND CONCLUSIONS

Patients with inherited lipidoses segregate clinically into two distinct groups: infants and young children with severe, progressive, neurological degeneration, and older individuals without serious nervous system involvement. The majority of patients belong to the first group and die within a few years after their disease is first recognized. Neither an effective treatment nor a cure is presently available for any of these diseases. The social and psychologic effects of these diseases and their drain on economic resources are immense.

Each lipidosis results from the deficiency of a lysosomal acid hydrolase that causes accumulation of a disease-specific lipid or group of lipids. Since each of the lipidoses, with the exception of Fabry's disease, is inherited according to an autosomal recessive pattern, the level of enzyme activity in the heterozygote should be reduced to approximately one-half normal values. This supposition provides the basis for the most commonly used tests in carrier detection for the lipidoses. In the case of Fabry's disease heterozygotes, modification of this principle is required because of the possibility of nonrandom X-chromosome inactivation.

The most frequently employed types of specimens for heterozygote detection in the lipidoses are serum, leukocytes, and cultured skin fibroblasts. Although the enzyme-specific activity is usually higher in leukocytes and cultured skin fibroblasts, the relatively time-consuming procedures involved in their preparation are a deterrent to their use in mass screening for heterozygotes. Because of the high activity of β-N-acetylhexosaminidase and α-galactosidase recently found in tears and the ease with which this fluid can be collected, it is likely that this will become an increasingly popular source of enzyme in heterozygote detection.

The use of artificial chromogenic and fluorigenic substrates has greatly simplified quantitation of lysosomal enzyme activities and has facilitated the development of automated instrumentation for these analyses—a necessary prerequisite for large-scale heterozygote screening. However, assay of lysosomal enzymes relevant to three of the sphingolipidoses still requires the use of their natural substrates.

A number of major problems must be solved before mass screening for heterozygotes can be instituted. In the case of several of the lysosomal enzymes, values of enzyme activity in heterozygotes overlap with values in normals and sometimes with values in affected homozygotes. This problem will probably be overcome by the use of methods that separate the component of interest from other molecular forms of the enzyme and by the use of comparisons between the activity of two enzymes with different actions.

Another problem is that enzyme assays currently in use do not distinguish between carriers of the infantile and adult forms of certain of the lipidoses. Since the different forms of these diseases have very different prognostic implications, responsible counseling requires that assay techniques be developed that will distinguish, unequivocally, the carriers of these different forms.

To economically screen for heterozygotes of the lipidoses, the handling of specimens must be simplified so that as much of the assay as possible can be automated. This simplification would presumably include a one-step procedure for whole blood fractionation to isolate leukocytes, and the use of micromethods for enzyme assay in single cells.

Currently, it is not economically feasible to screen the entire population, nor is it desirable. The major testing effort should be directed toward couples of reproductive age, with preference being given in this group to the relatives of known carriers, partners to consanguineous marriages, couples with a history of infertility or previous family history of genetic defects or mental retardation, and those of ethnic groups with a known predilection for one or another of the lipidoses.

A responsible heterozygote detection program for the lipidoses must be strictly voluntary and must include provisions for lay and professional education, confidentiality, genetic counseling, and follow-up testing for the relatives of heterozygotes. Widespread public acceptance of genetic screening programs can be expected in the next decade. Developments in laboratory methodology will undoubtedly keep pace. A significant reduction in new cases of the lipidoses is, therefore, within the reach of our own generation.

ACKNOWLEDGMENTS: Supported by USPHS Grants HD 05515 and HD 04147, and by the Tay-Sachs Foundation of New England, Inc.

References

1. Milunsky, A.: The Prenatal Diagnosis of Hereditary Disorders. Springfield, Illinois, Charles C Thomas, Publisher, 1973.
2. Nadler, H. L.: Prenatal detection of genetic disorders. *In* Harris, H., and Hirschhorn, K. (eds.): Advances in Human Genetics. Vol. 3. New York, Plenum Publishing Corp., 1972, p. 1.
3. Goto, I., Tabira, T., Nawa, A., Kurokawa, T., and Kuroiwa, Y.: Biochemical and genetic studies in two families with Fabry's disease. Arch. Neurol., 31:45, 1974.
4. Fried, K.: Population study of chronic Gaucher's disease. Isr. J. Med. Sci., 9:1396, 1973.
5. Aronson, S. M.: Epidemiology. *In* Volk, B. W. (ed.): Tay-Sachs Disease. New York, Grune & Stratton, Inc., 1964, p. 118.
6. Aronson, S. M., and Myrianthopoulos, M.: Epidemiology and genetics of the sphingolipidoses. *In* Vinken, P. J., and Bruyn, G. W. (eds.): Handbook of Clinical Neurology. Vol. 10. Amsterdam, North-Holland Publishing Company, 1970, p. 556.
7. Gustavson, K.-H., and Hagberg, B.: The incidence and genetics of metachromatic leukodystrophy in Northern Sweden. Acta Paediatr. Scand., 60:585, 1971.
8. Hagberg, B., Kollberg, H., Sourander, P., Åkesson, H. O.: Infantile globoid cell leukodystrophy (Krabbe's Disease). Neuropaediatrie, 1:74, 1969.
9. Sjögren, T.: Die juvenile amaurotische idiotie. Heredity (London), 14:197, 1931.
10. Zeman, W., Donahue, S., Hyken, P., and Green, J.: The neuronal ceroid-lipofuscinoses (Batten-Vogt syndrome). *In* Vinken, P. I., and Bruyn, G. W. (eds.): Handbook of Clinical Neurology. Vol. 10. Amsterdam, North-Holland Publishing Company, 1970, p. 588.
11. Statistical Abstract of the U.S.: U.S. National Center for Health Statistics, 1973.
12. Adachi, M., Torii, J., Schneck, L., and Volk, B. W.: The fine structure of fetal Tay-Sachs disease. Arch. Pathol., 91:48, 1971.
13. O'Brien, J. S., Okada, S., Fillerup, D. L., Veath, M. L., Adornato, B., Brenner, P. H., and Leroy, J. G.: Tay-Sachs disease: Prenatal diagnosis. Science, 172:61, 1971.
14. Lowden, J. A., Cutz, E., Conen, P. E., Rudd, N., and Doren, T. A.: Prenatal diagnosis of G$_{MI}$-gangliosidosis. N. Engl. J. Med., 288:225, 1973.
15. Schneider, E. L., Ellis, W. G., Brady, R. O., McCulloch, J. R., and Epstein, C. J.: Prenatal Niemann-Pick disease: Biochemical and histologic examination of a 19-gestational week fetus. Pediatr. Res., 6:720, 1972.
16. Ellis, W. G., Schneider, E. L., McCulloch, J. R., Suzuki, K., and Epstein, C. J.: Fetal globoid cell leukodystrophy (Krabbe's disease). Arch. Neurol., 29:253, 1973.
17. Groth, C. G., Blomstrand, R., Dreberg, S., Hagenfeldt, L., Lofstrom, B., Ockerman, P.-A., Samuelsson, K., and Svennerholm, L.: Splenic transplantation in Gaucher's disease. *In* Desnick, R. J., Bernlohr, R. W., and Krivit, W. (eds.): Enzyme Therapy in Genetic Disease. National Foundation Birth Defects: Orig. Art. Ser. Baltimore, Williams & Wilkins, 1973.
18. Philippart, M., Franklin, S. S., and Gorden, S.: Reversal of an inborn sphingolipidosis (Fabry's disease) by kidney transplantation. Ann. Intern. Med., 77:195, 1972.
19. Desnick, R. J., Simmons, R. L., and Allen, K. Y.: Correction of enzymatic deficiencies by renal transplantation: Fabry's disease. Surgery, 72:203, 1972.
20. Clarke, J. T. R., Guttmann, R. D., Wolfe, L. S., Beaudoin, J. G., and Morehouse, D. D.: Enzyme replacement therapy by renal allotransplantation in Fabry's disease. N. Engl. J. Med., 287:1215, 1972.
21. Skoog, W. A., and Beck, W. S.: Studies on the fibrinogen, dextran, and phytohemagglutinin methods of isolating leukocytes. Blood, 11:436, 1956.
22. Kampine, J. P., Brady, R. O., Kanfer, J. N., Feld, M., and Shapiro, D.: Diagnosis of Gaucher's disease and Niemann-Pick disease with small samples of venous blood. Science, 155:86, 1967.
23. Borrone, C., Gatti, R., Trias, X., and Durand, P.: Fucosidosis: Clinical, biochemical, immunologic, and genetic studies in two new cases. J. Pediatr., 84:727, 1974.
24. Singer, J. D., Cotlier, E., and Krimmer, R.: Hexosaminidase A in tears and saliva for rapid identification of Tay-Sachs disease and its carriers. Lancet, 2:1116, 1973.

25. Carmody, P. J., Rattazzi, M. C., and Davidson, R. G.: Tay-Sachs disease — The use of tears for the detection of heterozygotes. N. Engl. J. Med., 289:1072, 1973.

26. Del Monte, M. A., Johnson, D. C., Cotlier, E., Krivit, W., and Desnick, R. J.: Diagnosis of Fabry's disease by tear α-galactosidase A. N. Engl. J. Med., 290:57, 1974.

27. Navon, R., and Padeh, B.: Urinary test for identification of Tay-Sachs genotypes. J. Pediatr., 80:1026, 1972.

28. O'Brien, J. S., Okada, S., Chen, A., and Fillerup D. L.: Tay-Sachs disease. Detection of heterozygotes and homozygotes by serum hexosaminidase assay. N. Engl. J. Med., 283:15, 1970.

29. Friedland, J. Schneck, L., Saifer, A., Pourfar, M., and Volk, B. W.: Identification of Tay-Sachs disease carriers by acrylamide gel electrophoresis. Clin. Chim. Acta, 28:397, 1970.

30. Dance, N., Price, R. G., and Robinson, D.: Differential assay of human hexosaminidases A and B. Biochim. Biophys. Acta, 222:662, 1970.

31. Saifer, A., and Rosenthal, A. L.: Rapid test for the detection of Tay-Sachs disease heterozygotes and homozygotes by serum hexosaminidase assay. Clin. Chim. Acta, 43:417, 1973.

32. Matsuda, I., Arashima, S., Anakura, M., and Oka, T.: α-L-fucosidase and α-D-mannosidase activity in the white blood cells in the disease and carrier states of fucosidosis. Clin. Chim. Acta, 48:9, 1973.

33. Masson, P. K., and Lundblad, A.: Mannosidosis: Detection of the disease and of heterozygotes using serum and leukocytes. Biochim. Biophys. Res. Commun., 56:296, 1974.

34. Gold, R. J. M., Maag, U. R., Neal, J. L., and Scriver, C. R.: The use of biochemical data in screening for mutant alleles and in genetic counselling. Ann. Hum. Genet., 37:315, 1974.

35. Patrick, A. D., and Lake, B. D.: Wolman's disease. In Hers, H. G., and Van Hoof, F. (eds.): Lysosomes and Storage Diseases. New York, Academic Press, Inc., 1973, p. 466.

36. Wudl, L., and Paigen, K.: Enzyme measurements on single cells. Science, 184:992, 1974.

37. Lowden, J. A., Skomorowski, M. A., Henderson, F., and Kaback, M.: Automated assay of hexosaminidases in serum. Clin. Chem., 19:1345, 1973.

CHAPTER 8

SCREENING FOR THE HEMOGLOBINOPATHIES

Richard W. Erbe, M.D.*

The magnitude of the public health problems posed by the sickling disorders and the general previous neglect of these problems have been widely recognized in the United States since about 1970. Sickle cell anemia was emphasized as a national health problem in the President's Health Message to Congress in 1971. Since the passage in May 1972 of the National Sickle Cell Anemia Control Act, substantial amounts of federal funds have become available for sickle cell anemia programs[1] and have been used in part to establish large-scale hemoglobinopathy screening and counseling programs. In addition, many states hurriedly enacted laws, often mandating hemoglobinopathy screening, and hospitals, private groups, and others in numerous instances began their own screening programs. It seems likely that more people by far have been screened for sickle cell anemia and the related hemoglobinopathies than for any other simply-inherited disorders. Much attention has recently been focused on such screening programs as the result of the heightened professional and public interest in the hemoglobinopathies, the involvement of law-making bodies in aspects of screening, and the largely belated recognition of the many ethical, legal, and social issues involved in such screening. Unfortunately, to date there is still a shortage of data with which to test and evaluate the various alternative approaches that have been proposed. The purpose of this chapter is to review recent developments in screening and the current status of certain aspects of counseling that seem particularly relevant to the design of optimal hemoglobinopathy screening and genetic counseling programs.

HISTORICAL PERSPECTIVE

Since first described by Herrick in 1910,[2] a number of the advances in the understanding of sickle cell anemia have constituted milestones in

*Supported, in part, by NIH Grant HL 15157 to the Boston Sickle Cell Center.

human genetics. Previous suggestions that the disease was dominantly inherited were corrected in 1949 when Neel[3] and Beet[4] independently observed that it was carrier parents with sickle cell trait who gave birth to children with sickle cell anemia, thus indicating autosomal recessive inheritance of the disease even though the sickling phenomenon is expressed both in those with sickle cell anemia and in those with sickle cell trait. Pauling and associates[5] in the same year localized the defect in sickle cell anemia to the hemoglobin molecule itself, observing that hemoglobin S under appropriate conditions migrated more rapidly during electrophoretic analysis. In 1957 Ingram[6] identified the precise nature of the molecular lesion as consisting of the substitution of a glutamic acid for the normal valine in the sixth position of the globin beta chain, thus making sickle cell anemia the first human genetic disorder to be characterized at the molecular level. Subsequently, similar techniques of peptide analysis and protein sequencing have led to the identification of over 100 mutant hemoglobins, a minority of which are associated with serious disease.[7]

In contrast, the social and legal developments relevant to hemoglobinopathy screening have been more complex. In 1969 a survey in Richmond, Virginia, showed that only three out of ten adult Negroes had heard of sickle cell anemia.[8] Some 650 children in the Boston Head Start Program were screened by sickle preparations and 7.3 per cent were positive for hemoglobin S. Interviews with the parents of these children[9] revealed that about half had heard of sickle cell anemia, but of this group none were aware that the condition is transmitted from parent to child. About two thirds understood the difference between an inherited and a contagious or infectious disease. Yet nearly all of the parents agreed to be tested for the sickle trait, and one third requested a more detailed examination.[9] In 1970 Scott[10, 11] made a compelling case for increased attention to sickle cell anemia. He pointed out that, based on estimated incidence figures, approximately equal numbers of children were born annually with sickle cell anemia and cystic fibrosis, with far fewer having muscular dystrophy and phenylketonuria. Despite this evidence the financial support provided by volunteer organizations and the federal government was allocated overwhelmingly in favor of the less common genetic diseases. In July 1971, Massachusetts became the first state to pass a law requiring screening for sickle cell trait and sickle cell anemia (Chapter 491, Acts of 1971). This law mandated screening by unspecified methods prior to school attendance and failed to provide for genetic counseling or financial support among a number of other deficiencies. Based on what was perceived as great public interest in such legislation, a total of 17 states shortly thereafter passed similar laws. Yet, by mid-1973 eight states, including Massachusetts, had recognized the numerous problems in these laws and had repealed them outright. In May 1972, the National Sickle Cell Anemia Control Act (Public Law 92–

294) authorized over the subsequent three years the appropriation of $120 million, of which $85 million was ".....for the establishment and operation of voluntary sickle cell anemia screening and counseling programs, primarily through other existing health programs." Under this law, ten comprehensive sickle cell centers were established throughout the country with the responsibility for service (including screening, counseling, education, and training), and for basic research, hopefully increasing the clinical applications of fundamental research.

With the rapid proliferation of often hastily designed and incomplete screening programs, and with the appearance of anecdotal reports of serious abnormalities or even death in persons with sickle cell trait, problems and detrimental effects of screening and counseling were anticipated and actually observed. By 1971, the confusion of both lay and professional personnel about sickle cell trait and anemia, the anxiety of screenees and the inappropriate medical action that followed, problems with employers, and the implications of mandatory screening programs aimed at children prompted the beginning of a strong and vocal reaction against such screening programs as they were then evolving.[12] Certain ethical and social aspects, such as community participation in design, measures to insure the adequacy of testing procedures, provision of adequate informed consent, and protection of screenees and their privacy were not being met by many programs.[13] Undue alarm and incorrect impressions were being generated through misrepresentation of sickle cell trait as an undue health hazard, distorted depictions of sickle cell anemia, emphasis on the testing of young children who could neither adequately consent nor understand what the screening was about, and, in some instances, provision of highly directive counseling that conveyed the impression that abstinence from childbearing or selection of an alternative mate were the only reproductive courses justified when both members of a couple had sickle cell trait.[14, 15] In a study of 47 families from a total of 67 identified as having one or more children with sickle cell trait, a structured interview of the parents counseled previously indicated considerable confusion regarding the health implications of sickle cell trait.[16] Many parents were under the impression that it was necessary to limit the activity of persons with sickle cell trait and even to provide therapy.

Subsequently, major educational efforts have been launched by the National Sickle Cell Disease Program, and other educational programs have been undertaken by the various comprehensive centers as well as by numerous groups and individuals. Although recently there has generally been less discussion of the problems caused by misinformation and erroneous attitudes, it is difficult to know whether this is the result of modifications in the programs involved or simply the loss of interest. The social and ethical problems mentioned here are considered again in the section on program design.

FREQUENCIES OF THE HEMOGLOBINOPATHIES

Accurate measures of the frequency of the hemoglobinopathies in unselected populations at different ages are still unavailable, but a number of attempts have been made to estimate incidence and prevalence. Scott[10] suggested that sickle cell anemia occurs in about 1 in 500 Negro births and that the median survival of such individuals in the United States is only about 20 years. He estimated that some 1200 individuals with sickle cell anemia were born in the year 1969, suggesting that as many as 25,000 black individuals in the U.S. might have sickle cell anemia. Motulsky[17] combined the available data on frequencies of various hemoglobinopathy carrier states and estimates of lifespan to calculate frequencies of the hemoglobinopathies. He considered it likely that 8 per cent of blacks in this country have sickle cell trait, whereas about 3 per cent have hemoglobin C trait and 1.5 per cent have β-thalassemia trait. Similar frequencies have been found in the nearly 15,000 individuals screened to date in the Boston Sickle Cell Center. Based on these carrier frequencies, the incidence of sickle cell anemia would be

TABLE 8-1. ESTIMATED FREQUENCIES OF HEMOGLOBINOPATHIES AMONG UNITED STATES BLACKS *AT BIRTH**†

Feature	Maximum	Minimum	Most Likely
Trait frequencies (%):			
Hb AS	14(0.07)‡	6(0.03)	8(0.04)
Hb AC	4.6(0.023)	1(0.005)	3(0.015)
β thal§ trait	2(0.01)	0.5(0.0025)	1.5(0.0075)
Hereditary persistence			
of Hb F	—	—	0.1(0.0005)
Sickling disorders:			
Hb S-S (sickle cell			
anemia)	1:204	1:1111	1:625
Hb S-C	1:311	1:3333	1:833
Hb S-β thal	1:714	1:6667	1:1667
Hb S-persistence			
of Hb F	1:14,286	1:33,333	1:25,000
Other disorders:			
Hb C-C	1:1890	1:40,000	1:4444
β Thal major	1:10,000	1:160,000	1:17,778
Persistence of			
Hb F homozygote	—	—	1:4,000,000
Hb C-β thal	1:2174	1:40,000	1:4444
Hb C-persistence			
of Hb F	1:43,478	1:200,000	1:66,667
β thal-persistence			
of Hb F	1:100,000	1:400,000	1:133,333

*Reprinted by permission from Motulsky, A. G.: Frequency of sickling disorders in U.S. blacks. N. Engl. J. Med., 288:31, 1974.
†Calculated by Hardy-Weinberg statistics based on allelism of certain hemoglobinopathic traits (HbS, HbC, β-thal. hereditary persistence of fetal Hb).
‡Figures in parentheses signify gene frequencies.
§Thalassemia.

about 1 per 625 black births whereas that of S-C disease would be about 1 per 833, and that of S-β-thalassemia disease, about 1 per 1667. Assuming a two-thirds shortening of lifespan in sickle cell anemia as suggested by Scott and lesser degrees of shortening in the other two disorders, the prevalence of sickle cell anemia at all ages in American blacks would be about 1 per 1875, with hemoglobin S-C disease occurring in about 1 per 1250 and hemoglobin S-β-thalassemia disease occurring in about 1 per 3333.[17] Accordingly, by adulthood there would be more patients with hemoglobin S-C disease than with sickle cell anemia because of the fatality rate at a relatively early age in the latter disease. Sickle cell anemia is thus one of the major, if not the most frequent, serious autosomal recessively inherited diseases in the U.S. today, regardless of race. Although exact figures are lacking, there can be no doubt that many of these patients suffer greatly and incur substantial medical expenses during their lifetimes.

Although these disorders are relatively frequent, it is clear from these figures that numerically the carrier states for these disorders are vastly more frequent than the diseases themselves. Thus, at birth those with hemoglobin A-S outnumber those with hemoglobin S-S by 50:1, whereas over the entire age range there are 150 carriers for every person with sickle cell anemia. The designers of any hemoglobinopathy screening program will therefore need to consider that the preponderance of persons found to be positive by the tests employed will be carriers, whether the screening is done at birth or later. Most people screened will be found to have no hemoglobin abnormality, whereas the largest number of persons seen for counseling will be carriers and far smaller numbers will be identified as having a disease and therefore needing medical care as well as genetic counseling. Clearly, from the standpoint of genetic counseling, carrier-couple matings will be particularly important. It can be estimated that in American blacks both members will have sickle cell trait in about 1 per 156 couples, whereas in 1 per 208 couples one mate will have sickle cell trait while the other has hemoglobin C trait; in about 1 per 417 couples one mate will have sickle cell trait, while the other has β-thalassemia trait. It is these couples with a high risk of an affected child that are the focal point of those screening programs which concentrate on married adults and the avoidance of these carrier state combinations, which is one possible reproductive alternative available when unmarried individuals are screened.

MEDICAL ASPECTS OF THE HEMOGLOBINOPATHIES AND THEIR CARRIER STATES

Sickle Cell Anemia

Although a large variety of therapies have been used over the years for short- and long-term treatment of sickle cell anemia, there is still no cure or even effective therapy for this disorder. Instead, only palliative and supportive therapy is available.[7] Urea given intravenously as an anti-sickling agent for the treatment of sickle cell crisis was introduced by Nalbandian in 1961.[18] While its effectiveness and safety were widely debated, more extensive experience and carefully controlled clinical trials have provided no evidence that the administration of urea will shorten or terminate painful sickle crises.[19, 20, 21] Cyanate, found to be an effective inhibitor of erythrocyte sickling when used *in vitro*[22] and to be relatively nontoxic in initial clinical trials in humans,[23] has subsequently been shown to produce significant polyneuropathy that is more frequent with longer duration of therapy, higher daily cyanate dose, and greater carbamylation of hemoglobin,[24] all of which raise serious questions about the safety of cyanate therapy. This toxicity seems likely to limit the use of cyanate clinically even if subsequent double-blind studies demonstrate the effectiveness of cyanate in preventing sickle crises. Research on cyanate and other therapies is continuing, but at present there is no definite prospect of a safe, effective therapy for sickle cell anemia. This lack of therapy constitutes a major aspect of the rationale for mass screening and genetic counseling programs that have as one goal the prevention of sickle cell anemia through avoidance of the birth of individuals with the disease. It should be emphasized, however, that mass screening and counseling should not be viewed as competing with efforts to develop effective treatment, since the availability of therapy would obviate any need for genetic screening and require only the identification of affected individuals for appropriate medical care.

Other important and highly relevant research is aimed at the prenatal diagnosis of sickle cell anemia. It has been known for some time that sickle cell anemia, sickle cell trait, and homozygous normal adult hemoglobin can be distinguished in the 14 to 18 week fetus through the use of chromatographic analysis of radioactively labeled α, β, and γ chains synthesized *in vitro* in the presence[25] or absence[26] of an inhibitor of γ chain synthesis. By use of these methods it is possible to determine the adult hemoglobin genotype of the fetus with as little as 0.05 ml of fetal blood even in the presence of extensive contamination with maternal blood (the latter containing negligible quantities of reticulocytes and therefore incorporating little radioactivity). Although these initial studies depended on the use of blood obtained from fetuses aborted by hysterotomy for medical reasons, techniques have been developed more recently for obtaining blood samples from the fetus *in utero*[27] and for analyzing these samples directly or after concentration of the fetal cells.[28] These techniques, however, while showing great promise,[29] are still experimental, and many important questions about safety and feasibility must be answered before prenatal diagnosis of the hemoglobinopathies can be made widely available. This important avenue of research has been halted under the moratorium on fetal research imposed by the National Research Act, H.R. 7724.[30] While fears that efforts to limit reproduction of blacks through hemoglobinopathy screening and directive counseling have certainly been expressed,[14] it is this author's impression, based on discussions with black lay groups and screenees, that prenatal diagnosis would probably be used by many couples at-risk for having a child with sickle cell anemia as a means of having unaffected children, rather than as a directive to abstain from reproductive activity.

Sickle Cell Trait

Life-threatening consequences of sickle cell trait, contrary to at least one early anecdotal report in which causality was not established,[31] either are minimal or at least have yet to be demonstrated convincingly.[32, 33] Splenic infarction due to hypoxia, which should be avoidable, and mild renal abnormalities, particularly hyposthenuria and hematuria, which may require diagnostic evaluation, certainly occur, but they do not seriously compromise the health of the trait individual. It has been generally recognized that the finding at autopsy of massive intravascular sickling cannot alone be interpreted as the cause of death, since this response may well occur as an agonal change in anyone with sickle cell trait dying from any cause.

Indeed, sickle cell trait has been found in black professional football players with about the same frequency as in the general population, suggesting that sickle cell trait has not appreciably affected their physical capabilities.[34] Although longitudinal studies of morbidity and mortality of persons with sickle cell trait are still lacking, tests of large numbers of individuals of all ages have shown no decrease in the prevalence of sickle cell trait with increasing age and that sickle cell trait does not affect survival.[33, 35] Thus, in the absence of significantly increased morbidity or mortality in persons with sickle cell trait, and therefore in the absence of a medical basis for seeking to identify those with sickle cell trait, the purpose of screening for sickle cell trait remains, fundamentally, to provide genetic counseling.

Other Hemoglobinopathies

Recently, individuals who have been quite healthy and asymptomatic into adulthood have been identified as having apparent sickle cell anemia. Some of these have been shown to have other hemoglobinopathies of variable severity (*e.g.*, hemoglobin S-β-thalassemia) or of significantly lesser severity (*e.g.*, hemoglobin S-D), whereas others have inherited identifiable modifying genes (*e.g.*, hereditary persistence of fetal hemoglobin); still others seem to experience milder involvement without any apparent explanation.[36]

In general, it can be said that hemoglobin S-β-thalassemia disease, which occurs with a prevalence of about 1:3333 American blacks,[17] tends to be clinically less severe. However, significant heterogeneity has long been recognized in this disorder,[37, 38] and the prognosis can be estimated more accurately by correlating the clinical picture with the results of laboratory studies of hemoglobin and red blood cells in the affected individual and his or her family. In contrast, hemoglobin S-C disease, with a prevalence of about 1:1250 in American blacks,[17] is almost always milder in its manifestations than sickle cell anemia. Significant problems can occur, however, particularly during pregnancy, and can affect both mother and fetus.

CURRENT SCREENING PROGRAMS

Methods of Hemoglobin Detection

There has been a trend away from reliance solely on sickle preparations and hemoglobin solubility tests, earlier in widespread use, particularly in smaller screening programs. This has been due in part to recognition that solubility tests and sickle preparations fail to detect the carrier states for potentially serious hemoglobinopathies and in part to the widespread availability of simple, accurate, and inexpensive electrophoretic techniques.[39-41] In addition, sickle preparations and solubility tests do not distinguish sickle cell anemia from sickle cell trait, a more serious diagnostic problem in children than in asymptomatic adults (positives in the latter group probably have sickle cell trait).

The failure to detect nonsickling abnormal hemoglobin results in significant errors in genetic counseling. Sickling and solubility tests will be negative in the 3 per cent of American blacks with hemoglobin C trait and in the 1.5 per cent with β-thalassemia trait. Thus, for an individual with sickle cell trait there is *a priori* an 8 per cent risk that the mate will also have sickle cell trait, a 3 per cent risk that the mate will have hemoglobin C trait, and a 1.5 per cent risk of β-thalassemia trait. Although each of these matings results in a 25 per cent risk in each pregnancy of a significant hemoglobinopathy in the child, only the couple in which both partners have sickle cell trait will be identified by sickling or solubility tests, whereas the hemoglobin A-S × A-C couple and the hemoglobin A-S × A-β-thalassemia couple will be falsely reassured. As a result of this error alone, it can be calculated that over one third of the couples at-risk would be counseled incorrectly. It should be noted that an error of the same magnitude would result when sickling or solubility tests were used for initial screening followed by hemoglobin electrophoresis in those identified as positive. These errors are unacceptably large for a satisfactory screening and genetic counseling program.

Electrophoresis to identify hemoglobins has become the method of choice. Initial screening on cellulose acetate strips of cell lysates from small samples obtained by finger-stick will identify the more frequent hemoglobinopathies, as well as sickle cell trait, hemoglobin C trait, and some of those with β-thalassemia trait. Most of the individuals found to have a band at the position of hemoglobin S but none in the hemoglobin A region will prove to have sickle cell anemia. Further analysis by agar gel electrophoresis in acidic citrate buffer, however, should be carried out to identify the small proportion of individuals who have hemoglobin S-D and who are clinically indistinguishable from those with sickle cell trait.[40-42] Similarly, carriers of hemoglobin D, occurring about 1 per cent as often as sickle cell trait, will appear on routine electrophoresis to have sickle cell trait and can be properly identified if a simple solubility test[43] is performed on those found by electrophoretic screening to have the hemoglobin A-S pattern.[36]

The Center for Disease Control now offers workshops in hemoglobinopathy diagnosis as well as a standardization laboratory where the accuracy of techniques used in screening laboratories can be assessed. More complex problems encountered by less experienced programs should be referred to reference laboratories for further analysis. Cellulose acetate electrophoresis, as used in most screening programs,[40, 41] yields a low rate of false negatives, essentially confined to some of those with β-thalassemia trait who do not have elevations of hemoglobin A_2 detectable by this technique. This technique also yields a low rate of false positives that consist of the less common mutant hemoglobins with the same mobility as hemoglobin S under these conditions but that are resolved and identified under other appropriate conditions. The accurate identification of those with β-thalassemia trait requires more extensive testing of red blood cells as well as of hemoglobin.[41] In practice, this identification is generally carried out where this carrier state is suspected from family studies, or where the population being screened is enriched for β-thalassemia genes, such as in programs screening persons of Mediterranean origin.

Screening for sickle cell anemia in the neonatal period by the use of cord blood has been advocated by some[44] and is technically feasible with appropriate modification to allow identification of adult hemoglobin in the presence of large amounts of hemoglobin F.

Education

In spite of some gains in the education of the general public regarding the hemoglobinopathy screening and genetic counseling, it seems apparent that a widespread lack of valid information persists. Certainly, most individuals have now heard of sickle cell anemia as a result of printed materials and extensive radio and television presentations on the subject. Much potential harm has been done by the zealous misrepresentation of the magnitude and nature of the problems posed by sickle cell anemia in the more flagrant attempts to gain attention, financial support, and participation in screening programs. Yet, even the best efforts at careful and accurate presentations of background information have in many instances resulted in inaccurate and unbalanced information. Many of the informational materials available through 1972 were reviewed critically in a report by the Scientific Advisory Committee of the National Association for Sickle Cell Disease.[45] This report is a good reference for those attempting to develop valid educational materials.

Among the more frequent and persistent problems, failure to distinguish clearly between sickle cell trait and sickle cell anemia is particularly notable. Many presentations still state that "there are two forms of the disease," thereby failing to distinguish between sickle cell trait and sickle cell anemia. Similar errors are produced when the term "sickle cell" is used without further elaboration or definition. This unfortunate approach forms the basis for such dramatic statistics as the statement that 1.5 million Americans are "affected by" sickle cell anemia. As a result, education programs must be directed not only at providing basic medical and genetic information but also at anticipating and correcting certain increasingly common misconceptions about sickle cell trait and sickle cell anemia. Efforts to develop pamphlets, films, tape-slide presentations, and other materials are still rather fragmented, with much duplication of effort and no readily available, completely satisfactory educational program.

The lack of valid information on the part of potential screenees not only makes the subsequent counseling more difficult[46] but also seriously compromises the concept of informed consent.[47] With emphasis on the lack of experience in mass screening and genetic counseling programs, the uncertainty as to the best approach, and the seriousness of the potential consequences, the need for written informed consent prior to screening was recognized early.[13] However, it soon became evident that persons requesting screening for a variety of reasons often had only a superficial grasp of rather fundamental background information despite contact with the educational program. Even after considerable subsequent discussion, in many instances it was still not clear whether the person actually grasped sufficient information to give meaningful informed consent. As the number of persons seeking screening became large, it became practically impossible to assess the depth of understanding of each potential screenee. Here, as in many other instances, the question remains not whether meaningful informed consent *should* be obtained, but whether it *can* be obtained, in view of the general lack of a valid fund of information on the part of those seeking screening. In the opinion of this author a rigid prohibition of screening in cases where the validity of informed consent is in doubt is an unrealistic approach to this dilemma, particularly in view of the widespread nature of problems in obtaining meaningful consent. This problem emphasizes again, however, the great importance of prescreening educational programs that provide basic background information along with details of the screening procedures, possible outcomes, and resulting genetic alternatives.

Counseling

Initially, one of the most serious problems posed by hemoglobinopathy screening programs was that many provided no genetic counseling at all. Screenees were told the results of whatever test had been performed—usually a solubility test or sickle preparation—and were largely abandoned thereafter with little or no help in interpreting the significance of the result. This glaring error in program design was perpetuated by a number of the early laws pertaining to hemoglobinopathy screening.

Understandably, the greatest interest has focused on the "effectiveness" of counseling, most often defined in terms of the amount of information presented that is understood by the counselee and, less tangibly, the degree to which this information is used in making critical decisions about reproduction. Although the latter concept is admittedly ambiguous, it is not to be equated simply with compliance with a preconceived decision by the counselor on the course of reproductive action the counselees should take. Since prenatal diagnosis is not yet available in sickle cell anemia, the preventive aspects of the genetic counseling are at present limited to selection of a mate who is not also a carrier or, in the case of couples where both partners are carriers, self-imposed limitations on the number of pregnancies. To date there are no data available on the use of these genetic alternatives by any large group of persons at-risk within the United States.

Perhaps the most widely described problem in genetic counseling in the setting of hemoglobinopathy screening programs is the failure of those screened to appear for the scheduled counseling session. Often, this seems to result from a remediable problem of program design—that the results of the screening test and the counseling are given separately. Under these circumstances it is not readily apparent whether the screenees failed to return for counseling because of apathy, fear, or ignorance, or whether they did not return because they felt they had already grasped the relevant information and interpretation. Since the latter seems far less likely to be the case in most instances, many programs now provide counseling at the time the results are given with the screenee having the option of attending subsequent counseling sessions as needed. For those whose tests yield entirely normal results, the information received during counseling and a statement of its significance can be transmitted in writing and a return visit can be scheduled for those who request it.

No extensive studies have been published of counseling in hemoglobinopathy programs in the United States. A preliminary report[48] indicated that correct responses to a questionnaire after the session were positively correlated with the level of education of the counselees. There were major gaps in the understanding of risk.[46] A detailed study in Greece has yielded interesting and provocative results.[48a]

Available information suggests that the genetic counseling in most screening programs is provided exclusively by physicians. In the Boston Sickle Cell Center much of the genetic counseling that can be considered routine is provided by specially trained educator-counselors working under the supervision of physicians. Although the backgrounds of these persons vary considerably, most have had substantial experience with community-based health programs, although many have no advanced formal education. After completion of an initial intensive training and evaluation period, those who qualify are assigned to various community health centers where they coordinate the operation of the educational program, the submission of blood samples to the central laboratory for analysis, the transmission of the results of these tests to screenees, and the initial genetic counseling. A medical geneticist is readily available and is always brought into the counseling session whenever both members of a couple are found to be carriers of genes for potentially serious hemoglobinopathies, whenever any question of nonpaternity is raised by the results, or whenever any unusual problems arise. The judgment of the educator-counselors in these matters is supplemented by periodic reviews of the counselors' activities. This role for educator-counselors seems appropriate, for a number of reasons. A relatively circumscribed body of information that needs to be conveyed to potential screenees, to individuals found to have entirely normal hemoglobin and to couples where one or both have entirely normal hemoglobin can be identified. The educator-counselors often have the advantage of possessing much more extensive experience with the community in which they work, can be more immediately available (because of their geographic proximity) to screenees, and can serve as a focal point while they coordinate the diverse services of the center, including, as needed, those of geneticist, hematologist, general pediatrician and internist, gynecologist, nurse, social worker, and others.

OPTIMAL PROGRAM DESIGN

Based on the experience and considerations just reviewed, it is possible to formulate guidelines for the design of valid hemoglobinopathy screening and genetic counseling programs.

Principles

Certain essential principles must be incorporated into the design of the program. Participation in any such program must be voluntary. The decision to be screened must therefore be based on personal knowledge of the potential gains and risks as they are perceived by each individual to apply to his or her own situation. False and misleading information regarding the probability of having a disease, or of being a carrier, the risks of such a condition, the potential for treatment, and other similar considerations should not be used to promote screening. Some assurance of adequate background knowledge should be gained before screening is performed. In some instances this will be by personal interview or questionnaire, while probably less often a formal informed consent form will be appropriate. Similarly, an individual must be free to discontinue participation in the program at any stage and to select what, if any, information is applicable to his or her plans. Voluntarism also implies that the genetic counseling be nondirective in the sense that the counselor conveys all of the possible reproductive alternatives in the face of a given magnitude and type of risk to future offspring of a particular couple and does not attempt to say what the couple should do with this information. Although it is recognized that "nondirective" is a relative term, the special sensitivities of the black population toward eugenic genocide and other devious motives that have been incorrectly attributed to genetic screening programs make this approach essential.

The principle of voluntarism is also closely related to the concept of privacy and confidentiality. The latter helps not only to insure that couples will be free to choose from among all the possible options without fear of criticism or reprisal, but also to minimize the potential for stigmatization based on misinformation about these carrier states and disorders. Although the risk of loss of employment or increased insurance rates appears to have diminished recently, since these unfortunate practices have been addressed directly, the remaining risk can be minimized by strict confidentiality. Since medical records are unfortunately far from confidential, and because the medical consequences of having sickle cell trait are minimal, for the present this author is opposed to routine recording of hemoglobinopathy carrier states in the regular hospital record. Instead, records can be kept separate and more carefully protected by those responsible for the conduct of the screening program.

Community participation in the design and operation of any genetic screening and counseling program is essential. Through its leaders and representatives not only can the needs and sensitivities of potential screenees be expressed—a particularly important consideration when specific racial or ethnic groups are being screened—but also information regarding goals, policies, staffing, and other similar considerations can be conveyed more readily to the people the program aims to serve.

GOALS

The primary goal of hemoglobinopathy screening and genetic counseling programs is diminution of suffering through voluntary avoidance of the birth of children with serious hemoglobinopathies. Optimally, such screening and counseling are carried out prior to the birth of the first potentially affected child to a given couple. Young children have been screened in cases where the goal is the early diagnosis of those with hemoglobin diseases. Although there is no specific medical therapy yet available, prompt treatment of infections and of a crisis due to acute splenic sequestration of blood may be lifesaving in these children.[49] Additional goals include the conducting of valuable research aimed at developing effective therapy as well as improving screening and counseling programs, and the provision of further information regarding genetic disorders and of improved health care in general.

Methods

Screening should be offered only after or in the context of an accurate and effective educational program directed at both the community at large and individuals. Initial screening should be by hemoglobin electrophoresis and should be performed only in laboratories where accuracy is assured and where proper techniques are available for identifying all of the various hemoglobins that may complicate the diagnosis. With the present availability of training programs and reference laboratories,[40] most diagnostic errors, which necessarily lead to counseling errors, are avoidable.

The educational and screening programs should be directed particularly at high-school age persons who are old enough to grasp the significance of the findings and make critical decisions but who have generally not yet married. Even though the hemoglobinopathies in the United States are found primarily in blacks,[50] the general goal of increased genetic education and the possibility of tests for a variety of carrier states found in all racial and ethnic groups make it worthwhile to provide a broad educational program. While it should be clearly specified that the hemoglobinopathy carrier states and disorders are found primarily in blacks in the United States, it should also be stipulated that no person, regardless of apparent racial origin, be refused testing.

Genetic counseling must be readily available and can be provided by various qualified members of a team that includes medical geneticists and other physicians and paraprofessionals with special training as well as through supplementary written and illustrative material. There is at present considerable uncertainty as to what constitutes optimal counseling.[51] However, as a minimum, the counseling should provide information regarding the risks to offspring, the potential burden of the particular hemoglobinopathy, and the reproductive alternatives available to the couples and individuals at-risk. The detailed content of this counseling has been reviewed elsewhere.[36, 52, 53] (See also Chapter 3.) Clearly, the program must have the facilities to assist couples in implementing the reproductive alternatives selected, including provision of contraceptive information and materials, access to adoption agencies, referral to physicians who perform artificial insemination, and provision of diagnostic evaluation and any necessary medical care in cases where couples have elected to accept the risk of having an affected child.

Follow-up

It is highly desirable that screenees not be lost to follow-up but that an opportunity be provided for further contact. In this way the screenee can be furnished with further information, as needed, and can be supported in subsequent decisions and experiences. Valuable insights can be gained about the effectiveness of the program in conveying information and about the uses made of this information. Although valid assessments of the beneficial and adverse effects of these programs may be exceedingly difficult, the potential importance more than justifies the effort. It may well be some time yet before data are available to answer the key questions regarding whether the existing programs are progressing toward their goals.

CONCLUSION

Mass hemoglobinopathy screening and genetic counseling programs have been launched over the past five years largely because of the availability of funds and because of increased interest by professionals and the public. The relatively high frequencies of these diseases, the availability of accurate and simple testing procedures, and the present lack of specific, effective, safe therapy have provided the rationale for programs aimed at prevention. While many aspects of the existing programs remain controversial and objective data on their effects are still largely lacking, much critical reassessment has occurred over the past several years and many of the key unanswered questions have been identified. In light of these it is possible to make best present estimates about many technical, genetic, medical, ethical, and social aspects of program design. It continues to be essential that assessment of these programs be carried out at all levels to identify both the beneficial and adverse effects, to provide remedies for the latter where possible, and ultimately to allow evaluation of these programs in terms of their stated goals.

References

1. Jackson, R. E.: A perspective of the national sickle cell disease program. Arch. Intern. Med., 133:533, 1974.
2. Herrick, J. B.: Peculiar elongated and sickle-shaped red blood corpuscles in a case of severe anemia. Arch. Intern. Med., 6:517, 1910.
3. Neel, J. V.: The inheritance of sickle cell anemia. Science, 110:64, 1949.
4. Beet, E. A.: The genetics of the sickle cell trait in a Bantu tribe. Ann. Eugen., 14:229, 1949.
5. Pauling, L., Itano, H. A., Singer, S. J., et al.: Sickle-cell anemia: A molecular disease. Science, 110:543, 1949.
6. Ingram, V. M.: Gene mutations in human haemoglobin: The chemical difference between normal and sickle-cell haemoglobin. Nature (Lond.), 180:326, 1957.
7. Comings, D. E.: Sickle cell disease and related disorders. *In* Williams, W. J., Beutler, E., Erslev, A. J., et al. (eds.): Hematology. New York, McGraw-Hill Book Company, 1972, p. 413.
8. Lane, J. C., and Scott, R. B.: Awareness of sickle cell anemia among Negroes of Richmond, Va. Public Health Rep., 84:949, 1969.
9. Greenberg, M. S.: Need to identify sickle-cell trait. N. Engl. J. Med., 282:629, 1970.
10. Scott, R. B.: Health care priority and sickle cell anemia. J.A.M.A., 214:731, 1970.
11. Scott, R. B.: Sickle-cell anemia—high prevalence and low priority. N. Eng. J. Med., 282:164, 1970.
12. Beutler, E., Boggs, D. R., Hiller, P., et al.: Hazards of indiscriminate screening for sickling. N. Engl. J. Med., 285:1485, 1971.
13. Institute of Society, Ethics, and the Life Sciences: Ethical and social issues in screening for genetic disease. N. Engl. J. Med., 286:1129, 1972.
14. Whitten, C. F.: Sickle-cell programming—an imperiled promise. N. Engl. J. Med., 288:318, 1973.
15. Fost, N., and Kaback, M. M.: Why do sickle screening in children? Pediatrics, 51:742, 1973.
16. Hampton, M. L., Anderson, J., Lavizzo, B. S., et al.: Sickle cell "nondisease": A potentially serious public health problem. Am. J. Dis. Child., 128:58, 1974.
17. Motulsky, A. G.: Frequency of sickling disorders in U.S. blacks. N. Engl. J. Med., 288:31, 1974.
18. Nalbandian, R. M., Schultz, G., Lusher, J. M., et al.: Sickle cell crisis terminated by intravenous urea in sugar solutions: A preliminary report. Am. J. Med. Sci., 261:309, 1971.
19. Kraus, A. P., Robinson, H., Cooper, M. R., et al.: Clinical trials of therapy for sickle cell vaso-occlusive crises. J.A.M.A., 228:1120, 1974.
20. McCurdy, P. R., Binder, R. A., Mahmood, L., et al.: Treatment of sickle cell crisis with urea in invert sugar: A controlled trial. J.A.M.A., 228:1125, 1974.
21. Rhodes, R. S., Revo, L., Hara, S., et al.: Therapy for sickle cell vaso-occlusive crises: Controlled clinical trials and cooperative study of intravenously administered alkali. J.A.M.A., 228:1129, 1974.
22. Cerami, A., Manning, J. M.: Potassium cyanate as an inhibitor of the sickling of erythrocytes in vitro. Proc. Natl. Acad. Sci. U.S.A., 68:1180, 1971.
23. Gillette, P. N., Peterson, C. M., Lu, Y. S., et al.: Sodium cyanate as a potential treatment for sickle-cell disease. N. Engl. J. Med., 290:654, 1974.
24. Peterson, C. M., Tsairis, P., Ohnishi, A., et al.: Sodium cyanate induced polyneuropathy in patients with sickle-cell disease. Ann. Intern. Med., 81:152, 1974.
25. Hollenberg, M. D., Kaback, M. M., and Kazazian, H. H., Jr.: Adult hemoglobin synthesis by reticulocytes from the human fetus at midtrimester. Science, 174:698, 1971.
26. Kan, Y. W., Dozy, A. M., Alter, B. P., et al.: Detection of the sickle gene in the human fetus: Potential for intrauterine diagnosis of sickle-cell anemia. N. Engl. J. Med., 287:1, 1972.
27. Hobbins, J. C., and Mahoney, M. J.: In utero diagnosis of hemoglobinopathies: Technic for obtaining fetal blood. N. Engl. J. Med., 290:1065, 1974.

28. Chang, H., Hobbins, J. C., Cividalli, G., et al.: In utero diagnosis of hemoglobinopathies: Hemoglobin synthesis in fetal red cells. N. Engl. J. Med., 290:1067, 1974.
29. Mahoney, M. J., and Hobbins, J. C.: Experience with in utero fetal blood sampling in midtrimester pregnancies. Am. J. Hum. Genet., 26:58a, 1974.
30. Culliton, B. J.: National research act: Restores training, bans fetal research. Science, 185:426, 1974.
31. Jones, S. R., Binder, R. A., and Donowho, E. M., Jr.: Sudden death in sickle-cell trait. N. Engl. J. Med., 282:323, 1970.
32. National Academy of Sciences—National Research Council: The S-hemoglobinopathies: An Evaluation of Their Status in the Armed Forces. Washington, D.C., Government Printing Office, February, 1973.
33. Heller, P.: Once more: The pathogenic effects of the sickle cell trait. J.A.M.A., 225:987, 1973.
34. Murphy, J. R.: Sickle cell hemoglobin (Hb AS) in black football players. J.A.M.A., 225:981, 1973.
35. Petrakis, N. L., Wiesenfeld, S. L., Sams, B. J., et al.: Prevalence of sickle-cell trait and glucose-6-phosphate dehydrogenase deficiency: Decline with age in the frequency of G-6-PD-deficient Negro males. N. Engl. J. Med., 282:767, 1970.
36. Rucknagel, D. L.: The genetics of sickle cell anemia and related syndromes. Arch. Intern. Med., 133:595, 1974.
37. Weatherall, D. J., and Clegg, J. B.: The Thalassemia Syndromes. 2nd ed. Oxford, Blackwell Scientific Publications, 1972.
38. Comings, D. E.: Thalassemia. *In* Williams, W. J., Beutler, E., Erslev, A. J., et al. (eds.): Hematology. New York, McGraw-Hill Book Company, 1972, p. 328.
39. Barnes, M. G., Komarmy, L., and Novack, A. H.: A comprehensive screening program for hemoglobinopathies. J.A.M.A., 219:701, 1972.
40. Schmidt, R. M.: Laboratory diagnosis of hemoglobinopathies. J.A.M.A., 224:1276, 1973.
41. Schmidt, R. M., and Brosious, E. M.: Basic Laboratory Methods of Hemoglobinopathy Detection. 5th ed. Atlanta, Hematology Division, Bureau of Laboratories, Center for Disease Control, DHEW, HEW Pub. No. (CDC) 74–8266, 1974.
42. McCurdy, P. R., Lorkin, P. A., Casey, R., et al.: Hemoglobin S-G (S-D) syndrome. Am. J. Med., 57:665, 1974.
43. Greenberg, M. S., Harvey, H. A., and Morgan, C.: A simple and inexpensive test for sickle hemoglobin. N. Engl. J. Med., 286:1143, 1972.
44. Pearson, H. A., O'Brien, R. T., McIntosh, S., et al.: Routine screening of umbilical cord blood for sickle cell diseases. J.A.M.A., 227:420, 1974.
45. Scientific Advisory Committee, National Association for Sickle Cell Disease: A critical review of informational materials relating to sickle cell anemia and sickle cell trait. Los Angeles, National Association for Sickle Cell Disease, Inc., 1972.
46. Leonard, C. O., Chase, G. A., and Childs, B.: Genetic counseling: A consumer's view. N. Engl. J. Med., 287:433, 1972.
47. Ingelfinger, E. J.: Informed (but uneducated) consent. N. Engl. J. Med., 287:465, 1972.
48. Murray, R. F., Jr., Bolden, R., Headings, V. E., et al.: Information transfer in counseling for sickle cell trait. Am. J. Hum. Genet., 26:63a, 1974.
48a. Stamatoyannapoulos, G.: Problems of screening and counseling in the hemoglobinopathies. *In* Motulsky, A. G., and Lenz, W. (eds.): Birth Defects. Proceedings of the Fourth International Conference, Vienna, Austria, September 2–8, 1973. Amsterdam, Excerpta Medica, 1974, pp. 268–276.
49. Pearson, H. A., and Diamond, L. K.: The critically ill child: Sickle cell disease crises and their management. Pediatrics, 48:629, 1971.
50. Gelpi, A. P., and Perrine, R. P.: Sickle cell disease and trait in white populations. J.A.M.A., 224:605, 1973.
51. Fraser, F. C.: Genetic counseling. Am. J. Hum. Genet., 26:636, 1974.
52. Erbe, R. W.: Genetic counseling in sickle cell trait and sickle cell anemia. *In* Olafson, F., and Parker, A. W. (eds.): Sickle Cell Anemia—the Neglected Disease. Berkeley, University of California Health Center Seminar Program, Monograph Series No. 5, 1973, p. 41.
53. Motulsky, A. G.: Screening for sickle cell hemoglobinopathy and thalassemia. Isr. J. Med. Sci., 9:1341, 1973.

CHAPTER 9

PRENATAL DIAGNOSIS OF GENETIC DISORDERS

Aubrey Milunsky, MB.B.Ch., M.R.C.P., D.C.H.
and Leonard Atkins, M.D.

Almost ten years have passed since Steele and Breg and Klinger first cultured and karyotyped amniotic fluid cells.[1] Since that time we estimate that fewer than 5000 amniocenteses have been done (by mid-1975) in the United States for prenatal genetic diagnosis. Each year in the United States alone, there are more than 20,000 live births with chromosomal abnormalities.[2] It is therefore a sobering realization that during 1973, we estimated that chromosomal abnormalities were diagnosed in fewer than 75 fetuses *in utero* in the United States.[3] During this period, women aged 40 years and over (roughly 3 per cent of the child-bearing population) delivered approximately 2625 offspring with major chromosomal abnormalities (estimate based on the occurrence of maternal-age-dependent chromosomal abnormalities of 1 in 40 between 40 and 44 years of age).[4]

Various explanations can undoubtedly be offered to explain this remarkable delay in the application of a valuable but new technology. Consideration of a global perspective suggests that priorities other than genetic disease prevention occupy national and international attention. The havoc wreaked by natural disasters, starvation, and the problems concerned with the control of infection place considerations about genetic disease prevention at a deservedly lower priority level. In more affluent countries, concerns about the risks of amniocentesis in early pregnancy have been paramount. The evaluation of risk, however, should be resolved during 1975 as a result of the Collaborative Amniocentesis Registry Project under the auspices of the National Institutes of Child Health and Human Development (*vide infra*). A serious lack of awareness of the indications for prenatal genetic studies by profession and public alike probably constitutes the next major reason for delayed application of amniocentesis. Inadequate, unproven, and lack of facilities for amniotic fluid cell culture have been the reasons advanced by practitioners for not advocating prenatal genetic studies in those areas without such facilities. In view of the ability to transport amniotic fluid for study elsewhere, the latter objections have proved hard to sustain—at least in the United States.

EXPERIENCE WITH PRENATAL DIAGNOSIS

Boston

Amniocentesis

In Massachusetts, transabdominal amniocentesis is usually performed in the offices of private obstetricians. The vast majority of amniotic fluid samples are brought to our laboratory (a regional state facility) by hand. Successful results have been obtained from volumes of less than 1 ml, while failures have occurred with samples of 20 ml or over. Between 10 and 20 ml of amniotic fluid would still seem to be optimum for cell culture, obtained between 14 and 16 weeks of gestation.

Risks. Although the exact risks of mid-trimester amniocentesis are still under study, our experience may provide a useful guide. The outcome of pregnancy is known in 634 cases studied prenatally (Table 9–1). Major complications ("spontaneous" abortion, fetal death, stillbirth) occurred in 21 cases (3.3 per cent) (Table 9–2). It is conceivable that amniocentesis was possibly related to these major complications in ten cases (1.6 per cent) (it is highly probable that some cases of spontaneous abortion occurred without any relationship to amniocentesis). A spontaneous abortion rate of approximately 3 per cent between the fourteenth and eighteenth week of gestation has been reported in large surveys.[5, 6] Moreover, a general increase in the risk of spontaneous abortions with increasing maternal age and with a history of previous abortion has also been observed.[6, 7] These latter reports are especially pertinent to cases requiring prenatal genetic studies. In 330 "control" pregnancies we have followed in which *no* amniocenteses were done, a rate of 3 per cent for spontaneous abortion, fetal death, and stillbirth has been noted. In our experience with amniocenteses performed by many obstetricians of varying experience, the true risk of the procedure for causing fetal loss is probably less than 0.5 per cent. No maternal mortality has occurred and the maternal morbidity encountered has been minor.

We have no data on the frequency of fetal-maternal bleeding. Blajchman et al.,[8] using techniques that permitted detection of fetal bleeds greater than 0.25 ml, showed that in amniocenteses prior to 20

TABLE 9–1. OUTCOME OF PREGNANCY IN 634 CASES
STUDIED PRENATALLY

Outcome	Number of Cases
Live normal births (includes 9 twins)	582
Abortion:	
Spontaneous/missed	10
Elective	5
Therapeutic (affected fetus)	18
Stillbirth:	
Normal appearing fetus	4
Anencephaly	1
Trisomy 18	1
Fetal death	7
Live births with abnormalities:	15

Down's syndrome*	Claw deformity
Partial phocomelia	Multiple hypospadias
Craniosynostosis	Multiple congenital
Cleft lip	abnormalities
Cleft lip and palate	Hydrocephaly
Cystic fibrosis	Myelomeningocele
Adrenogenital syndrome	Mild hypospadias
Klinefelter's syndrome†	Tetralogy of Fallot

*Requested repeat amniocentesis not repeated
†Diagnosed prenatally

weeks gestation, 15 per cent of 26 procedures in 23 patients resulted in transplacental bleeding. Clear risks exist for isoimmunization in the Rh-negative mother, as well as for the possibility of fetal hemorrhage across the placenta. Recommendations have been made to administer anti-D gammaglobulin to mothers at-risk following early amniocentesis,[9] despite the absence of definitive data concerning the safety of this approach.

Results. Results have been provided in 90.8 per cent of cases (Table 9–3) from the first amniotic fluid sample received. In 78 samples where no such report was issued from the first sample cultured, mitigating circumstances figured prominently. More than 24 hours (2 to 9 days) elapsed before culture was initiated in almost 25 per cent (16) of these cases. Almost 50 per cent (37) of the samples were bloody and 15 per cent (9) were sent by air or regular mail. These figures are in striking contrast to our usual experience: over 95 per cent of samples initiated in culture within 1 to 8 hours of amniocentesis; markedly bloody samples are seen in less than 6.6 per cent of cases; except for mailed samples (15 per cent), the remainder are hand-carried to the laboratory.

A second amniocentesis has been necessary in 6.4 per cent of cases. Usually around the fifteenth day in culture, decisions are made about the need for a second amniocentesis. This cautious approach has been employed to avoid the difficult situations that could (and do) arise when

TABLE 9–2. UNTOWARD OUTCOME OF PREGNANCIES AFTER AMNIOCENTESIS IN 800 CASES*

Indications for Amniocentesis	Case No.	Outcome
Maternal age 35–39 years	A_1	Spontaneous abortion 3 weeks after tap
	A_2	Spontaneous abortion 2 weeks after tap
	A_3	Spontaneous abortion 3 days after tap
	A_4	Spontaneous abortion 20 days after tap
	A_5	Spontaneous abortion 26 days after tap
	A_6	Missed abortion 5 weeks after tap. Fetus had been dead several weeks
	A_7	No fetal heart by doptone at amniocentesis (15 weeks); spontaneous abortion 5 weeks after tap; very tight true umbilical cord knot around one leg; fetus and placenta grossly normal
Maternal age $\geq$ 40 years	B_1	Spontaneous abortion 7 weeks after tap
	B_2	Pregnancy complicated by hypertension, prediabetes, and excessive weight gain, ending with normal-appearing full-term stillborn
	B_3	Fetal death 2 days after tap with the abortus affected by erythroblastosis (as were two previous intrauterine deaths)
	B_4	Dark brown amniotic fluid at the time of tap (17 weeks) signifying probable fetal death, which was confirmed shortly thereafter
	B_5	Stillbirth in a pregnancy complicated by hypertension and placental insufficiency at 7 months gestation
Previous Down's syndrome	C_1	Fetal death occurred between 16–21 weeks gestation. Spontaneous abortion 7 weeks (23 weeks gest.) after tap. Amniocentesis conceivably related to fetal loss
	C_2	Missed abortion 3 weeks after tap
	C_3	Fetal death due to erythroblastosis fetalis at 20 weeks gestation

(*Table continued on opposite page.*)

TABLE 9–2. UNTOWARD OUTCOME OF PREGNANCIES AFTER AMNIOCENTESIS IN 800 CASES* (*Continued*)

Indications for Amniocentesis	Case No.	Outcome
	C_4	Cord constriction and intra-uterine death. Saline infusion at 23 weeks gestation fetus of 18 weeks size (amniocentesis at 16 weeks)
Family history of Down's syndrome	D_1	Vaginal bleeding from 4 weeks gestation until spontaneous abortion at 18 weeks gestation, 16 days after tap; abortus appeared normal and placenta circumvallate
Miscellaneous and previous neural tube defects	E_1	Spontaneous abortion at 32 weeks gestation; grossly mal-formed fetus; pregnancy complicated by hydramnios
	E_2	Premature twins at 27 weeks after a pregnancy compli-cated by hydramnios; one twin a macerated stillborn, other twin lived less than 2 hours.
	E_3	Fetal death between 18 and 19 weeks gestation; tap con-ceivably a cause
	E_4	Full-term stillborn male—nor-mal. Death due to strangula-tion

*Pregnancies completed in 634 cases

laboratories inform the obstetrician of failed cell culture four to five weeks after the first amniocentesis. The consequences of delayed communication may be the inability to provide the prenatal diagnosis sought. Not rarely, women have indeed become pregnant following the assurance by the obstetrician that prenatal diagnosis for the disorder in question is now possible. Therefore, early predictions are made, and in the presence of significant concern, a second amniocentesis is required. In 2.1 per cent of such cases, we subsequently have found that the second amniocentesis proved unnecessary. As a result of this practice, however, prenatal diagnoses have been provided in *all cases* in which the initial studies were performed at the recommended time of gestation (14 to 16 weeks).

Third amniocenteses have been necessary only rarely (0.4 per cent). The reasons for not repeating amniocenteses in 27 cases are shown in Table 9–4. In three cases, "spontaneous" abortion occurred after the first amniocentesis. Seven samples were received too late in gestation to

TABLE 9–3. DATA ON 800 CONSECUTIVE PRENATAL STUDIES

Data	Number of Cases	Percentage (%)
Total cases studied	800	100.0
Report on first sample*	726	90.8
Second amniocentesis needed and done	51†	6.4
Second amniocentesis done— eventually not needed‡	17	2.1
Third amniocentesis needed	3	0.4
Amniocentesis not repeated	27	3.4
Lost by contamination	4	0.5
Errors in diagnosis	0	0.0

*Excluding failures on delayed samples transported by air or by other means for 2 to 9 days, a report on first sample was provided in 92.6% of cases.
†Includes 4 cases where additional studies were recommended and done.
‡See text.

entertain repeat studies. Pain and discomfort or matters of conscience mainly dictated the decisions of ten patients not to have a repeat amniocentesis. In one instance, and during efforts to obtain a second amniotic fluid sample, the physician decided against proceeding because of difficulties in aspirating the fluid. That patient, who had previously had a child with trisomy 21, went on to have another similarly affected child.

Amniotic Fluid Cell Culture

Grossly bloody samples were received in 6.6 per cent of 600 cases.[3] Blood was cleared using the ammonium chloride lysis technique.[10] The average time for determining the karyotype was 17 days. By that day, 58 per cent of clear amniotic fluids were completed, in contrast to the 39 per cent of bloody samples. The commonest reason for requesting a second amniotic fluid sample and for failure in prenatal genetic diagnosis (Table 9–5) was poor cell growth. We have pointed previously to the greater frequency of problems with samples transported for periods in excess of 24 hours.[3]

TABLE 9–4. REASONS FOR NO REPEAT IN 27 CASES

Outcome	Number of Cases
Split sample	4
Spontaneous abortion	3
Initial amniocentesis too late in gestation	7
Patient refusal	10
MD decided against	3

TABLE 9–5. REASONS FOR REPEAT AMNIOCENTESES

	Number of Cases
Reasons for second amniocentesis in 68 cases	
Gross blood ? any amniotic fluid	6
Failure to harvest	5
Poor growth	42
? Insufficient cells for chemical assay	4
Transport failure	7
Verification of diagnosis	4
Reasons for third amniocentesis in 3 cases	
Grossly bloody ? any amniotic fluid	1
Poor growth – (bloody sample)	1
Rubella exposure	1

Fetal Abnormalities Diagnosed Prenatally

A summary of our experience with 800 cases (Table 9–6) shows that "affected" fetuses with chromosomal, X-linked, or metabolic disorders were observed in 28 cases. The details on some of those cases with abnormal karyotypes (Table 9–7) have been alluded to previously.[11] Some details on these and other cases follow.

Case 1. In one translocation carrier mother, the amniotic fluid cells in culture revealed one cell with a ring chromosome out of 90 metaphases analyzed. Leukocyte and skin karyotypes were normal, as were the observations made both in the newborn period and at 3 months of gestation.

TABLE 9–6. ANALYSIS OF CONSECUTIVE EXPERIENCE WITH 800 CASES FOR PRENATAL GENETIC DIAGNOSIS

Indication	Number	"Affected" Fetuses	Therapeutic Abortion	Diagnosis Confirmed Thus Far	Live Births Thus Far
Chromosomes					
Translocation carrier	11	3[a, j]	3	3	7
Mat. age 35–39	218	2[c]	1[d]	1	159
Mat. age ≥40	195	5[b, d, k]	3	4	143
Previous Down's Syndrome	120	—	—	—	96[d]
Fam. hist. Down's Syndrome	48	1[a]	1	1	37
Previous neural tube defect[f]	35	—	—	—	19
Miscellaneous	116	1[h]	—	—	98
X-Linked Disorders	16	8 (males)	3	6	11
Metabolic Disorders	41	7[c, e]	7	6	27
TOTAL	800	27	18	21	597[g]

a. Unbalanced D/G translocation (3)
b. Trisomy 18 (3)
c. 47 XXY (2)
d. Trisomy 21 (3)
e. Tay-Sachs (4); Hunter's (1); Niemann-Pick (1)
f. Including karyotyping
g. See Table 9–1
h. ? Radiation/viral damage (1)
j. Rubella (1)
k. Mosaicism 46, XY/47, XXY (1)

TABLE 9–7. ABNORMAL KARYOTYPES DETECTED DURING
CONSECUTIVE DIAGNOSTIC STUDIES OF 800 CASES

Abnormal Karyotypes	Maternal Age	Indication
Balanced t(DqGq)	25	D/G carrier
	38	Prev. D/G trans. D.S.*
	32	Prev. D/G trans. D.S.
	18	D/G carrier
Unbalanced t(DqGq)	18	D/G carrier
	28	D/G carrier
	18	Fam. hist. D.S.
Balanced t(DqDq)	21	Prev. D.S. (karyotype unknown)
	37	Adv. mat. age
	46	Adv. mat. age
Balanced t(5p−, 10q+)	40	Adv. mat. age
Balanced t(2p−, 4q+)	27	Prev. stillbirth
Balanced t(3p−, 4q+)	23	3/4 trans. carrier
Trisomy 18 (47, XY, +18)	45	Prev. trisomy 18
Trisomy 18 (47, XY, +18)	43	Adv. mat. age
Trisomy 18 (47, XX, +18)	40	Adv. mat. age
Trisomy 21 (47, XY, +21)	39	Adv. mat. age
	40	Adv. mat. age
47, XXY	38	Adv. mat. age
	34	Prev. Niemann-Pick dis.
46, XY (Familial abnormal Y, ? isochromosome, ? pericentric inversion)	21	Prev. Niemann-Pick dis.
Chromosomal breakage, ring formation and dicentric chromosomes (10–14% cells)†	25	Irradiation at 3 weeks gestation
Mosaicism (46, XY/47 Xip, XiqY)†	47	Adv. mat. age
Ring Chromosome (1/90 cells)†	19	D/G carrier
Ring Chromosome (1/60 cells)†	38	Prev. Hurler's
Pericentric Inv. 3/6 pi (6/30 cells)†	36	Adv. mat. age
46, XY Abnormal 20† (? duplication, ? dicentric)	36	Adv. mat. age
46, XY (5/74 cells with missing G)†	25	Prev. cong. malf.
Mosaicism 46, XY/47 XXY‡	36	Prev. anencephalic
46, XY (Familial pericentric inv. Y)	19	5/13 trans. carrier
TOTAL	30	

*D.S. = Down's syndrome
†Newborn karyotype and phenotype normal
‡On first amniotic fluid sample; second sample yielded 46 XY

Case 2. In a fetus at-risk for Hurler's syndrome, a single ring chromosome was observed in one cell out of 60 cells analyzed. The outcome after birth was again a normal child with a normal karyotype.

Case 3. What appeared to be a pericentric inversion of chromosome 3 and isochromosome of number 6 on the X chromosome were abnormalities observed in 6 cells out of 30 metaphases studied in a 38-year-old mother. Karyotypes of cultured leukocytes of both mother and father were normal. Repeat amniocentesis and karyotyping revealed normal karyotypes in all cells analyzed. The reason for the abnormalities observed is not known, but an abnormality arising in culture may well have been the cause. The child seemed perfectly normal at 4 months of age and blood karyotypes were normal.

Case 4. In a patient who had previously had a child with multiple malformations similar to the Meckel's syndrome, 5 out of 74 cells analyzed were shown to have a missing G chromosome. It was too late in pregnancy for a repeat amniotic fluid sample and the parents elected to continue the pregnancy in the face of recognized uncertainty about the implications of the observations made. The karyotypes of both parents were normal, as were the phenotype and karyotype of the newborn child.

Case 5.   In a 36-year-old primipara, an abnormal chromosome number 20 was observed in cultured amniotic fluid cells. Trypsin banding showed an increase in the darkly stained centromeric zone of this chromosome. C-banding demonstrated this zone to be an abnormally large block of heterochromatin. Possible explanations entertained included duplication or a dicentric chromosome. A similar abnormality was seen both with trypsin banding and C-banding of the maternal blood. Although a subtle abnormality could not be excluded, evidence pointed to a fetus without expected phenotypic abnormality relating to this chromosomal aberration. A phenotypically normal child with the described karyotype was born.

Case 6.   The prenatal diagnosis of Niemann-Pick disease was sought in one patient who had previously had an affected child. While sphingomyelinase studies revealed an unaffected fetus, simultaneous chromosomal analysis showed a 47, XXY karyotype. The parents elected to terminate that pregnancy.

Case 7.   In a 46-year-old mother, whose amniotic fluid was received at 29 weeks gestation, most unusual mosaicism (46, XY/47, Xip XiqY) was found. Karyotypes on both blood and skin after birth were entirely normal, as was the phenotypic appearance of the child. It is perhaps most likely that this unusual mosaicism arose in cell culture. Conceivably, this could have been true mosaicism with gradual elimination of the abnormal cell line. It would be necessary that long-term follow-up be provided in this child to determine whether any stigmata of Klinefelter's syndrome appear after puberty. This was a most unusual abnormality, *i.e.,* a combination of Xip and Xiq in the same cell, and has not heretofore been described.

We have pursued a uniform approach to cases with unusual karyotypes and which therefore pose diagnostic dilemmas. The obstetrician and parents are informed about the details of the dilemma. Karyotypes from cultured leukocytes of both parents are usually obtained, banding and fluorescent studies (where indicated) are performed on the first amniotic cell samples as well as the repeat amniotic fluid studies invariably requested. Notwithstanding any normal results obtained after the first sample, the parents are clearly informed about the inability to determine with 100 per cent certainty the relationship of the first observations to the ultimate fetal status. No justification could be advanced to withhold information from the parents when unusual karyotypes are being detected. The options of terminating any such pregnancies must be communicated to the parents immediately. Although experience is beginning to indicate a

TABLE 9-8. CHROMOSOME FRAGMENTS IN 500 CASES

No. Cells/Case with Fragments	1			2			3		4
Total No. Fragments*	1	2	3	2	3	4	3	4	4
No. of Cases	60	7	4	10	6	1	2	1	2

Irradiation (?) case excluded. See Table 9-7.

normal outcome in some cases with unusual findings as described above, the geneticist is counseled *not* to make predictions in specific cases.

We have previously considered the questions of chromosomal fragments, polyploidy, and hypo- and hyperdiploidy.[11, 12] The frequency of chromosome fragments observed in amniotic cells during efforts to make prenatal genetic diagnoses is reflected in Table 9-8. Our observations on polyploidy continue to indicate that even high degrees of polyploidy are consistent and compatible with the birth of a karyotypically normal offspring. The occurrence of hypo- or hyperdiploidy in some cultured amniotic fluid cells could be most disturbing. Questions concerning mosaicism arise with an increasing number of involved cells. Almost invariably, however, with marked hypo-diploidy (in four to eight cells) loss appears to be random, *i.e.*, it is not the same chromosome missing from each cell (Table 9-9).

Reported Experience with Prenatal Diagnosis

The cumulative experience among geneticists in the U.S.A. and Canada up to the early months of 1973 has been extensively reviewed.[13] There has been a relatively modest increase in the number of cases being submitted for study (Table 9-10). Generally in the larger series published, failure to culture successfully the first amniotic fluid sample received has been experienced in less than 10 per cent of cases (Table 9-10). The average time in most laboratories ranged between 17 to 28 days for determination of the fetal karyotype.

The risk of bearing offspring with chromosomal abnormalities for mothers between 40 and 44 years of age remains, as previously calculated,

TABLE 9-9. HYPO- AND HYPERDIPLOIDY IN 482 CASES*

Chromosome No.	43		44			45						47		48
No. of Cells/Case	1	2	1	2	3	1	2	3	4	6	8	1	2	1
No. Cases	2	3	23	3	1	99	28	12	5	1	1	15	2	1

*All karyotypes listed in Table 9-7 excluded.

TABLE 9–10. RECENTLY PUBLISHED EXPERIENCE WITH PRENATAL DIAGNOSIS

Authors	No. Cases	Failed Culture-first Sample (%)	Days in Culture (Mean)	Reference
Nadler and Gerbie (1970)	160	4.0	3–28 (14.2)‡ 15–40 (30.1)§	14
Ferguson-Smith et al. (1971)	30	16.0	7–31 (18.4)	15
Gerbie et al. (1971)	238	5.0	Not stated	16
Epstein et al. (1972)	28	10.0	17–38 (28.0)	17
Therkelsen et al. (1972)	18	0.0	– (10.5)	18
Prescott et al. (1973)	50	8.0	14–44 (25.9)	19
Mulcahy and Jenkyn (1973)	18	0.0	17–34 (23.0)	20
Aula et al. (1973)	16	25.0	8–30 (15.5)	21
Hsu et al.* (1973)	>200	9.0	Not stated	22
Turnbull et al.† (1973)	80	42.0	Not stated	23
Robinson et al. (1973)	128	12.0–30.0	Not stated	24
Wahlstrom (1973)	75	6.7	Not stated	25
Hsu and Hirschhorn* (1974)	325	6.0	Not stated	26
Cox et al. (1974)	45	8.5	8–18 (10.0)	27
Collaborative Series* (1974)	507	6.0	Not stated	28
Doran et al. (1974)	55	9.0	Not stated	29
Allen et al. (1974)	100	10.0	Not stated	30
Laurence† (1974)	181	9.0	Not stated	31
Golbus et al. (1974)	100	6.0	(25.5)	32
Philip et al. (1974)	93	10.8	8–42	33
Milunsky and Atkins (1972, '73, '74)	800	7.4–9.2‖	12–34 (17.0)	34, 13, 3, 11

*Overlapping cases
†Overlapping cases
‡For cytogenetics
§For biochemistry
‖See text and Table 9–3.

at roughly 1:40.[4] In his cumulative survey, Milunsky noted a roughly 1:60 risk for mothers between 35 and 39 years of age.[13] Although there are still an insufficient number of cases for any definitive statements, accumulating data still point to roughly this same risk figure for this age range[26] (Table 9–11). In Boston, prenatal genetic studies are routinely recommended to women who are 35 years and over.

Ultrasound studies prior to amniocentesis have been performed in less than 10 per cent of cases whose amniotic fluids we have studied. Ideally, every patient undergoing amniocentesis should have prior ultra-

TABLE 9–11. CUMULATIVE PRENATAL DIAGNOSTIC CHROMOSOMAL STUDIES: ANALYSIS OF COSTS INCURRED TO DIAGNOSE A SINGLE CASE IN EACH "INDICATION" CATEGORY. (DATA FROM ACCUMULATED U.S. AND CANADIAN EXPERIENCE[13] WITH ADDITIONAL CASES STUDIED BY US)

Indication for Amniocentesis	No. Cases	Fetuses with Major Abnormalities	Frequency of Abnormalities	Cost ($)* per Abnormal Case Detected
Translocation carrier	101	19	1/5.3	1,857
Mat. age 35 to 39	418	6	1/69.7	19,567
Mat. age 40 and over	481	12	1/40.0	11,400
Previous trisomy 21	566	6	1/94.3	26,332
Miscellaneous	326	2	1/163.0	45,225
TOTAL	1892	45	1/42.0	$11,950

*Number of cases × $275 + $400

sound studies. An earlier recommendation that patients in high-risk categories (10 per cent or more) have ultrasound studies[13] is still often forgotten. This absolute recommendation for prior ultrasound studies is also applicable to those embarking upon a prenatal diagnosis of the neural tube defects. Even the smallest fetal blood contamination of an amniotic fluid sample will alter the alpha-fetoprotein content of amniotic fluid—a consequence invariably followed by consideration of pregnancy termination, which may be needless.

Doran *et al.* report that the use of ultrasound prior to amniocentesis clearly diminishes the frequency of blood taps.[29] Placental localization prior to amniocentesis may be valuable in decreasing the chance of feto-maternal transfusion, as well as bloody taps. Furthermore, detection of more than one fetus prior to amniocentesis may have critical consequences. Prenatal genetic studies in high-risk patients (*e.g.*, at risk for Tay-Sachs disease) who subsequently are found to be carrying twins provide major dilemmas for the physician and emotional chaos for the parents. Even though amniotic fluid has successfully been aspirated from two amniotic sacs in sequence after the instillation of contrast medium, this procedure remains a most difficult exercise and is complicated by an increased inability to provide certain reassurance to the very anxious parents at-risk. Should this technique be pursued in the presence of twins (and it would seem extraordinarily hazardous even to offer this approach in the presence of triplets), then a written request from the parents should be obtained following oral and written disclosure (consent form) by the obstetrician. Certainly, parents in such high-risk situations may elect— and should be offered—termination of the pregnancy. We have seen sufficient examples (*e.g.*, mothers at-risk for having offspring with Hurler's syndrome) in which twin pregnancy has been diagnosed more than 10 weeks after the diagnostic amniocentesis. Parenthetically, we have been approached a number of times when in early diagnosed twin pregnancy, anencephaly has been found in one of the twin pair after ultrasound examination. Since concordance for anencephaly in twins is most unusual,[35] and in view of the hazards of needling twin sacs, parents in such situations may elect to continue such pregnancies in the hope that at least one normal twin will survive.

Since multiple fetuses will be found in at least 1 in 80 pregnancies, a reasonable case could be made for routine ultrasound studies prior to amniocentesis. It is true, however, that while chromosomal abnormalities from brief pulsed diagnostic ultrasound is most unlikely,[36, 37] absence of fetal cochlea damage has yet to be documented.

RECENT PROGRESS IN PRENATAL DIAGNOSIS

Neural Tube Defects

Amniotic Fluid Studies

Brock and Sutcliffe[38] first demonstrated elevated alpha-fetoprotein concentrations in the amniotic fluid in pregnancies in which the fetus had an open neural tube defect. Our own subsequent studies[39, 40] and those of others[41-45] confirmed the original observations made by Brock and Sutcliffe.

Human alpha-fetoprotein (AFP) has been recognized as a fetal-specific alpha-1-globulin since 1956.[46] AFP is synthesized by normal embryonal liver cells, the yolk sac, and the gastrointestinal tract of the human conceptus.[47-49] The fetal serum concentrations of AFP rise from about the sixth week of gestation to a peak between 12 and 14 weeks and then steadily fall toward term.[38, 47] The concentration gradient between fetal serum and amniotic fluid AFP is about 200/1. Unlike other proteins, the major source of amniotic fluid AFP seems to be fetal urine,[50] where the concentration is higher than in amniotic fluid. This apparently holds for early but not late pregnancy.[50]

In amniotic fluid the AFP concentrations decrease from the fifteenth week of gestation, and *at term* the concentrations are similar to those in maternal serum[50, 51] where very small amounts of AFP are present normally.[52-54] Hence, the optimal time for amniocentesis for AFP studies should be between 14 and 16 weeks gestation. This point has been exemplified by cases we have seen in which a striking fall in AFP concentration (16.5 to 4.55 mg/100 ml with an anencephalic) was witnessed between 16 and 21 weeks[40] (Fig. 9–1).

Our current studies of over 1000 cases (Table 9–12 and Fig. 9–1) indicate that there is a narrow normal range of AFP in the amniotic fluid in normal pregnancy. In at least 11 cases, levels of AFP that were 2 standard deviations above the mean were observed in pregnancies in

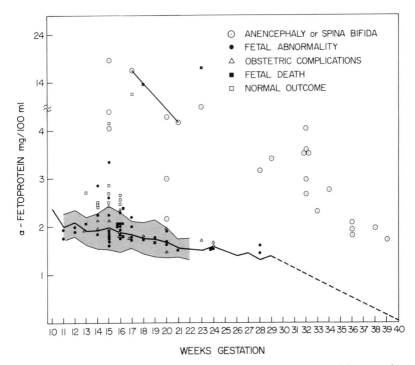

Figure 9–1. Alpha-fetoprotein concentrations (± 2 S.D. shaded area) in normal amniotic fluid (918 cases) and in anencephaly or spina bifida (24 cases), various fetal abnormalities (37), fetal death (10), obstetric complications (15), and normal outcomes (11). Note the striking fall in AFP concentration (16.5 to 4.55 mg/100 ml) between 16 and 21 weeks in one case of anencephaly.

which the liveborn child was normal. Almost all these samples had at least macroscopically visible blood after centrifugation, indicating the presence of *fetal* blood. In a disturbing example studied at 20 weeks gestation, the AFP value was found to be 2.16 mg per cent — just 0.16 mg per cent above the 2 standard deviation point above the mean. This case had been studied prospectively, and the amniotic fluid had been subjected to routine assay, the indication for the amniocentesis being advanced maternal age. The patient subsequently delivered a karyotypically normal infant with a large *open* myelomeningocele. This case serves to emphasize the need for amniocentesis prior to 16 weeks.

Elevated amniotic fluid AFP concentration was also noted in a number of cases where fetal abnormality had been diagnosed[39] and almost certainly reflects the presence of fetal blood.

The mechanism causing raised amniotic fluid AFP concentration in anencephaly and the other neural tube defects is not completely understood. Present evidence suggests that leakage of fetal cerebrospinal fluid containing AFP leads to the elevated levels of amniotic fluid AFP, since

TABLE 9–12. AMNIOTIC FLUID ALPHA-FETOPROTEIN DATA FROM 882 NORMAL FETUSES

Gestational Age (weeks)	13	14	15	16	17	18	19	20	21	22
No. of Cases	42	83	131	232	160	101	51	43	15	24
Alpha-fetoprotein mg% (mean)	1.92	1.92	1.98	1.89	1.84	1.76	1.75	1.67	1.56	1.53
± 2 S.D. mg%	1.64–2.20	1.55–2.29	1.54–2.42	1.48–2.30	1.56–2.12	1.43–2.09	1.37–2.13	1.36–1.98	1.38–1.74	1.31–1.75

fetal serum AFP concentrations have been found to be normal in fetuses affected with neural tube defects.[55] Further, since a gradient of about 1:30 has been noted between the concentrations of other proteins in maternal serum and amniotic fluid, a significant contribution by maternal AFP seems unlikely.[56] Egress of AFP through the cervical cystic hygroma has been offered as the explanation for elevated amniotic fluid AFP in Turner's syndrome.[57]

Since about 90 per cent of neural tube defects are "open," AFP assay would seem to be of valuable diagnostic use in that group of disorders. In "closed" neural tube defects, or those in which the defect is very small, though open, elevated AFP concentrations are usually not found. In all pregnancies at-risk for neural tube defects, ultrasound prior to amniocentesis followed by AFP assay on the amniotic fluid can now be regarded as standard recommendations. At the Boston Hospital for Women[58] and elsewhere, amniography prior to 20 weeks gestation is now also being offered as a third diagnostic dimension. Fetoscopy in which current fiberoptic instrumentation is used is not likely to succeed in the presence of a rigid instrument that makes negotiation around the fetal back nearly impossible.

Besides the neural tube defects, there have been other disorders in which elevated amniotic fluid AFP has been noted. Seppälä reported a third trimester case in which the fetus had esophageal atresia.[59] He has made similar observations in cases where the fetus had hydrocephalus — but in later pregnancy.[60] Congenital nephrosis,[61] occasional chromosomal abnormalities, fetal death or imminent fetal death, and severe Rh immunization — among others — have all been documented. Hence, the non-specific nature of elevated AFP levels in the amniotic fluid has become apparent.[40]

A search for other more specific markers in the amniotic fluid has been pursued. Milunsky and Alpert (unpublished observations) assayed amniotic fluid samples for S-100 protein. We thought this specific neuronal brain protein[62] might also leak into the amniotic fluid via open neural tube defects. It was, however, not possible to demonstrate the presence of S-100 protein in amniotic fluid obtained from pregnancies with anencephalic fetuses studied prior to 24 weeks gestation.

Macri et al.,[63] with a similar aim in mind, initiated studies using cerebrospinal fluid β-trace-protein (βTP). Their preliminary results indicated that βTP is present in the amniotic fluid of fetuses with open neural tube defects.[64] This potentially important development primed our efforts to perform a collaborative study in which we compared the use of AFP and βTP in the prenatal detection of neural tube defects.[65]

AFP was assayed by immunoelectrodiffusion.[66] The Oucherlony double diffusion technique was employed for βTP assay.[67] A quantitative measurement of βTP by electroimmunodiffusion is still in development, and hence it was possible only to recognize the presence or absence of

βTP or in some cases to express uncertainty. From these interim studies it was concluded that AFP must still be considered the most reliable prenatal detection technique for neural tube defects until such time that a quantitative assay for βTP will be developed.

Sutherland et al.[68] and, subsequently, others[69] have commented on the predictive value of amniotic fluid macrophages when the fetus has a gross neural tube defect. Presently, AFP analysis is likely to be more reliable for the prenatal diagnosis of neural tube defects. However, those amniotic fluid samples obtained for chromosomal or biochemical analyses may on occasion be seen to have macrophages. It would seem judicious to do AFP studies on all such cases. We have found high values in at least two samples sent for AFP studies, where macrophages had been seen, and where the subsequently aborted fetuses had anencephaly.

Maternal Serum Studies

In 1973 Leek et al.[70] and others[71] reported that maternal serum AFP may be elevated in pregnancy when the fetus is affected with an open neural tube defect. Subsequent studies by collaborating centers in the United Kingdom have shown that maternal serum AFP was elevated in 10 out of 17 fetuses with spina bifida and in 12 of 15 anencephalics all studied between 15 and 20 weeks of gestation.[72]

The British data open up what could become a most appealing approach — namely, testing maternal serum routinely during the early antenatal visits. Presently, amniocenteses are performed only for those patients who have had prior offspring with a neural tube defect. Up to now, the first tragedy could not be averted. With the advent of screening of maternal serum for AFP prospectively, this pattern may change. Certainly, the ability to seek out prospectively 80 per cent of anencephalic fetuses and almost 60 per cent of fetuses affected with spina bifida/myelomeningocele, as the preliminary data suggest,[72] represents a most dramatic advance in the prenatal detection of one of the commonest major congenital malformations. There are, however, certain reservations that need to be borne in mind.

Elevated maternal serum AFP has been noted in a variety of conditions, including threatened abortion,[73] impending fetal death,[73] actual fetal death,[74] Rh immunization,[75] and conditions associated with increased feto-maternal transfusion (placental abruption). In addition, disorders causing liver dysfunction in pregnancy (e.g., viral hepatitis, hepatic carcinoma, and diabetes mellitus) may also be associated with elevated maternal serum AFP.[76]

The reason for elevated maternal serum AFP when the fetus has a neural tube defect remains unclear. In some cases elevated maternal serum values may be related to impending or actual fetal death. Indeed, results of a collaborative prospective study in the United Kingdom and

Helsinki[77] support this possibility. Maternal serum AFP was found to be raised in 1 out of 14 pregnancies in which the fetuses had neural tube defects. That single case was associated with fetal death *in utero*. Finally, the *screening* nature of the maternal serum AFP testing should be emphasized, since cases with markedly elevated amniotic fluid AFP have had associated normal maternal serum AFP levels. For example, Seppälä and Ruoslahti in their extensive experience noted a case of fetal meningomyelocele in which the maternal AFP remained normal in the face of death *in utero* and extremely high levels of amniotic fluid AFP levels.[76]

Biochemical and Other Disorders (Table 9-13)

Deficient β-galactosidase, necessary for the degradation of galactosyl ceramide, is characteristic of Krabbe's disease.[78] Most recently, however, it would appear that patients with Krabbe's disease have a marked deficiency in the β-galactosidase activity necessary to degrade galactosyl ceramide and lactosyl ceramide,[79] as well as monogalactosyl diglyceride and psychosine. Detection of Krabbe's disease heterozygotes has been complicated by the overlapping values obtained for galactosyl ceramide β-galactosidase activity in controls and obligate carriers.[80] Wenger *et al.*[81] have described a very sensitive assay using lactosyl ceramide in which they found no overlap between carriers and controls. Because of the greater sensitivity of this assay (already successfully used for prenatal detection of an affected fetus),[81] much fewer cultured amniotic fluid cells are required.

Wenger (personal communication) and others have restudied cells of the patient originally described as having lactosyl ceramidosis.[82] They observed normal activity for the two enzymes that degrade lactosyl ceramide. The implication is, therefore, that lactosyl ceramidosis is not a separate disease.

Max *et al.*[83] have recently described the first synthetic defect in ganglioside metabolism. The affected child had phenotypic features reminiscent of the mucopolysaccharidoses and an accumulation of ganglioside GM_3 in brain and liver with a deficiency of more complex gangliosides. In further tissue studies, deficient activity of GM_3-UDP-N-acetylgalactosaminyl transferase was demonstrated. Since complex gangliosides are found in cultured skin fibroblasts and amniotic cells, it can be anticipated that the synthetic enzymes are also present and that prenatal diagnosis should ultimately be possible.

Ceramidase activity is deficient in Farber's disease tissues.[84] This enzyme is present in normal cultured skin fibroblasts and amniotic cells (J. T. Dulaney—personal communication), thereby making the prenatal diagnosis of Farber's disease possible. Close attention, however, should be paid to the observation that ceramidase is active at both pH 4.0 and

TABLE 9–13. INBORN ERRORS OF METABOLISM DIAGNOSABLE
PRENATALLY

Disorders	Prenatal Diagnosis	References
LIPIDOSES		
Cholesterol ester storage disease	Possible	117
Fabry's disease	Made	118, 229
Farber's disease	Possible	84
Gaucher's disease	Made	119, 119b
Generalized gangliosidosis (GM$_1$ gangliosidosis type 1)	Made	120
Juvenile GM$_1$ gangliosidosis (GM$_1$ gangliosidosis type 2)	Made	121
Tay-Sachs disease (GM$_2$ gangliosidosis type 1)	Made	122
Sandhoff's disease (GM$_2$ gangliosidosis type 2)	Made	123
Juvenile GM$_2$ gangliosidosis (GM$_2$ gangliosidosis type 3)	Possible	124
GM$_3$ sphingolipidystrophy	Potentially possible	83
Krabbe's disease (globoid cell leukodystrophy)	Made	139
Metachromatic leukodystrophy	Made	14, 223, 224, 229
Niemann-Pick disease type A	Made	125
Niemann-Pick disease type B	Possible	227
Refsum's disease	Possible	126
Wolman's disease	Possible	127
MUCOPOLYSACCHARIDOSES		
MPS I-Hurler	Made	128, 131
MPS I-Scheie	Possible	129
MPS — Hurler/Scheie	Possible	129
MPS II A — Hunter	Made	130, 131
MPS II B — Hunter	Possible	130
MPS III — Sanfilippo A	Made	133
— Sanfilippo B	Possible	134
MPS IV — Morquio's syndrome	Possible	86
MPS VI A — Maroteaux-Lamy syndrome	Possible	135, 185
MPS VI B — Maroteaux-Lamy syndrome	Possible	135
MPS VII — β-glucuronidase deficiency	Possible	136
AMINO ACID AND RELATED DISORDERS		
Argininosuccinic aciduria	Made	138, 225
Aspartylglucosaminuria	Possible	226
Citrullinemia	Made	140
Congenital hyperammonemia	Possible	141
Cystathionine synthase deficiency (homocystinuria)	Possible	142
Cystathioninuria	Possible	143
Cystinuria	Made	144, 145
Hartnup Disease	Possible	145, 146

(*Table continued on following page.*)

TABLE 9–13. INBORN ERRORS OF METABOLISM DIAGNOSABLE PRENATALLY (*Continued*)

Disorders	Prenatal Diagnosis	References
Histidinemia	Possible	183
Hypervalinemia	Possible	148
Iminoglycinuria	Possible	145, 147
Isoleucine catabolism disorder	Possible	149
Isovaleric acidemia	Possible	150
Maple syrup urine disease:		
Severe infantile	Made	193
Intermittent	Possible	152
Methylmalonic aciduria		
Unresponsive to vitamin B_{12}	Possible	153
Responsive to vitamin B_{12}	Made	85
Methylenetetrahydrofolate		
reductase deficiency	Possible	154
Methyltetrahydrofolate		
Methyltransferase deficiency	Possible	155
Ornithine-α-ketoacid		
transaminase deficiency	Possible	156
Propionyl CoA carboxylase		
deficiency (ketotic hyper-		
glycinemia	Possible	157
Succinyl-CoA: 3 ketoacid		
CoA-transferase deficiency	Possible	158
Vitamin B_{12} metabolic defect	Possible	159

DISORDERS OF CARBOHYDRATE METABOLISM

Fucosidosis	Possible	160
Galactokinase deficiency	Potentially possible	161
Galactosemia	Made	162
Glucose-6-phosphate		
dehydrogenase deficiency	Possible	163
Glycogen storage disease		
(type II)	Made	151, 164, 229
Glycogen storage disease		
(type III)	Possible	165
Glycogen storage disease		
(type IV)	Made	166
Mannosidosis	Possible	167
Phosphohexose isomerase		
deficiency	Possible	168
Pyruvate decarboxylase		
deficiency	Possible	169
Pyruvate dehydrogenase		
deficiency	Possible	170

MISCELLANEOUS HEREDITARY DISORDERS

Acatalasemia	Possible	171
Adenosine deaminase		
deficiency	Made	172, 231
Chediak-Higashi syndrome	Potentially possible	173
Congenital erythropoietic		
porphyria	Potentially possible	174
Congenital nephrosis	Made	232
Cystinosis	Made	175

(*Table continued on opposite page.*)

TABLE 9–13. INBORN ERRORS OF METABOLISM DIAGNOSABLE
PRENATALLY (*Continued*)

Disorders	Prenatal Diagnosis	References
Familial hypercholesterolemia	Potentially possible	103–106
Hypophosphatasia	Made	114
I-cell disease	Made	92
Leigh's encephalopathy	Possible	95
Lesch-Nyhan syndrome	Made	176
Lysosomal acid phosphatase deficiency	Made	177
Lysyl-protocollagen hydroxylase deficiency	Possible	178
Myotonic muscular dystrophy	Possible	179
Nail-patella syndrome	Possible	132
Orotic aciduria	Possible	181
Saccharopinuria	Possible	182
Sickle cell anemia	Possible	137, 185
Testicular feminization	Potentially possible	184
Thalassemia	Possible	180
Xeroderma pigmentosum	Made	186

pH 9.0 in normal kidney and cerebellar tissues. In Farber's disease, however, ceramidase activity is deficient in these tissues only at pH 4.0.[84] We have monitored one pregnancy at risk for this disease – the outcome being an unaffected child, as predicted.

Ampola et al.[85] made the prenatal diagnosis of an affected fetus with methylmalonic aciduria. For the first time in hereditary metabolic disease, they followed their prenatal diagnosis with successful *in utero* therapy. When massive doses of vitamin B_{12} were administered intramuscularly to the mother, the affected child appeared well and normal when born. Now at 12 months of age, development of the child has progressed normally on a low protein diet (M. Ampola – personal communication). Thus far, vitamin B_{12} therapy has not been required, probably reflecting hepatic storage from the *in utero* treatment.

Neufeld's pioneering efforts have made for impressive progress in the understanding of the basic defects in the mucopolysaccharidoses (see Chapter 5). Delineation of the specific enzymatic deficiencies and recognition of both heterogeneity and compound genetic disease have been achieved for the major occurring mucopolysaccharidoses. Most recent is the description by Matalon et al.[86] of chondroitin sulfate N-acetylhexosamine sulfate sulfatase as the deficient enzyme in Morquio's syndrome. Further confirmatory studies are required before prenatal diagnosis of Morquio's syndrome can be offered. Diagnostic use of cell-free amniotic fluid for the mucopolysaccharidoses has not been recommended[13] – the observations of Lee and Schafer[87] having confirmed the limitations of that approach. In elegant studies using microtechniques for lysosomal enzyme assays, Galjaard et al.[228–230] have made prenatal diagnoses of Fabry's disease, metachromatic leukodystrophy, and Pompe's disease.

The mucolipidoses (or lipomucopolysaccharidoses) represent a group of disorders with abnormalities in glycoprotein, lipid and/or mucopolysaccharide metabolism. Three types have been recognized,[88-92] all of which are inherited in an autosomal recessive manner. While their exact interrelationships remain to be discerned (heterogeneity alone may explain many of the observations), their manifestations in cultured skin fibroblasts have raised the possibilities for prenatal diagnosis.

Indeed, mucolipidosis II (I-cell disease) has been diagnosed prenatally[92] despite lack of knowledge concerning the basic defect. A marked intracellular reduction of six lysosomal enzymes has been demonstrated in cultured skin fibroblasts from patients with I-cell disease.[93] The culture medium of fibroblasts from obligate heterozygotes contains elevated lysosomal enzyme activities, but normal intracellular enzyme activities.[93] This cumbersome technique would appear to yield more reliable determination of heterozygosity for I-cell disease than does measurement of serum lysosomal enzymes.[94]

Leigh's subacute necrotizing encephalomyelopathy does manifest in the cultured skin fibroblasts.[95] Thus far, only one biochemical factor has been common to all untreated and autopsy-proved patients with this autosomal recessive genetic disorder. An inhibitor to the phosphotransferase responsible for the synthesis of cerebral thiamine triphosphate has been found. This inhibitor has been observed not only in homozygotes but also in all obligate heterozygous carriers of Leigh's disease.[96] Current studies suggest that this inhibitor is a low molecular weight protein — possibly a glycoprotein.[97] Murphy et al. in their elegant studies have shown that this inhibitor is not normally detectable in cultured skin fibroblasts. When these fibroblasts are cultured in low thiamine media, the inhibitor is demonstrable. It would seem that both carrier detection and prenatal diagnosis are now at least potentially possible.

Cystic fibrosis cannot yet be diagnosed in utero, despite clear evidence that the ciliary inhibition factor is made by cultured skin fibroblasts[98] and amniotic fluid cells.[99] Possibly, after quantitation of the ciliary inhibitor, differentiation of the homozygous from the heterozygous fetus may be achieved. Refinement of current assay techniques is still being sought to attain concordance of results on identical samples of both amniotic fluid and serum.[100, 101]

Histologic studies of muscle from fetuses (16 to 21 weeks gestation) at-risk for Duchenne muscular dystrophy have shown an increase in variability of fiber size, mean fiber diameter, and the presence of a significant number of hyaline fibers.[102] Toop and Emery[102] have therefore suggested that these findings may represent the earliest histologic signs of muscular dystrophy already manifest in utero. Fetal muscle biopsy may ultimately be required for the prenatal diagnosis of muscular dystrophy.

Recent studies using cultured fibroblasts from patients and carriers of familial hypercholesterolemia suggest that there is a derangement in the negative feedback inhibition of cholesterol biosynthesis.[103-106] Prenatal diagnosis would therefore seem to be potentially possible for this disorder characterized by juvenile xanthomatosis and very rapidly developing atherosclerosis. Reliable carrier detection would seem still to require further studies, in view of the overlap observed between heterozygotes and normals.[106]

Perfection of the fetoscopy technique into a safe and reliable technology would have an enormous impact on prenatal genetic diagnosis. Aspiration of pure fetal blood would allow karyotyping within 5 days, in contrast to the 2 to 6 weeks now required. Biochemical disorders of metabolism—particularly those with lysosomal enzyme deficiencies—might be diagnosable through direct fetal blood sampling in less than 24 hours, as opposed to the usual 3 to 6 weeks. Many presently undiagnosable disorders, in addition to the hemoglobinopathies, could be approached prenatally (see Chapter 1, Table 1-8). This table of selected genetic malformation syndromes all associated with mental retardation provides some insight into the anticipated possibilities for prenatal diagnosis by direct fetoscopy, ultrasound, radiography, or amniography. We have experienced difficulties with direct fetoscopic visualization, as have others.[107] In view of Scrimgeour's experience (with a diagnostic error) it would seem judicious not to offer further diagnostic studies until the technology is sufficiently safe and accurate to provide reliable answers.

In the face of recent legislation in many states in the U.S. and the fetal research moratorium, progress with fetoscopy has been slowed. Nevertheless, direct fetal blood sampling via a fiberoptic amnioscope has been achieved.[108-111] Since Hollenberg et al.[112] demonstrated appreciable synthesis of hemoglobin A by reticulocytes of human fetuses from about the ninth week of gestation, the prenatal diagnosis of the hemoglobinopathies has been anticipated. Kan and his coworkers[137] in Boston demonstrated that the β-chain of hemoglobin S is synthesized in reticulocytes of the human fetus at least at the fifteenth week of gestation. In addition, they showed that a mixed maternal fetal blood sample obtained from the placenta contained adequate numbers of fetal reticulocytes. They therefore suggested that placental aspiration after localization of the placenta by ultrasonic techniques may serve as a sufficient source of fetal blood by providing the opportunity to make the prenatal diagnosis of certain hemoglobinopathies. Since hemoglobin S is produced by fetuses early in gestation,[113] the prenatal diagnosis of sickle cell anemia, thalassemia, and other hemoglobinopathies simply awaits a safer technology for fetal blood aspiration. Placental aspiration for mixed fetal-maternal blood samples is often not successful. Moreover, the possibility of inducing a retroplacental hemorrhage or maternal sensitization is as yet an undefinable risk.

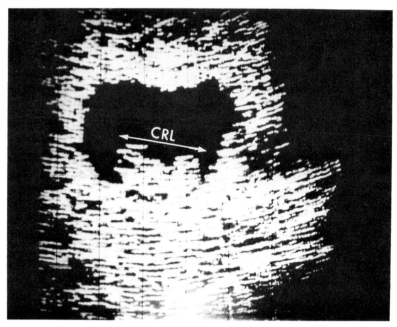

Figure 9–2. Echogram taken at 9 weeks menstrual age, showing the correct longitudinal section of the embryo for crown-rump length (CRL) measurement. (From Campbell, S.: The assessment of fetal development by diagnostic ultrasound. *In* Milunsky, A. (ed.): Clinics in Perinatology. Vol. 1, No. 2. Philadelphia, W. B. Saunders Company, Sept. 1974, p. 510.)

Knowledge that alkaline phosphatase is normally present in the amniotic fluid was used in efforts to make the prenatal diagnosis of hypophosphatasia. Rudd *et al.*[114] demonstrated the absence of alkaline phosphatase in the amniotic fluid of a fetus suspected as having hypophosphatasia. Prior ultrasound studies had drawn attention to apparent abnormalities of the fetal head. The value of ultrasound studies in prenatal diagnosis of congenital malformations has been discussed previously.[13] Campbell, in his continuing sophisticated ultrasound studies, has presented graphic evidence of his ability to measure fetal growth from as early as the ninth week of gestation (Fig. 9–2, 9–3).[115] Measurements of the crown rump length and the fetal biparietal diameter of the head, together now provide extremely helpful information concerning the growth pattern of the fetus. Hereditary disorders in which microcephaly and intrauterine growth retardation are features may easily be diagnosed by such combined serial measurements.

Holmberg *et al.*[116] have demonstrated that the fetus from the twelfth week of gestation has already detectable coagulation factors. Since factor VIII activity and antigen appear in substantial amounts in early fetal life, the prenatal diagnosis of hemophilia (and possibly other coagulation disorders) may become possible in the not-too-distant future.

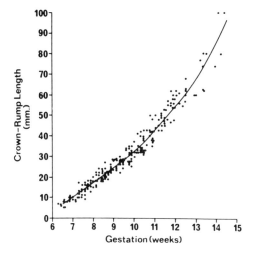

Figure 9–3. Relationship between embryonic CRL measurement and menstrual age from the sixth to the fourteenth week of normal pregnancy. (From Robinson, H. D.: Sonar measurement of fetal crown-rump length to assess maturity in first trimester of pregnancy. Br. Med. J., 4:28, 1973.)

PROBLEMS IN PRENATAL DIAGNOSIS

Chromosomal

Chromosomal Mosaicism

The interpretation of cytogenetic observations from amniotic fluid cell cultures may on occasion be difficult and may raise unexpected dilemmas.[13, 22, 27, 187, 188] Our earlier observations with polyploidy in such cultures simply heralded an increasing sequence of karyotypic dilemmas involving aneuploidy, translocation, and mosaicism. In one example reported by Kardon et al.[22, 189] a diagnosis of 45, X was made on the first amniotic fluid sample. Subsequent culture of an amniotic fluid sample obtained at the time of abortion revealed a 46, XY karyotype. In another case, Kajii[188] observed 5 out of 64 cultured amniotic fluid cells with 47 chromosomes in each. Cytogenetic observations on the skin, lung, and amnion of the abortus revealed a normal karyotype.

Cox et al.[27] tackled the problem of mosaicism in amniotic fluid cell cultures involving tetraploidy, aneuploidy, or translocation. They performed chromosomal analyses on individual amniotic fluid cell colonies. In their careful studies they demonstrated that 50 per cent of successful cell cultures had one or more colonies in which the chromosomes were entirely tetraploid. In subsequent follow-up of the outcome of those cases, they corroborated conclusions, which we had made earlier,[12] that high degrees of polyploidy are compatible with the birth of chromosomally normal offspring.

More important, in four amniotic fluid samples they found one or two colonies in which all the cells studied showed an aberrant chromosome pattern involving either aneuploidy or translocation. Colonies with both trisomy and translocation either separately or together in the same culture were observed. These findings suggest that the occurrence of entire colonies with constant aneuploidy (tetraploidy, aneuploidy, or translocation) must be attributable either to chromosome change very early in culture or to the presence of an abnormal cell type in the amniotic fluid itself. Distortion in chromosomal analysis may occur because of over-representation of a more vigorously growing abnormal cell type from the amniotic fluid.[192]

There have indeed been very disturbing reports of *spontaneous translocations* occurring in cell culture.[190, 191] The distinct possibility exists of such an unhappy eventuality during prenatal diagnostic studies. Routine analysis of at least 30 metaphases[13] from amniotic cells grown in 2 or more containers should greatly help diminish errors and resolve dilemmas. Hence, prenatal diagnosis of true mosaicism is most difficult and can never be excluded with absolute certainty. Nevertheless, Bloom *et al.* made the prenatal diagnosis of 46, XX/47, XX, + D mosaicism, which was subsequently confirmed by analysis of fetal tissues obtained at hysterotomy.

The occurrence of hypo- or hyperdiploidy in a few cultured amniotic fluid cells may give rise to some anxiety. In our experience (Table 9–9) hypo- or hyperdiploid cells are not infrequent in such cultures. Almost invariably, however, chromosomal loss or gain in a few cells appears to be random, and most often in the culture the aberrations are different in the "affected" cells. The possibilities of missing mosaicism adds a hazardous dimension to the prenatal diagnosis of chromosomal abnormalities. Cases in which apparent mosaicism is present in amniotic fluid cell cultures are best managed by immediately seeking a repeat amniocentesis while simultaneously studying blood samples from both parents. Notwithstanding a normal result obtained on the second sample, it would seem to be most judicious for a genetic counselor to inform the parents of the results of the first study, carefully delineating the available explanations for the mosaicism found and leaving the option of termination of the pregnancy open to such parents.

Mycoplasma Contamination

The possible contamination of amniotic fluid cell culture by a variety of agents — bacteria, fungi, bacteriophages, viruses, and mycoplasma — fosters chronic anxiety in the tortured heart of the cell biologist. When such contamination ruins the entire amniotic fluid cell sample, this torture is shared by the already anxious parents. Despite close attention to the approaches used to eliminate contamination — propagation of cell cultures in the absence of antibiotics, good handling techniques, and continuous monitoring of cell cultures — mycoplasma contamination may still become a problem.

Mycoplasma organisms may infect cells thereby leading to growth retardation or cytopathogenicity, or they may cause no obvious alteration in cell growth or morphology.[194, 195] They may, however, cause chromosomal abnormalities in cultured human diploid fibroblasts.[195] Hence, a potential hazard exists if such contamination occurs during efforts to establish a prenatal genetic diagnosis. Indeed, Schneider et al.[196] found a significant four-fold increase in chromosomal breaks and gaps in mycoplasma-infected amniotic fluid cell cultures when these were compared with uninfected cultures. Additional observations included aneuploidy — as well as unusual mosaicism, in one case — with half the cells examined possessing consistent multiple translocations.

The generally accepted reason for the increase in the frequency of chromosomal aberrations caused by the presence of mycoplasmas is that these organisms deplete the medium of certain nutrients that are essential to the cells.[197] It has, for example, been shown that cells grown in arginine-depleted medium show an increased level of chromosomal breakage.[195] This mechanism is probably operative in the majority of cell cultures infected with mycoplasmas, since they are known to convert arginine to ornithine via citrulline.[198]

Even while exercising careful control of techniques used, including monitoring for mycoplasma, cell biologists find that occasional mycoplasma contaminants may not readily be isolated by conventional microbiological techniques. Ideally, therefore, it would be safer to monitor cell cultures for mycoplasma by both microbiological[199] and biochemical[200, 201] techniques. Certainly, any amniotic fluid cell culture revealing unusual chromosomal mosaicism or an increased frequency of chromosomal breakage should be carefully examined for mycoplasma contamination.

Chromosomal Aberrations and pH Disturbances

Ford observed chromosomal abnormalities in actively dividing human fetal fibroblasts cultured even for a short time in an alkaline medium.[202] She noted that chromosomal errors were induced in up to 60 per cent of the cells after exposure in alkaline medium for merely 15 to 30 minutes. The errors included not only polyploidy but also aneuploidy with trisomy 21, trisomy 18, and monosomy X, as well as other abnormal karyotypes.

Other investigators[203] have also reported chromosomal aberrations in short-term cultured lymphocytes exposed to small pH ranges between 6.7 and 6.9. They noted a broad spectrum of anomalies, including monosomies, trisomies, double trisomies, translocations, quadriradial chromosomes, and various hypo- and hyperdiploidies. Indeed, they noted such aneuploidies in up to 30 per cent of plates examined after exposure to pH 6.9.

The importance of these observations for routine diagnostic amniotic fluid cell cultures cannot be assessed now. Nevertheless, they provide sufficient warning to those involved with tissue culture to warrant the maintenance of the pH of the culture medium within a narrow range throughout the duration of the culture. Furthermore, in the presence of unusual karyotypic findings — including trisomic mosaicism — the possible effect of pH disturbances in culture should be considered.

Satellites, Markers, and Subtle Chromosomal Abnormalities

An uncommon indication for genetic amniocentesis is the presence of satellite associations. Lubs and Ruddle have found that if the karyotype of either parent shows an increased number of satellite associations, the risk of having aneuploid offspring is increased.[204] The recommendation has also been made in cases with enlarged satellites that cells of late prophase or early metaphase be used for analysis to eliminate confusion with translocations.[26] Indeed, in one such case with enlarged satellites, the initial interpretation had been a D/G translocation.[26]

"Marker" chromosomes may on occasion provide a diagnostic dilemma during chromosomal analysis for prenatal diagnosis. A few laboratories routinely obtain blood samples from both parents at the time of the amniocentesis to be able to resolve more quickly problems concerning satellites, markers, and translocations. Since there is significant time, effort, and cost involved in such endeavors, we prefer to obtain blood samples for karyotyping and banding studies when the observations on the amniotic cells indicate this course of action. We have thus far had no reason to change this philosophy.

Some years ago Lejeune and Berger suggested that some apparently balanced reciprocal translocations may represent aneusomy by recombination, which may in turn lead to multiple congenital abnormalities.[205] With the advent of chromosomal banding techniques, cases of true aneusomy by recombination could be detected prenatally. Certainly, the occurrence of subtle chromosomal abnormalities represents potential heartbreak for geneticists and parents alike. Indeed, such subtle translocations have been missed by routine staining and karyotyping in living patients with mental retardation.[206] Routine use of chromosomal banding techniques would obviate the likelihood of such mishaps in many cases. Routine banding of all amniotic fluid cell samples would be ideal.

Prenatal sex determination is best provided by full karyotyping.[13] The use of uncultured amniotic fluid cells for rapid identification of the Y chromosome, although reported,[207] is explicitly *not* recommended. Examination by light microscopy of cells with various constitutions with one, two, or no Y chromosomes has been shown to be subject to error in the identification of the Y chromosomes.[208] Even fluorescent staining techniques for the Y chromosome cannot provide sufficiently reliable prenatal sex determination. For example, large Y chromosomes may have terminal nonfluorescent segments of the long arms[209] or other unusual staining patterns.[210] There are also brightly fluorescent bands that may be found in chromosomes 13 and 3 that might give false positive results in fluorescent staining of interphase nuclei.

Biochemical

Definitive Prenatal Diagnosis of Biochemical Disorders of Metabolism

One of the present built-in limitations of prenatal diagnosis is the inability to perform assays for many different enzymes from the same amniotic fluid cell sample. While micromodification of current assay techniques are being successfully pursued,[211, 212] there remains a critical need to know *exactly* which form or type of biochemical disorder is to be diagnosed prenatally. For example, there is a need to specify exactly which form of Sanfilippo syndrome (A or B) is the problem. Hence, ideally either leukocyte or cultured skin fibroblast studies on the propositus — where available — should be performed and underlined prior to subsequent pregnancy. Heterozygote detection for the specific disorder in question should ideally be completed prior to initiation of a new pregnancy.

Cell Culture Problems

Extreme caution is necessary in attempts to establish the prenatal diagnosis of a biochemical disorder. Multiple variables may affect enzyme activities in cultured amniotic cells;[13] these include pH variations of the medium, the frequency with which cultures are fed, the type of medium used and the percentage of fetal calf serum, the age of the cells in culture, the gestational age of the fetus, the nature of the cells used (epithelial or fibroblastic), the stage of cell growth at which cells are harvested, the degree of cell confluency, the proteolytic enzyme used in harvesting, and the presence of mycoplasma in the cultured cells. To this disturbing array of variables must be added known or suspected heterogeneity of the particular disease.

Finally, the closest attention must be paid to controls used in these comparative assays. The employment of at least two controls is always recommended, since it is possible that the control can also be lost to contamination. When selecting controls, one should take into account gestational age, age of the cells in culture, and the cell type (epithelial or fibroblastic), as well as the other factors already mentioned. Inevitably, there will be instances where some abnormality is fortuitously detected in the control sample. Wenger and Riccardi[213] had one such disturbing experience when the cells of a healthy nurse appeared to be homozygous for Krabbe's disease! In one of our cases studied for the prenatal diagnosis of

Tay-Sach's disease, the control sample, especially selected for its non-Jewish origin, turned out to be heterozygous for Tay-Sachs disease.

False-Positive Diagnoses in Biochemical Disorders of Metabolism

Recently, Vidgoff et al.[214] reported a family in which they observed a woman with a marked deficiency of hexosaminidase A – the enzyme deficient in Tay-Sachs disease. The enzymatic deficiency was present in all tissues studied, although the woman was asymptomatic. They interpreted her hexosaminidase A deficiency as representing the carrier state for Tay-Sachs disease. The prenatal diagnosis of Sandhoff's disease (hexosaminidases A and B deficient) has now been complicated by the description of a new variant[215] in which hexosaminidase A activities in serum, plasma, leukocytes, and cultured fibroblasts were found to be higher than previously reported.

Immunologic and amino acid analytic studies on hexosaminidases A and B suggest that these enzymes possess polypeptide subunits,[216–218] which could be subject to different mutations. Much has yet to be learned about possible heterogeneity, amino acid differences, locus control, and phenotypic outcome for disorders characterized by hexosaminidase deficiency. Prenatal diagnosis is now serving as the painful anvil of progress.

More recently, Wenger and Riccardi,[213] while studying Krabbe's disease in two unrelated families, made observations similar to those of Vidgoff. Patients with Krabbe's disease have 0 to 10 per cent of normal activity of β-galactosidase (active against galactosyl ceramide, psychosine, monogalactosyldiglyceride, and galactosyl ceramide.[78, 79, 219] Carriers of Krabbe's disease ordinarily have about 50 per cent of the normal activity of β-galactosidase.[81] Wenger et al.[81] demonstrated a new mutant for this enzyme that is characterized by having activity in the heterozygote that overlaps with levels typical of homozygous Krabbe's disease. These investigators noted that β-galactosidase activity is consistently below 10 per cent of the control values in totally asymptomatic individuals.[213] One patient was the father – and therefore the obligate heterozygote – of a child with typical Krabbe's disease. This father, with only 10 per cent of normal activity for leukocyte galactosyl ceramide β-galactosidase and 6 per cent of normal for leukocyte lactosyl ceramide β-galactosidase, was nevertheless entirely normal.

The implications of these observations, both in Krabbe's disease and in Tay-Sachs disease, are that false-positive diagnoses of both disorders could be made *in utero* as well as after birth. Undoubtedly, similar observations will be made, in time, in other lysosomal storage disorders. Meanwhile, the exhortation that heterozygote status of prospective parents be accurately delineated has been embellished with added credence.

ORGANIZATIONAL PRINCIPLES
OF PRENATAL GENETIC DIAGNOSIS

Philosophy

The basic philosophy, aim, and essential thrust of prenatal genetic diagnosis has been previously emphasized.[13] *It is the assurance to parents of selectively having unaffected offspring when the procreative risk for having defective children becomes unacceptably high.* The real emphasis has *not* been on the removal of defective offspring, but rather on the provision of life for children who may never even have been born. Through the use of the currently accepted indications for genetic amniocentesis, about 95 per cent of all diagnostic studies conclude with a reassuring answer concerning the fetal status.

In the light of our expressed philosophy, parents would not be required to make commitments prior to amniocentesis to seek abortion, should the fetus be found to be defective. All parents have a basic right to know, not only about their own personal health, but also about the health status of their child-to-be. Some parents may wish to exercise this right prior to any decision to terminate a pregnancy. Indeed, experience has shown that most parents in these high-risk situations under study do not need to confront the abortion decision. There are certain geneticists, however, who will deny prenatal genetic studies to those patients not agreeing beforehand to terminate pregnancies where the fetus is found to be abnormal. They point to the time, effort, hazards, and expense involved in prenatal genetic studies and do not feel impelled to provide answers that will be ignored.

Many laboratories have been approached for prenatal sex determination on the basis that if the fetus is not of a particular sex, then pregnancy termination would be sought by the parents. Notwithstanding the foregoing remarks, we have taken the position that prenatal sex determination for "family-planning" reasons represents an inappropriate use of a scarce and expensive technology and have therefore declined to provide any such service.

Education

Initial and continuing education of the physician involved in seeing the pregnant patient and in performing the amniocentesis is vital. Indeed, simultaneous education of the patient serves as an excellent stimulant to education of the physician. The indications for genetic amniocentesis will continue to broaden slowly in the light of new techniques and technologies that make additional disorders diagnosable prenatally. It may not be possible — or even necessary — to familiarize the physician practitioner about detailed tissue culture problems involved in achieving the prenatal diagnosis of complex biochemical disorders. It is critical, however, that he remain apprised of the possibilities and impossibilities of prenatal genetic diagnosis.

Ideally, it might seem wiser to have each patient attend genetic counseling prior to amniocentesis. Although this idea may be presently achievable in some centers, it is likely that only the more difficult problems will be referred to the geneticist for prior consultation. Certainly, genetic counseling would seem to be mandatory after the prenatal diagnosis of fetal abnormality where some explanations would be necessary, e.g., XYY syndrome, XXY syndrome, and other similar conditions.

The physician practitioner should obtain written consent from the patient prior to amniocentesis and after full verbal disclosure of the possible hazards, the need for repeat amniocentesis, the possibilities of failure to obtain fluid or a result, the difficulties of a bloody tap (complicates prenatal diagnosis of neural tube defects), and the rare possibilities of missing the diagnosis (e.g., subtle unbalanced translocation).

The Laboratory

Patients are rarely requested to provide written informed consent for laboratory tests, when the risk is not related to bodily injury. Prenatal genetic studies provide a clear and important exception to this general rule, since written informed consent should be a prerequisite to the initiation of studies. The need for such safeguards is emphasized further by our experience with patients who have not been apprised of the risks of failure due, for example, to lack of cell growth, to contamination of cell cultures, and to other causes previously enumerated. Frequently a patient is encountered who hears from us about the limitations of prenatal diagnosis after the amniocentesis. Such a patient has had no counseling to indicate the universal risks that exist for all parents (2 to 3 per cent) for having offspring with major birth defects.

For a laboratory to initiate a prenatal genetic diagnostic service, a certain level of expertise should have been attained by staff members. It would seem reasonable that the minimum prerequisite for initiation be a success rate in excess of 85 per cent for successful cultures on the first amniotic fluid sample received. All major laboratories providing such services are now clearly in the 90 to 96 per cent success range for results from the first sample. Until such a standard is reached it would seem more prudent that the developing laboratory split all samples with another established service. (Sample splitting should occur at the time of amniocentesis when *two aliquots in separate vials* are collected. Further meddling with a single sample causes very real threats of contamination and subsequent loss of the entire sample in both laboratories.) A final success rate with cell culture following receipt of two or three consecutive amniotic fluid samples from the same patient should ultimately be between 98 and 100 per cent. These figures represent no idle hope or speculation, since many patients may have undertaken pregnancy following the specific reassurance that prenatal diagnosis could assure them of not having an affected child with a serious or fatal genetic disease. In those rare (hopefully) instances where no answers are obtained even after three amniotic fluid samples have been studied, termination of the pregnancy simply on the basis of risk should be offered. The greatest care should be exercised by both obstetrician and geneticist to ensure that such an offer is provided in time for parents to elect abortion. We have thus far been fortunate in having always provided answers to patients (in 800 cases), provided that the first amniotic fluid sample was obtained between 14 and 16 weeks of pregnancy. There are still, however, too many sad occasions where the first antenatal visit occurs after 20 weeks gestation, or where the obstetrician has omitted mentioning amniocentesis and prenatal genetic studies until many weeks after the recommended time for study.

Prenatal genetic studies are unique in that there may only be a brief opportunity to make a serious diagnosis. Hence, the highest standards of excellence must be demanded. Meticulous attention to cell culture technique, continuous monitoring for mycoplasma, connection of incubators to emergency power lines (to avoid catastrophe following power failure), and the application of various measures aimed at anticipating and preventing possible errors, all constitute critical aspects of the care and attention envisaged in such laboratories. Still required are state or federal agencies issuing definitive guidelines for cytogenetic laboratories that provide prenatal diagnostic studies, for surveillance, and for operation of such facilities to ensure or enforce compliance. Meanwhile the single salutary external guideline is the threat of possible litigation. The American Society of Human Genetics is currently working towards resolution of some of these problems. Meanwhile, and as part of routine laboratory surveillance, each group should make it a mandatory rule to confirm their prenatal diagnoses by study of the abortus, cord blood, and newborn skin or blood, as well as by the phenotypic appearance and subsequent short-term development of the child in question. In this context, Kohn *et al.* have cautioned that cord blood — because of maternal lymphocytes — is not an ideal tissue for verification of a prenatal diagnosis.[220] Termination of pregnancy by the use of intra-amniotic prostaglandin F_{2a}, in contrast to the use of hypertonic saline, yields viable fetal tissues that are easily utilized for confirmatory diagnostic studies.[221]

Society

We have previously taken the position that the prevention of mental retardation and genetic disease is a matter for the public health authorities. Hence, it would seem reasonable for society, through its governmental agencies, to ensure that economic considerations not interdict the use by the patient of the new technology available to prevent the occurrence of serious/fatal genetic disease. To condemn the poor to having children with genetic disease or mental retardation is unconscionable. Sadly, these could become the facts only too easily.

Careful cost-benefit analyses for prenatal genetic diagnosis are discussed in Chapter 20. Through the use of the same formula ($275 as the cost of an amniocentesis and possible repeat sample in about 8 per cent of cases + $400 for therapeutic abortion), experience with 1892 cases where chromosomal analysis was performed has been analyzed (Table 9–11). Analysis of the costs incurred to diagnose a single case in the varying categories in which amniocentesis was indicated is shown. As would be expected, the high-risk cases occur in mothers who are aged 40 years and over and where the cost of detecting one abnormal case was $11,400. In the maternal age group 35 to 39 years, the costs were surprisingly low at $19,567. It should be recalled that in previous considerations of the economics of prenatal diagnosis[13] the anticipated lifetime costs of the care of one child with Down's syndrome approximated a quarter of a million dollars.

Milunsky and Reilly have discussed the many medicolegal problems that now abound in prenatal genetic diagnosis.[222] The absence of legal precedent in this new field of medicine has made clear perception of the issues extremely important. In this way, orderly development of guidelines could be pursued, without the complications of poorly considered legislation. Indeed, the tendency to enactment of hasty legislation when an important technology of prevention becomes available has become only too apparent when one views recent legislation governing genetic screening in the United States. Should society see fit to introduce legislation concerning prenatal diagnosis, it is hoped that its essential thrust would be to catalyze the further development of technology and its easy application to those who need it most. Any such legislation must be noncoercive and nondiscriminatory, paying special heed to the respect for the freedom and rights of the individual. The detection of carriers of hereditary disorders is often the first step in prenatal diagnosis, and societal support for such efforts is of paramount importance.

References

1. Steele, M. W., and Breg, W. R., Jr.: Chromosome analysis of human amniotic-fluid cells. Lancet, 1:383, 1966.
2. Lubs, H. A., and Ruddle, F. H.: Chromosomal abnormalities in the human population: Estimation of rates based on New Haven newborn study. Science, 169:495, 1970.
3. Milunsky, A., and Atkins, L.: Prenatal diagnosis of genetic disorders. An analysis of experience with 600 cases. J.A.M.A., 230:232, 1974.
4. Milunsky, A., Littlefield, J. W., Kanfer, J. N., et al.: Prenatal genetic diagnosis. N. Engl. J. Med., 283:1370, 1441, 1498, 1970.

5. Carr, D. H.: Chromosome studies in spontaneous abortions. Obstet. Gynecol., 26:308, 1965.
6. Shapiro, S., Levine, H. S., and Abramowicz, M.: Factors associated with early and late fetal loss. Excerpta Med. Int. Congr. Ser., 224:45, 1970.
7. Warburton, D., and Fraser, F. C.: Spontaneous abortion risks in man: Data from reproductive histories collected in a medical genetics unit. Am. J. Hum. Genet., 16:1, 1964.
8. Blajchman, M. A., Maudsley, R. F., Uchida, I. et al.: Diagnostic amniocentesis and fetal-maternal bleeding. Lancet, 1:993, 1974.
9. Guidelines for antenatal diagnosis of genetic disease. MRC (Canada) Prenatal Diagnosis Newsletter, 3:9, 1974.
10. Dioguardi, N., Agostani, A., Fiorelli, G. et al.: Characterization of lactic dehydrogenase of normal human granulocytes. J. Lab. Clin. Med., 61:713, 1963.
11. Atkins, L., Milunsky, A., and Shahood, J. M.: Prenatal diagnosis: Detailed chromosomal analysis in 500 cases. Clin. Genet., 6:317, 1974.
12. Milunsky, A., Atkins, L., and Littlefield, J. M.: Polyploidy in prenatal genetic diagnosis. J. Pediatr., 79:303, 1971.
13. Milunsky, A.: The Prenatal Diagnosis of Hereditary Disorders. Springfield, Ill., Charles C Thomas, Publisher, 1973.
14. Nadler, H. L., and Gerbie, A. B.: Role of amniocentesis in the intrauterine detection of genetic disorders. N. Engl. J. Med., 282:596, 1970.
15. Ferguson-Smith, M. E., Ferguson-Smith, M. A., Nevin, N. C. et al.: Chromosome analysis before birth and its value in genetic counselling. Br. Med. J., 9:69, 1971.
16. Gerbie, A. B., Nadler, H. L., and Gerbie, M. V.: Amniocentesis in genetic counseling. Am. J. Obstet. Gynecol., 109:765, 1971.
17. Epstein, C. J., Schneider, E. L., Conte, F. A. et al.: Prenatal detection of genetic disorders. Am. J. Hum. Genet., 24:214, 1972.
18. Therkelsen, A. J., Petersen, G. B., Steenstrip, O. R. et al.: Prenatal diagnosis of chromosome abnormalities. Acta Paediatr. Scand., 61:397, 1972.
19. Prescott, G. H., Pernoll, M. L., Hecht, F. et al.: A prenatal diagnostic clinic: An initial report. Am. J. Obstet. Gynecol., 116:942, 1973.
20. Mulcahy, M. T., and Jenkyn, J.: Prenatal diagnosis: Results of cytogenetic analysis of amniotic fluid cell cultures. Med. J. Aust., 1:979, 1973.
21. Aula, P., and Karjalainen, O.: Prenatal karyotype analysis in high risk families. Ann. Clin. Res., 5:142, 1973.
22. Hsu, Y. F., Dubin, E. C., Kerenyi, T. et al.: Results and pitfalls in prenatal cytogenetic diagnosis. J. Med. Genet., 10:112, 1973.
23. Turnbull, A. C., Gregory, P. J., and Laurence, K. M.: Antenatal diagnosis of fetal abnormality with special reference to amniocentesis. Proc. R. Soc. Med., 66:1115, 1973.
24. Robinson, A., Bowes, W., Droegemueller, W. et al.: Intrauterine diagnosis: Potential complications. Am. J. Obstet. Gynecol., 116:937, 1973.
25. Wahlstrom, J.: Prenatal analysis of the chromosome constitution. In Forssman, H., and Åkesson, H. O. (eds.): Reports from the Psychiatric Research Center, University of Göteborg, Sweden, Göteborg Sweden, Orstadius Boktryckeri AB, 1973.
26. Hsu, L. Y. F., and Hirschhorn, K.: Prenatal diagnosis of genetic disease. Life Sci., 14:2311, 1974.
27. Cox, D. M., Niewczas-Late, V., Riffell, I. et al.: Chromosomal mosaicism in diagnostic amniotic fluid cell cultures. Pediatr. Res., 8:679, 1974.
28. Burton, B. K., Gerbie, A. B., and Nadler, H. L.: Present status of intrauterine diagnosis of genetic defects. Am. J. Obstet. Gynecol., 118:718, 1974.
29. Doran, T. A., Rudd, N. L., Gardner, H. A. et al.: The antenatal diagnosis of genetic disease. Am. J. Obstet. Gynecol., 118:310, 1974.
30. Allen, H. H., Sergovich, F., Stuart, E. M. et al.: Infants undergoing antenatal genetic diagnosis: A preliminary report. Am. J. Obstet. Gynecol., 118:310, 1974.
31. Laurence, K. M.: Fetal malformations and abnormalities. Lancet, 2:939, 1974.
32. Golbus, M. S., Conte, F. A., Schneider, E. L., et al.: Intrauterine diagnosis of genetic defects: Results, problems, and follow-up of one hundred cases in a prenatal genetic detection center. Am. J. Obstet. Gynecol., 118:897, 1974.
33. Philip, J., Bang, J., Hahnemann, N. et al.: Chromosome analysis of fetuses in risk pregnancies. Acta Obstet. Gynecol. Scand., 53:9, 1974.

34. Milunsky, A., Atkins, L., and Littlefield, J. W.: Amniocentesis for prenatal genetic studies. Obstet. Gynecol., 40:104, 1972.
35. Yen, S., and MacMahon, B.: Genetics of anencephaly and spina bifida. Lancet, 2:623, 1968.
36. Bobrow, M., Blackwell, N., Unrau, A. E. et al.: Absence of any observed effect of ultrasonic irradiation on human chromosomes J. Obstet. Gynaecol. Br. Commonw., 78:730, 1971.
37. Boyd, E., Abdulla, U., Donald, I. et al.: Chromosome breakage and ultrasound. Br. Med. J., 2:501, 1971.
38. Brock, D. J. H., and Sutcliffe, R. G.: Alpha-fetoprotein in the antenatal diagnosis of anencephaly and spina bifida. Lancet, 2:197, 1972.
39. Milunsky, A., Alpert, E., and Charles, D.: Amniotic fluid alpha-fetoprotein in anencephaly. Obstet. Gynecol., 43:592, 1974.
40. Milunsky, A., and Alpert, E.: The value of alpha-fetoprotein in the prenatal diagnosis of neural tube defects. J. Pediatr., 84:889, 1974.
41. Lorber, J., Stewart, C. R., and Ward, A. M.: Alpha-fetoprotein in antenatal diagnosis of anencephaly and spina bifida. Lancet, 1:1187, 1973.
42. Nevin, N. C., Nesbitt, S., and Thompson, W.: Myelocele and alpha-fetoprotein in amniotic fluid. Lancet, 1:1383, 1973.
43. Allan, L. D., Ferguson-Smith, M. A., Donald, I. et al.: Amniotic fluid alpha-fetoprotein in the antenatal diagnosis of spina bifida. Lancet, 2:522, 1973.
44. Field, B., Mitchell, G., Garrett, W. et al.: Amniotic alpha-fetoprotein levels and anencephaly. Lancet, 2:798, 1973.
45. Brock, D. J. H., and Scrimgeour, J. B.: Early prenatal diagnosis of anencephaly. Lancet, 2:1252, 1972.
46. Bergstrand, C. G., and Czar, B.: Demonstration of a new protein fraction in serum from the human fetus. Scand. J. Clin. Lab. Invest., 8:174, 1956.
47. Gitlin, D., and Boesman, M.: Serum alpha-fetoprotein, albumin, and gamma-G-globulin in the human conceptus. J. Clin. Invest., 45:1826, 1966.
48. Gitlin, D., and Boesman, M.: Sites of serum alpha-fetoprotein synthesis in the human and in the rat. J. Clin. Invest., 46:1010, 1967.
49. Gitlin, D., Perricelli, A., and Gitlin, G. M.: Synthesis of alpha-fetoprotein by liver, yolk sac, and gastrointestinal tract of the human conceptus. Cancer Res., 32:979, 1972.
50. Seppälä, M., and Ruoslahti, E.: Alpha-fetoprotein in amniotic fluid: An index of gestational age. Am. J. Obstet. Gynecol., 114:595, 1972.
51. Seppälä, M., and Ruoslahti, E.: Radioimmunoassay of maternal serum alpha-fetoprotein during pregnancy and delivery. Am. J. Obstet. Gynecol., 112:208, 1972.
52. Ruoslahti, E., and Seppälä, M.: Studies of carcino-fetal proteins. III. Development of a radioimmunoassay for alpha-fetoprotein. Demonstration of alpha-fetoprotein in serum of healthy human adults. Int. J. Cancer, 8:374, 1971.
53. Ruoslahti, E., and Seppälä, M.: Alpha-foetoprotein in normal human serum. Nature (Lond.), 235:161, 1972.
54. Purves, L. R., and Purves, M.: Serum alpha-fetoprotein. VI. The radioimmunoassay evidence for the presence of AFP in the serum of normal people and during pregnancy. S. Afr. Med. J., 46:1290, 1972.
55. Seppälä, M., and Ruoslahti, E.: C. R. Conf. Int. L'Alphafoeto-protein. In Masseyeff, R. (ed.): INSERM, Paris, 1974, p. 387.
56. Bonsnes, R. W.: Composition of amniotic fluid. Clin. Obstet. Gynecol., 9:440, 1966.
57. Seller, M. J., Creasy, M. R., and Alberman, E. D.: Alpha-fetoprotein levels in amniotic fluids from spontaneous abortions. Br. Med. J., 1:524, 1974.
58. Frigoletto, F. D., and Griscom, N. T.: Amniography for the detection of myelomeningocele. Obstet. Gynecol., 44:286, 1974.
59. Seppälä, M.: Increased alpha-fetoprotein in amniotic fluid associated with a congenital esophageal atresia of the fetus. Obstet. Gynecol., 42:613, 1973.
60. Seppälä, M., and Unnerus, H.-A.: Elevated amniotic fluid alpha-fetoprotein in fetal hydrocephaly. Am. J. Obstet. Gynecol., 119:270, 1974.
61. Seppälä, M., Tallberg, T., and Ehnholm, C.: Studies on embryo-specific proteins. Physiological characteristics of embryo-specific alpha-globulin. Ann. Med. Exper. Fenn., 45:16, 1967.

62. Hyden, H., and McEwen, B.: A glial protein specific for the nervous system. Proc. Natl. Acad. Sci. U.S.A., 55:354, 1966.
63. Macri, J. N., Weiss, R. R., Joshi, M. S. et al.: Antenatal diagnosis of neural tube defects using cerebrospinal fluid proteins. Lancet, 1:14, 1974.
64. Weiss, R. R., Macri, J. N., Tejani, N. et al.: Antenatal diagnosis and lung maturation in anencephaly. Obstet. Gynecol., 44:368, 1974.
65. Milunsky, A., Macri, J. N., Weiss, R. R. et al.: Prenatal detection of neural tube defects. Comparative studies between alpha-fetoprotein and β-trace protein. Am. J. Obstet. Gynecol. (In press.)
66. Laurell, C. B.: Quantitative estimation of proteins by electrophoresis in agarose gel-containing antibodies. An. Biochem., 15:45, 1966.
67. Macri, J. N., Weiss, R. R., and Joshi, M. S.: Beta trace protein and neural tube defects. Lancet, 1:1109, 1974.
68. Sutherland, G. R., Brick, D. J. H., and Scrimgeour, J. B.: Amniotic-fluid macrophages and anencephaly. Lancet, 2:1098, 1973.
69. Nelson, M. M., Ruttiman, R.-T., and Brock, D. J. H.: Predictive value of amniotic fluid macrophages in gross C.N.S. defects. Lancet, 1:504, 1974.
70. Leek, A. E., Ruoss, C. F., Kitaw, M. J. et al.: Raised α-fetoprotein in maternal serum with anencephalic pregnancy. Lancet, 2:385, 1973.
71. Brock, D. J. H., Bolton, A. E., and Monaghan, J. M.: Prenatal diagnosis of anencephaly through maternal serum alpha-fetoprotein measurement. Lancet, 2:923, 1973.
72. Editorial: Towards the prevention of spina bifida. Lancet, 1:907, 1974.
73. Seppälä, M.: Alpha-fetoprotein in the management of high-risk pregnancies. In Milunsky, A. (ed.): Clinics in Perinatology. Symposium on Management of the High-Risk Pregnancy. Vol. 1. Philadelphia, W. B. Saunders Company, 1974, p. 293.
74. Seppälä, M., and Ruoslahti, E.: Alpha-fetoprotein in maternal serum: A new marker for detection of fetal distress and intrauterine death. Am. J. Obstet. Gynecol., 115:48, 1973.
75. Seppälä, M., and Ruoslahti, E.: Alpha-fetoprotein in Rh-immunized pregnancies. Obstet. Gynecol., 42:701, 1973.
76. Seppälä, M., and Ruoslahti, E.: Alpha-fetoprotein: Physiology and pathology during pregnancy and application to antenatal diagnosis. J. Perinat. Med. 1:104, 1973.
77. Harris, R., Jennison, R. F., Barson, A. J. et al.: A comparison of amniotic fluid and maternal serum alpha feto-protein levels in the early antenatal diagnosis of spina bifida and anencephaly. Lancet, 1:429, 1974.
78. Suzuki, K., and Suzuki, Y.: Globoid cell leukodystrophy (Krabbe's disease): Deficiency of galactocerebroside beta-galactosidase. Proc. Natl. Acad. Sci. U.S.A., 66:302, 1970.
79. Wenger, D. A., Sattler, M., and Hiatt, W.: Globoid cell leukodystrophy: Deficiency of lactosyl ceramide beta-galactosidase. Proc. Natl. Acad. Sci. U.S.A., 71:854, 1974.
80. Suzuki, Y., and Suzuki, K.: Krabbe's globoid cell leukodystrophy: Deficiency of galactocerebrosidase in serum, leukocytes, and fibroblasts. Science, 171:73, 1971.
81. Wenger, D. A., Sattler, M., Clark, C. et al.: An improved method for the identification of patients and carriers of Krabbe's disease. Clin. Chim. Acta, 56:199, 1974.
82. Dawson, G., and Stein, A. O.: Lactosyl ceramidosis: Catabolic enzyme defect of glycosphingolipid metabolism. Science, 170:556, 1970.
83. Max, S. R., Maclaren, N. K., Brady, R. O. et al.: GM_3 (hematoside) sphingolipodystrophy. N. Engl. J. Med., 291:929, 1974.
84. Sugita, M., Dulaney, J. T., and Moser, H. W.: Ceramidase deficiency in Farber's disease (lipogranulomatosis). Science, 178:1100, 1972.
85. Ampola, M. G., Mahoney, M. J., Nakamura, E. et al.: In utero treatment of methylmalonic acidemia (MMA-EMIA) with vitamin B_{12}. Pediatr. Res., 8:113, 1974 (Abstract).
86. Matalon, R., Arbogast, B., and Dorfman, A.: Morquio's syndrome: A deficiency of chondroitin sulfate N-acetylhexosamine sulfate sulfatase. Pediatr. Res., 8:162, 1974 (Abstract).
87. Lee, T.-Y., and Schafer, I. A.: Glycosaminoglycan composition of human amniotic fluid. Biochim. Biopshys. Acta, 354:264, 1974.

88. Spranger, J. W., and Wiedemann, H. R.: The genetic mucolipidoses. Humangenetik, 9:113, 1970.

89. Leroy, J. G., and De Mars, R. I.: Mutant enzymatic and cytological phenotypes in cultured human fibroblasts. Science, 157:804, 1967.

90. Tondeur, M., Vamos-Hurwitz, E., Mockel-Pohl, S. et al.: Clinical, biochemical and ultrastructural studies in cases of chondrodystrophy presenting the I-cell phenotype in tissue culture. J. Pediatr., 79:366, 1971.

91. Maroteaux, P., and Lamy, M.: La pseudo-polydystrophie de Hurler. Presse Med., 74:2889, 1966.

92. Huijing, F., Warren, R. J., and McLeod, A. G. W.: Elevated activity of lysosomal enzymes in amniotic fluid of a fetus with mucolipidosis II (I-cell disease). Clin. Chim. Acta, 44:453, 1973.

93. Wiesmann, U. N., and Herschkowitz, N. N.: Studies on the pathogenetic mechanism of I-cell disease in cultured fibroblasts. Pediatr. Res., 8:865, 1974.

94. Leroy, J. G., and van Elsen, A. F.: I-cell disease (mucolipidosis type II) serum hydrolases in obligate heterozygotes. Humangenetik, 20:119, 1973.

95. Murphy, J. V., Diven, W. F., and Craig, L.: Detection of Leigh's disease in fibroblasts. Pediatr. Res., 8:119, 1974 (Abstract).

96. Murphy, J. V.: Subacute necrotizing encephalomyelopathy (Leigh's disease). Detection of the heterozygous carrier state. Pediatrics, 51:710, 1973.

97. Murphy, J. V., Craig, L., and Glew, R.: Leigh's disease: Biochemical nature of the inhibitor. Pediatr. Res., 8:118, 1974 (Abstract).

98. Barnett, D. R., Barranco, S. C., Lockhart, L. H. et al.: Cystic fibrosis: Growth kinetics and production of the ciliary inhibitor by cultured fibroblasts. Tex. Rep. Biol. Med. 31:4, 1973.

99. Bowman, B. J., Lockhart, L. H., Herzberg, V. L. et al.: Cystic fibrosis: Synthesis of ciliary inhibitor by amniotic cells. Clin. Genet., 4:461, 1973.

100. Conover, J. H., Bonforte, R. J., Hathaway, P. et al.: Studies on ciliary dyskinesia factor in cystic fibrosis. I. Bioassay and heterozygote detection in serum. Pediatr. Res., 7:220, 1973.

101. Bowman, B. H., Hirschhorn, K., and Bearn, A. G.: Bioassays of cystic-fibrosis factor. Lancet, 1:404, 1974.

102. Toop, J., and Emery, A. E. H.: Muscle histology in fetuses at risk for Duchenne muscular dystrophy. Clin. Genet., 5:230, 1974.

103. Brown, M. S., and Goldstein, J. L.: Familial hypercholesterolemia: Defective binding of lipoproteins to cultured fibroblasts associated with impaired regulation of 3-hydroxy-3-methylglutaryl coenzyme—a reductase activity. Proc. Natl. Acad. Sci., 71:788, 1974.

104. Goldstein, J. L., and Brown, M. S.: Familial hypercholesterolemia: Identification of a defect in the regulation of 3-hydroxy-3-methylglutaryl coenzyme. A reductase activity associated with overproduction of cholesterol. Proc. Natl. Acad. Sci. U.S.A., 70:2804, 1973.

105. Goldstein, J. L., and Brown, M. S.: Binding and degradation of low density lipoproteins by cultured human fibroblasts. J. Biol. Chem., 249:5153, 1973.

106. Kachadurian, A. K., and Kawahara, F. S.: Cholesterol synthesis by cultured fibroblasts: Decreased feedback inhibition in familial hypercholesterolemia. J. Lab. Clin. Med., 83:7, 1974.

107. Scrimgeour, J. B.: Other techniques for antenatal diagnosis. In Emery, A. E. H. (ed.): Antenatal Diagnosis of Genetic Disease. Baltimore, Md., The Williams & Wilkins Co., 1973, p. 40.

108. Valenti, C.: Endoamnioscopy and fetal biopsy: A new technique Am. J. Obstet. Gynecol., 114:561, 1972.

109. Valenti, C.: Antenatal detection of hemoglobinopathies: A preliminary report. Am. J. Obstet. Gynecol., 115:851, 1973.

110. Hobbins, J. C., Mahoney, M. J., and Goldstein, L. A.: New method of intrauterine evaluation by the combined use of fetoscopy and ultrasound. Am. J. Obstet. Gynecol., 118:1069, 1974.

111. Patrick, J. E., Perry, T. B., and Kinch, R. A. H.: Fetoscopy and fetal blood sampling: A percutaneous approach. Am. J. Obstet. Gynecol., 119:539, 1974.

112. Hollenberg, M. D., Kaback, M. M., and Kazazian, H. H., Jr.: Adult hemoglobin synthesis by reticulocytes from the human fetus at midtrimester. Science, 174:698, 1971.

113. Pataryas, H. A., and Stamatoyannopoulos, G.: Hemoglobins in human fetuses: Evidence for adult hemoglobin production after the 11th gestational week. Blood, 39:688, 1972.

114. Rudd, N.: Personal communication.

115. Campbell, S.: The assessment of fetal development by diagnostic ultrasound. *In* Milunsky, A. (ed.): Clinics in Perinatology. Symposium on Management of the High-Risk Pregnancy. Vol. 1. Philadelphia, W. B. Saunders Company, 1974, p. 507.

116. Holmberg, L., Henriksson, P., Ekelund, H. et al.: Coagulation in the human fetus. Comparison with term newborn infants. J. Pediatr., 85:860, 1974.

117. Beaudet, A. L., Lipson, M. H., Ferry, G. D. et al.: Acid lipase in cultured fibroblasts: Cholesterol ester storage disease. J. Lab. Clin. Med., 84:54, 1974.

118. Brady, R. O., Uhlendorf, B. W., and Jacobson, C. B.: Fabry's disease: Antenatal detection. Science, 172:175, 1971.

119. Epstein, C. J., Schneider, E. L., Conte, F. A. et al.: Prenatal detection of genetic disorders. Am. J. Hum. Genet., 24:214, 1972.

119b. Schneider, E. L., Ellis, W. G., Brady, R. O. et al.: Infantile (type II) Gaucher's disease: In utero diagnosis and fetal pathology. J. Pediatr., 81:1134, 1972.

120. Lowden, J. A., Cutz, E., Conen, P. E. et al.: Prenatal diagnosis of GM_1 gangliosidosis. N. Engl. J. Med., 288:225, 1973.

121. O'Brien, J. S., Ho, M. W., Veath, M. L. et al.: Juvenile GM_1 gangliosidosis: Clinical, pathological, chemical and enzymatic studies. Clin. Genet., 3:411, 1972.

122. O'Brien, J. S., Okada, S., Fillerup, D. L. et al.: Tay-Sachs disease: Prenatal diagnosis. Science, 172:61, 1971.

123. Desnick, R. J., Raman, M. K., Bendel, R. P. et al.: Prenatal diagnosis of glycosphingolipidoses: Sandhoff's and Fabry's diseases. J. Pediatr., 83:149, 1973.

124. O'Brien, J. S., Okada, S., Ho, M. W. et al.: Ganglioside-storage diseases. Fed. Proc., 30:956, 1971.

125. Epstein, C. J., Brady, R. O., Schneider, R. M. et al.: In utero diagnosis of Niemann-Pick disease. Am. J. Hum. Genet., 23:533, 1971.

126. Herndon, J. H., Jr., Steinberg, D., Uhlendorf, B. W. et al.: Refsum's disease: Characterization of the enzyme defect in cell culture. J. Clin. Invest., 48:1017, 1969.

127. Kyriakides, E. C., Filippone, N., Paul, B. et al.: Lipid studies in Wolman's disease. Pediatrics, 46:431, 1970.

128. Crawfurd, M. d'A., Dean, M. F., Hunt, D. M. et al.: Early prenatal diagnosis of Hurler's syndrome with termination of pregnancy and confirmatory findings on the fetus. J. Med. Genet., 10:144, 1973.

129. Bach, G., Friedman, R., Weismann, B. et al.: The defect in the Hurler and Scheie syndromes: Deficiency of an α-L-iduronidase. Proc. Natl. Acad. Sci. U.S.A., 69:2048, 1972.

130. Bach, G., Eisenberg, F., Jr., Cantz, M. et al.: The defect in the Hunter syndrome: Deficiency of sulfoiduronate sulfatase. Proc. Natl. Acad. Sci. U.S.A., 70:2134, 1973.

131. Fratantoni, J. C., Neufeld, E. F., Uhlendorf, B. W. et al.: Intrauterine diagnosis of the Hurler and Hunter syndromes. N. Engl. J. Med., 280:686, 1969.

132. Renwick, J. H.: Progress in mapping human autosomes. Br. Med. Bull., 25:65, 1969.

133. Harper, P. S., Laurence, K. M., Parkes, A. et al.: Sanfilippo A disease in the fetus. J. Med. Genet., 11:123, 1974.

134. O'Brien, J. S.: Sanfilippo syndrome: Profound deficiency of alpha-acetyl glucosaminidase activity in organs and skin fibroblasts from type B patients. Proc. Natl. Acad. Sci. U.S.A., 69:1720, 1972.

135. Stumph, D. A., Austin, J. H., Crocker, A. C. et al.: Mucopolysaccharidosis type VI (Maroteaux-Lamy syndrome). I sulfatase B deficiency in tissues. Am. J. Dis. Child., 126:747, 1973.

136. Hall, C. W., Cantz, M., and Neufeld, E. H.: A β-glucuronidase deficiency mucopolysaccharidosis: Studies in cultured fibroblasts. Arch. Biochem. Biophys., 155:32, 1973.

137. Kan, Y. W., Dozy, A. M., Alter, B. P., et al.: Detection of the sickle gene in the human fetus. Potential for intrauterine diagnosis of sickle cell anemia. N. Engl. J. Med., 287:1, 1972.
138. Jacoby, L. B., Littlefield, J. W., Milunsky, A. et al.: A microassay for argininosuccinase in cultured cells. Am. J. Hum. Genet., 24:321, 1972.
139. Suzuki, K., Schneider, E. L. and Epstein, C. J.: In utero diagnosis of globoid cell leucodystrophy (Krabbe's disease). Biochem. Biophys. Res. Commun., 45:1363, 1971.
140. Roerdink, F. H., Gouw, W. L. M., Okken, A. et al.: Citrullinemia. Report of a case with studies on antenatal diagnosis. Pediatr. Res., 7:863, 1973.
141. Russell, A., Levin, B., Oberholzer, V. G. et al.: Hyperammonaemia: A new instance of an inborn enzymatic defect of the biosynthesis of urea. Lancet, 2:699, 1962.
142. Fleisher, L. D., Longhi, R. C., Tallan, H. H. et al.: Homocystinuria: Investigations of cystathionine synthase in cultured fetal cells and the prenatal determination of genetic status. J. Pediatr., 85:677, 1974.
143. Frimpter, G. W., Greenberg, A. J., Hilgartner, M. et al.: Cystathioninuria: Management. Am. J. Dis. Child., 113:115, 1967.
144. Komrower, G. M.: The philosophy and practice of screening for inherited diseases. Pediatrics, 53:182, 1974.
145. Emery, A. E. H., Burt, D., Nelson, M. M. et al.: Antenatal diagnosis and amino-acid composition of amniotic fluid. Lancet, 1:1307, 1970.
146. Levy, H. L.: Hartnup disease. In Goldensohn, E. S., and Appel, S. H. (eds.): Cellular and Molecular Basis of Neurologic Disease. Philadelphia, Lea & Febiger (In press).
147. Bostrom, H., and Hambraeus, L.: Cystinuria in Sweden VII. Clinical histopathological and medico-social aspects of the disease. Acta Med. Scand., 175:411, 1964.
148. Tada, K., Wada, Y., and Arakawa, T.: Hypervalinemia: Its metabolic lesion and therapeutic approach. Am. J. Dis. Child., 113:64, 1967.
149. Daum, R. S., Mamer, O. A., Lamm, P. H. et al.: A "new" disorder of isoleucine catabolism. Lancet, 2:1289, 1971.
150. Shih, V., and Tanaka, K.: Personal communication.
151. Nadler, H. L., and Messina, A. M.: In-utero detection of type-II glucogenosis (Pompe's disease). Lancet, 2:1277, 1969.
152. Dancis, J., Hutzler, J., and Cox, R. O.: Enzyme defect in skin fibroblasts in intermittent branched-chain ketonuria and in maple syrup urine disease. Biochem. Med., 2:408, 1969.
153. Mahoney, M. J., and Rosenberg, J. E.: Defective metabolism of vitamin B_{12} in fibroblasts from children with methylmalonicaciduria. Biochem. Biophys. Res. Commun., 44:375, 1971.
154. Mudd, S. H., Uhlendorf, B. W., Freeman, J. M. et al.: Homocystinuria associated with decreased methylenetetrahydrofolate reductase activity. Biochem. Biophys. Res. Commun., 46:905, 1972.
155. Mudd, S. H.: Homocystinuria: The known causes in inherited disorders of sulphur metabolism. In Carson, N. A. J., and Raine, D. N. (eds.): Proceedings of the 8th Symposium of the Society for the Study of Inborn Errors of Metabolism, Belfast, 1970. London, E. & S. Livingston, Ltd., 1971.
156. Shih, V. E., and Schulman, J. D.: Ornithine-ketoacid transaminase activity in human skin and amniotic fluid cell culture. Clin. Chim. Acta, 27:73, 1970.
157. Logan, R. W.: Antenatal diagnosis of propionicacidaemia. Lancet, 1:1128, 1973.
158. Tildon, J. T., Leffler, A. T., Cornblath, M. et al.: Abnormal glucose metabolism in skin fibroblasts cultured from a patient with a new syndrome of ketoacidemia. Pediatr. Res., 5:518, 1971.
159. Levy, H. L., Mudd, S. H., Schulman, J. D. et al.: A derangement in B_{12} metabolism associated with homocystinemia, cystathioninemia, hypomethioninemia and methylmalonic aciduria. Am. J. Med., 48:390, 1970.
160. Zielke, K., Okada, S., and O'Brien, J. S.: Fucosidosis: Diagnosis by serum assay of alpha-L-fucosidase. J. Lab. Clin. Med., 79:164, 1972.
161. Gitzelmann, R.: Hereditary galactokinase deficiency, a newly recognized cause of juvenile cataracts. Pediatr. Res., 1:14, 1967.
162. Nadler, H. L.: Antenatal detection of hereditary disorders. Pediatrics, 42:912, 1968.
163. Nadler, H. L.: Patterns of enzyme development utilizing cultivated human fetal cells derived from amniotic fluid. Biochem. Genet., 2:119, 1968.

164. Galjaard, H., Mekes, R., De Josselin De Jong, J. E. et al.: A method for rapid prenatal diagnosis of glycogenosis II (Pompe's disease). Clin. Chim. Acta, 49:361, 1973.

165. Justice, P., Ryan, C., Hsia, D. Y.-Y.: Amylo-1, 6-glucosidase in human fibroblasts: Studies in type III glycogen storage disease. Biochem. Biophys. Res. Commun., 39:301, 1970.

166. Howell, R. R., Kaback, M. M., and Brown, B. I.: Glycogen storage disease type IV. Branching enzyme deficiency in skin fibroblasts and possibly heterozygote detection. J. Pediatr., 78:638, 1971.

167. Kjellman, B., Gamstorp, I., Brun, A. et al.: Mannosidosis: A clinical and histopathologic study. J. Pediatr., 75:366, 1969.

168. Krone, W., Schneider, G., and Schulz, D.: Detection of phosphohexose isomerase deficiency in human fibroblast cultures. Humangenetik, 10:224, 1970.

169. Blass, J. P., Avigan, J., and Uhlendorf, B. W.: A defect in pyruvate decarboxylase in a child with an intermittent movement disorder. J. Clin. Invest., 49:423, 1970.

170. Blass, J. P., Schulman, J. D., Young, D. S. et al.: An inherited defect affecting the tricarboxylic acid cycle in a patient with congenital lactic acidosis. J. Clin. Invest., 51:1845, 1972.

171. Krooth, R. S., Howell, R. R., and Hamilton, H. B.: Properties of acatalasic cells growing in vitro. J. Exp. Med., 115:313, 1962.

172. Giblett, E. R., Anderson, J. E., Cohen, F. et al.: Adenosine-deaminase deficiency in two patients with severely impaired cellular immunity. Lancet, 2:1067, 1972.

173. Danes, B. S., and Bearn, A. G.: Cell culture and the Chediak-Higashi syndrome. Lancet, 2:65, 1967.

174. Romeo, G., Kaback, M. M., and Levin, E. Y.: Uroporphyrinogen III. Cosynthetase activity in fibroblasts from patients with congenital erythropoietic porphyria. Biochem. Genet., 4:659, 1970.

175. Schneider, J. A., Verroust, F. M., Kroll, W. A. et al.: Prenatal diagnosis of cystinosis. N. Engl. J. Med., 290:878, 1974.

176. Boyle, J. A., Raivio, K. O., Schulman, J. D., et al.: Lesch-Nyhan syndrome: Preventive control by prenatal diagnosis. Science, 169:688, 1970.

177. Nadler, H. L., and Egan, T. J.: Deficiency of lysosomal acid phosphatase: A new familial metabolic disorder. N. Engl. J. Med., 282:302, 1970.

178. Krane, S. M., Pinnell, S. R., and Erbe, R. W.: Lysyl-protocollagen hydroxylase deficiency in fibroblasts from siblings with hydroxylysine-deficient collagen. Proc. Natl. Acad. Sci. U.S.A., 69:2899, 1972.

179. Schrott, H. G., Karp, L., and Omenn, G. S.: Prenatal prediction in myotonic dystrophy: Guidelines for genetic counseling. Clin. Genet., 4:38, 1973.

180. Kan, Y. W., Dozy, A. M., Alter, B. T. et al.: Intrauterine diagnosis of thalassemia. Ann. N.Y. Acad. Sci., 232:145, 1974.

181. Krooth, R. S.: Properties of diploid cell strains developed from patients with an inherited abnormality of uridine biosynthesis. Symp. Quant. Biol., 29:189, 1964.

182. Simell, O., Johansson, T., and Aula, P.: Enzyme defect in saccharopinuria. J. Pediatr., 82:54, 1973.

183. Melancon, S. B., Lee, S. Y., and Nadler, H. L.: Histidase activity in cultivated human amniotic fluid cells. Science, 173:627, 1971.

184. Shanie, D. D., Hirschhorn, K., and New, M. I.: Metabolism of testosterone-^{14}C by cultured human cells. J. Clin. Invest., 51:1459, 1972.

185. Fluharty, A. L., Stevens, R. L., Sanders, D. L. et al.: Arylsulfatase B deficiency in Maroteaux-Lamy syndrome cultured fibroblasts. Biochem. Biophys. Res. Commun., 59:455, 1974.

186. Ramsay, C. A., Coltart, T. M., Blunt, S. et al.: Prenatal diagnosis of xeroderma pigmentosum. Lancet, 2:1109, 1974.

187. Bloom, A. D., Schmickel, R., Barr, M. et al.: Prenatal detection of autosomal mosaicism. J. Pediatr., 84:732, 1974.

188. Kajii, T.: Pseudomosaicism in cultured amniotic fluid cells. Lancet, 2:1037, 1971.

189. Kardon, N. B., Chernay, P. R., Hsu, L. Y. et al.: Pitfalls in prenatal diagnosis resulting from chromosomal mosaicism. J. Pediatr., 80:297, 1972.

190. Ellis, J. R.: Spontaneous translocation in a cell culture. Ann. Hum. Genet., 26:287, 1963.

191. Kohn, G., Aronson, M., and Mellman, W. J.: Spontaneous D/G translocation in fibroblasts derived from a patient with trisomy 21 mosaicism. Clin. Genet., 5:113, 1974.
192. Mannanal, G. S.: The issue of chromosomal mosaicism in prenatal diagnosis. J. Pediatr., 81:422, 1972.
193. Wendel, U., Rudiger, H. W., Passarge, E. et al.: Maple syrup urine disease: Rapid prenatal diagnosis by enzyme assay. Humangenetik, 19:127, 1973.
194. Fogh, J. (ed.): Contamination in cell cultures. New York, Academic Press, Inc., 1973.
195. Stanbridge, E.: Mycoplasmas and cell cultures. Bacteriol. Rev., 35:206, 1971.
196. Schneider, E. L., Stanbridge, E. J., Epstein, C. J. et al.: Mycoplasma contamination of cultured amniotic fluid cells: Potential hazard to prenatal chromosomal diagnosis. Science, 184:477, 1974.
197. Aula, P., and Nichols, W. W.: The cytogenetic effects of mycoplasma in human leukocyte cultures. J. Cell. Physiol., 70:281, 1967.
198. Smith, P. F.: Amino acid metabolism of PPLO. Ann. N.Y. Acad. Sci., 79:543, 1960.
199. Hayflick, L.: Tissue cultures and mycoplasmas. Tex. Rep. Biol. Med., 23(suppl.) 285, 1965.
200. Schneider, E. L., Epstein, C. J., Epstein, W. J. et al.: Detection of mycoplasma contamination in cultured human fibroblasts. Comparison of biochemical and microbiological techniques. Exp. Cell. Res., 79:343, 1973.
201. Schneider, E. L., Stanbridge, E. J., and Epstein, C. J.: Incorporation of ^{3}H-uridine ^{3}H-uracil into RNA. Exp. Cell. Res., 84:311, 1974.
202. Ford, J.: Induction of chromosomal errors. Lancet, 1:54, 1973.
203. Ingalls, T. H., and Shimada, T.: pH disturbances and chromosomal anomalies. Lancet, 1:872, 1974.
204. Lubs, H. A., and Ruddle, F. H.: Applications of quantitative karyotypy to chromosome variation in 4400 consecutive newborns. In Jacobs, P. A., Price, W. H., and Law, P. (eds.): Human Population Cytogenetics. Edinburgh, University Press, 1970.
205. Lejeune, J., and Berger, R.: Sur deux observations familiales de translocations complexes. Ann. Genet. (Paris), 8:21, 1965.
206. Uchida, I. A., Holunga, R., and Lawler, C.: Maternal radiation and chromosomal aberrations. Lancet, 2:1045, 1968.
207. Khudr, G., and Benirschke, K.: Fluorescence of the Y chromosome: A rapid test to determine fetal sex. Am. J. Obstet. Gynecol., 110:1091, 1971.
208. Sulica, L. O., Borgaonkar, D. S., and Shah, S. A.: Accurate identification of the human Y chromosome. Clin. Genet., 5:17, 1974.
209. Wilson, M. G., Towner, J. W., Lipshin, J. et al.: Identification of an unusual Y chromosome in YY mosaicism by quinacrine fluorescence. Nature (Lond.), 231:388, 1971.
210. Buhler, E. M., Muller, H., Muller, J. et al.: Variant of the Y-fluorescence pattern in an abnormal human Y chromosome. Nature (Lond.), 234:348, 1971.
211. Galjaard, H.: Possibilities of quantitative histochemical analyses in prenatal detection and genetic classification of hereditary metabolic disorders. In Proceedings. Fourth International Congress of Histo- & Cytochemistry, Kyoto, 1972, p. 185.
212. Richardson, B. J., and Cox, D. M.: Rapid tissue culture and microbiochemical methods for analyzing colonially grown fibroblasts from normal, Lesch-Nyhan and Tay-Sachs patients and amniotic fluid cells. Clin. Genet., 4:376, 1973.
213. Wenger, D. A., and Riccardi, V. M.: False positive diagnosis of Krabbe's disease. J. Pediatr., (In press).
214. Vidgoff, J., Buist, N. R. M., and O'Brien, J. S.: Absence of beta-N-acetyl-D-hexosaminidase A activity in a healthy woman. Am. J. Hum. Genet., 25:372, 1973.
215. Spence, M. W., Ripley, B. A., Embil, J. A. et al.: A new variant of Sandhoff's disease. Pediatr. Res., 8:628, 1974.
216. Srivastava, S. K., Awasthi, Y. C., Yoshida, A. et al.: Studies on human β-D-N-acetyl-hexosaminidases. I: Purification and properties. J. Biol. Chem., 249:2043, 1974.
217. Srivastava, S. K., and Beutler, E.: Studies on human β-D-N-acetylhexosaminidases. III. Biochemical genetics of Tay-Sachs and Sandhoff's diseases. J. Biol. Chem., 249:2054, 1974.
218. Lalley, P. A., Rattazzi, M. C., and Shows, T. B.: Human β-D-N-acetylhexosaminidases A and B: Expression and linkage relationships in somatic cell hybrids. Proc. Natl. Acad. Sci. U.S.A., 71:1569, 1974.

219. Wenger, D. A., Sattler, M., and Markey, S. P.: Deficiency of monogalactosyl diglyceride beta-galactosidase activity in Krabbe's disease. Biochem. Biophys. Res. Commun., 53:680, 1973.
220. Kohn, G., Ornoy, A., and Cohen, M. M.: Prenatal diagnosis: A problem in verification. J. Med. Genet., 11:247, 1974.
221. Golbus, M. S., and Erickson, R. P.: Mid-trimester abortion induced by intraamniotic prostaglandin F_{2a}: Fetal tissue viability. Am. J. Obstet. Gynecol., 119:268, 1974.
222. Milunsky, A., and Reilly, P.: Medico-legal issues in the prenatal diagnosis of hereditary disorders. Am. J. Law. Med. Vol. 1, 1975 (In press).
223. Leroy, J. G., Van Elsen, A. F., Martin, J.-J. et al.: Infantile metachromatic leukodystrophy — confirmation of a prenatal diagnosis. N. Engl. J. Med., 288:1365, 1973.
224. Van der Hagen, C. B., Borresen, A.-L., Molne, K. et al.: Metachromatic leukodystrophy I. Prenatal detection of arylsulphatase A deficiency. Clin. Genet., 4:256, 1973.
225. Goodman, S. I., Mace, J. W., Turner, B. et al.: Antenatal diagnosis of argininosuccinic aciduria. Clin. Genet., 4:236, 1973.
226. Aula, P., Nanto, V., Laipio, M.-L. et al.: Aspartylglucosaminuria: Deficiency of aspartylglucosaminidase in cultured fibroblasts. Clin. Genet., 4:297, 1973.
227. Stanbury, J. B., Wyngaarden, J. B., and Fredrickson, D. S. (eds.): The Metabolic Basis of Inherited Disease, 3rd ed. New York, McGraw-Hill Book Co., 1972.
228. Galjaard, H., Van Hoogstraten, J. J., DeJosselin, J. E. et al.: Methodology of the quantitative cytochemical analysis of single or small numbers of cultured cells. Histochem. J., 6:409, 1974.
229. Niermeijer, M. F., Fortuin, J. J. H., Koster, J. F. et al.: Prenatal diagnosis of some lysosomal storage diseases. In Tager, J. M., Hooghwinkel, G. J. M., and Daems, W. Th. (eds.): Enzyme Therapy in Lysosomal Storage Diseases. Amsterdam, North-Holland Publishing Company, 1974, p. 25.
230. Galjaard, H., Hoogeveen, A., Keijzer, W. et al.: The use of quantitative cytochemical analyses in rapid prenatal detection and somatic cell genetic studies of metabolic diseases. Histochem. J., 6:491, 1974.
231. Hirschhorn, R., Beratis, N., Rosen, F. S. et al.: Adenosine-deaminase deficiency in a child diagnosed prenatally. Lancet, 1:73, 1975.
232. Kjessler, B., Johansson, S. G. O., Sherman, M. et al.: Alpha-fetoprotein in antenatal diagnosis of congenital nephrosis. Lancet, 1:432, 1975.

CHAPTER 10

HOMOZYGOTE SCREENING IN THE DISORDERS OF AMINO ACID METABOLISM

Vivian E. Shih, M.D.

Inborn errors of amino acid metabolism are among the few genetic disorders in which mental retardation is preventable by early diagnosis and early treatment. It was not possible to estimate the frequency of inborn errors of amino acid metabolism in the general population until results of mass screening of newborn infants became available. Table 10-1 gives the worldwide results of routine newborn screening, as compiled by Levy.[1, 2]

There are considerable variations among different countries in the incidence of some of the disorders. These variations may be related to genetic differences in the population. In addition, the diagnostic criteria and the completeness of follow-up surveys may influence the reported incidence. Furthermore, reports of rare disorders, on the basis of one or two cases, cannot be considered reliable. Phenylketonuria (PKU) has been screened in over 13 million newborns and continues to be, by far, the most common of the amino acid disorders that cause mental retardation.

Experience in the Massachusetts Newborn Metabolic Screening Program has shown that, as a group, inborn errors of amino acid metabolism occur in about one out of 2500 neonates screened.[2] In other words, at least 30 to 35 infants born each year in Massachusetts are expected to have an aminoacidopathy. One third of these newborns will be mentally retarded or will die early if the condition is undiagnosed and untreated.

TABLE 10–1. WORLDWIDE INCIDENCE OF AMINOACIDOPATHY BY ROUTINE NEWBORN SCREENING*

Disorder	Number Screened	Number Detected (Overall Incidence)	Range of Incidence HIGH	Range of Incidence LOW
Hyperphenylalaninemias	13,665,644			
Phenylketonuria		1190 (1:11,500)	1:4,500	1:20,000
Other hyperphenylalaninemias		133 (1:30,000)	1:11,000	1:200,000
Maple syrup urine disease	3,372,363	17 (1:200,000)	1:42,000	1:350,000
Hypermethioninemia (Homocystinuria)	2,781,042	12 (1:230,000)	1:36,000	1:310,000
Hereditary tyrosinemia	1,401,777	0		
Histidinemia	1,098,215	47 (1:23,000)	1:12,000	1:77,000
Hyperprolinemia	631,025	4 (1:150,000)		
Hartnup disease	651,323	26 (1:25,000)	1:15,000	1:46,000
Cystinuria	651,323	83 (1:8,000)	1:2,000	1:15,000
Renal familial iminoglycinuria	651,323	53 (1:12,000)	1:10,000	1:18,000
Argininosuccinic acidemia	364,025	4 (1:90,000)		
Cystathionuria	364,025	5 (1:73,000)		
Hyperglycinemia, nonketotic	364,025	2 (1:180,000)		
Hyperlysinemia	364,025	1 (4:300,000)		
Hyperglycinemia, ketotic (Propionic acidemia)	364,025	1 < (1:300,000)		

*Adapted from Levy, H. L.: Genetic screening. *In* Harris, H., and Hirschhorn, K. (eds.): Advances in Human Genetics. Vol. 4. New York, Plenum Publishing Corp., 1974, p. 1, and Levy, H. L.: Neonatal screening for inborn errors of amino acid metabolism. Clin. Endocr. Metab., 3:153, 1974.

Thus, each year in Massachusetts alone, there will be 8 to 10 potential candidates for institutional care. Certain homocystinuric patients (having cystathionine synthase deficiency) may escape mental retardation, but they may be considerably disabled by poor vision resulting from dislocated lens or glaucoma and by stroke and myocardial infarction resulting from thromboembolism. In terms of the financial burden to society, the savings from early diagnosis by screening is considerable. For instance, the cost for diagnosis and treatment of phenylketonuria in North America is only about one third of the corresponding cost for institutional care.[3] Furthermore, benefits to the affected persons and society, in terms of quality of life and productivity, can hardly be argued.

HISTORY

Homozygote screening started with phenylketonuria (PKU). PKU is the first amino acid disorder that has been found to cause mental retardation. It was discovered in Norway by Fölling, who identified phenylpyruvic acid in the urine of two retarded siblings.[4] He then hypothesized that a disturbance in the metabolism of phenylalanine might be causing the retardation in these patients and in other retarded people, as well. In 1954, Jervis[5] conducted a survey of over 48,000 mentally defective institutionalized patients from 12 countries. He reported the incidence of PKU to be 0.64 per cent of this population, thereby calling attention to the role of PKU in mental retardation.

This area of research was further advanced by the discovery that biochemical abnormalities in PKU could be controlled by a low-phenylalanine diet.[6] It was subsequently learned that when such a diet was instituted early in life, it could prevent, as well as control, mental retardation. Understandably, the realization by investigators that mental retardation associated with PKU could be prevented by early treatment underscored the importance of early diagnosis in the population-at-large. Ferric chloride tests of the urine in infant screening proved to be unsatisfactory because the excretion of phenylpyruvic acid, which is responsible for the positive ferric chloride test, is variable, and the level is usually low in PKU neonates. Large-scale screening was made possible by the development of a simple, reliable, and inexpensive test by Guthrie.[7] This test is a bacterial assay, performed on blood dried on filter paper. The availability of this test and the success of its field trial in Massachusetts[8] ushered in the modern era of screening. Within a few years, similar neonatal screening programs for PKU were under way in many states in this country, as well as in other countries. The increasing awareness of inborn errors of amino acid metablism as causes of mental retardation, and the availability of dietary therapy for disorders other than PKU, led to expansion of the screening program to include many other metabolic disorders of amino acids and related compounds.

CURRENT SCREENING PROGRAM

At present, the greater majority of the screening programs in over 20 countries test only newborn blood specimens; approximately one half of these programs have included disorders other than PKU. Only a few laboratories do urine screening. The most comprehensive neonatal screening program is the one in Massachusetts.

With a few exceptions, Guthrie's bacterial inhibition assays (GBIA) are used for blood screening. The two most widely used paper chromatographic methods are those developed by Efron et al.[9] and by Scriver et al;[10] the Efron technique utilizes blood dried on filter paper, whereas the Scriver technique utilizes capillary serum. Aside from the differences in the collection of the sample, these two techniques are comparable. The paper chromatographic methods are not as sensitive as the GBIA for detecting mild elevations in blood; however, they are generally the most valuable technique for examining the whole amino acid pattern in urine.

Shortly after the initiation of screening programs, it became evident that heterogeneity of a disorder is frequently encountered and that not all amino acid abnormalities are due to genetic defects. Elevation of phenylalanine, for example, is not equivalent to phenylketonuria, and elevation of methionine is not always indicative of cystathionine synthase deficiency. More commonly, these changes are transient in nature. A recent survey of screening for PKU in the United States showed that for every infant found to have PKU, there were 18 "false-positives"—mostly transient elevations.[11] As has been found in Massachusetts, there is a similar ratio of hypermethioninemia (due to cystathionine synthase deficiency) to transient hypermethioninemia (secondary to high protein intake) when the blood specimen is obtained in the second month of life. Only through follow-up surveys and further investigation may the various causes be distinguished.

Much has been learned from mass screening; otherwise unobtainable information has changed the researcher's perspective of inborn errors of amino acid metabolism. A number of the disorders previously thought to be associated with neurological deficit are now considered to be benign conditions, thereby causing the list of "benign genetic disorders" to grow (see Table 10–2). A notable example is Hartnup disease. The "characteristic" symptom-complex of intermittent ataxia, mental retardation, and photosensitive rashes has not been observed in

TABLE 10–2. INBORN ERRORS OF AMINO ACID METABOLISM GROUPED ACCORDING TO THE FREQUENCY OF ASSOCIATION WITH MENTAL RETARDATION AND/OR OTHER CLINICAL ABNORMALITIES*

Most Frequent	Infrequent	"Benign"
Phenylketonuria	Severe hyperphenylalaninemia	Mild persistent
Maple syrup urine disease	Carnosinase deficiency	hyperphenyl-
Homocystinuria	Hyperprolinemia, Type I	alaninemia
Hereditary tyrosinemia with	Hyperprolinemia, Type II	Histidinemia
and without liver disease	Hyperlysinemia	Cystathioninemia
Carbamyl phosphate	Cystinuria†	Hydroxyprolinemia
synthetase deficiency		Familial renal
Ornithine carbamyltransferase		iminoglycinuria
deficiency		Hartnup disease
Citrullinemia		
Argininosuccinic aciduria		
Hyperargininemia		
Aspartylglycosaminuria		
Hyperglycinemia, nonketotic		
Hyperglycinemia, ketotic		
Propionic acidemia		
Methylmalonic acidemia		
Isovaleric acidemia		
Hyperalaninemia with pyruvic		
acidemia and lactic acidemia		

*Only those disorders that have been reported in more than 1 or 2 cases are included.
†Urolithiasis, but no mental retardation.

any of the infants or their similarly "affected" older siblings who showed the typical urinary amino acid pattern by screening.

In any large-scale screening program, artifacts can present significant problems. Abnormalities secondary to diet are most common. For instance, infants fed a soy-protein–based formula supplemented with DL-methionine often excrete D-methionine in the urine. Contamination of the specimens with feces or medicinal ointment for diaper dermatitis may also cause unusual amino acid patterns.[12] Drugs, especially antibiotics, have become an increasing problem in the interpretation of a urine chromatogram. The presence of ampicillin and Keflex (cephalexin monohydrate) in the urine may mimic the findings in maple syrup urine disease and phenylketonuria, respectively.[13] At the screening program in Massachusetts there have been several unusual and anecdotal experiences with artifacts. These experiences show that interpretation of chromatograms is more an art, than a science.

COMMENTS

Urine Screening

Many of the screening programs have expanded from PKU testing alone to include testing for other amino acid metabolic disorders. Frequently, the same newborn blood specimens are used for all such tests; only a few laboratories use urine samples in their screening. Several years ago, Massachusetts started urine screening as a pilot program, and it has been found to be invaluable. It is now an integral part of the screening program. The diagnosis of many disorders not detectable by blood screening is now feasible with urine screening. Furthermore, the patient "missed" by initial blood screening will be detected by urine screening. The advantages of urine screening combined with blood testing is therefore, self-evident. Blood testing alone is only partially useful in diagnostic screening. Parents — and even physicians — may gain a false sense of security by believing that inborn errors of amino acid metabolism have been ruled out by blood screening. Therefore, I would strongly advocate the incorporation of urine screening into current screening programs.

Types of Disorders

Since resources and manpower are limited, disorders that are known to cause serious clinical abnormalities and for which treatment is available should be given high priority for screening. It is also important to detect the disorders that cause clinical abnormalities for which treatment is not yet available, so that genetic counseling and the potential development of effective treatment can be considered. The group of disorders listed in the left-hand column of Table 10–2 should all be given high priority for screening. There are a number of disorders in which the causal relationship of the clinical disease and biochemical defect has not yet been established. Because these disorders were discovered during amino acid screening of mentally retarded patients, the association could very well be due to sampling bias. Thus, information gathered from routine screening is necessary to clarify the issue. Information about the natural history of those amino acid disorders is vital to the evaluation of the indication and the results of treatment. The emotional and financial burden to the family resulting from unnecessary treatment cannot be ignored.

In over one million infants screened for hereditary tyrosinemias, not one case was found (Table 10–1).[1] Only recently was one case detected in Great Britain.[15] It is not clear whether this finding represents the rarity of the disorder or the ineffectiveness of the testing. On the other hand, the most frequent finding in newborn screening is tyrosine elevation due to the transient type of neonatal tyrosinemia. This type of tyrosinemia was considered to be a benign entity until recently, when a report of a long-term study by Menkes et al.[14] showed that high tyrosine levels in the neonatal period had an adverse effect on the intellect of larger low–birth weight (2.0 to 2.5 kg) infants. If a causal relationship between high blood levels of tyrosine and impairment of intellectual performance can be proven from long-term follow-up studies, routine screening of newborns for tyrosine elevation would certainly be warranted.[14a]

Several screening laboratories use GBIA for histidine to detect histidinemia. Although some histidinemic children have had speech defects, mental retardation, and seizures, it is far from certain that this disorder produces clinical disease and requires treatment.[16] Thus, the justification for such testing depends largely upon one's view of the necessity for treatment for histidinemia. As long as there is no consensus in this regard, there will be no agreement about the need for a separate test for histidinemia. In Massachusetts, histidinemia is not specifically tested for, but is detected as a result of urine chromatographic screening, and patients with this disorder are not treated.[16a]

In addition to inborn errors in amino acid metabolism, disorders in organic acid metabolism (such as methylmalonic acidemia, propionic acidemia, isovaleric acidemia, congenital lactic acidosis, and pyruvic acidemia) and in monosaccharide metabolism (such as classical galactosemia and galactokinase deficiency) result in serious clinical diseases of early onset and for which treatment is available. With little added effort and expense, screening for these disorders can be done by using the same blood and urine specimens collected for amino acid screening. Two excellent tests are available for galactose screening: Beutler's fluorescence spot test for galactose-1-phosphate uridyltransferase deficiency[17] and Paigen's bacteriophage test for galactose elevation.[18]

For organic acid screening, field testing in Massachusetts of a paper chromatographic technique failed to detect any single case in over 100,000 urine specimens. A most distressing finding is that a patient later diagnosed as having methylmalonic acidemia had a "normal" result with the screening. Therefore, there are serious doubts about the efficacy of this technique. The reasons for failure to detect organic acid disorders may be twofold: (1) the test may be too insensitive to detect the amounts of organic acids in the urine of affected infants; and (2) substantial accumulation of the abnormal metabolite(s) may not have occurred during the asymptomatic period. Thus, the development of an improved staining reagent for paper chromatography or a simplified gas-liquid chromatographic technique is a prerequisite for organic acid screening.

Age for Testing

The appropriate age for screening is determined by several factors. In most instances, the screening test is based upon finding an abnormal amount of a metabolite; therefore, the blood should be taken after the infant has had at least 8 milk feedings and the accumulation of metabolites becomes detectable. From the practical point of view, blood sampling should be done before discharge from the nursery. In the United States, babies are usually discharged when they are 3 and 5 days of age. The blood specimen can be obtained just prior to discharge. If the infant is discharged before this time, repeat testing on a second blood specimen — preferably at 1 or 2 weeks of life — is advisable. For a premature infant remaining in the nursery for an extended period of time, the blood sampling can be done at about 10 days to 2 weeks of age. In countries such as Great Britain, where it is common for babies to be born at home, blood can be collected by health visitors on the first visit, which would normally be during the second or third week of life.[19]

One question occasionally asked is whether a single blood test in the newborn period is sufficient to detect PKU, or if an additional blood test at a later age is necessary to prevent missing any case. Holtzman and his colleagues critically reviewed the data from many of the large PKU screening programs in North America[20] and noted that one fourth of the classical PKU infants had blood phenylalanine levels ranging from only minimal elevation to less than 10 mg per 100 ml. In approximately 7 per cent of all PKU infants, the initial test results were between 2 and 4 mg per 100 ml. Most of these infants were tested on or before the third day. Holtzman et al. concluded that the findings suggest that as long as most infants in the United States continue to be screened on or before 4 days of age, some infants with PKU will be missed. However, this conclusion has not been borne out by experiences from screening in Massachusetts[21] and in Oregon.[22] Nearly 700,000 blood specimens taken at 4 and 6 weeks of age were tested for PKU in these states, and no additional case was identified by the second testing. Therefore, it seems unlikely that any case would be missed as long as a repeat blood specimen is obtained from all infants whose blood phenylalanine is greater than 2 mg per 100 ml and also from those whose blood specimens were taken at less than 2 days of age. The same can be said about blood methionine and leucine. It is advisable to request a repeat specimen when there is any deviation from normal values in the initial testing.

Diagnosis by enzyme assay has one obvious advantage — the test can be done on cord blood, which obviates a heel prick on the newborn. In addition, the results may be obtained before the neonate is discharged from the nursery and, more importantly, before symptoms become manifest. The two diseases for which enzyme assays are available, argininosuccinic aciduria (argininosuccinase deficiency)[23] and classical galactosemia (galactose-1-phosphate uridyltransferase deficiency)[17] may both result in a fulminating fatal neonatal course. It has been our experience, as well as that of others, that when results of the screening of blood taken at discharge from nursery became available, the infant had either died or was already critically ill.[24, 25] For these infants, at least, screening at birth, instead of at 3 days of age, may be life-saving.

One observation made from working with the Massachusetts screening program is that a sick infant, who may be the very infant who needs to be screened, often is not. It is an unfortunate misconception that there is no time for this routine screening procedure when the baby is having problems. Another condition under which an infant may "miss" screening is when he or she is transferred to another hospital, and no one at either hospital remembers to obtain the specimen. These factors may partially account for the apparent rarity of disorders such as maple syrup urine disease (MSUD), galactosemia, and nonketotic hyperglycinemia, all of which cause acute illness in the neonate. Some physicians feel that because these metabolic disorders cause such fulminating disease and are very costly to treat, it is best to let nature take its course; thus, they do not screen for them. We take a brighter view of it. First, we have had good results in treating such patients. Children with MSUD and galactosemia, who were given the special diet from the neonatal period, have had normal growth and development at 3 years of age. Second, although the detection of a metabolic defect as the cause of the clinical disease may not be enough to save the life of this patient, the information will be useful in genetic counseling and will help with the management of a subsequent pregnancy.

At present, there is not enough information to determine the optimal age for urine screening, and, particularly, to determine how early urine screening can be done without missing important aminoacidurias. In Massachusetts, 3 to 4 weeks of age is chosen somewhat arbitrarily. Because most infants will have their first pediatric checkup at approximately 6 weeks of age, the hope is that screening results will be available before that time and that specimens for additional investigation can be conveniently obtained during the checkup.

Family Survey

When the diagnosis of a metabolic disorder is made in one member of the family, a survey of siblings and parents is indicated. Other relatives should be included, when their medical history suggests that they may be similarly affected. A sibling of a child affected with a hereditary metabolic disorder is at high risk, usually with a one-in-four chance of being similarly affected. At the time of birth of such a high-risk infant, the screening laboratory should be notified immediately. When the activity of the enzyme involved can be measured, an assay could be performed on cord blood for definitive diagnosis; otherwise, quantitative measurement for the specific metabolite should be performed on serial blood and urine specimens in the neonatal period until such time as the disorder can be ruled out. It is not known, with any certainty, whether those diseases that have not yet been detected by newborn screening can actually be identified in the first several weeks of life. In such cases, the follow-up period should be extended to 1 or 2 months of age, or beyond the time when clinical symptoms usually appear.

Maternal Disease

Maternal disease as a cause of mental retardation in the offspring who do not have the metabolic defect has recently been recognized. Thus far, mental defect and malformations have been reported only in children of women with untreated PKU.[26] Women with homocystinuria,[27] histidinemia,[28] and Hartnup disease[29] have successfully produced normal children. Search for biochemical diseases in mothers of children who had mental retardation revealed no cause other than PKU.[26, 30, 31] Thus, in any family in which all children are retarded without a specific cause, maternal disease — particularly PKU — should be considered.

There is evidence to suggest that the mental retardation in children of mothers having untreated PKU may be due to intrauterine brain damage caused by exposure to the unfavorable chemical environment (i.e., high concentration of phenylalanine). The degree of brain damage is more or less directly proportional to the maternal blood level of phenylalanine.[33] Preliminary results of low phenylalanine dietary therapy are encouraging.[33, 34] Thus, early diagnosis of maternal PKU is essential for the welfare of the fetus. One approach to this problem is to use ferric chloride testing (Phenistix test) routinely at the first prenatal visit. A positive result usually indicates that the blood phenylalanine is greater than 12 mg per dl. Since a blood phenylalanine level of less than 12 mg per dl seems to have no obvious adverse effect on the fetus,[32] this urine test is simple and adequate for detecting any hyperphenylalaninemic pregnant woman who needs treatment. Recently, this practice has received attention from obstetricians,[35] and it is hoped that testing for PKU will soon be a routine procedure included in the initial prenatal workup. If the manufacturer of the popular test strip for urine, sugar, and protein would add a square of the ferric chloride reagent, the test for

PKU could be carried out without additional effort on the part of the obstetrician. It is of utmost importance that the diagnosis be confirmed by a quantitative measurement of blood phenylalanine.

Dietary management of these patients is discussed in Chapter 11. It has yet to be determined at what period the fetal brain is most susceptible to the "toxicity" of phenylalanine and its metabolites. If the most vulnerable period is the first trimester, then screening at the initial prenatal checkup, which is often in the second month of gestation, may be too late to prevent brain damage. There has been a report, however, of a PKU woman who was treated through the last 5 months of gestation and gave birth to a child who was normal as of 9 months of age.[33] This finding would appear to indicate that dietary restriction from the second trimester on can be effective in preventing mental retardation.

At the present time in Massachusetts, there is screening for maternal PKU by GBIA for phenylalanine of the cord blood. The phenylalanine level in the cord blood reflects that of the mother and is greatly elevated in an infant born of a PKU mother. The level drops rapidly to a normal range within 48 hours. Therefore, the diagnosis of maternal PKU can best be made by testing cord blood, conversely that of PKU in infants can best be made by blood obtained at 3 or 4 days of age. Obviously, such screening will not prevent the brain damage of the firstborn of the PKU mother, but it will benefit the future offspring of this woman. It is hoped that in the near future such testing will be replaced by "prenatal" screening.

Intrauterine Screening

Intrauterine screening is different from intrauterine or prenatal diagnosis. The latter is the detection of a metabolic defect suspected in a specific fetus. The former is similar to neonatal screening in that the group of amino acid disorders, as the potential cause of mental retardation, is searched for—and amniotic fluid seems to be suitable for this purpose. Data on the normal amino acids in amniotic fluid[36-38] have been collected as background information for comparison with values from the amniotic fluid of a fetus at risk. Little is known, however, about the amino acid pattern in the amniotic fluid of an affected infant with a metabolic defect. Since blood amino acid patterns of infants with PKU, MSUD, and other similar conditions are normal at birth, it is unlikely that the abnormalities of these amino acid disorders can be detected from amniotic fluid. On the other hand, argininosuccinic acid[39] and methylmalonic acid[40, 41] have been found in the cord blood of the affected newborn, as well as in the amniotic fluid. Prenatal screening for these disorders is potentially possible. The scarcity of information regarding many other amino acid disorders prevents the drawing of any conclusions as to the usefulness of measurement of amino acids in amniotic fluid as a screening procedure.

Recent developments in the techniques of fetoscopy make possible the sampling of fetal blood for specific diagnostic purposes, but the use of this approach for routine screening awaits further investigation and depends largely upon the development of tests for various enzyme activities in a small quantity of blood.

Analysis of the intracellular amino acid pattern of the cultivated amniotic fluid cell as a screening test has been investigated, but the results have been rather disappointing. The amino acid patterns of normal cells and of cells known to have an enzyme defect appeared to be similar.[42]

Record Keeping

It is evident that in a program of mass screening a large volume of data can be collected in a relatively short period. To make this information useful, it should be carefully cross-indexed and computerized for easy retrieval. Judicious use of this information within the context of confidentiality and ethics is, nevertheless, important.

Patient Referral Facility

A facility to which patients can be referred for further evaluation, treatment, and genetic counseling is an integral part of any screening program, without which the purpose of screening would be defeated. The laboratory personnel should be able to communicate freely with the clinical physician. Feedback information about the final diagnosis and the progress of the patient is essential—both in monitoring and in planning the future of the screening program. As new information in the field of inborn errors of metabolism becomes available, it is essential that effort be directed toward the areas most relevant to the prevention of mental retardation and other disabling diseases.

ACKNOWLEDGMENT: I am indebted to Dr. Harvey L. Levy, Assistant Director of the Diagnostic Laboratories, State Laboratory Institute, and Principal Consultant of the Massachusetts Metabolic Disorders Screening Program, for allowing me to include some of the information obtained through this screening program.

References

1. Levy, H. L.: Genetic screening. *In* Harris, H., and Hirschhorn, K. (eds.): Advances in Human Genetics. Vol. 4. New York, Plenum Publishing Corp., 1974, p. 1.

2. Levy, H. L.: Neonatal screening for inborn errors of amino acid metabolism. Clin. Endocr. Metab., 3:153, 1974.
3. Clow, C. L., Fraser, F. C., Laberge, C., et al.: On the application of knowledge to the patient with genetic disease. Progr. Med. Genet., 9:159, 1973.
4. Fölling, A.: Phenylpyruvic acid as a metabolic anomaly in connection with imbecility. Z. Physiol. Chem., 227:169, 1934.
5. Jervis, G. A.: Phenylpyruvic oligophrenia (phenylketonuria). Proc. Assoc. Res. Nerv. Ment. Dis., 33:259, 1954.
6. Bickel, H., Gerrard, J., and Hickmans, E. M.: Influence of phenylalanine intake on phenylketonuria. Lancet, 2:812, 1953.
7. Guthrie, R., and Susi, A.: A simple phenylalanine method for detecting phenylketonuria in large populations of newborn infants. Pediatrics, 32:338, 1963.
8. MacCready, R. A., and Hussey, M. G.: Newborn phenylketonuria detection program in Massachusetts. Am. J. Public Health, 54:2075, 1964.
9. Efron, M. L., Young, D., Moser, H. W., et al.: A simple chromatographic screening test for the detection of disorders of amino acid metabolism: A technique using whole blood or urine collected on filter paper. N. Engl. J. Med., 270:1378, 1964.
10. Scriver, C. R., Davies, E., and Cullen, A. M.: Application of a simple method to the screening of plasma for a variety of aminoacidopathies. Lancet, 2:230, 1964.
11. Holtzman, N. A., Meek, A. G., and Mellits, E. D.: Neonatal screening for phenylketonuria I. Effectiveness. J.A.M.A., 229:667, 1974.
12. Levy, H. L., Madigan, P. M., and Shih, V. E.: Massachusetts metabolic disorders screening program. I. Technics and results of urine screening. Pediatrics, 49:825, 1972.
13. Shih, V. E.: Laboratory Techniques for the Detection of Hereditary Metabolic Disorders. Cleveland, Ohio, CRC Press, 1973.
14. Menkes, J. H., Welcher, D. W., Levi, H., et al.: Relationship of elevated blood tyrosine to the ultimate intellectual performance of premature infants. Pediatrics, 49:218, 1972.
14a. Hunt, P. A., and Hitchcock, E. S.: Intellectual deficits after transient tyrosinemia in term neonates. Pediatr. Res., 8:344, 1974.
15. Beckers, R. G., Wamberg, E., Bickel, H., et al.: Collective results of mass screening for inborn metabolic errors in eight European countries. Acta Paediatr. Scand., 62:413, 1973.
16. Editorial: Histidinaemia: To treat or not to treat. Lancet, 1:719, 1974.
16a. Levy, H. L., Shih, V. E., and Madigan, P. M.: Routine newborn screening for histidinemia. Clinical and biochemical results. N. Engl. J. Med. 291:1214, 1974.
17. Beutler, E., Baluda, M. C.: A simple spot test for galactosemia. J. Lab. Clin. Med., 68:137, 1966.
18. Paigen, K.: Personal communication.
19. Komrower, G. M., Fowler, B., Griffiths, M. J., et al.: A prospective community survey for aminoacidaemias. Proc. R. Soc. Med., 61:294, 1968.
20. Holtzman, N. A., Mellits, E. D., and Kallman, C. H.: Neonatal screening for phenylketonuria: II: Age dependence of initial phenylalanine in infants with PKU. Pediatrics, 53:353, 1974.
21. Levy, H. L., Shih, V. E., Karolkewics, V., et al: Screening for phenylketonuria. Lancet, 2:522, 1970.
22. Buist, N. R. M., Brandon, G. T., and Penn, R. L.: Follow-up screening for phenylketonuria. N. Engl. J. Med., 290:577, 1974.
23. Murphey, W. H., Patchen, L., Guthrie, R.: Screening tests for argininosuccinic aciduria, orotic aciduria, and other inherited enzyme deficiencies using dried blood specimens. Biochem. Genet., 6:51, 1972.
24. Shih, V. E., Levy, H. L., Karolkewicz, V., et al.: Galactosemia screening of newborns in Massachusetts. N. Engl. J. Med., 284:753, 1971.
25. Murphey, W. H.: Personal communication.
26. MacCready, R. A., Levy, H. L.: The problem of maternal phenylketonuria. Am. J. Obstet. Gynecol., 113:121, 1972.
27. McKusick, V. A.: Heritable Disorders of Connective Tissue. St. Louis, The C. V. Mosby Company, 1972, pp. 224–281.

28. Neville, B. G. R., Harris, R. F., Stern, D. J., et al.: Maternal histidinaemia. Arch. Dis. Child., 46:119, 1971.
29. Seakins, J. W. T.: Personal communication.
30. Hansen, H.: Epidemiological considerations on maternal hyperphenylalaninemia. Am. J. Ment. Defic., 75:22, 1970.
31. Perry, T. L., Bunting, R., Tischler, B., et al.: Unrecognized maternal biochemical disease: An uncommon cause of mental retardation in children. J. Ment. Defic. Res., 14:44, 1970.
32. Levy, H. L., Shih, V. E.: Maternal phenylketonuria and hyperphenylalanemia. A prospective study. Pediatr. Res., 8:391, 1974.
33. Allan, J. D., Brown, J. K.: Maternal phenylketonuria and fetal brain damage, an attempt at prevention by dietary control. *In* Holt, K. S., and Coffey, V. P. (eds.): Some Recent Advances in Inborn Errors of Metabolism. Edinburgh, E. & S. Livingstone, Ltd., 1968, pp. 14–35.
34. Arthur, L. J. H., and Hulme, J. D.: Intelligent, small for dates baby born to oligophrenic phenylketonuric mother after low phenylalanine diet during pregnancy. Pediatrics, 46:235, 1970.
35. Johnson, C. C.: Phenylketonuria and the obstetrician. Obstet. Gynecol., 39:942, 1972.
36. Emery, A. E. H., Burt, D., Nelson, M. M.: Antenatal diagnosis and amino acid composition of amniotic fluid. Lancet, 1:1307, 1970.
37. Reid, D. W. J., Campbell, D. J., and Yakymyshyn, L. Y.: Quantitative amino acids in amniotic fluid and maternal plasma in early and late pregnancy. Am. J. Obstet. Gynecol., 111:251, 1971.
38. Levy, H. L., Easterday, C. L., Montag, P. P., et al.: Amino acids in amniotic fluid. *In* Dorfman, A., (ed.): Antenatal Diagnosis. Chicago, University of Chicago Press, 1972, pp. 109–113.
39. Goodman, S. I., Mace, J. W., Turner, B., et al.: Antenatal diagnosis of argininosuccinic aciduria. Clin. Genet., 4:236, 1973.
40. Morrow, G., Schwarz, R. H., Hallock, J. A., et al.: Prenatal detection of methylmalonic acidemia. J. Pediatr., 77:120, 1970.
41. Ampola, M. G., Mahoney, M. J., Nakamura, E., et al.: In utero treatment of methylmalonic acidemia (MMA-EM1A) with vitamin B_{12}. Pediatr. Res., 8:387, 1974.
42. Shih, V. E., Mandel, R., Levy, H. L., and Littlefield, J. W.: Free amino acids in extracts of cultured skin fibroblasts from patients with various amino acid metabolic disorders. Clin. Genet., 1975. (In press.)

CHAPTER 11

TREATMENT IN GENETIC DISEASES

Yujen Edward Hsia,* B.M., D.C.H., M.R.C.P.

To cure, sometimes
To relieve, often
To comfort, always

Anonymous

INTRODUCTION

When confronted with a child totally incapacitated by one of the neurolipidoses, or an infant grotesquely malformed because of a major chromosomal anomaly, one may feel that no treatment is possible for genetic diseases. This belief is unjustified, for not only is some form of treatment always possible, but recent advances have provided increasingly effective rational therapeutic approaches for genetic diseases.[1-6]

The *goal* of treatment for genetic disorders is to modify or prevent the natural history of expression of genetic traits, so that an affected individual may develop optimally within his or her inherent potential, despite the possession of disadvantageous genes. To achieve this goal, the *focus* of treatment should be on the immediate products of gene expression, rather than on the more diffuse levels of eventual clinical expression.

Modification of the natural history of the disease by supportive measures,[7] although necessary and important, is make-shift and costly without being curative. Thomas has pointed out[8] that most medical treatment is based on "notechnology" or "halfway technology." In this category, he includes the increased emotional demands on a physician for continuing moral support of a family that is burdened by an incurable

*Supported by the National Foundation—March of Dimes Medical Service Grant #C-41, Human Genetics Center Grant GM-20124-02, USPHS Grants AM-09527 and HD 00198-07.

disease. Prevention of the expression of diseases, on the other hand, requires generous investment in scientific research to develop "high technology" for rational treatments (see Chapter 13). For example, orthopedic and post-traumatic treatments of the complications of hemophilia are on a "notechnology," stand-by clinical level that fails to correct the basic pathology, and are only stop-gap measures applied in response to each complication. These treatments are often unsuccessful in avoiding severe crippling handicaps and are costly in terms of both human suffering and medical resources. Anticipatory infusions of antihemophilic factor are on a "halfway technology" level that will correct the coagulation defect and prevent bleeding complications. This treatment is "designed to make up for disease," has dramatic appeal for the public, but is still a drain on the supply and processing of human plasma. This type of therapy is "primitive," despite its sophistication, because it maintains a patient's clinical status without being curative.[8] The potential benefit of unrestricted normal activity for these patients, freeing them from reliance on "notechnology" treatment, nonetheless is a major advance in effective therapy for this condition (see Chapter 13). Ultimately, the possibility of biochemical manipulations that will cure the coagulation defect, perhaps by introduction of missing genetic material, must await "high technology" advances necessary for understanding the biological regulation of coagulation factors, or altering the genetic constitution of an individual. In the meantime, genetic counseling should be given to carrier females to forewarn them of their risks of bearing an affected son, and to offer them the option of prenatal selection on the basis of sex (see Chapters 3, 6, 9). This counseling may help families to avoid the birth of a hemophilic son. The problem of hemophilia (its prevalence is estimated to be about 20,000 in the United States) could be somewhat reduced by this means of lowering the number of births of hemophilic babies.

Decisive curative treatment, based on genuine understanding of disease mechanisms, is possible or foreseeable for many of the inherited biochemical diseases that are single-gene disorders, primarily affecting a single enzyme or membrane protein. Correction of the abnormal structure or quantity of this protein would produce a complete cure. Whenever this is accomplished, Thomas points out, this form of "high technology" cure will be less expensive and simpler to administer than the complex procedures of "halfway technology," such as organ transplantation. The compromise of evading the problem, by reducing the number of births of affected individuals, is a prudent interim objective, until curative treatment becomes a reality.

Levels of Treatment

Every normal gene has a replicative and a functional role. Its replicative role directs the transcription of copies of itself for somatic daughter cells (horizontal arrows in Figure 11-1) or for sexual germ cells (curved arrows in Figure 11-1), which transmit its genetic information to progeny. Its functional role directs formation of gene products, which act to produce the eventual overt expression of genetic activity (vertical arrows in Figure 11-1). Genetic disease is detectable only from the expression of an abnormal (or missing) gene product. The function of a given gene may be restricted in time or in location; for instance, some genes are functional only for embryogenesis, others, only in a single differentiated tissue. Also, every gene product interacts with the products of other genes, and the coordinated function of gene products is subject to external influences. Hence, expression of genetic activity is modulated by endogenous factors and is modified by exogenous factors. An example is the doubly heterozygous state for the sickle hemoglobin gene and the β-thalassemia gene. Interaction between the two genes causes one form of sickle cell disease, although either heterozygous state by itself is innocuous; furthermore, a patient with sickle cell disease is rendered vulnerable to environmental stresses that cause hypoxia or dehydration.

Rational treatment for genetic diseases aims to control the effects of endogenous or exogenous factors on any of the arrows depicted in Figure 11-1, but is best focused in place and time to act on the appropriate target organ before clinical expression has become irreversible. Treatment at multiple levels can often be more effective than any single approach by itself.

Genetic diseases can be treated by (1) controlling the external environment; (2) regulating the foods that are ingested; (3) modifying the internal environment with chemicals, drugs, or hormones; (4) removing

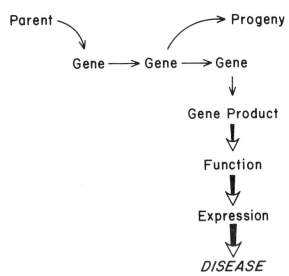

Figure 11-1. Replicative and functional role of the gene. (See the text for a more complete explanation.)

diseased tissue or introducing normal donor tissue (organs, cells, proteins — even genes); (5) stimulating genes to form more of a needed product, or suppressing genes from forming as much of a harmful product; and (6) preventing the conception or birth of affected individuals.

Complete treatment for any genetic disease must include sympathetic attention to a family's psychological needs and to a patient's educational, occupational, and recreational needs. A family should be given a clear understanding of the nature of its problem, treatment possibilities, prognosis, and recurrence risks. Treatment must be accompanied by care and comfort.

WAYS TO TREAT GENETIC DISEASES

The External Environment

Human frailty is such that each of us needs protection from the hazards of the world around us, and we all utilize many manufactured devices to improve the prowess with which we control our environment. Genetic disorders that render affected individuals unduly vulnerable to external hazards increase their need to live in a controlled environment.

Protection from mechanical trauma is obviously required for patients with fragile bones, weak muscles, lax joints, or bleeding tendencies; and for patients with disorders of connective tissue, like the Ehlers-Danlos syndrome, which impairs healing. Protection from thermal injury is necessary in a few rare genetic conditions: in anhidrotic ectodermal dysplasia, hyperthermia may kill or cause irreversible brain damage; in Menkes' syndrome, hypothermia is a well-recognized complication (pathological fractures also occur in Menkes' syndrome); in sickle cell disease, there may be increased vascular susceptibility to cold injury.

Ultraviolet (UV) irradiation is particularly harmful in xeroderma pigmentosa, where there is genetic loss of a DNA repair enzyme, with high risk of malignant change in irradiated skin cells. UV irradiation is also harmful in the albinisms where pigmentary protection from ultraviolet rays is lacking and there is serious risk of sunburn and eventual malignant change in exposed skin. On the other hand, ultraviolet light activates vitamin D, to enhance its potency, and visible light detoxifies bilirubin, thereby allowing phototherapy to be used for neonatal hyperbilirubinemia.

Protection from environmental pollutants and allergens is an important aspect of treatment for chronic lung diseases and allergic conditions — whether or not they are genetically determined. Protection from infectious agents is the first line of defense for patients with deficiencies of the cellular or humoral immune systems.[9] One or two children with a-γ-globulinemia are being reared in isolated germ-free conditions — an extreme example of tailoring the physical environment to suit an individual's requirements.

Optical correction is underestimated as a treatment for genetic disorders, because any genetically determined cause of refractory errors or visual impairment would be a major handicap without optical or surgical correction. For instance, a patient with Marfan's syndrome or homocystinuria requires watchful ophthalmological supervision because the complications of dislocation of the lenses may threaten vision. Two new promising approaches for treating blindness are being developed — one, the magnification by electronic photomultiplier systems of visual signals impinging on the retina; and the other, the direct electrical stimulation of the occipital visual cortex via a network of implanted electrodes. Similarly, auditory amplification by use of hearing aids will reduce the handicap of patients with conductive or partial neural deafness.

Dental care is also relevant to genetic diseases.[7] There is significant dental pathology in several systemic genetic diseases: poor alveolar tooth support in hypophosphatemic rickets; abnormal enamel in dentinogenesis imperfecta and the cholestatic syndromes; malocclusion in some of the chondrodystrophies; and absent or malformed teeth in anhidrotic ectodermal dysplasia. Prophylactic dental care for these conditions should include protecting the teeth in hypophosphatemia from the trauma of mastication on tough foods, using reconstructive and prosthetic procedures for anomalous dentition, and paying attention to routine dental hygiene. Good dental prostheses may contribute significantly to nutrition and to an individual's cosmetic self-image.

Physical supportive devices are indicated for habilitation of patients with locomotor instability or weakness.[7] Previously cumbersome orthopedic supports, such as body casts and braces, are becoming less and less unwieldy with technological advances in materials and design. Their better acceptability is enhancing their usefulness. For the future, servo-assisted mechanical joints are being developed that may eventually enable paralyzed or severely weakened patients with neuromuscular problems to move about under self-controlled power, thereby prolonging their mobility and their ability to engage in normal social activities.

Food and Nutrition

The differences between individuals — are seen — in the end products of metabolism. Even those idiosyncrasies — that what is one man's meat is another man's poison, presumably have a chemical basis.

Sir A. Garrod, 1909[10]

All food is not equally palatable for all men. As one example, a large proportion of the human population has a genetically scheduled loss of intestinal lactase activity after weaning. In these individuals, the lactose in any subsequently ingested milk will cause diarrhea, but this

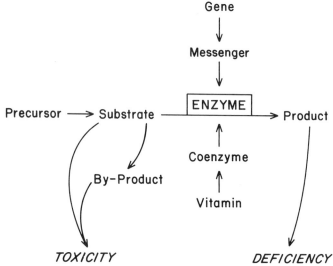

Figure 11–2. Schematic illustration of possible consequences of defective enzyme function, indicating potential principles of therapy.

milk intolerance is of no consequence unless milk is introduced into their postweaning diet. This illustrates that some genetic deficiencies are common; that an enzyme deficiency need not be detrimental unless a given food is ingested; and that a major principle of diet treatment is to avoid a toxic food. (Food is defined here as any ingested foodstuff, including water, calories, constituents of carbohydrates, protein or fat, vitamins, minerals, and other essential nutrients.)

Genetic abnormalities that cause metabolic disturbances often are affected by food intake. Certain foods can be vitally protective in some of these disorders, or can be lethally toxic in others. These disturbances constitute the most important genetic causes of mental retardation because they are preventable by treatment.

When an enzyme is quantitatively deficient or qualitatively defective, impaired activity can result in toxic accumulation of the substrate or a by-product or may lead to product deficiency. Secondarily reduced enzyme activity can arise from defective interaction of the enzyme with a coenzyme, or deficient production of a coenzyme from its precursor vitamin (Fig. 11–2). In metabolic diseases that can be modified by food, the aim of treatment is to sustain nutrition, while suppressing toxicity and preventing deficiency. Principles of diet therapy have focused on each step represented by an arrow in Figure 11–2. These include: (1) augmented intake of protective foods; (2) total exclusion of nonessential foods that are toxic; (3) restricted intake of essential foods that are toxic; (4) supplementation of foods that are needed in greater than normal amounts; (5) substitution of synthetic foods to fulfill the nutritional needs of individuals for whom a natural food is toxic; and (6) oral replacement of missing enzymes, when this is feasible.

Augmentation

Increased intake of certain foods indirectly protects against the complications of some metabolic diseases.

Water protects patients with sickle cell disease from dehydration, which increases the tendency of red cells to sickle; in cystinuria water reduces the danger of cystine crystallization in the urine. Changes in hydrogen ion concentration can also be therapeutic. Alkalinization will prevent the formation of cystine stones, although at the increased risk of precipitating calcium salts in the urinary tract. Acidosis is protective against the neurological complications of hyperammonemia.[11] Sodium chloride helps protect against the mineralocorticoid deficiency of the congenital adrenal hyperplasia syndromes;[12] phosphate and magnesium may increase the solubility of urine oxalate in hyperoxaluria, protecting against stone formation and renal damage.

Glucose, polysaccharides, and gluconeogenic foods protect against the hypoglycemia that complicates many metabolic diseases.[13] In fructose-1,6-diphosphatase deficiency, gluconeogenesis is blocked, and any carbohydrate food will be beneficial for maintaining blood glucose levels; in deficiencies of the phosphorylase system, only glycogenolysis is blocked, thereby allowing gluconeogenic foods to be as beneficial as carbohydrates.

A very remarkable concept of biochemical protection is the use of glycine for isovalericacidemia. In deficiency of isovaleryl CoA dehydrogenase, excess leucine causes accumulation of free isovalerate or its presumably less toxic glycine-conjugate. When given glycine supplements, a patient has been shown to have improved tolerance to a leucine load,[79] the improved tolerance being accompanied by a corresponding shift in the ratio of free to conjugated isovalerate. Similar therapy may prove to be applicable for detoxication of unconjugated metabolic intermediates in other disorders.

Exclusion

The body is able to synthesize all of the carbohydrates it needs, but only some of the amino acids and fats. When a food is nonessential, it can be safely eliminated from the diet, provided full caloric and other needs are met by compensatory supplies in the diet.

Dietary glucose and galactose in glucose-galactose malabsorption, fructose in fructose intolerance, and galactose in galactosemia can cause severe metabolic disturbances that may lead to brain damage. This is preventable by strict dietary exclusion—from early infancy—of the sugar responsible for the condition. In fructose intolerance, a characteristic acquired aversion to sweet-tasting foods[14] simplifies the management of older patients. In galactosemia, strict avoidance of galactose is more difficult because of the commercial practice of adding milk or lac-

TABLE 11–1. EXAMPLES OF CONDITIONS TREATABLE BY
EXCLUSION OR RESTRICTION OF TOXIC FOODS

Exclusion	Condition
Carbohydrate	
Fructose	Fructose intolerance
Galactose	Galactosemia
Glucose and galactose	Glucose-galactose malabsorption
Lactose	Lactase deficiency
Fat	
Neutral fats	Steatorrheic syndromes
	Type I hyperlipidemia
Cholesterol and saturated fats	Type II hyperlipidemia
Alcohol	Alcoholism
Protein	
Gluten	Celiac syndromes
Glycine*	Nonketotic hyperglycinemia
Histidine*	Histidinemia
Proline*	Hyperprolinemia
Purines (uric acid)	Gout
Chlorophyll (phytanic acid)	Refsum's disease
Fava beans	Favism

Restriction	Condition
Protein	Urea cycle enzyme defects
Branched-chain amino acids	Maple syrup urine disease
Isoleucine, methionine, threonine, valine	Methylmalonicacidemias and propionicacidemias
Leucine	Isovalericacidemia
Methionine and homocystine*	Homocystinuria
Phenylalanine	Phenylketonuria
Tyrosine	Hereditary tyrosinemia
	Hypertyrosinemia
Minerals	
Copper	Wilson's disease
Iron*	Hemochromatosis

*Of questionable or limited value

tose to processed foods; hence expert dietetic guidance is necessary in the management of galactosemia.

Other examples (Table 11–1) include cholesterol and saturated fats, which are nonessential foods that exacerbate the hypercholesterolemia of type II hyperlipidemia; and the phytanic acid moiety of chlorophyll, which is probably responsible for all of the features of Refsum's disease, including the polyneuritis and retinitis pigmentosa.

Among the nonessential amino acids, glycine restriction will reduce blood glycine levels in the nonketotic hyperglycinemia syndromes,[15] but with inconsistent clinical improvement;[16] proline restriction can lower the blood levels of this substance in the hyperprolinemia syndromes, also with no clinical benefit.

Restriction

When the toxic food in a metabolic disease is also an essential nutrient, completely depriving a patient of the nutrient would be complicated by malnutrition. Hence, ideally, the prescribed intake of the potentially toxic food should be sufficient for growth and maintenance of health, but should not exceed its toxic threshold. When there is a safe margin of tolerance between nutritional needs and toxic levels, management is simpler, but when toxicity appears before nutritional needs can be met, diet therapy alone will be inadequate or impractical.

Treatment for *phenylketonuria* is the prototype for this concept of diet therapy.[15, 16] Phenylalanine is an essential amino acid, ubiquitously present in all high-grade protein and required for synthesis of all body proteins. The enzyme phenylalanine hydroxylase, present exclusively in the liver, converts phenylalanine to tyrosine (which is not an essential amino acid only because phenylalanine is its source of supply). In phenylketonuria, deficiency of this hydroxylase blocks phenylalanine metabolism, with formation of phenylketone by-products. There is failure of normal myelination and brain development in the phenylketonuric infant (the pathogenic mechanism of which is still poorly understood), but avoidance of phenylalanine accumulation appears to arrest this mechanism. Because presymptomatic detection of phenylketonuric infants is possible by newborn screening programs (see Chapter 10), restrictive dietary treatment can be started in time to prevent brain damage.[15, 16, 17] A carefully regulated diet, based on a synthetic low phenylalanine milk formula, must provide supplemental tyrosine (which has become an essential amino acid for these patients), while progressively reducing the allocation of phenylalanine according to the decelerating growth and amino acid requirements of the infant. The details of regulation still have not been mastered, but, in principle, they aim at keeping the blood phenylalanine level within safe limits.

Safe limits are those that allow an unimpaired intellectual and physical growth. In practice, these limits have been difficult to delineate because of the long lag before minimal brain damage becomes measureable — even with the most precise psychometric tests. As soon as the diagnosis is confirmed, these infants can be given a very low phenylalanine diet for a few days to allow the blood phenylalanine concentration to fall to acceptable limits. After this, the diet should be composed of a judicious mixture of the low-phenylalanine formula with a regular high phenylalanine milk, adjusted to meet growth needs. With the introduction of other foods, regular milk is phased out, and only the special milk is given together with predetermined amounts of the other foods, under the supervision of dietitians and physicians experienced in the management of these infants. This diet is balanced to provide a large proportion of the infant's protein requirements in the form of the low phenylalanine–high tyrosine formula, mixed as appetizingly as possible with low phenylalanine foods. After the initial infant growth spurt, particularly as physical activity increases when the child begins to walk, phenylalanine tolerance decreases, while caloric demand increases. Together with the capricious taste preferences of this age group, these changing diet requirements become a difficult challenge for the parents and the dietetic team. Low protein flour, and use of the special formula as a milk substitute in ingenious recipes, can help to ease the monotony of the diet.

The age at which the diet can be liberalized is not at all clear,[18] but by school age no risk of further brain damage has been demonstrated, and restriction of the diet has often been permitted to lapse when an affected child starts school. Many treated phenylketonuric children have been found at the age of 4 or 5 to have some degree of hyperactivity, with some lack of coordination of fine visuomotor movements.[19] Further investigations are required to determine whether these minor abnormalities are intrinsically irreversible consequences of the disorder, are due to lapses in dietary control, or are psychological sequelae of the intrafamilial stresses of coping with such severe food restrictions. The cost-benefit of treatment for phenylketonuria is convincingly attested to by the experience of centers caring for these patients,[20] but the justification for phenylketonuria screening programs must be viewed in the context of the overall health needs of a society.[6]

In personal experience in a Genetics Clinic with two to five patients per annum, since the beginning of a statewide screening program in Connecticut in 1965, 48 diagnosed patients have been treated (including at least 15 with hyperphenylalaninemia); two patients have been missed; one treated patient has mental retardation due to an unrelated condition; six treated patients have marginal intelligence, probably due to unsatisfactory diet control; and 24 children are now of preschool or school age, with normal developmental quotients by standard tests. During this period at least 15 to 20 of these patients would probably have suffered from severe brain damage if they had not been diagnosed and treated.[17]

The undoubted benefit of diet restriction for children who would otherwise suffer grave brain damage must be weighed against the strain on the family of the unnatural diet that deprives a child of many normal eating experiences; and against the potential harm—physical, as well as psychological—of treating some children unnecessarily. Throughout the duration of diet treatment, attention must be paid to the psychological health of the child and the family; full supportive services should be readily available for the mother of a child with phenylketonuria, to guide her in her child's upbringing and diet supervision. Hyperphenylalaninemic states that mimic presymptomatic phenylketonuria pose a therapeutic dilemma that is resolved by treating all suspected cases with low phenylalanine diets, while constantly monitoring blood phenylalanine levels, and with periodic challenges of protein loads for those children tolerant of more diet phenylalanine than would be expected in classic phenylketonuria. Diet restriction can then be safely abandoned as soon as phenylalanine tolerance is shown to have reached permissible levels.

The fear that successful treatment will increase the problem of phenylketonuria for future generations is partly justified. The offspring of women with phenylketonuria have a high risk of intrauterine brain damage,[16] which might be preventable by reinstituting a restricted diet during pregnancy. All successfully treated phenylketonuric girls must be forewarned about this risk and be offered the options of refraining from procreation, or of resubmitting to a difficult diet which has not yet been proven to protect fully the brain of their unborn child. The preservation of most homozygous phenylketonuric children to become normal reproductive adults, however, and any increased family size of heterozygous parents, will not raise the population incidence of the phenylketonuric gene significantly for many generations. It can be calculated that the incidence would only be increased about 20 per cent in 100 generations.[5]

Other methods of treatment attempted for phenylketonuria include the use of (1) β-2-thienylalanine to compete with and impede intestinal absorption and renal reabsorption of phenylalanine; (2) serotonin congeners to counteract the neurochemical effects of phenylketonuria;[21] (3) folic acid to promote formation of the pteridine cofactor for phenylalanine hydroxylase;[16] and (4) anabolic androgens to promote growth and accelerate the utilization of phenylalanine for protein synthesis.[22] None of these alternative therapies has yet shown promise of replacing or even complementing diet therapy for this disorder.

Once brain damage has already appeared in phenylketonuria, there is no evidence that it can be reversed by later diet restriction, although diet might have some effect on the seizures and eczema, will reduce the formation of malodorous phenylacetate, and should darken skin and hair pigmentation. Hyperactivity can be ameliorated by low phenylalanine diets, but may be aggravated by the resentment retarded patients might feel when limitations are set on their choice of foods.[16] Intensive behavior modification training might somewhat improve a patient's social responsiveness, but this is obviously an extravagant "notechnology" stopgap measure compared to the "half-way technology" of successful diet restriction. Replacement of the defective enzyme at present can be only by liver transplantation, which is not yet a practical option for this condition.

Several other metabolic diseases are treatable by following the same principle of restricting dietary intake to reduce blood levels of a toxic substance, but not every hyperaminoacidemia or organicacidemia has to be treated. Many of the disorders of branched-chain amino acid catabolism and urea cycle metabolism will require stringent diet regulation, but whether to restrict histidine in histidinemia,[23, 24] or methionine and homocysteine in homocystinuria,[16] is still being evaluated. Experimental treatment by restriction for hyperaminoacidemias must proceed with caution, since attempting to lower the plasma concentrations to normal levels conceivably can be detrimental, as it is in phenylketonuria where growth and developmental retardation occur unless the phenylalanine level exceeds the normal range by at least 50 per cent.

Mineral restriction of dietary copper in Wilson's disease is an important part of its treatment, but restriction of dietary iron in hemochromatosis does not contribute significantly to its treatment, although avoidance of excessive oral iron intake is a necessary corollary to depletion therapy.

Supplementation

Supplementation of specific foods will counteract the deficiency states that complicate certain genetic disorders (see Fig. 11–2).

Blocked transport or metabolism of essential nutrients produces these specific deficiency states. Table 11–2 lists some examples of conditions that require supplementation. Increased dietary tyrosine for patients with phenylketonuria and the possibly protective effect of increased glycine in isovalericacidemia have already been discussed. In oroticaciduria there is defective pyrimidine biosynthesis with megaloblastic anemia and mental retardation. Supplying the body's need for pyrimidines by administration of exogenous uridine will totally correct the anemia, and if treatment is started early enough, full intellectual function may be preserved.[25] In citrullinemia, argininosuccinicaciduria and lysinuric familial protein intolerance, at least some of the pathology seems to be from arginine deficiency.[11] Treatment with supplements of arginine may prime the Krebs-Henseleit urea cycle to accelerate ammonia detoxification, and also may promote protein synthesis to restore normal growth rates. Phosphate supplements are an integral part of the treatment of hypophosphatemia. If blood phosphate levels are raised in this condition, the vitamin D–resistant rickets will be responsive to less enormous doses of the vitamin, and if hypophosphatemia is completely corrected, stunting of physical growth has been successfully avoided.[26] Water supplementation

TABLE 11–2. EXAMPLES OF CONDITIONS TREATABLE, IN PART, BY SUPPLEMENTATION OF DIET FOR DEFICIENCY STATES

Supplement	Condition
Glucose	Hypoglycemic states
Glycine	Isovalericacidemia
Arginine	Citrullinemia, argininosuccinicaciduria
	Lysinuric dibasic aminoaciduria
	(familial protein intolerance)
Tyrosine	Phenylketonuria
Uridine	Oroticaciduria
Phosphate	Fructose intolerance
	Hypophosphatemic rickets
Water	Diabetes insipidus

is essential to prevent dehydration in nephrogenic diabetes insipidus, but in conditions such as this, with grossly increased drinking of water, a cautionary note is introduced by the unforeseen complication of fluorosis from excessive imbibing of fluorinated water.[27]

Supplementation: Vitamins. (1) Abnormalities of transport of specific vitamins can lead to deficiency states that are treatable by giving larger doses of the vitamin, and by giving the vitamin parenterally or in an altered chemical form. (2) Some biochemical derangements can be countered by administering doses of vitamins to stimulate compensatory metabolic reactions. (3) Genetic mutations that affect either biosynthesis of cofactors from vitamins or biochemical interaction of a cofactor with an apoenzyme are occasionally responsive to treatment with vitamin supplements (Fig. 11–2 and Table 11–3).

In any of the causes of fat malabsorption, the steatorrheas, including genetic causes ranging from cholestasis to cystic fibrosis, deficiency of the fat soluble vitamins may follow. Supplementation with appropriate doses of the water soluble forms of vitamins A, D, E, and K will prevent these deficiency states from appearing. In Hartnup disease, the abnormality in transport of tryptophan blocks this source of nicotinic acid, so that maintenance doses of the vitamin will abolish the rash, and may improve the neurological abnormalities of this disorder. The retinitis pigmentosa complicating a-β-lipoproteinemia might be preventable, by administering supplements of vitamin A sufficient to maintain normal blood levels of this vitamin.[28]

Large doses of vitamin D are required to counteract the abnormalities of calcium metabolism resulting from failure of bicarbonate reabsorption in Lowe's syndrome, and of phosphate reabsorption in conditions such as cystinosis and hypophosphatemia. Large doses of ascorbic acid have been proposed, among other things, for treatment of oxalosis,[16] alcaptonuria,[29] and the hypertyrosinemia of immature infants.

Vitamin dependency is a term reserved for conditions improved by vitamin therapy in which there is (1) biochemical impairment of coenzyme formation from vitamin precursors or (2) altered coenzyme interaction with a structurally abnormal apoenzyme. Disorders of folic acid metabolism, of coenzyme formation from vitamin B_{12},[16] and of vitamin D metabolism[30] are typical examples of the first group. Some of these disorders will respond only to massive doses of the precursor vitamin, but can respond to physiological doses of the active derivative.[30]

Table 11–3 lists examples of vitamin dependency in which an apoenzyme seems to be stimulated to greater activity by an administered vitamin. Thiamine has been uniquely beneficial in one form of hemolytic anemia, one form of maple syrup urine disease, and one form of hyperalaninemic lactic acidosis associated with an intermittent movement disorder. Pyridoxine has been effective in a host of disorders, and may be of value in many others, since pyridoxal phosphate is a coenzyme for many enzymes of intermediary metabolism. The most note-

TABLE 11-3. EXAMPLES OF CONDITIONS PROVEN OR POSTULATED TO BE RESPONSIVE TO VITAMIN SUPPLEMENTATION

Correction of Deficiency	Condition
Fat soluble vitamins	Fat malabsorption syndromes
Vitamin A	A-β-lipoproteinemia
Nicotinic acid	Hartnup disease
Folic acid	Folic acid malabsorption
Vitamin B_{12}	B_{12} malabsorption
	Transcobalamin deficiencies

Counteraction of Metabolic Disturbance	Condition
Ascorbic acid	Hypertyrosinemia of immaturity
	? Alcaptonuria
	? Cystinosis
Nicotinic acid	Type II hyperlipidemia
Pyridoxine	Hyperoxalosis
Vitamin D	Rickets and osteomalacia of
	Lowe's syndrome
	Cystinosis
	Hypophosphatemia
Vitamin K	Hypoprothrombinemia of liver disease

Vitamin Dependency Syndromes	Condition
Thiamine	Thiamine-responsive anemia
	Thiamine-responsive branched-chain ketoacidosis
	? Thiamine-responsive hyperalaninemia
Lipoic acid	? Subacute necrotizing encephalomyelopathy
Pyridoxine	Pyridoxine-responsive anemia
	Pyridoxine-dependent seizures
	Pyridoxine-responsive homocystinuria
	Cystathioninuria
Folic acid	Fructose-1,6-diphosphatase deficiency
Biotin	Biotin-responsive propionicacidemia
	Biotin-responsive β-methylcrotonylglycinuria

Inborn Errors of Vitamin Metabolism	Condition
Folic acid	Formimino transferase deficiency
	Methyltetrahydrofolate transferase deficiency
	Methylene tetrahydrofolate reductase deficiency
B_{12}	B_{12}-responsive methylmalonicacidemias
	B_{12}-responsive homocystinuria with methylmalonicaciduria
Vitamin D	Defective 1-hydroxylation of 25-hydroxyvitamin D

worthy of these disorders is pyridoxine-dependent seizures, in which otherwise intractable seizures with progressive brain damage are dramatically arrested by this vitamin. Apart from abnormalities of folate transport and metabolism, supplemental folic acid has stimulated the defective enzyme in one patient with fructose-1,6-diphosphatase deficiency[31] and has been tried for the stimulation of glycine cleavage enzyme in nonketotic hyperglycinemia.[32] Biotin-responsiveness has been reported for propionicacidemia and β-methylcrotonylglycinuria,[16] but it is possible that the effect of biotin in these conditions is nonspecific, since the molecular activation of apoenzyme *in vitro* has yet to be demonstrated in these biotin-responsive patients.

Substitution

The semisynthetic formula used for phenylketonuria has already been discussed. For strict regulation of amino acid intake in other metabolic disorders, such as maple syrup urine disease or propionicacidemia, recourse to synthetic amino acid mixtures as a substitute for protein has preserved life, health, and intellect successfully in several patients.[16, 33]

In the hyperammonemia syndromes, which can produce acute and chronic neurological complications, self-imposed low protein intake is the only treatment required for the milder degrees of hyperammonemia. When protein intolerance is extreme, however, many other therapeutic measures have to be instituted,[11] including the use of keto-acid derivatives of the amino acids to provide the carbon skeleton of the essential amino acids without overloading the body's capacity to detoxify nitrogen.[34] This experimental approach, shown to be helpful in uremia, has yet to be tried for any infant with a severe congenital hyperammonemia syndrome. Chemical treatments for hyperammonemia are discussed in a following section.

A completely synthetic diet is used in total parenteral alimentation, when all nutritional sustenance is given intravenously. Although generally used for patients with severe small bowel malfunction, this treatment has improved the metabolic balance and diminished the enlarged liver of patients with glycogenoses.[35]

Replacement

Restoring enzymes by oral therapy is possible only for the enteric digestive enzymes, such as in cystic fibrosis, and in pepsinogen or lactase deficiency, because few protein molecules are absorbed into the body from the intestine before they are degraded by digestive enzymes. Insufflation of antidiuretic hormone has allowed nasal absorption of this polypeptide to be used in the treatment of diabetes insipidus.

The Internal Environment

Chemical Modification

Correction of electrolyte and fluid disturbances is not unique to genetic diseases, but is a critical aspect of treatment for several of these disorders. Restoring tissue concentrations of glucose, amino acids, minerals, and vitamins by diet supplements has already been discussed. Correction of abnormal concentrations is not always beneficial; for example, raising copper levels by parenteral copper administration has not helped the profound retardation or seizure disorder in Menkes' disease,[36] perhaps because irreversible brain damage had already occurred before therapy was started.

Lowering concentrations of metabolites by chemical modification of the internal environment is the treatment of choice for some conditions, with or without simultaneous dietary restrictions. *Hyperammonemia* is aggravated by the ammonia released by urease-forming microorganisms in the colon, so prevention of constipation is an obvious precaution in any condition leading to hyperammonemia. Abolishing colonic microorganisms with antibiotics can be useful for short-term treatment; use of chemical or immunological urease inhibitors has not gained clinical acceptance.[37] Altering the bowel flora by giving the nonassimilable sugar lactulose (or use of lactose in lactase-deficient individuals)[38] will promote growth of fermentative bacteria, producing acidic, loose stools. This traps ammonium ions in the fecal contents, and has been used successfully to complement dietary management of children with hyperammonemia.[11] Detoxification of ammonia has been attempted by acid infusions to stimulate renal excretion of ammonium salts; by administration of glutamate to increase glutamine formation; and by use of other metabolites to stimulate the Krebs-Henseleit cycle and accelerate urea synthesis. None of these approaches, however, has been convincingly helpful.[11, 37] Other treatments for hyperammonemia include the use of exchange transfusion, peritoneal dialysis, colonic by-pass surgery and liver transplantation as desperate measures for extreme hyperammonemia. Sterilization of the bowel has reduced methylmalonicacidemia in one patient, and has been proposed for treatment of acute exacerbations of metabolic acidosis in this disease.[39]

Exchange transfusions have been used for decades to remove bilirubin from infants with hyperbilirubinemia, and can be used to remove other toxic substances from the blood. Hemodialysis or peritoneal dialysis is more efficient than exchange transfusion for removing small molecules from the extracellular fluid and has been used occasionally in metabolic diseases.[11, 40] Charcoal hemoperfusion may also find a place in severe biochemical disturbances of genetic etiology.[41]

Chemical removal of the copper retained in Wilson's disease has been very successful in preventing the liver and brain damage complicating this disorder. Decreasing food copper by diet restriction and by use of potassium sulfide or resins to precipitate copper in the intestinal lumen are of marginal benefit. Chelating agents, of which D-penicillamine is the most effective, will flush sufficient copper out into the urine to deplete the excess copper stored in these patients' tissues. D-Penicillamine is also used in cystinuria, since it forms a soluble mixed disulfide with cysteine that will prevent or reverse stone formation.[16]

Unfortunately, D-penicillamine does have toxic side effects, some of which can be quite serious, which means it is not the final solution for treatment of these conditions. Removal of iron by chelating agents in hemochromatosis has not been very useful, but repeated bloodlettings will deplete iron stores very efficiently, to prevent the complications of this disease. In treating hypercholesterolemia, the enterohepatic circulation of the bile acids, essential for fat absorption, can be interrupted by chelating them with cholestyramine. If this procedure does not suffice, it can be treated with nicotinic acid, or with clofibrate, which inhibits enzymatic biosynthesis of cholesterol.[42]

Inhibition of enzyme activity is applicable to a minority of genetic disorders where overproduction of a metabolite is responsible for clinical expression of the disease. In acute intermittent porphyria, feedback inhibition of porphyrin synthesis by heme infusions has lowered blood levels of porphyrin precursors.[43] In hyperuricemia, allopurinol will inhibit xanthine oxidase and block formation of uric acid from its more soluble precursor, xanthine. This is effective in preventing attacks of gout, but it has had no discernible benefit on the severe neurological abnormalities of the Lesch-Nyhan syndrome.

Pharmacological Modification

All drugs have potentially toxic side effects, but some drugs may be uniquely toxic to persons with certain genetic disorders. Glucose-6-phosphate dehydrogenase deficiencies render affected individuals prone to hemolytic crises whenever they are exposed to drugs that are strong oxidizing agents. In affected newborn infants this could lead to hyperbilirubinemia, kernicterus and irreversible brain damage. Barbiturates precipitate systemic or neuropsychiatric attacks in patients with acute intermittent porphyria. Anesthetic agents precipitate a malignant hyperthermia syndrome in genetically susceptible individuals. Thus, drugs may unmask the clinical expression of otherwise benign genetic disorders, or may exacerbate the clinical expression of some genetic diseases. For the many conditions in which drugs are harmful, the responsible drugs should be avoided.[44]

Pharmacological treatment of genetic disorders can be by stimulation or suppression of physiological processes and by altering neurological sensitivity to toxic substances. Stimulation of the cellular immune system by transfer factor in some of the genetic immunodeficiency diseases has met with mixed success.[45] Suppression by antihistamines, kallikrein inhibitors,[46] antifibrinolytic agents[47] and perhaps an anticomplement agent[48] apparently have reduced attacks of angioneurotic edema due to C $\overline{1}$ esterase deficiency.

The use of *neurochemical agents* to oppose the toxic effect of abnormal metabolic products in genetic diseases is an alternative concept of preventive therapy that aims to protect the target organ. The toxic effect of ammonia on central nervous function far outweighs its effect on systemic metabolism, and moderate hyperammonemia would be innocuous if an effective chemical means were available to prevent its neurotoxicity. Unconjugated hyperbilirubinemia is most harmful during the first week of life, because after that the establishment of a blood brain barrier blocks entry of bilirubin into the nervous system, except in some patients with severe Crigler-Najjar syndrome. L-Dihydroxyphenylalanine (L-dopa), a precursor of the neurotransmitter dopamine, has both an antirigidity and an antiakinesic effect. It has been useful in the treatment of parkinsonism, and has improved the basal ganglion dysfunction of some patients with Wilson's disease.[49] Huntington's chorea, which causes severe degenerative brain disease in midadult life, is generally associated with an increased sensitivity to the effects of dopaminergic and serotoninergic drugs. It is possible that the disease is caused by decreased glutamate decarboxylase in certain regions of the brain, resulting in defective production of the neuroinhibitor γ-aminobutyrate (or that this degenerative disease is associated with a selective loss of neurons that contain glutamate decarboxylase). Attempting to treat Huntington's chorea with γ-aminobutyrate, however, appears to be somewhat hazardous.[50] Apart from these few sketchy examples, neurological remission by use of neuroactive chemicals has not been convincingly demonstrated in genetic disorders leading to mental retardation, although symptomatic use of anticonvulsants, for instance, is standard neurochemical therapy. Many other attempts have been made in conditions such as phenylketonuria to find neurochemical antidotes, but with little success, so far.[15, 16] In Down's syndrome, the postulate that low concentration of brain serotonin is responsible for failure to develop normal intelligence has been refuted by proving that there is a lack of response to 5-hydroxytryptophan treatment.[51]

Endocrinological Modification

Replacement therapy of primary hormone deficiencies is well recognized for genetic abnormalities of endocrine function. Deficiencies of hypothalamic releasing factors, of pituitary trophic hormones, and of other hormones will all respond to replacement therapy with the target hormone. Timely administration of thyroid hormone is essential for preservation of normal brain development in congenital hypothyroid states (in some instances apparent intrauterine brain damage has been recalcitrant to postnatal but perhaps not prenatal therapy). Inborn errors of adrenal corticosteroid biosynthesis can lead to interruption of feedback control of corticosteroid metabolism with overproduction of androgenic byproducts; therefore, in these disorders replacement therapy also serves to suppress virilization.[12] Re-establishing feedback inhibition is exemplified in the suppression of hyperparathyroidism by chemical correction of disturbances of calcium and phosphate metabolism.

Contraception by hormone suppression of ovulation is relevant as a treatment for genetic disorders in which parents at risk wish to avoid or defer pregnancies. Here the treatment is directed, not at a patient, but at a normal, fertile woman.

Donation and Excision of Gene Products

Enzyme or protein replacement therapy and organ transplantation are the subjects of Chapter 13. Protein replacement for the coagulation defects and immunoprotein deficiencies is well established, but its role in treatment for the mucopolysaccharidoses and other lysosomal disorders is still highly experimental. Cell replacement is useful, albeit only temporarily, for blood cell or blood platelet abnormalities; it has been successful in marrow transplants for hematopoietic and immune deficiencies. Organ replacement is established therapy only for renal disorders, despite the fact that transplants of other organs have been attempted for serious genetic diseases (see Chapter 13).

Excision of mutant genes or aneuploid chromosomes is not technically possible yet. When gene expression produces malignant changes in any tissue, there is little choice but to use excision or ablation therapy. In addition to use in malignant or premalignant conditions, excision can be used for cosmetic, temporizing, or symptomatic treatment; then, however, it is only indirect "low technology" therapy for genetic disorders.

Repair and Reconstruction

Anatomical malformation syndromes can sometimes be totally repaired by surgical reconstruction; both structural abnormalities and metabolic disturbances may respond to surgical rearrangements of normal anatomy. Among the conditions that threaten mental retardation, hydrocephalus can often be prevented by the increasingly successful neurosurgical shunt procedures. Surgical shunting operations on the gastrointestinal tract and portal venous system include colonic by-pass for the hyperammonemia syndromes and ileal by-pass to decrease fat absorption in type II hyperlipidemia. Portacaval shunting has been used to treat portal hypertension complicating some of the glycogenoses, or to divert alimentary glucose to the systemic circulation before it is trapped in the liver as glycogen. Not only has there been modest prolongation of postprandial blood glucose levels, but some patients have had dramatic clearance of hyperlipidemia, reduced hepatomegaly, and growth spurts.[52] That liver metabolism is radically benefited by portal diversion has also been demonstrated in homozygous type II hyperlipidemia, where, surprisingly, remission of hypercholesterolemia and xanthomata followed a portacaval shunt.[53]

Manipulation of Gene Expression

Stabilization of a defective protein molecule, to sustain its activity, is illustrated by the prophylactic treatments used for sickle hemoglobinopathy. Red cell life is prolonged by removal of an overactive spleen; tendency to sickle is reduced by avoidance of dehydration and hypoxia; and molecular stabilization might be achieved by carbamylation of the amino-terminal ends of the β-polypeptide with cyanate.[54] Cofactor potentiation of enzyme function in the vitamin dependency syndromes might also be by molecular stabilization of the apoenzyme molecule, in some instances.

Stimulation of increased formation of a deficient enzyme is possible for some inducible microsomal enzymes. In Gilbert's unconjugated bilirubinemia, barbiturates induce measurably increased glucuronyl transferase activity,[55] but in the more serious Crigler-Najjar hyperbilirubinemia, barbiturates are ineffective. Phenobarbital also causes symptomatic and chemical improvement in some of the familial cholestases. It is not clear whether the conditions are errors primarily of bile salt metabolism or of bile salt transport across the canalicular membrane. Consequently, it is also not clear whether phenobarbital relieves some variants by inducing microsomal enzyme activity, or by stimulating canalicular bile excretion.[56]

Rhesus-incompatible hemolytic disease in the fetus is prevented by curtailing the circulation of Rh antigens in the genetically susceptible mother. Injections of Rh-specific antibodies are used to coat and inactivate the antigens to interrupt the immunization process. *Inhibition* of enzyme expression by poisoning its activity with substrate analogues, or by feedback suppression with enzyme products, has already been cited for treatment of hyperuricemia with allopurinol, of hypercholesterolemia with clofibrate, and of congenital adrenal hyperplasia with mineralocorticoids. Nonspecific blocking of gene product formation by folic acid antagonists, purine analogues, actinomycin, and other antimetabolites that block DNA replication or RNA translation and protein synthesis are useful in genetic disorders for immunosuppressive or antineoplastic therapy. In one rare hypermetabolic state, chloramphenicol seemed to suppress the excessive formation of mitochondria.[57]

Donor genes are of course transplanted along with any cell or tissue graft, but to donate isolated human genes, technical advances in gene purification are required that might not be safe to pursue.[58] Gene therapy by deliberate introduction of viral genes in man has been tried only once, in two sisters with argininemia, in an attempt to transduce a viral gene for arginase.[11, 59] This virus appeared to transduce viral arginase accidentally in laboratory workers, and was reported to have some *in vitro* effect on cultured cells from one of the argininemic patients.[60] The experiment produced no detectable biochemical or chemical response, and generated considerable controversy about its advisability.[59]

SELECTION

The choosing of genes for one's offspring is generally practiced only by selective mating, in the choosing of marriage partners on social or romantic criteria. Because most discernible traits are an expression of contributions from many genetic factors, choice of a mate by social, physical, or behavioral attributes has a very indirect relation to gene selection. For instance, mating of racehorses is based on ancestry and form, from a highly inbred stock; yet prediction of successful progeny is still as much of a gamble as prediction of racetrack performance. In all of animal husbandry, selection for certain characteristics is often at the cost of losing other traits; therefore, with inbreeding there is considerable constriction of the variability of the gene pool. In human society, the positive or negative value of a given trait is strongly colored by changing cultural mores. This means that the endowment of the future human gene pool is being determined by uncontrolled sociological forces of little selective genetic benefit. Such a situation is not to the detriment of society, for the ill-conceived doctrines of eugenicists in the past have rightly discredited attempts at selective breeding for supposedly desirable traits.

An unfair selective disadvantage may be imposed on heterozygote carriers detected by screening programs, because their identification may stigmatize them as being unacceptable candidates for marriage (see Chapter 4).

For multifactorial and recessive genetic disorders, the reproductive practices of couples at-risk will have negligible effects on the human gene pool within many generations, although for each couple the birth of a genetically handicapped child would still be of paramount importance. For dominant genetic disorders, if affected adults were to refrain from having children, the responsible gene would be removed from the gene pool except for replenishment by new mutations. In this way conditions such as achondroplasia would be only somewhat reduced in incidence, whereas conditions such as Huntington's chorea might be almost completely eradicated.

The *reproductive options* open to an individual couple include abstention, adoption, artificial insemination by donor, natural reproduction, and the solely negative option of prenatal selection against fetuses found wanting by intrauterine tests (Chapter 9). A family can select against undesirable traits only by refraining from procreation; using donor sperm; or destruction of the whole genome, *i.e.,* by aborting the fetus. Prenatal testing will, when applicable, allow parents to bear children with the assurance that they will be born free of a dreaded disorder — provided that any affected fetus is aborted.

Selection against the commoner polygenic traits is limited by present inability to predict occurrence of these diseases other than by empirical observations of recurrence frequencies. The neural tube malformations, meningomyelocele and anencephaly, and the chromosome aneuploidy syndromes form the only groups of such diseases that, so far, are detectable by prenatal tests (see Chapter 9). Mutagenic and teratogenic hazards for the fetus can be selected against by careful adherence to high standards of nutrition and of antenatal care, and by avoidance of potential toxins in the expectant mother's external and internal environment.[61] Developments in prenatal diagnosis that permit visual inspection of fetal contours[62, 63, 64] promise that other major malformation syndromes will soon be detectable by amnioscopy or fetography. Selection against disorders caused by single genes of large effect is more definitive, because risk predictions can be given with greater precision, especially if genetic carriers can be accurately identified. The option of prenatal diagnosis by biochemical tests is possible for many of these disorders (see Chapter 9), if this option is morally and emotionally acceptable to the prospective parents. The availability of prenatal tests presents the possibility that a family might select progeny for survival on the basis of criteria other than that of absence of untreatable disease. If parents could freely choose the sex of their children, fear has been expressed that gross imbalance of the sex ratio of children would soon appear,[65] but at least one expert opinion is that this fear is unfounded in a society such as that in North America.[66]

Selection of gametes is now practiced by artificial insemination from a donor. Experiments with selection of X- or Y-carrying human sperm by differential motility or buoyant density have not yet been successful. Selection of oocytes, or even transfer of human zygotes, has already been attempted experimentally,[67] but not without arousing critical ethical debate[65, 68, 69] (see Chapter 21). Singling out a gamete according to its specific genetic composition, for fertilization *in vitro*, is beyond present technical capabilities.

Postnatal selection by withholding corrective treatment for children with incurable mental or physical handicaps might be considered a logical extension of prenatal selection by abortion. While it is inhumane to subject an incurably malformed infant to major surgery – sometimes repeatedly – it is inhuman to deny therapy that might make possible the hope of meaningful survival. The boundary between ethically acceptable and unacceptable policies must be determined cautiously for each case, preferably by a broader representation from society than the immediate family and its physicians, although the rights and responsibilities of those most closely involved must be respected. The problem of whether, if ever, quality of life outweighs the absolute duty to preserve life should be of concern for all.[65, 69, 70]

FUTURE PROSPECTS

The immediate future holds predictable advances in treatment for genetic disorders, but the more distant future is less foreseeable, and utopian forecasts of genetic engineering triumphs may be premature, if not unrealistic. Expansion of present facilities would give more families access to currently available treatment possibilities: if public education and genetic counseling services were more effective; if diagnostic heterozygote and homozygote screening programs were more widespread and encompassed more disorders; and if prenatal testing centers and genetic treatment clinics were more widely distributed. The question of societal priorities in choosing whether to invest in preventive genetic programs must be considered in the context of overall societal health needs, but the long-range benefits of forestalling mental retardation,[20, 69] or reducing the burden of recurrent genetic diseases, must claim some priority in the allocation of health resources.[6, 65] The impetus of current research activities can be expected to expand knowledge about the pathogenesis and response to therapy of individual genetic diseases and about fundamental mechanisms of molecular genetics. Particularly promising are anticipated developments in immunogenetics, embryology and teratogenesis, metabolic diseases, and neurochemistry.

Undoubtedly, major advances will soon occur in defining the genetic mechanisms of immune reactions. These findings will allow new approaches to be developed in treatment of the immune deficiency diseases, and refined immunosuppressive therapy will hopefully protect against rejection of heterologous protein or tissue transplants, with fewer toxic side effects.

Developmental biology is just beginning to detail the physiological changes that govern the constancy of the intrauterine environment in the growing embryo and fetus, while controlling cell differentiation and fetal maturation. This new information, together with the more precise identification of teratogens and their mode of action, will allow better protection of pregnancy from exogenous noxious influences and will lead to new approaches to intrauterine therapy.

Intrauterine therapy for genetic diseases has already been achieved in three distinct ways. The first is by prevention of toxic changes in maternal physiology as illustrated by desensitization in rhesus incompatibility, and by dietary control of metabolic imbalance in maternal phenylketonuria.[16] The next is by intrauterine transfusions, also used for rhesus-incompatible hemolytic disease, but with a possibility that infusion of medications, proteins, or cells directly into the fetus might be used for other problems. The last is by correction of toxic changes in fetal physiology, by administration of intrauterine corticosteroids for congenital adrenal hyperplasia,[12] and by massive maternal injections of vitamin B_{12} in vitamin B_{12}-responsive methylmalonicaciduria.[71] This third approach, or a combination of the second and third approaches, may become of vital significance in the treatment of certain genetic disorders in which toxic effects are already treatable postnatally, but in which irreversible changes may have occurred by the time of birth, as in some congenital hypothyroid states.[72]

More accurate chemical data about biochemical disturbances in metabolic diseases and their responses to therapy will be obtained with improved sophistication of analytical techniques. Many new metabolic diseases and new variants of known metabolic diseases will be diagnosed and investigated; effective variations on the themes of diet restriction and vitamin supplementation will be applied to many of these conditions; and presymptomatic newborn screening programs will become possible for more diseases as diagnostic tests become both more accurate and more economical. Similarly, improved methodology will allow carriers of more genetic disorders to be detectable, and prenatal testing to be available for affected fetuses in many more disorders. Prenatal testing will be extended to diseases identifiable only in blood or tissue cells rather than in amniotic fluid cells, when safe methods for obtaining fetal tissue become established.[62] Reasonably safe amnioscopy and noninvasive ultrasound[64] will also enable any gross malformation syndrome to be detected antenatally.

Neurochemistry is an important burgeoning new specialty that is defining the chemical basis for normal and abnormal neural activity, including factors controlling brain growth and nerve cell differentiation.[73] The pathophysiology of neurotoxicity and of arrested brain development in metabolic diseases will soon reach the point where rational neurochemical treatment may become possible. For instance, as in many metabolic diseases, the biochemical lesions in phenylketonuria and the urea cycle enzyme defects are in the liver, but the major important consequences are neurological. If an effective barrier could be raised to obstruct entry of the responsible toxin into nervous tissue, if imbalanced intraneuronal cell metabolism could be re-equilibrated, or if an effective neurochemical antagonist or antidote to the toxin could be introduced, then treatment for this group of disorders might focus on normalization of brain function rather than on correction of systemic disturbances. This new approach to treatment for genetic diseases causing mental retardation should be explored and exploited as soon as neurochemical developments permit.

Fundamental advances in molecular genetics, although dramatic, are unlikely to be applicable to treatment of human disease until many intermediate studies have been thoroughly evaluated. *In vitro* cell culture, and particularly cell hybridization experiments, have been impressively productive in the study of eukaryotic molecular genetics and human enzyme defects. Advantage has been taken of progressive loss of human chromosomes from man-rodent heterokaryons to localize genes on individual chromosomes.[74] This information will enable gene-linkage analyses to be used in prediction of how genetic disorders will segregate within families. As an example, the known linkage of the gene for myotonic dystrophy with the secretor locus can be helpful in genetic counseling, and, at least in theory, could be used to predict some cases of myotonic dystrophy antenatally by testing for the secretor marker. The possibility of cloning a new individual from cultured somatic cells has been demonstrated for plants and amphibia. Although a popular topic of alarmists, cloning of human cells is unlikely to lead to the creation of manmade men, because of virtually insuperable biological problems of *in vitro* tissue culture requirements and because natural procreation is far simpler to encourage.[6, 65, 69] Cloning of differentiated human cells for organ transplantation is a less unlikely prospect that has therapeutic possibilities. Research into gene structure and gene function has progressed to the stage where the structural portion of human genes has been sequenced and replicated *in vitro* for the hemoglobin polypeptides;[75] a functional bacterial gene has been introduced into cultured galactosemic human cells;[76] and transduction of a viral gene has been deliberately attempted for two patients with argininemia.[59] Despite these isolated accomplishments, the immediate prospects of gene therapy by introduction of functional donor genes for the vast majority of single gene diseases are not bright. The approach of seeking to modify gene regulation, however, may be a promising interim objective. For instance, for abnormalities of the β-polypeptide of hemoglobin, therapeutic measures that can cause persistence of fetal hemoglobin, repressing product formation from the β-hemoglobin gene, inactivating the developmental genetic switch, would suspend hemoglobin synthesis in a permanently fetal state. This would effectively prevent any clinical expression of these hemoglobinopathies.

Until such time as effective "high technology" prevention of mental retardation from genetic diseases becomes a practical reality, genetic counseling and zygote selection by antenatal diagnosis are still the mainstays of prevention for these diseases. Alternative reproductive options, such as artificial insemination, might eventually be expanded to include the possibilities of oocyte donation, zygote or embryo implantation and surrogate motherhood[67] but experimental testing of these procedures may exceed the bounds both of technical feasibility and of ethical acceptability[65, 69, 77] (see Chapter 21). Public outcry against uninhibited fetal experimentation has recently mandated a moratorium on fetal research, but scientifically sound humane investigations must be allowed to continue, although they must be supervised by responsible regulatory agencies. Research in embryology and fetology is essential if causes of fetal morbidity and mortality are to be understood and then eradicated. Whether long-range prospects for effective treatment of genetic disease will come to fruition depends on continued exploration of fundamental genetic mechanisms and investigations of each genetic disorder. If active basic and applied genetic research is not encouraged, the present momentum in these fields will be lost. It is not extravagant to invest in training and research which promise no immediate dividends, since future prospects of treatment for genetic disorders will not materialize without the scientific talent and facilities to pursue them.[78]

Future therapeutic prospects for mental retardation in genetic disorders are quite promising for many metabolic diseases, with excellent probability of completely preventing brain damage in some. Continuing research should develop new possibilities beyond those that are readily predictable. For the neurological and neuropsychiatric disorders as a group, therapeutic prospects depend on clarification of pathogenic mechanisms. For chromosomal anomalies, and other malformation syndromes associated with abnormal brain structure and function, therapeutic prospects would seem to be limited to prevention by genetic counseling and antenatal diagnosis. It is evident there are a multitude of approaches for treatment in genetic disease. Emphasis is progressing toward "high technology" preventive therapy for some disorders, and "halfway technology" for many others, thereby lowering the proportion of genetic problems for which only "notechnology" care and comfort can be given.

References

1. Scriver, C. R.: Treatment of inherited disease: Realized and potential. Med. Clin. North Am., 53:941, 1969.
2. Davis, B. D.: Prospects for genetic intervention in man. Science, 170:1279, 1970.
3. Lederberg, J.: Genetic engineering, or the amelioration of genetic defect. Pharos, 34:9, 1971.
4. WHO Technical Report: Genetic Disorders: Prevention, Treatment, and Rehabilitation. Technical Report No. 497, 1972.
5. Howell, R. R.: Genetic disease: The present status of treatment. In McKusick, V. A., and Claiborne, R. (eds.): Medical Genetics. New York, HP Publishing Co., Inc., 1973, pp. 271–280.
6. Motulsky, A. G.: Brave new world? Science, 186:653, 1974.
7. Haslam, R. H. A. (ed.): Symposium on habilitation of the handicapped child. Pediatr. Clin. North Am., 20:1, 1973.
8. Thomas, L.: The Lives of a Cell: Notes of a Biology Watcher. New York, Viking Press, Inc., 1974, pp. 33–35.
9. Stiehm, E. R., and Fulginiti, V. A. (eds.): Immunologic Disorders in Infants and Children. Philadelphia, W. B. Saunders Company, 1973.
10. Garrod, A. E.: In Harris, H. (ed.): Garrod's Inborn Errors of Metabolism. London, Oxford University Press, 1963.
11. Hsia, Y. E.: Inherited hyperammonemic syndromes. Gastroenterology, 67:347, 1974.
12. New, M. I., and Levine, L. S.: Congenital adrenal hyperplasia. Adv. Hum. Genet., 4:251, 1973.
13. Pagliara, A. S., Karl, I. E., Haymond, M., and Kipnis, D. M.: Hypoglycemia in infancy and childhood. J. Pediatr., 82:365, 558, 1973.
14. Garcia, J., Hankins, W. G., and Rusiniak, K. W.: Behavioral regulation of the *milieu interne* in man and rat. Science, 185:824, 1974.
15. Bickel, H., Hudson, F. P., and Woolf, L. I. (eds.): Phenylketonuria and Some Other Inborn Errors of Aminoacid Metabolism. Stuttgart, Georg Thieme Verlag, 1970.
16. Scriver, C. R., and Rosenberg, L. E.: Amino Acid Metabolism and Its Disorders. Philadelphia, W. B. Saunders Company, 1973.
17. Smith, I., and Wolff, O. H.: Natural history of phenylketonuria and influence of early treatment. Lancet, 2:540, 1974.
18. Anonymous: The duration of treatment of phenylketonuria. Lancet, 1:971, 1974.
19. Berman, J. L., and Ford, R.: Intelligence quotients and intelligence loss in patients with phenylketonuria and some variant states. J. Pediatr., 77:764, 1970.
20. Bush, J. W., Chen, M. M., and Patrick, D. L.: Analysis of the New York State PKU screening program using a health status index. In Coddington, D. (ed.): Health Status Indexes. Chicago, Hospital Research & Educational Trust, 1973.

21. Wooley, D. W., and van der Hoeven, T.: Prevention of a mental defect in phenylketonuria with serotonin congeners such as melatonin and 5-hydroxytryptophan. Science, 144:1593, 1964.

22. Brambilla, F., Giardini, M., and Russo, R.: Prospects for a pharmacological treatment of phenylketonuria. Berkeley, Calif., Conference on Phenylketonuria, 1973.

23. Holmgren, G., Hambraeus, L., and deChateau, P.: Histidinemia and "normohistidinemic histidinuria." Acta Paediatr. Scand., 63:220, 1974.

24. Anonymous: Histidinaemia: To treat or not to treat? Lancet, 1:719, 1974.

25. Smith, L. H., Huguley, C. M., and Bain, J. A.: Hereditary orotic aciduria. *In* Stanbury, J. B., Wyngaarden, J. B., and Fredrickson, D. S. (eds.): The Metabolic Basis of Inherited Disease. 3rd ed. New York, McGraw-Hill Book Company, 1972, pp. 1003–1029.

26. Glorieux, F. H., Scriver, C. R., Reade, T. M., Goldman, H., and Roseborough, A.: Use of phosphate and vitamin D to prevent dwarfism and rickets in X-linked hypophosphatemia. N. Engl. J. Med., 287:481, 1972.

27. Greenberg, L. W., Nelsen, C. E., and Kramer, N.: Nephrogenic diabetes insipidus with fluorosis. Pediatrics, 54:320, 1974.

28. Gouras, P., Carr, R. E., and Gunkel, R. D.: Retinitis pigmentosa in abetalipoproteinemia: Effects of vitamin A. Invest. Ophthalmol., 10:784, 1971.

29. Lustberg, T. J., Schulman, J. D., and Seegmiller, J. E.: Decreased binding of ^{14}C-homogentisic acid induced by ascorbic acid in connective tissues of rats with experimental alcaptonuria. Nature, 228:770, 1970.

30. Fraser, D., Kooh, S. W., Kind, H. P., Holick, M. F., Tanaka, Y., and DeLuca, H. F.: Pathogenesis of hereditary vitamin D–dependent rickets: An inborn error of vitamin D metabolism involving defective conversion of 25-hydroxyvitamin D to 1α,25-dihydroxyvitamin D. N. Engl. J. Med., 289:817, 1973.

31. Greene, H. L., Stifel, E. B., and Herman, R. H.: "Ketotic hypoglycemia" due to hepatic fructose-1,6-diphosphatase deficiency: Treatment with folic acid. Am. J. Dis. Child., 124:415, 1972.

32. Trijbels, J. M. F., Monnens, L. A. H., van der Zee, S. P. M., Vrenken, J. A. T., Sengers, R. C. A., and Schretlen, E. D. A. M.: A patient with nonketotic hyperglycinemia: Biochemical findings and therapeutic approaches. Pediatr. Res., 8:598, 1974.

33. Brandt, I. K., Hsia, Y. E., Clement, D. H., and Provence, S. A.: Propionicacidemia (ketotic hyperglycinemia): Dietary treatment resulting in normal growth and development. Pediatrics, 53:391, 1974.

34. Close, J. H.: The use of amino acid precursors in nitrogen-accumulation diseases. N. Engl. J. Med., 290:663, 1974.

35. Folkman, J., Philippart, A., Tze, W. J., and Crigler, J.: Portacaval shunt for glycogen storage disease: Value of prolonged intravenous hyperalimentation before surgery. Surgery, 72:306, 1972.

36. Bucknall, W. E., Haslam, R. H. A., and Holtzman, N. A.: Kinky hair syndrome: Response to copper therapy. Pediatrics, 52:653, 1973.

37. Schenker, S., Breen, K. J., and Hoyumpa, A. M.: Hepatic encephalopathy: Current status. Gastroenterology, 66:121, 1974.

38. Welsh, J. D., Cassidy, D., Prigatano, G. P., and Gunn, C. G.: Chronic hepatic encephalopathy treated with oral lactose in a patient with lactose malabsorption. N. Engl. J. Med., 291:240, 1974.

39. Snyderman, S. E., Sansaricq, C., Norton, P., and Phansalkar, S. V.: The use of neomycin in the treatment of methylmalonic aciduria. Pediatrics, 50:925, 1972.

40. Russell, G., Thom, H., Tarlow, M. J., and Gompertz, D. J.: Reduction of plasma propionate by peritoneal dialysis. Pediatrics, 53:281, 1974.

41. Gazzard, B. G., Portmann, B., Weston, M. J., Langley, P. G., Murray-Lyon, I. M., Dunlop, E. H., Flax, H., Mellon, P. J., Record, C. O., Ward, M. B., and Williams, R.: Charcoal haemoperfusion in the treatment of fulminant hepatic failure. Lancet, 1:1301, 1974.

42. Segall, M. M., Fosbrooke, A. S., Lloyd, J. K., and Wolff, O. H.: Treatment of familial hypercholesterolemia in children. Am. Heart J., 82:707, 1971.

43. Bonkowsky, H. L., Tschudy, D. P., Collins, A., Doherty, J., Bossenmaier, I., Cardinal, R., and Watson, C. J.: Repression of the overproduction of porphyrin precursors in acute intermittent porphyria by intravenous infusions of hematin. Proc. Natl. Acad. Sci. U.S.A., 68:2725, 1971.

44. LaDu, B. N.: Pharmacogenetics: Defective enzymes in relation to reactions to drugs. Annu. Rev. Med., 23:453, 1972.
45. Pachman, L. M., Kirkpatrick, C. H., Kaufman, D. B., and Rothberg, R. M.: The lack of effect of transfer factor in thymic dysplasia with immunoglobin synthesis. J. Pediatr., 84:681, 1974.
46. Juhlin, L., and Michaëlsson, G.: Use of a kallikrein inhibitor in the treatment of urticaria and hereditary angioneurotic edema. Acta Derm.-Venereol., 49:37, 1969.
47. Sheffer, A. L., Austen, K. F., Rosen, F. S.: Tranexamic acid therapy in hereditary angioneurotic edema. N. Engl. J. Med., 287:452, 1972.
48. Erill, S., Cabezas, R., and Ausina, V.: Hereditary angioneurotic edema. Lancet, 1:169, 1974.
49. Morgan, J. P., Preziosi, T. J., and Bianchine, J. R.: Ineffectiveness of L-dopa as supplement to penicillamine in a case of Wilson's disease. Lancet, 2:659, 1970.
50. Perry, T. L., Hansen, S., and Urquhart, N.: GABA in Huntington's chorea. Lancet, 1:995, 1974.
51. Weise, P., Koch, R., Shaw, K. N. F., and Rosenfeld, M. J.: The use of 5-HTP in the treatment of Down's syndrome. Pediatrics, 54:165, 1974.
52. Starzl, T. E., Putnam, C. W., Porter, K. A., Halgrimson, C. G., Corman, J., Brown, B. I., Gotlin, R. W., Rodgerson, D. O., and Greene, H. L.: Portal diversion for the treatment of glycogen storage disease in humans. Ann. Surg., 178:525, 1973.
53. Starzl, T. E., Chase, H. P., Putnam, C. W., and Porter, K. A.: Portacaval shunt in hyperlipoproteinaemia. Lancet, 2:940, 1973.
54. Gillette, P. N., Manning, J. M., and Cerami, A.: Increased survival of sickle-cell erythrocytes after treatment *in vitro* with sodium cyanate. Proc. Natl. Acad. Sci. U. S. A., 68:2791, 1971.
55. Black, M., Fevery, J., Parker, D., Jacobson, J., Billing, B. H., and Carson, E. R.: Effect of phenobarbitone on plasma (^{14}C)bilirubin clearance in patients with unconjugated hyperbilirubinemia. Clin. Sci. Molec. Med., 46:1, 1974.
56. Ballow, M., Margolis, C. Z., Schachtel, B., and Hsia, Y. E.: Progressive familial intrahepatic cholestasis. Pediatrics, 51:998, 1973.
57. Haydar, N. A., Conn, H. L., Afifi, A., Wakid, N., Ballas, S., and Fawaz, K.: Severe hypermetabolism with primary abnormality of skeletal muscle mitochondria: Functional and therapeutic effects of choramphenicol treatment. Ann. Intern. Med., 74:548, 1971.
58. Berg, P., Baltimore, D., Boyer, H. W., Cohen, S. N., Davis, R. W., Hogness, D. S., Nathans, D., Roblin, R., Watson, J. D., Weissman, S., and Zinder, N. D.: Potential biohazards of recombinant DNA molecules. Science, 185:303, 1974.
59. Colombo, J. P., Terheggen, H. G., and Lowenthal, A.: Argininaemia. *In* Hommes, F. A., and van den Berg, C. J. (eds.): Inborn Errors of Metabolism. New York, Academic Press, Inc., 1973, pp. 239–248.
60. Rogers, S., Lowenthal, A., and Terheggen, H. G.: Induction of arginase activity with the Shope papilloma virus in tissue culture cells from an argininemic patient. J. Exp. Med., 137:1091, 1973.
61. Wilson, J. G.: Environment and Birth Defects. New York, Academic Press, Inc., 1973.
62. Hobbins, J. C., and Mahoney, M. J.: In utero diagnosis of hemoglobinopathies: Technic for obtaining fetal blood. N. Engl. J. Med., 290:1065, 1974.
63. Ogita, S., Hasegawa, H., Matsumoto, M., Kamei, T., Shimamoto, T., Ohnishi, M., and Sugawa, T.: Prenatal diagnosis of E trisomy syndrome by fetography. Obstet. Gynecol., 43:887, 1974.
64. Leopold, G. R., and Asher, W. M.: Ultrasound in obstetrics and gynecology. Radiol. Clin. North Am., 12:127, 1974.
65. Lipkin, M., and Rowley, P. T. (eds.): Genetic Responsibility: On Choosing Our Children's Genes. New York, Plenum Publishing Corp., 1974.
66. Westoff, C. F., and Rindfuss, R. R.: Sex preselection in the United States: Some implications. Science, 184:633, 1974.
67. deKretzer, D., Dennis, P., Hudson, B., Leeton, J., Lopata, A., Outch, K., Talbot, J., and Wood, C.: Transfer of a human zygote. Lancet, 2:728, 1973.
68. Fletcher, J.: Ethical aspects of genetic controls: Designated genetic changes in man. N. Engl. J. Med., 285:776, 1971.

69. Hilton, B., Callahan, D., Harris, M., Condliffe, P., and Berkley, B. (eds.): Ethical Issues in Human Genetics: Genetic Counseling and the Use of Genetic Knowledge. New York, Plenum Publishing Corporation, 1973.
70. Duff, R. S., and Campbell, A. G. M.: Moral and ethical dilemmas in the special-care nursery. N. Engl. J. Med., 289:890, 1973.
71. Ampola, M. G., Mahoney, M. J., Nakamura, E., and Tanaka, K.: *In utero* treatment of methylmalonic acidemia with vitamin B_{12}. Pediatr. Res., 8:387, 1974 (Abstr).
72. Deol, M. S.: An experimental approach to the understanding and treatment of hereditary syndromes with congenital deafness and hypothyroidism. J. Med. Genet., 10:235, 1973.
73. Albers, R. W., Siegel, G. J., Katzman, R., and Agranoff, B. W. (eds.): Basic Neurochemistry. Boston, Little, Brown, and Company, 1972.
74. Bergsma, D. (ed.): Human Gene Mapping: New Haven Conference, 1973. Miami, Symposia Specialists, 1974.
75. Kacian, D. L., Spiegelman, S., Bank, A., Terada, M., Metafora, S., Dow, L., and Marks, P. A.: *In vitro* synthesis of DNA components of human genes for globins. Nature [New Biol.], 235:167, 1972.
76. Merrill, C. R., Geier, M. R., and Petricciani, J. C.: Bacterial virus gene expression in human cells. Nature, 233:398, 1971.
77. Motulsky, A. G., Lenz, W., and Ebling, F. J. G. (eds.): Birth Defects: Proceedings of Fourth International Conference. Amsterdam, Exerpta Medica, 1974.
78. Scriver, C. R.: Mutants: Consumers with special needs. Nutr. Rev., 29:155, 1971
79. Krieger, I., and Tanaka, K.: Personal communication.

CHAPTER 12

SCREENING FOR THE HYPERLIPIDEMIAS

Arno G. Motulsky, M.D., and Helge Boman, M.D.

Coronary heart disease (CHD) is a major public health problem in the industrialized countries; in 1967, there were 626,000 deaths attributable to CHD in the United States. A quarter of these deaths which accounted for one third of all deaths, occurred in persons in the age group 35 to 64 years. In addition, several million Americans are disabled from CHD.[39] A strong association between hyperlipidemia and the risk to develop coronary heart disease can be demonstrated.[1, 3, 24, 39] The degree of risk is continuously related to the cholesterol level. There is no cholesterol level below which a person is protected, but persons with low levels have a decreased risk of developing CHD. Data from an epidemiologic study in Framingham (Massachusetts) indicated that men with plasma cholesterol levels above the 95th percentile may comprise a quarter of the myocardial infarcts.[10] In a Seattle study of consecutive survivors of myocardial infarction, about a third below the age of 60 years had hyperlipidemia,[20] and of these, about two thirds had a simply inherited hyperlipidemia.[21]

Low-density lipoprotein (LDL) carries the major part of plasma cholesterol in the normal and hyperlipidemic state. Very low density lipoprotein (VLDL), as a rule, contains most of the triglycerides in hypertriglyceridemia. Since LDL contains some triglyceride, and VLDL carries some cholesterol, a positive correlation ($r = 0.3$)[21] between cholesterol and triglyceride values is not surprising. Thus, whereas hypercholesterolemia definitely predisposes to CHD, hypertriglyceridemia may or may not be an independent risk factor.[39]

Supported in part by NIH Grant GM 15253 and by a Public Health Service International Research Fellowship (1 FO5 TWO 1905) to H.B.

Heterogeneity of the Hyperlipidemias

The term "hyperlipidemia" is used when a fasting plasma cholesterol or triglyceride, or both lipid values, exceed a certain level. Cholesterol and triglyceride levels are distributed in the population on a continuum; the cut-off point, which designates some individuals as hyperlipidemic, is arbitrary. Age- and sex-corrected lipid levels above the 95th percentile of an assumed normal distribution, derived from a specified population, are frequently-used criteria for determining the presence of hyperlipidemia.[21] Sometimes, hyperlipidemia in the United States has been arbitrarily defined as the condition existing "when cholesterol values exceed 250 mg per 100 ml and triglyceride levels exceed 150 mg per 100 ml."[39] Evidence for the existence of three monogenic forms, as well as of polygenic and nongenetic forms, of hyperlipidemia has been presented.[21] In familial hypercholesterolemia and familial hypertriglyceridemia, the abnormal cholesterol and triglyceride levels, respectively, were transmitted as autosomal dominant traits.[12, 16, 21]

Familial hypercholesterolemia is a well-established entity.[12] The affected individuals are at a high risk to develop premature CHD. Thus, in an extensive study of 116 such families,[40] the hypercholesterolemic males incurred a 16 per cent chance of nonfatal or fatal CHD by the age of 40. By the age of 60, the cumulative risk for CHD was 52 per cent, compared with 12.7 per cent among their normolipidemic male relatives. The risk of developing CHD by the age of 60 was 32.8 per cent in hyperlipidemic females, compared with 9.1 per cent in their normolipidemic female relatives.

The risk of coronary heart disease in familial hypertriglyceridemia is unknown. The data of Goldstein et al.[21] suggest an increased risk, but more extensive direct data are required.

In familial combined hyperlipidemia the genetic defect is also an autosomal dominant trait and expresses with elevated cholesterol, with high triglyceride levels, or with increases of both lipids.[21] Family studies supporting the existence of combined hyperlipidemia as a genetic entity have been published,[14] but with varying interpretations.[31, 34] Further work is needed to clearly delineate this condition.

It is likely that several different disorders may make up the currently defined entities. Further proof of heterogeneity of a monogenic disorder may come from demonstration of linkage of a given type of hyperlipidemia to a marker gene or from the identification of a unique biochemical defect. Thus, recent work makes it likely that a membrane defect is the basic lesion in the autosomally dominant transmitted familial hypercholesterolemia.[2]

At the present time, there are no simple methods, such as studies of a blood sample from a single hyperlipidemic individual, available to identify monogenic forms of hyperlipidemia. Extended family studies are necessary to gain information about the cause of hyperlipidemia.

General Criteria for Screening

In general, screening programs are undertaken for several reasons:

1. To identify individuals of a certain genotype to provide medical advice and initiate treatment of a potentially preventable or curable disease. Most often, the tested individual is unaware of the medical problem.

2. To identify individuals heterozygous for a given trait for reproductive advice. For instance, it may be possible to detect homozygosity for familial hypercholesterolemia, with its feared childhood mortality, by intrauterine diagnosis.[19] Matings of two affected parents, however, are exceedingly rare, except in areas where consanguineous marriages are common.[25]

3. To collect epidemiologic data.

4. To perform research studies.

If prevention were possible and detection simple, it would be urgent to launch population screening programs to identify hyperlipidemic individuals who are at high risk to develop CHD. Simultaneously, useful information could be gained for epidemiologic and research studies. However, recent experience with screening programs suggests caution in wholesale application of screening.[29]

Problems in Heterozygote Identification

In screening programs, a simple laboratory test should distinguish between the different phenotypes. When "abnormal" lipid levels are defined as above the 95th percentile for cholesterol, triglyceride, or both, approximately 10 per cent of the population will be classified as hyperlipidemic. Thus, millions of individuals will be generated for diagnosis, classification, treatment, and follow-up studies, if the program were extended to most of the population. Since the monogenic hyperlipidemias are relatively rare (at most, 1 per cent of the general population),[21] only a small portion of those identified will have a simply inherited abnormality. Although increased risk for CHD with increasing lipid values is indisputable, the extent to which the risk is related to the lipid values alone or to the various inherited forms of hyperlipidemia is not yet clear.

Following identification of abnormal individuals by the screening procedure, a well-established treatment should be available for them. The hyperlipidemias have no single cause, and treatment differs for the different forms.[28] Low-fat diet and drugs are the two main forms of treatment at this time, and their use is based on empirical observations. Commonly used drugs, such as clofibrate and cholestyramine, interfere with both the formation and the removal of VLDL and LDL from plasma; the empirical net effect is generally a decrease in triglyceride and cholesterol values, respectively.[28] No definite proof yet exists to indicate that lowered lipid levels reduce the risk of CHD, but several large studies are under way to elucidate this crucial point.[28] Furthermore, it is reasonable to assume, from basic pharmacogenetic principles, that different types of hyperlipidemia are caused by different mechanisms and therefore are likely to respond to different modes of treatment.

Frequency of Hyperlipidemia

The trait under study should be neither too rare nor too common. If cholesterol levels above 250 mg per 100 ml or triglyceride levels above 160 mg per 100 ml are considered "abnormal," at least one third of the population needs treatment.[4] It might therefore be most logical to treat the whole population. The same reasoning was used for the widespread addition of fluoride to drinking water. Dental caries is found in more than half the population, and a simple, general prophylaxis was possible. The eating habits of persons of North America or Europe might contribute to elevated blood lipid levels, and widescale prevention might be possible by changing the national diet—a difficult task.

It is conceivable that some of the simply inherited (monogenic) hyperlipidemias respond to diet and drug treatment more readily than the more complex hyperlipidemias. Since persons in the monogenic categories have the highest blood lipid values and therefore, perhaps, are at the highest risk, attention might be focused initially on the relatively small proportion of individuals with these metabolic abnormalities. The total effort would certainly be much reduced and more intensive measures could be concentrated on these individuals. In any case, before any major undertaking is started, careful consideration must be given to the large effort and the psychological and economic impact of the medical investigations and family studies required.

PRESENT STATUS OF SCREENING FOR THE HYPERLIPIDEMIAS

Laboratory Screening Methods

Determination of fasting plasma cholesterol and triglyceride levels in patients *and* their families provided sufficient information to identify several familial forms of hyperlipidemia.[21] The two determinations may even suffice as a basis for treatment.[4, 28] Unfortunately, the determination of plasma cholesterol and triglyceride, however widespread and readily available, is, in fact, not simple.[28] High quality laboratories standardized on a national and, hopefully, international basis are required. Widely disparate results from different laboratories are being obtained at the present time.[7, 37, 38] The determination of cholesterol carried in the low-density lipoprotein fraction (LDL-cholesterol) has been shown to be more effective than the determination of total cholesterol in separating affected from nonaffected individuals in familial hypercholesterolemia.[26, 27]

Lipoprotein electrophoresis is the basis for the Fredrickson classification.[12] By this method the different lipoprotein classes are identified; this type of categorization, however, has not proved useful in categorizing genetically defined types of hyperlipidemia.[22] Of high potential interest is the "atypical pre-β-lipoprotein" band recently identified by electrophoresis, and possibly associated with CHD.[5] This lipoprotein band may be inherited as an autosomal dominant trait,[5] and may represent an independently inherited risk factor for CHD, since its presence is not directly related to lipid levels.[13] This trait segregates independently from familial hypercholesterolemia.[23] If the prognostic value of this particular electrophoretic pattern can be validated, lipoprotein electrophoresis as a screening procedure might be justified.

Family Screening

In families where a patient has been clearly identified as suffering from familial hypercholesterolemia, family members should be screened.[35] On the average, 50 per cent of the siblings and children of index cases and 25 per cent of the uncles, aunts, nephews, and nieces will also be affected. Screening of such first-degree and second-degree family members would be indicated. Hyperlipidemic family members can also be identified among relatives of hyperlipidemic survivors of myocardial infarction. Such screening, moreover, may determine the form of hyperlipidemia in the index patient.[21, 32]

Since lesions leading to CHD may start in early childhood[26, 41] the identification and preventive treatment of the condition should probably start early in life. It is therefore logical that great interest has focused on cord blood screening, since cord blood represents a newborn specimen and is readily obtainable. Kwiterovich et al.[27] have successfully used such samples to identify, at birth, hypercholesterolemic infants who had one parent affected with familial hypercholesterolemia. Unfortunately, monogenic hypertriglyceridemia and combined hyperlipidemia may not manifest until the late twenties, and cord blood screening will miss many affected persons (see Cord Blood Screening).

Population Screening

The entire population is at risk for developing CHD; men more frequently and earlier than women. No single ethnic group can be singled out as at a particularly high risk. Several different approaches for hyperlipidemia screening have been discussed.

Cord Blood Screening.   Screening for hyperlipidemia in cord blood samples of all newborns has been proposed. The prime interest has focused on cholesterol determination. In a study of 1800 cord blood samples, 65 infants were determined to have hypercholesterolemia (above mean + 2SD).[15] Based on family studies, eight of these infants had familial hypercholesterolemia. After one year, only three of these children had abnormal cholesterol levels, but some of them had been on a low-cholesterol diet.[42] Darmady et al.[6] found 34 of 302 infants to be hypercholesterolemic (above 100 mg per 100 ml) at birth. However at 1 year of age, the cholesterol levels of the hyperlipidemic infants were distributed throughout the normal range. Goldstein et al.[18] analyzed 2000 cord blood samples. These investigators found 71 infants with elevated cholesterol levels (above the 95th percentile), including 16 with coexisting hypertriglyceridemia. By studying parents, grandparents, and other relatives of these infants, five families with familial hyperlipidemia were found. Of these, three families had familial hypercholesterolemia (all ascertained through a purely hypercholesterolemic index case), and two had familial combined hyperlipidemia (one ascertained through a hypercholesterolemic index case, the other, through a hypertriglyceridemic index case. The mean cholesterol level of the parents and grandparents of all hypercholesterolemic infants, however, was not significantly different from that of parents and grandparents of 133 control infants.

Family studies on 71 hypertriglyceridemic (above the 95th percentile) infants, including 16 with abnormal cholesterol levels, failed to reveal any cases of familial hypertriglyceridemia. The distribution of triglyceride levels among the parents and grandparents of the hypertriglyceridemic index cases and normolipidemic control families was similar.[18] The distribution of triglyceride values among 57 neonates in which one of the parents had primary hypertriglyceridemia was identical with that of the control population.[42]

These data are in keeping with the finding that triglyceride levels in the newborn are influenced largely by perinatal factors. Thus, the triglyceride level in the neonate is linearly correlated with the number of significant antepartum or intrapartum complications[42] and also correlates with the maturity of pregnancy and the nutritional status of the fetus.[9]

On the basis of presently available data, it can be concluded that most hyperlipidemias in neonates are not genetically determined, and that the usefulness of this screening procedure in the general population for identifying individuals at high risk in developing CHD later in life is questionable. Familial hypertriglyceridemia cannot be identified through cord blood screening, and it seems unlikely that this form of screening will be helpful in the identification of the familial combined hyperlipidemia. The identification of familial hypercholesterolemia requires a full-scale family study of all infants with hypercholesterolemia; even then, it may present difficulties. As soon as a test system better than determination of total or LDL cholesterol is devised, large-scale trials of population screening of cord blood, followed by dietary treatment, may be feasible. However, before advocating mass screening, one must prove that there is a harmless effective treatment for familial hypercholesterolemia to decrease substantially the high risk of CHD.

Screening in Children. Cholesterol levels double within the first week of extrauterine life and thereafter remain relatively stable throughout childhood.[11] Thus, it seems possible to conduct screening for hypercholesterolemia in children.

A positive correlation exists between cholesterol levels of school children and their parents. It has been assumed that a high cholesterol level in childhood is correlated with hypercholesterolemia in adulthood and with subsequent development of CHD.[8, 17] How much of this increased risk is attributable to the hereditary forms of the hyperlipidemias is not known. Since a large proportion of children affected with familial hypertriglyceridemia and combined hyperlipidemia may not express this abnormality until later in life, it is difficult to assess fully the significance of these correlations. Knowledge is also lacking about the relationship of the sporadic and polygenic forms of hypercholesterolemia in children to adult hyperlipidemia and the risk of developing CHD. Concern over potential adverse effects, following treatment of hyperlipidemia in childhood, has also been expressed.[36] Until these problems are solved, mass screening for hyperlipidemia in childhood appears unjustified.

Screening of Adults. By the age of 30, the monogenic forms of hyperlipidemia are fully expressed. Studies of relatives of hyperlipidemic individuals identified by population screening in adulthood are thus likely to reveal most of the monogenic hyperlipidemias. However, a monogenic mechanism was found in only one half the hyperlipidemic survivors of myocardial infarction.[21] The proportion of monogenic disorders in hyperlipidemic individuals derived from a healthy population is not known, but it is expected that the majority of hyperlipidemic individuals detected by population screening will not have a familial form of hyperlipidemia.

The correlation between lipid levels in adults and the risk of developing CHD does not specify the contribution of the hereditary forms of hyperlipidemias to this risk. Familial hypercholesterolemia and familial combined hyperlipidemia are associated with coronary heart disease at an earlier mean age in males (46 and 52 years, respectively) than familial hypertriglyceridemia (57 years), polygenic hypercholesterolemia (58 years), sporadic hypertriglyceridemia (59 years), and normolipidemia (63 years).[30] This may reflect the higher risk associated with the monogenic forms, but a similar trend in females was not found. In a necropsy study of 42 patients with normal (14) and abnormal (28, at least 13 inheritable) lipoprotein patterns, 20 had symptomatic heart disease. Early and accelerated atherosclerotic damage was seen in patients with hyperlipoproteinemia (Types II, III, and IV), but the hyperlipidemia was not a prerequisite for such damage.[33] Again, it is not known how much of the atherosclerotic damage is caused directly by the elevated lipid level or by other factors related to the underlying genetic defect.

Thus, while there are some clues that individuals with monogenic forms of hyperlipidemia may be more susceptible to premature atherosclerotic disease, the present state of knowledge is such that all hyperlipidemic persons must be considered at risk.

The prophylactic effect of treatment of hyperlipidemic individuals in several population trials is not obvious. There is some evidence[28] that treatment of hyperlipidemia in patients, even with overt CHD, may have beneficial influence on the survival rate. There are now several programs under way to determine the effect of treatment of hyperlipidemia in asymptomatic individuals in the age prone to coronary heart disease.[28] The results of such trials may give answers in only a few years, whereas the interpretation of trials involving children or neonates must wait a generation. It is likely that treatment groups selected for hyperlipidemia are heterogeneous from a genetic point of view, and that persons having the sporadic, polygenic, and monogenic forms of hyperlipidemia[21] may show different response to diets and drugs. Therefore, attempts must be made to subdivide the hyperlipidemic index cases into their genetic forms, and to study the response to treatment in these groups separately. If a good response to a certain treatment regimen is obtained in one individual with a monogenic form of hyperlipidemia, pharmacogenetic principles indicate that other hyperlipidemic members of the same family have the same defect and will therefore respond to the same treatment.

CONCLUSION

Hypercholesterolemia is a significant risk factor for coronary heart disease, as probably is hypertriglyceridemia. Several simply inherited disorders cause elevation of cholesterol, of triglyceride, or of both lipids. However, more complex genetic interaction and nongenetic factors may also cause lipid elevations. Simple tests to differentiate the various types of hyperlipidemia do not exist at our current state of knowledge. Treatment of the hyperlipidemias has not been standardized, and the long-term effects of antihyperlipidemic therapy remain unknown. It is likely that different treatments will be required for the various genetic and nongenetic types of hyperlipidemia. Whether such treatments decrease the occurrence of the clinical manifestations of coronary heart disease is not yet known.

Under these circumstances, large-scale mass screening for the hyperlipidemias, using newborn, juvenile, or adult populations, cannot be recommended. Well-planned, extensive pilot studies, using all existing biochemical and genetical knowledge, are required to assess the utility of mass screening for the hyperlipidemias. Families of hyperlipidemic patients with coronary heart disease are a high-risk group, and lipid studies of such persons should be encouraged on a research basis. Investigations of family members of individuals clearly affected with a monogenic disorder, such as familial hypercholesterolemia—whether or not affected with coronary heart disease—should be urged. Such family members have a greater chance of having hyperlipidemia, and detection of these individuals causes fewer difficulties than in the general population. Screening for any disorder has widespread psychological, social, economic, and medical implications. Before initiating programs affecting large segments of the population, one must carefully assess the total impact of such screening.

References

1. Albrink, M. J.: Serum lipids, diet, and cardiovascular disease. Postgrad. Med., 55:86, 1974.
2. Brown, M. S., and Goldstein, J. L.: Familial hypercholesterolemia: Defective binding of lipoproteins to cultured fibroblasts associated with impaired regulation of 3-hydroxy-3-methylglutaryl Coenzyme A reductase activity. Proc. Natl. Acad. Sci. U.S.A., 71:788, 1974.
3. Carlson, L. A., and Böttiger, L. E.: Ischaemic heart disease in relation to fasting values of plasma triglycerides and cholesterol. Lancet, 1:865, 1972.
4. Committee on Diet and Heart Disease of the National Heart Foundation of Australia: Dietary fat and coronary heart disease. A review. Med. J. Aust., 1:575–579, 616–620, 663–668, 1974.
5. Dahlén, G., Ericson, C., Furberg, C., Lundkvist, L., and Svärdsudd, K.: Studies on an extra pre-beta lipoprotein fraction. Acta Med. Scand. Suppl. 531, pp. 1–29, 1972.
6. Darmady, J. M., Fosbrooke, A. S., and Lloyd, J. K.: Prospective study of serum cholesterol levels during first year of life. Br. Med. J., 2:685, 1972.
7. de Beer, A., and Gibbs, C. C. J.: Unreliability of serum cholesterol estimations. S. Afr. Med. J., 47:1569, 1973.
8. Drash, A., and Hengstenberg, F.: The identification of risk factors in normal children in the development of arteriosclerosis. Ann. Clin. Lab. Sci., 2:348, 1972.
9. Fosbrooke, A. S., and Wharton, B. A.: Plasma lipids in umbilical cord blood from infants of normal and low birth weight. Biol. Neonate, 23:330, 1973.
10. Fredrickson, D. S.: Mutants, hyperlipoproteinaemia, and coronary artery disease. Br. Med. J., 2:187, 1971.
11. Fredrickson, D. S., and Breslow, J. L.: Primary hyperlipoproteinemia in infants. Annu. Rev. Med., 24:315, 1973.
12. Fredrickson, D. S., and Levy, R. I.: Familial hyperlipoproteinemia. In Stanbury, J. B., Wyngaarden, J. B., and Fredrickson, D. S. (eds.): The Metabolic Basis of Inherited Disease. 3rd ed., New York, McGraw-Hill Book Company, 1972, pp. 545–614.
13. Frick, M. H., Dahlén, G., Furberg, C., Ericson, C., and Wiljasalo, M.: Serum pre-β-1 lipoprotein fraction in coronary atherosclerosis. Acta Med. Scand., 195:337, 1974.
14. Glueck, C. J., Fallat, R., Buncher, C. R., Tsang, R., and Steiner, P.: Familial combined hyperlipoproteinemia: Studies in 91 adults and 95 children from 33 kindreds. Metabolism, 22:1403, 1973.
15. Glueck, C. J., Heckman, F., Schoenfeld, M., Steiner, P., and Pearce, W.: Neonatal

familial type II hyperlipoproteinemia: Cord blood cholesterol in 1800 births. Metabolism, 20:597, 1971.

16. Glueck, C. J., Tsang, R., Fallat, R., Buncher, C. R., Evans, G., and Steiner, P.: Familial hypertriglyceridemia: Studies in 130 children and 45 siblings of 36 index cases. Metabolism, 22:1287, 1973.

17. Godfrey, R. C., Stenhouse, N. S., Cullen, K. J., and Blackman, V.: Cholesterol and the child: Studies of the cholesterol levels of Busselton school children and their parents. Austr. Paediatr. J., 8:72, 1972.

18. Goldstein, J. L., Albers, J. J., Schrott, H. G., Hazzard, W. R., Bierman, E. L., and Motulsky, A. G.: Plasma lipid levels and coronary heart disease in adult relatives of newborns with normal and elevated cord blood lipids. Am. J. Hum. Genet., 26:727, 1974.

19. Goldstein, J. L., Harrod, M. J. E., and Brown, M. S.: Homozygous familial hypercholesterolemia: Specificity of the biochemical defect in cultured cells and feasibility of prenatal detection. Am. J. Hum. Genet., 26:199, 1974.

20. Goldstein, J. L., Hazzard, W. R., Schrott, H. G., Bierman, E. L., and Motulsky, A. G.: Hyperlipidemia in coronary heart disease. I. Lipid levels in 500 survivors of myocardial infarction. J. Clin. Invest., 52:1533, 1973.

21. Goldstein, J. L., Schrott, H. G., Hazzard, W. R., Bierman, E. L., and Motulsky, A. G.: Hyperlipidemia in coronary heart disease. II. Genetic analysis of lipid levels in 176 families and delineation of a new inherited disorder, combined hyperlipidemia. J. Clin. Invest., 52:1544, 1973.

22. Hazzard, W. R., Goldstein, J. L., Schrott, H. G., Motulsky, A. G., and Bierman, E. L.: Hyperlipidemia in coronary heart disease. III. Evaluation of lipoprotein phenotypes of 156 genetically defined survivors of myocardial infarction. J. Clin. Invest., 52:1569, 1973.

23. Heiberg, A., and Berg, K.: On the relationship between Lp(a) lipoprotein, "sinking pre-β-lipoprotein" and inherited hyper-β-lipoproteinaemia. Clin. Genet., 5:144, 1974.

24. Kannel, W. B.: The role of cholesterol in coronary atherogenesis. Med. Clin. North Am., 58:363, 1974.

25. Khachadurian, A. K.: The inheritance of essential familial hypercholesterolemia. Am. J. Med., 37:402, 1964.

26. Kwiterovich, P. O., Frederickson, D. S., and Levy, R. I.: Familial hypercholesterolemia (One form of familial type II hyperlipoproteinemia). A study of its biochemical, genetic, and clinical presentation in childhood. J. Clin. Invest., 53:1237, 1974.

27. Kwiterovich, P. O., Levy, R. I., and Fredrickson, D. S.: Neonatal diagnosis of familial type-II hyperlipoproteinaemia. Lancet, 1:118, 1973.

28. Levy, R. I., Morganroth, J., and Rifkin, B. M.: Drug therapy. Treatment of hyperlipidemia. N. Engl. J. Med., 290:1295, 1974.

29. Motulsky, A. G.: Screening for sickle cell hemoglobinopathy and thalassemia. Isr. J. Med. Sci., 9:1341, 1973.

30. Motulsky, A. G., and Boman, H.: Genetics and atherosclerosis. In Schettler, G. (ed.): Second International Symposium on Atherosclerosis. Heidelberg, Springer-Verlag, (In press, 1973.)

31. Nikkilä, E. A., and Aro, A.: Family study of serum lipids and lipoproteins in coronary heart disease. Lancet, 1:954, 1973.

32. Patterson, D., and Slack, J.: Lipid abnormalities in male and female survivors of myocardial infarction and their first-degree relatives. Lancet, 1:393, 1972.

33. Roberts, W. C., Ferrans, V. J., Levy, R. I., and Fredrickson, D. S.: Cardiovascular pathology in hyperlipoproteinemia. Anatomic observations in 42 necropsy patients with normal or abnormal serum lipoprotein patterns. Am. J. Cardiol., 31:557, 1973.

34. Rose, H. G., Kranz, P., Weinstock, M., Juliano, J., and Haft, J. I.: Inheritance of combined hyperlipoproteinemia: Evidence for a new lipoprotein phenotype. Am. J. Med., 54:148, 1973.

35. Schrott, H. G., Goldstein, J. L., Hazzard, W. R., McGoodwin, M. M., and Motulsky, A. G.: Familial hypercholesterolemia in a large kindred. Evidence for a monogenic mechanism. Ann. Intern. Med., 76:711, 1972.

36. Schubert, W. K.: Fat nutrition and diet in childhood. Am. J. Cardiol., 31:581, 1973.
37. Shapiro, B. G.: Unreliability of serum cholesterol estimations. S. Afr. Med. J., 47:1092, 1973.
38. Simson, E.: Unreliability of serum cholesterol estimations. S. Afr. Med. J., 47:1337, 1973.
39. Stamler, J.: Epidemiology of coronary heart disease. Med. Clin. North Am., 57:5, 1973.
40. Stone, N. J., Levy, R. I., Fredrickson, D. S., and Verter, J.: Coronary artery disease in 116 kindred with familial type II hyperlipoproteinemia. Circulation, 49:476, 1974.
41. Strong, J. P., and McGill, H. C.: The pediatric aspects of atherosclerosis. J. Atheroscl. Res., 9:251, 1969.
42. Tsang, R., and Glueck, C. J.: Cord blood hypertriglyceridemia. Am. J. Dis. Child., 127:78, 1974.

CHAPTER 13

ENZYME THERAPY IN GENETIC DISEASES: PROGRESS, PRINCIPLES, AND PROSPECTS*

R. J. Desnick, Ph.D., M.D., W. Krivit, M.D., Ph.D., and M. B. Fiddler, B.S.

INTRODUCTION

Dramatic progress has been made in the elucidation of the molecular pathology of inherited metabolic diseases during the past two decades. In an ever increasing number of these disorders, the clinical and pathophysiologic manifestations have been delineated, and the metabolic derangements have been characterized. Sophisticated chemical and enzymatic techniques as well as somatic cell systems have been developed to identify the specific enzymatic defects in more than 120 of the over 400 catalogued, recessively inherited, inborn errors of metabolism.[1-5] Already, many of these mutant structural alleles have been assigned or localized to specific human chromosomes.[6] Implementation of these techniques in major centers has made the diagnosis of these disorders a reality. Indeed, the demonstration of a specific enzymatic deficiency has

*This review was supported in part by grants (CBRS-273 and C-174) from the National Foundation—March of Dimes, a grant (74-915) from the American Heart Association, grants (AM 15174 and AM 14470) from the National Institutes of Health, a grant (RR-400) from the Clinical Research Centers Program of the Division of Research Resources. National Institutes of Health, and a predoctoral fellowship to MBF from the National Science Foundation. RJD is a recipient of an NIH Research Career Development Award (1 KO4 AM 00042).

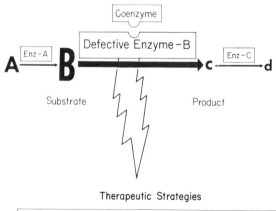

Therapeutic Strategies

SUBSTRATE:
1. Limit toxic metabolites – dietary therapy
2. Deplete stored metabolites – chelators
3. Metabolic inhibitors
PRODUCT:
4. Supply deficient metabolic product
ENZYME:
5. Supply coenzyme – cofactor, vitamin
6. Enzyme induction - ↑ residual activity/alternate pathway
7. DIRECT GENE PRODUCT REPLACEMENT - ENZYME THERAPY

Figure 13–1. Treatment of inherited metabolic diseases, including various therapeutic strategies, at the level of the substrate, product, or enzyme.

provided for the accurate diagnosis of affected homozygotes or hemizygotes, the detection of heterozygous carriers, and the capability to diagnose prenatally and prevent the birth of affected fetuses. However, in spite of these major diagnostic achievements, patients with these debilitating disorders and their families have become increasingly disappointed by the absence of specific therapy for most of these diseases.

During the past decade, considerable attention has been focused on the development of strategies to treat patients with inherited metabolic diseases, as illustrated in Figure 13–1. Early therapeutic endeavors primarily involved attempts to alter the disease course by manipulations at the level of the metabolic or biochemical defects, by either decreasing the accumulated substrate concentration or replacing the deficient product of the metabolic block. In selected diseases, investigators attempted to reduce the levels of the accumulated substrate or precursors proximal to the metabolic block by dietary restriction, chelation, or administration of appropriate metabolic inhibitors. Alternatively, the deficient metabolic product has been supplied. The biochemical as well as clinical effectiveness of specific metabolic manipulations in amenable disorders has been discussed in recent reviews.[7-9]

Therapeutic trials directed at the level of the enzymatic defect have included attempts to (1) supply the appropriate cofactor or vitamin in coenzyme-responsive mutations, (2) increase a residual activity or induce an alternative metabolic pathway by the use of appropriate drugs, and (3) replace the gene product directly by enzyme replacement therapy or by the transplantation of allografts capable of producing the normal gene product. The encouraging experiences as well as the limitations of these strategies for enzyme therapy in genetic diseases have been the subject of recent symposia and reviews.[10-12]

The thrust of current research is directed at the treatment of inborn errors of metabolism, particularly the lysosomal storage diseases that have become a major focus of exploratory enzyme therapy endeavors. It is the purpose of this chapter to discuss the progress, principles, and future prospects for the treatment of genetic diseases, specifically by enzyme replacement therapy for selected lysosomal storage diseases. Clearly, the development of therapeutic strategies will continue to be dependent on further progress in the elucidation and understanding of the basic molecular pathology of each inherited enzymatic deficiency. Perhaps the past progress and future knowledge, gained from studies of the lysosomal storage diseases as prototypes, may provide the basis for innovative therapeutic endeavors, specifically modified for other inherited metabolic diseases.

ENZYME THERAPY: PROGRESS

The identification of the specific pathogenic enzymatic defects in several lysosomal storage diseases immediately intrigued investigators with the therapeutic possibilities of enzyme replacement. De Duve's suggestion in 1964[13] that enzyme replacement may be of therapeutic value, since exogenous substances taken into cells by endocytosis are likely to end up within lysosomes, stimulated investigators to attempt direct enzyme replacement. Table 13–1 summarizes the human trials of enzyme therapy that have been attempted in patients with various lysosomal storage diseases. The initial experiments, performed in the mid-1960s, involved the administration of crude preparations of heterologous enzymes into severely debilitated patients. In 1964, Baudhuin and associates[14] were the first to administer intravenously an enzyme-rich extract from *Aspergillus niger* as a source of α-glucosidase to a patient with glycogenosis type II (Pompe's disease). Further trials using fungal enzyme preparations were attempted in patients with glycogenosis types II[15-17] and IV[18, 19]; the results indicated that the active exogenous enzyme could gain access to visceral tissues for the catabolism of the accumulated glycogen substrates. However, these studies were complicated by pyrogenic contaminants and immunologic reactions to the fungal proteins.

In 1966, enzyme replacement in patients with metachromatic leuko-

TABLE 13-1. HUMAN TRIALS OF ENZYME THERAPY IN LYSOSOMAL STORAGE DISEASES

Disease (defective enzyme)	Source of Exogenous Enzyme	Route of Administration	Tissue Uptake	Substrate Metabolism	Comments or Complications	Reference
GLYCOGENOSES						
Glycogenosis type II (α-glucosidase)	A. niger	I.V.*	Liver	Liver −		14
	A. niger	I.V.	Liver	Liver + E.M.†	Immunologic reaction	15, 16
	A. niger	I.M.‡	Liver Leukocytes	Liver −		17
	Human placenta	I.V.	Liver	Liver −	Hepatic activity T½ of 4–5 days	32
Glycogenosis type IV (glycogen branching enzyme)	A. niger	I.V.	Liver	Liver +	α-Glucosidase activity monitored	18, 19
GLYCOSPHINGOLIPIDOSES						
Metachromatic leukodystrophy (arylsulfatase A)	Bovine brain	I.V.	Liver		No neural tissue catabolism by E.M.	20
	Human urine	I.T.§ I.T.	CSF		Immunologic reaction	21
Sandhoff's disease (hexosaminidase A & B)	Plasma concentrates	I.V.		Plasma +	Immunologic reaction	24
	Human urine	I.V.	Liver		Hexosaminidase A injected, no activity detected in neural tissue or CSF	31
Fabry's disease (α-galactosidase A)	Plasma	I.V.		Plasma +		22, 23
	Plasma	I.V.		Plasma −		33
	Leukocytes & platelets	I.V.		Plasma +		33

Disease (enzyme)	Source	Route	Liver		Comments	Ref.
Gaucher's disease (β-glucocerebrosidase)	Human placenta	I.V.		Plasma +		33
	Human placenta	I.V.		Liver + Plasma + Erythrocytes +		34
MUCOPOLYSACCHARIDOSES						
Mucopolysaccharidosis type I (α-L-iduronidase)	Plasma	I.V.		Urine +||		25
Mucopolysaccharidosis type II (iduronosulfate sulfatase)	Plasma			Urine −¶		26
	Plasma	I.V.		Urine +		25
	Plasma			Urine −		27
Mucopolysaccharidosis type III	Leukocytes	I.V.		Urine +	Compatible donors required	30
	Whole blood	I.V.		Urine −		26
	Plasma			Urine +		28
(Type IIIa: heparan sulfate sulfamidase)	Plasma			Urine −		29, 26
(Type IIIb: α-N-acetylglucosaminidase)	Whole blood			Urine −		26
Mucopolysaccharidosis type VII (β-glucuronidase)	Plasma	I.V.			Blood clearance of platelet > plasma enzyme (minutes vs. days)	35
	Platelets	I.V.				

*I.V. = Intravenous
†E.M. = Electron microscopic observation of decreased substrate
‡I.M. = Intramuscular
§I.T. = Intrathecal
||Urine + = Increased urinary excretion of glycosaminoglycans and degradative products
¶Urine − = Unchanged urinary excretion of glycosaminoglycans and degradative products

dystrophy was accomplished by intravenous and intrathecal administration of arylsulfatase A, partially purified from beef brain.[20] Following intravenous administration, the levels of enzymatic activity were increased significantly in serum and biopsied hepatic tissue. After intrathecal administration, the arylsulfatase A activity was increased in cerebrospinal fluid for 20 hours, but it could not be demonstrated in biopsied brain tissue. These studies indicated that although apparently feasible, future trials of enzyme therapy would clearly require the use of nonimmunogenic enzymes, presumably from human sources.

Austin,[21] recognizing the need for homologous enzyme, was the first to use a human enzyme source for therapy. He administered human urinary arylsulfatase A intrathecally to a patient with metachromatic leukodystrophy. Seven and one-half hours after injection, the arylsulfatase A activity in the cerebrospinal fluid was increased approximately fiftyfold. Concomitantly, the patient developed an immunologic hypersensitivity reaction, a febrile episode (105° F), and a CSF polymorphonuclear leukocytosis; there was no evidence of chemical or clinical therapeutic effect.

Stymied by the previous clinically unsuccessful attempts to administer partially purified exogenous enzymes, investigators turned to the use of normal human plasma and leukocyte preparations for therapy. In 1970, Mapes and coworkers[22, 23] administered fresh heparinized plasma containing active α-galactosidase A to patients with Fabry's disease. Not only was the enzymatic activity in the plasma of the recipients increased, but also the level of the specific plasma substrate, trihexosyl ceramide, was significantly decreased, providing evidence of the catabolic activity of the exogenously supplied enzyme. Similar results were observed when plasma concentrates were administered to a patient with Sandhoff's disease;[24] the hexosaminidase levels were increased in the plasma, and the concentration of plasma globoside, a substrate of the enzyme, was significantly reduced postinjection.

Shortly thereafter, several investigators reported enzyme replacement studies in the mucopolysaccharidoses types I, II, and III by infusion of normal plasma,[25-28] leukocytes,[30] and whole blood.[26] The chemical results of these studies were highly variable; some authors reported evidence for mucopolysaccharide catabolism,[25, 28, 30] whereas others found none.[26, 27, 29] The current concensus is that this approach for the treatment of the mucopolysaccharidoses is unlikely to have significant therapeutic effect.

Based on previous observations, recent trials of enzyme replacement therapy have utilized highly purified enzymes isolated from human sources. Pilot intravenous administrations of the appropriate enzymatic activity have been accomplished in patients with Sandhoff's disease (urinary hexosaminidase A),[31] glycogenosis type II (placental α-glucosidase),[32] Fabry's disease (placental α-galactosidase A),[33] and most recently, Gaucher's disease type I (placental β-glucosidase).[34] In each case the highly purified human enzyme was shown to hydrolyze its natural substrate *in vitro* prior to *in vivo* trials. The plasma clearance of the injected enzymes was rapid, and exogenous enzymatic activity was recovered in biopsied liver samples from each of the patients studied. The approximate plasma half-lives of the injected enzymatic activity were 10 minutes in Sandhoff's disease, 20 minutes in Pompe's disease, 10 to 12 minutes in Fabry's disease, and 18 minutes in Gaucher's disease. Evidence for concomitant substrate catabolism was demonstrated in three of these trials; decreased concentrations of plasma globoside (Sandhoff's disease), hepatic glycogen (Pompe's disease), plasma trihexosyl ceramide (Fabry's disease), and plasma, erythrocytic, and hepatic glucocerebroside (Gaucher's disease) were found following enzyme administration. These preliminary but encouraging results support the feasibility of enzyme therapy using highly purified human enzyme preparations. However, the major limitations of these endeavors were the short half-lives of the circulating exogenous enzymatic activities (instability? bioinactivation?) and the lack of evidence for substrate catabolism in the target tissue and subcellular sites of pathologic deposition; demonstrable clinical improvement was not expected or observed in these pilot trials.

Although the past decade of progress in enzyme therapy has been encouraging, the optimistic expectations for this therapeutic strategy have not been realized. As a result of the experiences of the early endeavors investigators have identified the major problem areas to be resolved before direct enzyme replacement will be therapeutically beneficial. There is a clear need to develop and apply enzyme and cellular engineering techniques to introduce active enzyme into the target pathologic sites in order to metabolize accumulated substrate as well as to

TABLE 13-2. REQUISITES FOR EFFECTIVE ENZYME THERAPY

Enzyme Technology
   Sufficient quantities of highly active, stable, nonimmunogenic, sterile enzyme

Enzyme Administration
   DELIVERY of sufficient enzymatic activity, by appropriate ROUTES, to target tissue and subcellular sites for effective substrate metabolism at reasonable intervals
   PROTECTION of enzyme from inactivation, degradation, and immunologic surveillance

In Vivo **Test System**
   Mammalian model systems to evaluate and maximize enzyme stability, protection, tissue distribution and uptake, subcellular localization, and substrate metabolism prior to human trials

Therapeutic Evaluation
   Demonstration of biochemical and clinical improvement

---

retard its further deposition; in addition, these techniques must be designed to circumvent potential immunologic complications, including inactivation of the administered activity. A summary of the requisites for effective enzyme therapy, identified by the early endeavors, is presented in Table 13–2.

Requisite 1 – Enzyme Technology

Sufficient Quantities of Maximally Stable Enzymes with High Specific Activities Must Be Available for Routine Administration

Enzyme modification techniques to stabilize the enzyme against potential bioinactivation or instability under physiologic conditions (pH 7.4, 37° C, proteases, inhibitors, etc.) and to prolong the intracellular half-life of exogenous activity will maximize the enzyme duration and dose effectiveness, resulting in optimal utilization and increased intervals between enzyme administrations.

Nonimmunogenic, Sterile Enzyme Preparations

Enzyme purification and administration strategies must be designed to avoid possible pathologic and molecular immunologic complications resulting from the administration of exogenous enzymes. Extensive experience with replacement therapy in patients with hemophilia and diabetes mellitus has indicated the limited, but clinically manageable, immunologic reactions to administered proteins. Since enzyme replacement involves the direct administration of a protein that may be foreign to the recipient, possibly triggering the synthesis of antibodies against the "enzyme-antigen," it is important to determine whether a particular patient's or family's genetic mutation produces a cross-reacting material (CRM).[36]

The majority of enzymatic defects characterized to date are CRM-positive; these recipients may not recognize the exogenous enzyme as foreign, since they produce an inactive enzyme with one amino acid difference. Use of human enzymes should minimize the potential immunologic complications in CRM-positive patients. Specific human isozymes, obtained perhaps from the normal tissues or fluids corresponding to the pathologic target tissue, may further minimize the possibility of these complications (*e.g.,* the specific, normal splenic, or hepatic β-glucosidase isozyme for replacement therapy in type 1 Gaucher's disease, or similarly, the specific normal renal or endothelial α-galactosidase A isozyme for Fabry's disease). In addition, selected isozymes may promote uptake by the pathologic target tissues via recognition of specific tissue and perhaps subcellular receptors and may also provide additional protection from endogenous catheptic destruction. However, a single base-pair substitution may alter a CRM-positive enzyme conformation sufficiently so that even normal homologous isozyme is recognized as foreign. Similarly, if the recipient is CRM-negative, the administered enzyme may be recognized as foreign; the antibody produced against the enzyme may inactivate or precipitate the enzyme causing immunologic consequences, although it is also possible that the antibody may stabilize the enzyme as has been demonstrated *in vitro* for normal and defective arylsulfatase A.[37, 38] Therefore, strategies must be developed to avoid the immunologic surveillance system and to circumvent potential immunologic inactivation of administered activity in CRM-negative patients by modifications that render the enzyme nonimmunogenic while retaining enzymatic activity or by the possible passive immunization of the recipient.

Requisite 2 — Enzyme Administration

Route of Administration

Although intramuscular and intrathecal administrations have been attempted (Table 13–1), presumably, intravenous injection will provide the most accessible and direct route to the widest variety of tissues. However, as has been demonstrated in Sandhoff's disease,[31] it is not likely that appreciable quantities of intravenously administered enzyme can cross the blood brain barrier to gain access to neural tissues or cerebrospinal fluid. Alternative routes of administration may be required to provide neural tissue and perhaps other target sites with therapeutic quantities of exogenous enzyme.

Enzyme Protection

Methods must be developed that will protect exogenous enzymes from the potentially hazardous physiologic conditions of the circulation, including the immunologic surveillance system, thereby maximizing the enzymatic activity that reaches target subcellular sites. The entrapment of enzyme preparations in biodegradable microvesicles may provide effective protection of the activity and facilitate enzyme delivery.

Enzyme Delivery

Critical to effective therapy is the delivery of active, stable enzyme to the specific tissues and subcellular sites of pathologic substrate accumulation (*e.g.,* secondary lysosomes in lysosomal storage diseases). Targeting to specific sites may require either purification of an isoenzyme from specific tissue sources and/or chemical modification of the enzyme; alternatively, the enzyme preparation may be immobilized in various microvesicles that can be chemically or physically modified to promote uptake by specific cells or tissues.

Requisite 3— *In Vivo* Test Systems

The inherent limitations of human experimentation mandate the development of mammalian model systems to appraise the application of various enzyme engineering techniques necessary to fulfill the requisites for effective therapy. Suitable models that permit the evaluation and manipulation of the factors that maximize enzyme stability, protection, target tissue distribution, subcellular localization, and substrate metabolism prior to human trials have become a prime requisite for the systematic development of enzyme replacement techniques. These model systems should also provide data regarding initial enzyme dosages that will result in both catabolism of pathogenic metabolites and significant decrease in the rate of their deposition.

In vitro test systems have been exploited to determine whether exogenous enzyme incubated in the media of cultured skin fibroblasts could gain access to the cell's interior. These studies demonstrated that enzymes from a variety of sources could be taken up by fibroblasts cultured from patients with various lysosomal storage diseases and that accumulated substrates could be metabolized. However, enzyme uptake by fibroblasts provides only an indication that exogenously supplied enzymes may gain access to living cells in a metabolically active state. The limited physiology of tissue culture systems requires that *in vivo* animal models now be used for the critical investigations confronting the complex physiology of mammals prior to the design of therapeutic endeavors.

Requisite 4— Demonstration of Biochemical and Clinical Effectiveness

The results of human therapeutic trials must be carefully documented, both biochemically and clinically, to evaluate therapeutic effectiveness and to pinpoint problems to be resolved. Prior to enzyme administration, multiple determinations of the endogenous levels of defective enzymatic activity and accumulated substrate, as well as precursors and metabolites distal to the enzymatic block, should be obtained. If any drugs or other agents are to be administered with the exogenous enzyme, the patient should be placed on a preadministration control regimen so that enzyme and metabolite levels can be determined under appropriate control conditions. The fate of the administered enzymatic activity and the levels of the pathogenic metabolite(s) should be serially determined in all easily obtainable sources (e.g., plasma, serum, urine, tears) and in appropriate biopsied tissues (e.g., CSF, bone marrow, liver, kidney) by percutaneous techniques. The optimal intervals for these determinations should be based initially on the results of the *in vivo* test system studies. In addition, physical, kinetic and/or electrophoretic characterization of the administered enzyme in the above sources will further document the biochemical effectiveness of therapeutic endeavors. Ultimately, the fulfillment of all the requisites will be determined by the demonstration of clinical improvement in appropriately selected patients. Clinical assessment should include, when possible, the objective and quantitative evaluation of selected clinical parameters that must be improved if the therapeutic endeavor is to be considered beneficial.

ENZYME THERAPY: PRINCIPLES

The development of a successful therapeutic modality as complex as enzyme replacement therapy requires a framework of general principles both to guide the evolution of rational therapeutic strategies and to provide criteria for evaluating their effectiveness. Therefore, the following principles are intended to provide guidelines for future investigations of enzyme therapy in general, and for enzymatic deficiencies characterized by the accumulation of a pathogenic substrate(s) in particular.

Principle 1 — Ethical Aspects of Human Therapeutic Experimentation

Every new method of treatment requires, in the final analysis, proof or disproof of its efficacy in man — human experimentation. The first and major principle of any exploratory medical endeavor involving human subjects is DO NO HARM. Clinical trials of newly developed therapeutic techniques must be rigidly controlled and cautiously pursued to protect the volunteer subjects from any potential complications or detrimental effects; their legal and moral rights must be respected and safeguarded.[39-42] In keeping with these concepts, the ethical and medical principles for experimental trials include: (1) genetic and family counseling relative to the specific diagnosis, prognosis, and therapeutic options; (2) full and informed consent to participation by volunteer subjects or guardians, including complete disclosure of the risks, limitations, and likelihood of therapeutic success or failure; (3) rational therapeutic strategies designed to obtain sufficient data, including appropriate control studies, to assess adequately the short- and long-term chemical and clinical effects, and (4) protection of the mental, as well as physical, welfare of the participant subjects.

Principle 2—Characterization of the Molecular Pathology

The development of effective strategies for the treatment of patients with genetic diseases requires thorough understanding of the molecular pathology of the specific disorder or disease variant. Therefore, the second principle stresses the need to characterize molecular and pathologic alterations for the design of rational therapeutic strategies. The essential experimental logic for therapeutic design and assessment protocols clearly requires careful documentation of the specific nature of the primary genetic defect and pathology in each disease.

The identification of the specific defective enzyme or isozyme does not provide sufficient information to design appropriate therapeutic strategies; further elucidation of the molecular nature of the putative enzymatic lesion is required. Indeed, similar clinical phenotypes may result from molecular heterogeneity in the primary genetic lesion. Table 13–3 lists the potential causes of an enzymatic defect resulting from various types of gene mutations. For example, a genetic defect significantly decreasing the rate of enzyme synthesis, resulting in extremely small quantities of a catalytically active protein, might be therapeutically increased to metabolically effective levels by methods different from those that would be applied to an enzymatic lesion resulting from a structural mutation altering its kinetic properties and/or stability. Similarly, a specific deficient isozyme activity may result from a defective post-translational modification process, possibly a mutant glycosyltransferase or phosphorylase activity; thus, therapeutic strategies might be

TABLE 13–3.   POTENTIAL CAUSES OF AN ENZYMATIC DEFICIENCY

Regulatory Gene Mutation
    Low rate of normal enzyme synthesis

Structural Gene Mutation
    Normal rate of enzyme synthesis, *but*
        Altered (raised) K_m value
        Altered (lowered) V_{max} value
        Altered allosteric characteristics
    Elevated rate of enzyme degradation
        Physical instability
        Increased susceptibility to endogenous proteases
    Other structural defects
        Altered pH optima
        Altered secondary binding sites
        Altered interactions (*e.g.*, subunits)

Post-Translational Defects
    Normal structural gene product, but
        Altered enzyme modification
            (*e.g.*, glycosylation, phosphorylation, peptide cleavage or
            addition, etc.)
        Altered tissue or subcellular localization

more effective if designed to correct the modifying process rather than to directly replace the deficient isozyme.

Analogous to the delineation of the molecular nature of the enzymatic lesion, characterization of the molecular pathology includes the critical assessment of all the metabolic processes that are altered by the enzymatic defect. The concentrations of the specific substrate(s) and product(s), the metabolites proximal and distal to the enzymatic block, as well as metabolites that may be altered via secondary metabolic interactions, must be quantitated in the appropriate tissues and fluids and their pathogenicity, if any, determined. Concomitantly, the localization of the tissue and the subcellular sites of the defective enzyme (isozyme) as well as pathogenic metabolite(s) accumulation (or deprivation) is required to determine the targets for enzyme therapy. In addition, the metabolic origin of the normal enzyme and product must be considered, since a metabolite may normally be synthesized in the liver, for example, and be secreted into the circulation for delivery to another tissue site for its physiologic action;[43] thus, enzyme therapy directed at the primary site of origin may be more effective than enzyme therapy delivered to the secondary site of substrate deposition. Finally, the kinetics of the molecular pathology are important to consider; the rate of accumulation of the pathologic metabolite is critical for determining the feasibility of therapy as well as the frequency and dosage of enzyme administrations. In certain disorders, the *in utero* substrate deposition may have already caused significant, possibly irreversible, pathology prior to birth,[48] so that enzyme therapy may not be feasible. Furthermore, in amenable disorders, appropriate patients should be selected for treatment; for example, the progressive age and condition, in terms of "substrate years" of deposition, will be a major factor in determining the therapeutically reasonable chemical and clinical expectations of enzyme therapy. Clearly, enzyme therapy for mucopolysaccharidosis type I (Hurler's disease) should not be attempted unless strategies to replace the deficient enzyme in neural cells can be achieved. Thus, the thorough characterization of the molecular pathology of each familial mutation is a necessary principle for the development of effective strategies for enzyme therapy.

Principle 3 — Rational Therapeutic Design and Adequate Assessment

Based on the molecular pathology of each disease, therapeutic strategies must be designed using enzyme and cellular engineering techniques to fulfill the previously described therapeutic requisites (Table 13–2) to warrant human trials and to assess their effectiveness. Initial trials are necessarily cautious experiments aimed at establishing that the therapeutic strategy merits further evaluation. Therapeutic assessment must be objective; demonstration of chemical effectiveness (*i.e.,* active enzyme recovered from pathologic target sites, or evidence of substrate metabolism, etc.) is a reasonable expectation for initial trials. Investigators should be cautious not to implicate subjective clinical improvement as justification for further trials, since clinical benefit, particularly in disorders characterized by years of substrate deposition, is an unlikely expectation from single-dose or short-term therapy. Failure to demonstrate chemical and/or morphologic improvement should be fully evaluated and all possible explanations should be explored in animal model systems. If the initial trials provide promise of therapeutic benefit, further carefully controlled, chemically and clinically monitored studies must be accomplished before the efficacy and safety of an enzyme therapeutic endeavor can be documented.

It is of note that enzyme therapy is potentially feasible only in selected diseases. Attempts must be made to identify the inherent therapeutic restraints in the design of enzyme replacement endeavors. Certainly, the constraints of the blood-brain barrier would limit the effectiveness of intravenous enzyme replacement in disorders with severe neurologic involvement. For example, in the family of lysosomal glycosphingolipid storage diseases, only Gaucher's type 1 and Fabry's diseases can be considered amenable to enzyme therapy at present. Effective therapy awaits the development of chemical and physical techniques to permit enzymes to gain access to neural sites of lipid deposition, a requisite for the treatment of the other glycosphingolipidoses. Similar therapeutic restraints must be considered for each disorder. Although enzyme replacement endeavors should be limited to selected diseases, future advances may provide the cellular engineering technology necessary to overcome these limitations.

ENZYME THERAPY: PROSPECTS

Prospects for the development of effective enzyme therapy rely on the integration and application of basic contributions from many disciplines, including genetics, biochemistry, immunology, and cell biology. The thrust of current research efforts in these areas, involving the development of enzyme purification and modification techniques, elucidation of the chemistry and immunology of cell membranes, and the characterization of the molecular pathology of genetic diseases, will be applicable to future enzyme therapy endeavors. The following section presents an overview of the most recent and innovative developments that may provide strategies to overcome the current obstacles to successful enzyme therapy in genetic diseases.

Enzyme Technology

An essential requisite for enzyme therapy is the availability of sufficient quantities of purified, highly active, human enzymes that can be modified to increase their stability and maximize their metabolic effectiveness. Figure 13–2 illustrates various methods for the isolation, purification, and modification of human enzymes as well as for the production of antienzyme antibodies. In addition to conventional techniques for enzyme purification, the application of affinity chromatography has already demonstrated its usefulness for the simplified and rapid purification of normal human enzymes.[44-46] Similarly, the value of lectins has been demonstrated recently for the isolation of glycoprotein enzymes including various lysosomal hydrolases.[47, 48] Preparative gel-electrophoresis and isoelectric focusing procedures can further resolve the enzymatic activity into its various isozymes for further characterization and can provide homogenous sources of specific antigens for antibody production. In the future, chemical engineering techniques will be applied to scale-up these purification procedures for the production of large quantities of specific enzymes required for long-term enzyme replacement.

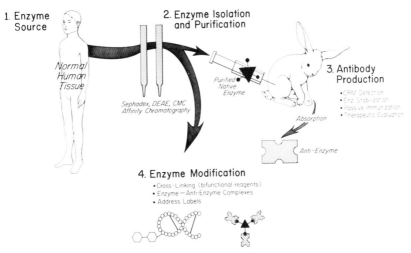

Figure 13–2. Strategies for the application of enzyme technology to the isolation, purification, antibody production, and modification of human enzymes.

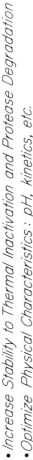

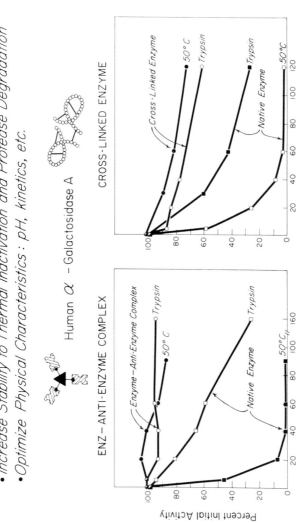

Figure 13-3. Modification of native purified α-galactosidase A designed to increase the stability of the enzymatic activity to thermal inactivation and protease degradation.

Various chemical modification procedures for the production of enzyme derivatives designed to act optimally as therapeutic agents have recently been reviewed by Wold.[49] Application of these enzyme engineering techniques potentially can produce modified enzymes altered to maximize *in vivo* catalytic properties (K_m, V_{max}, and pH optimum), stability, and compartmentalization. Figure 13–3 illustrates two approaches to the modification of the lysosomal hydrolase deficient in Fabry's disease, α-galactosidase A. The enzyme, either complexed with anti-α-galactosidase antibody or cross-linked with an inorganic bifunctional reagent, was significantly more stable than the purified native activity; the modified α-galactosidase A demonstrated markedly increased stability to thermal inactivation and protease degradation without loss of activity as determined with artificial[50] or natural substrates.[51] Furthermore, incubation of cross-linked and native α-galactosidase A activities in plasma from a hemizygote with Fabry's disease, to simulate physiologic conditions (pH 7.3, 37° C), demonstrated that the half-life of the cross-linked activity ($t_{1/2} = 46$ minutes) was significantly increased compared to the native activity ($t_{1/2} = 9$ minutes)[51] (Fig. 13–4). These results suggest that chemical modification of lysosomal hydrolases will be of value to maximize the stability of administered enzymes, at least in the circulation, as well as to stabilize them, potentially, to intracellular catheptic destruction, thus prolonging their catalytic lives. Recently, Gregoriadis[52] reported that administration of glutaraldehyde cross-linked invertase from *Saccharomyces cerevisiae* entrapped in liposomes and injected into rats had a longer intracellular half-life than liposome-en-

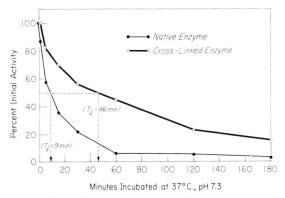

Figure 13–4. Comparison of the half-lives of native versus cross-linked α-galactosidase A activity incubated under physiologic conditions in plasma from a hemizygote with Fabry's disease.

trapped, native invertase. Further evaluations of modified enzymes in *in vivo* test systems must be accomplished to determine the fate of these enzyme derivatives under physiologic conditions; in addition, the rate of clearance due to protease degradation, or to immunologic inactivation, as well as the potential toxicity of these modified enzymes, must be evaluated.

Although the likelihood, frequency, and severity of immunologic reactions to administered exogenous enzymes in CRM-positive or -negative recipients are unknown, the potential immunologic complications of enzyme therapy must be circumvented. Production of monospecific antibodies to the normal active enzyme will provide the means to test an individual patient's or family's mutation for CRM; in addition, these antibodies, when radiolabeled, will provide for sensitive radioimmunoassays that may be valuable for demonstrating the presence of injected enzyme protein in the recipient. However, as previously discussed, it is desirable to protect administered enzymes from immunologic surveillance in both CRM-positive and -negative patients. Various strategies, including the chemical modification of exogenous enzymes, rendering them less or nonimmunogenic and/or entrapment of enzymes in biodegradable vesicles (see following section, Enzyme Administration) to protect the enzyme from the immunologic surveillance system, should be evaluated in animal test systems.

Chemical modification may provide an important means to render enzymes nonimmunogenic; the covalent binding of polyethylene glycol to catalase has been reported to reduce its antigenicity in mice.[53] Desired immunologic protection to avoid both immune responses and inactivation of enzymes would also be expected when enzymes are camouflaged by entrapment in microcapsules, erythrocytes, or liposomes. For example, repeated injections of microencapsulated catalase into mice revealed no immunologic reactions;[54] administration of liposome-entrapped antigens (diphtheria toxoid, asparaginase) to mice or rats demonstrated that entrapped enzymes are less likely to elicit antibody-mediated responses than the unentrapped proteins.[52] Further developments in the chemical modification of enzymes and enzyme immobilization, combined with advances in immunosuppression therapy, may be expected to enhance the ability to attenuate immune responses possibly induced by enzyme administrations.

Enzyme Administration

The results of early attempts to replace enzymes directly underscore the essential requisite for administration methods that promote the delivery of enzyme to target tissue and subcellular sites while protecting it from bioinactivation and other degradative and immunologic processes. Several strategies to accomplish these objectives are depicted in Figure 13–5. The observations that intravenously injected native enzymes are recovered primarily from liver (Table 13–1) indicate that targeting of enzymes to other critical sites of pathology is a priority. The intriguing possibility that specific isozymes may be biologically coded to promote their differential distribution and uptake may prove invaluable for the controlled and specific delivery of exogenous enzymes.

Several recent observations suggest that selective targeting may be achieved by exploiting the inherent properties of proteins to signal their uptake by specific tissues and/or cells. For example, desialylation of certain circulating glycoproteins has been shown to result in their rapid clearance from the circulation and selective uptake by hepatocytes;[55, 56] these and similar studies[57-59] implicate the role of carbohydrates on proteins in signaling specific membrane uptake. Indeed, investigators are currently searching to determine if an extensive carbohydrate code exists to specifically address proteins ("address labels") for uptake by particular tissues and cells.[60, 61]

The tissue or subcellular source from which a specific enzyme is isolated may also prove valuable. Our studies,[57] demonstrating a differential *in vivo* tissue distribution of administered isogenic, murine serum, hepatic, and renal proteins, suggest that the delivery of enzymes to a wide variety of tissues may require the administration of isozymes purified from multiple, selected tissue sources. Analogous observations have been made *in vitro;* investigators have found striking differences in

ADMINISTRATION OF ENZYME PREPARATION

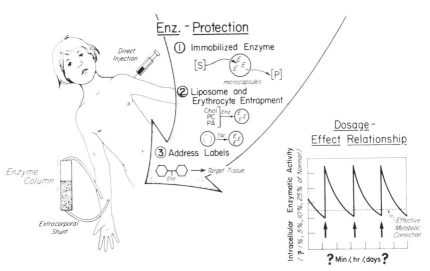

Figure 13–5. Strategies for the protection and delivery of enzyme preparations for *in vivo* administration.

the rate of enzyme uptake by enzyme-deficient fibroblasts dependent on the source of the enzyme.[44, 62]

Other chemical modifications of enzymes may promote their differential uptake by target tissues. For example, partial degradation of a high molecular weight enzyme, in a manner that permits retention of its catalytic activity, may result in tissue-specific uptake (e.g., kidney) by virtue of its smaller size.

An essential adjunct to the delivery of administered enzymes is their protection. The entrapment of enzymes in synthetic microcapsules[63] or, preferably, in biodegradable vesicles should prove to be valuable for both enzyme delivery and protection. Of particular interest is the immobilization of enzymes in lipid vesicles, or liposomes,[64, 65] and erythrocyte ghosts[66] isolated from the perspective enzyme recipient to avoid immunologic complications. Both of these techniques have been utilized to deliver active enzyme to hepatic lysosomes,[52, 67] target subcellular sites for lysosomal storage diseases. Liposomes tend to be taken up primarily by hepatic parenchymal cells,[68] whereas enzyme-loaded erythrocyte ghosts are presumably cleared by cells of the reticuloendothelial system, permitting a differential cellular distribution for administered enzymes.

Chemical and/or physical modifications of the membrane surfaces of erythrocytes and liposomes, similar to those discussed above for unentrapped enzymes, may direct them to target sites; for example, the incorporation of specific carbohydrate, glycolipid, lipopolysaccharide, or glycopeptide "address labels" may provide the appropriate surface marker for specific tissue recognition and selective uptake. Innovative approaches to "educate" liposomes by the incorporation of specific antibodies directed against selected cell or tissue membrane antigens into the liposomal (or erythrocyte) membrane may also address these enzyme carriers.[52] Furthermore, appropriately selected lipid compositions for liposomes may induce their cellular uptake by either fusion (e.g., lysolecithin) or endocytosis;[69] these approaches offer exciting possibilities for the delivery of enzymes into the cytoplasm as well as into lysosomes. The surface of enzyme-loaded erythrocytes can be similarly manipulated to expand their tissue distribution as, for example, by alteration of surface sulfhydryl groups.[70]

An alternative approach for disorders characterized by metabolite accumulation in the blood is the use of enzymes entrapped in extracorporeal shunts. In this case, blood is diverted through an extracorporeal shunt containing enzyme immobilized within gels or capsules for metabolism of the circulating toxic metabolite.[71] Clearance of the circulating pathogenic metabolite would presumably establish equilibrium gradients between the plasma and tissue sites of metabolite accumulation, eventually providing a mechanism for the continual clearance and metabolism of the systemic metabolite load.

Thus, enzyme carriers will be useful not only for the delivery of enzymes but also for the protection of entrapped enzymatic activity from *in vivo* degradative processes. Utilization of methodologies to stabilize, protect, deliver, and target administered enzymes will presumably yield the most fruitful strategy for enzyme therapy.

In Vivo Test Systems

Animal model systems are essential for the *in vivo* evaluation and maximization of the enzyme and cellular engineering strategies required for enzyme therapy prior to human trials. Preferred model systems are those in which the test animals have the same enzyme defect as those found in human disorders. Various animal models of human metabolic diseases have been described.[72-76] Among the lysosomal storage diseases several models have been identified that have similar clinical and pathologic manifestations caused by the same biochemical defect as their human counterpart, including cats with G_{M1} gangliosidosis,[77] English setters with Batten disease,[78] Cairn and West Highland terriers with Krabbe's disease,[79] and cows with mannosidosis.[80] Further support should be provided to our colleagues in veterinary medicine to continue their efforts to search for new animal models of human inborn errors of metabolism and other diseases, preferably in smaller animals that are easily bred and conveniently studied in the laboratory.

Previous experience with enzyme replacement in selected animal models has already demonstrated the value of these experimental systems. Feinstein[81] purposely sought and identified an animal model of the human disorder acatalasemia in mutagenized mice. Attempted enzyme replacement with native[82] and microencapsulated catalase[54] demonstrated the relative effectiveness of these approaches to supply a defective circulating enzymatic activity; these studies documented the feasibility of enzyme therapy for disorders that may be ameliorated by metabolism of circulating pathogenic metabolites. Normal animals also have been useful for studies of enzyme therapy. Huijing and coworkers[19] have characterized the *in vivo* fate of α-glucosidase, partially purified from *A. niger,* following intravenous administration of various dosages to normal rats. Although no significant effects of dosage were observed, a rapid uptake and clearance of the injected activity in hepatic lysosomes were demonstrated, similar to early findings in human trials (Table 13–1). Recently, Gregoriadas[52] used normal rats to evaluate the fate and antigenicity of liposome-entrapped, native, and cross-linked enzymes; these studies demonstrated the effectiveness of enzyme immobilization for the protection and delivery of native and cross-linked enzymatic activities.

Recently, we have developed a mammalian model system to deter-

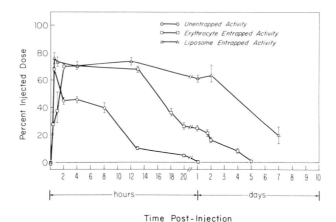

Time Post-Injection

Figure 13–6. Comparison of the hepatic recoveries of unentrapped, erythrocyte entrapped and liposome entrapped bovine β-glucuronidase intravenously administered to β-glucuronidase deficient mice.

mine the fate of intravenously administered β-glucuronidase in β-glucuronidase deficient mice.[83] This system provides the means to assess the various biochemical and physiologic processes affecting the fate of administered enzymes prior to human trials, including (1) methods of administration and protection; (2) uptake, distribution, and metabolic half-life; and (3) potential bioinactivation, immunologic complications, and toxicity of the exogenously supplied enzyme. Through use of this system, the effectiveness of various enzyme delivery systems has been compared, as shown in Figure 13–6. Preliminary results have shown that the half-lives of recovered exogenous activity in the liver were significantly prolonged when administered enzyme was entrapped in liposomes or erythrocytes compared to unentrapped enzyme.[84] Furthermore, enzyme delivery by each of these methods resulted in the recovery of exogenous activity primarily in the lysosomally enriched subcellular fraction, the target subcellular site for enzyme replacement in lysosomal storage diseases. Future studies using animal model systems should be directed toward determining (1) the *in vivo* metabolism of the pathogenic metabolite; (2) the cell types to which administered enzymes are delivered; (3) potential immunologic complications; and (4) the relative uptake and tissue and subcellular distribution of chemically and immunologically stabilized enzymes from various heterologous and homologous sources.

SUMMARY

The integration of prior experiences with recent developments and future ingenuity portends exciting prospects for the treatment of patients with enzyme deficiency diseases. Advances in enzyme technology are already enhancing the feasibility and potential practicality of enzyme replacement. In addition, the rapid and anticipated progress in cellular engineering to deliver highly purified, chemically modified enzymes to target sites for maximal therapeutic effectiveness makes the future of this therapeutic modality promising. Indeed, the significant contributions to the understanding of normal human biochemistry and cell biology made by the patients with inherited metabolic diseases and their families mandate our continued efforts to develop specific therapies for these severely debilitating disorders.

References

1. McKusick, V. A.: Mendelian Inheritance In Man. Baltimore, Md., Johns Hopkins Press, 1971.
2. Stanbury, J. G., Wyngaarden, J. B., and Frederickson, D. S. (eds.): The Metabolic Basis of Inherited Disease. 3rd ed. New York, McGraw-Hill Book Company, 1972.
3. Nyhan, W. L.: Heritable Disorders of Amino Acid Metabolism. New York, John Wiley & Sons, Inc., 1974.
4. Scriver, C. R., and Rosenberg, L. E.: Amino Acid Metabolism and Its Disorders. Philadelphia, W. B. Saunders Company, 1973.
5. Raivio, K. O., and Seegmiller, J. E.: Genetic diseases of metabolism. Ann. Rev. Biochem., 41:543, 1972.
6. New Haven Conference (1973): First International Workshop on Human Gene Mapping. Birth Defects: Original Articles Series. Vol. 9. New York, The National Foundation, 1974; and Cytogenetics 13, 1974.
7. Desnick, R. J., and Fiddler, M. B.: Advances in the treatment of genetic disease: An overview. In Kelly, S., Hook, E. B., Janerich, D. T., and Porter, I. H. (eds.): Birth Defects: Risks and Consequences. New York, Academic Press, Inc. (in press).
8. Seakins, J. W. T., Saunders, R. A., and Toothill, C. (eds.): Treatment of Inborn Errors of Metabolism. Edinburgh, Churchill Livingstone, 1973.
9. Howell, R.: Genetic disease: The present status of treatment. In McKusick, V. A., and Claiborne, R. (eds.): Medical Genetics. New York, Hospital Practice Publishing Co., 1973, p. 271.
10. Desnick, R. J., Bernlohr, R. W., and Krivit, W. (eds.): Enzyme Therapy in Genetic Disease. Birth Defects: Original Article Series. Vol. 9. Baltimore, Md., Williams & Wilkins Co., 1973.
11. Rietra, P. J. G. M., van den Bergh, F. A. J. T. M., and Tager, J. M.: Recent developments in enzyme replacement therapy of lysosomal storage disease. In Tager, J. M., Hooghwinkel, G. J. M., and Daems, W. T. (eds.): Enzyme Therapy in Lysosomal Storage Diseases. Amsterdam, North-Holland Publishing Co., 1974.
12. Kirkman, H. N.: Enzyme defects. Progr. Med. Genet., 8:125, 1972.
13. de Duve, C.: From cytases to lysosomes. Fed. Proc., 23:1045, 1964.
14. Baudhuin, P., Hers, H. G., and Loeb, H.: An electron microscopic and biochemical study of type II glycogenosis. Lab. Invest., 13:1139, 1964.
15. Hug, G., and Schubert, W. K.: Lysosomes in type II glycogenosis: Changes during administration of extract from Aspergillus niger. J. Cell Biol., 35:1, 1967.
16. Hug, G., and Schubert, W. K.: Hepatic lysosomes in Pompe's disease: Disappearance during glucosidase administration. J. Clin. Invest., 46:1073, 1967.
17. Lauer, R. M., Mascarinas, T., Racela, A. S., Diehl, A. M., and Brown, B. I.: Administration of a mixture of fungal glucosidases to a patient with type II glycogenosis (Pompe's disease). Pediatrics, 42:672, 1968.
18. Fernandes, J., and Huijing, F.: Branching enzyme-deficiency glycogenosis: Studies in therapy. Arch. Dis. Child., 43:347, 1968.
19. Huijing, F., Waltuck, B. L., and Whelan, W. J.: α-Glucosidase administration: Experiences in two patients with glycogen storage disease compared with animal experi-

ments. *In* Desnick, R. J., Bernlohr, R. W., and Krivit, W. (eds.): Enzyme Therapy in Genetic Diseases. Birth Defects: Original Article Series. Vol. 9. Baltimore, Md., Williams & Wilkins Co., 1973, p. 191.

20. Greene, H. L., Hug, G., and Schubert, W. K.: Metachromatic leucodystrophy: Treatment with arylsulfatase-A. Arch. Neurol., 20:147, 1969.

21. Austin, J. H.: Some recent findings in leukodystrophies and in gargoylism. *In* Aronson, S. M., and Volk, B. W. (eds.): Inborn Disorders of Sphingolipid Metabolism. Oxford, Pergamon Press, 1967, p. 359.

22. Mapes, C. A., Anderson, R. L., Sweeley, C. C., Desnick, R. J., and Krivit, W.: Enzyme replacement in Fabry's disease, an inborn error of metabolism. Science, 169:987, 1970.

23. Sweeley, C. C., Mapes, C. A., Krivit, W., and Desnick, R. J.: Chemistry and metabolism of glycosphingolipids in Fabry's disease. *In* Volk, B. W., and Aronson, S. M. (eds.): Sphingolipids, Sphingolipidoses and Allied Disorders. New York, Plenum Publishing Corp., 1972, p. 287.

24. Desnick, R. J., Krivit, W., Snyder, P. D., Desnick, S. J., and Sharp, H. L.: Sandhoff's disease: Ultrastructural and biochemical studies. *In* Volk, B. W., and Aronson, S. M. (eds.): Sphingolipids, Sphingolipidoses and Allied Disorders. New York, Plenum Publishing Corp., 1972, p. 351.

25. DiFerrante, N., Nichols, B. L., Donnelly, P. V., Neri, G., Hrgovcic, R., and Berglund, R. K.: Induced degradation of glycosaminoglycans in Hurler's and Hunter's syndromes by plasma infusion. Proc. Natl. Acad. Sci. U.S.A., 68:303, 1971.

26. Dekaban, A. S., Holden, K. R., and Constantopoulos, G.: Effects of fresh plasma or whole blood transfusions on patients with various types of mucopolysaccharidosis. Pediatrics, 50:688, 1972.

27. Erickson, R. P., Sandman, R., Robertson, W. van B., and Epstein, C. J.: Inefficacy of fresh frozen plasma therapy of mucopolysaccharidosis II. Pediatrics, 50:693, 1972.

28. Dean, M. F., Muir, H., and Benson, P. F.: Mobilization of glycosaminoglycans by plasma infusion in mucopolysaccharidosis type III — two types of response. Nature [New Biol.], 243:143, 1973.

29. Kolodny, E. H.: Discussion. *In* Desnick, R. J., Bernlohr, R. W., and Krivit, W. (eds.): Enzyme Therapy in Genetic Diseases. Birth Defects: Original Article Series. Vol. 9. Baltimore, Md., Williams & Wilkins Co., 1973, p. 40.

30. Knudson, A. G., Jr., DiFerrante, N., and Curtis, J. E.: Effect of leucocyte transfusion in a child with type II mucopolysaccharidosis. Proc. Natl. Acad. Sci. U.S.A., 68:1738, 1971.

31. Johnson, W. G., Desnick, R. J., Long, D. M., Sharp, H. L., Krivit, W., Brady, B., and Brady, R. O.: Intravenous injection of purified hexosaminidase A into a patient with Tay-Sachs disease. *In* Desnick, R. J., Bernlohr, R. W., and Krivit, W. (eds.): Enzyme Therapy in Genetic Diseases. Birth Defects: Original Article Series. Vol. 9. Baltimore, Md., Williams & Wilkins Co., 1973, p. 120.

32. De Barsy, T., Jacquemin, P., van Hoof, F., and Hers, H. G.: Enzyme replacement in Pompe disease: An attempt with purified human acid α-glucosidase. *In* Desnick, R. J., Bernlohr, R. W., and Krivit, W. (eds.): Enzyme Therapy in Genetic Diseases. Birth Defects: Original Articles Series. Vol. 9. Baltimore, Md., Williams and Wilkins Co., 1973, p. 184.

33. Brady, R. O., Tallman, J. F., Johnson, W. G., Gal, A. E., Leahy, W. R., Quirk, J. M., and Dekaban, A. S.: Replacement therapy for inherited enzyme deficiency: Use of purified ceramidetrihexosidase in Fabry's disease. N. Engl. J. Med., 289:9, 1973.

34. Brady, R. O., Penchev, P. G., Gal, A. E., Hibbert, S. R., and Dekaban, A. S.: Replacement therapy for inherited enzyme deficiency: Use of purified glucocerebrosidase in Gaucher's disease. N. Engl. J. Med., 291:989, 1974.

35. Sly, W. S., Glaser, J. G., Roogen, K., Brot, F., and Stahl, P.: Enzyme replacement studies with β-glucuronidase deficiency. *In* Tager, J. M., Hooghwinkel, G. J. M., and Daems, W. T. (eds.): Enzyme Therapy in Lysosomal Storage Diseases. Amsterdam, North-Holland Publishing Co., 1974, p. 288.

36. Boyer, S. H., Siggers, D. C., and Krueger, L. J.: Caveat to protein replacement therapy for genetic disease. Immunological implications of accurate molecular diagnosis. Lancet, 2:654, 1973.

37. Neuwelt, E., Stumpf, D., Austin, J., and Kohler, P.: A monospecific antibody to human sulfatase A preparation, characterization and significance. Biochim. Biophys. Acta, 236:333, 1971.

38. Stempf, D., Neuwelt, E., Austin, J., and Kohler, P.: Metachromatic Leukodystrophy (MLD) X. Immunological studies of the abnormal sulfatase A. Arch. Neurol., 25:427, 1971.
39. Rutstein, D. D.: The ethical design of human experiments. Daedalus, 98:523, 1969.
40. Strauss, M. B.: Ethics of experimental therapeutics. N. Engl. J. Med., 288:1183, 1973.
41. Motulsky, A. G.: Brave new world? Science, 185:653, 1974.
42. Federal Register (Protector of Human Subjects. Proposed Policy), 39:30648, August 23, 1974.
43. Sharp, H. L.: α_1 Anti-trypsin deficiency. Hosp. Pract., 6:83, 1971.
44. Brot, F. E., Glaser, J. H., Roozen, K. J., and Sly, W. S.: In vitro correction of deficient human fibroblasts by β-glucuronidase from different human sources. Biochem. Biophys. Res. Commun., 57:1, 1974.
45. Radin, N. S., Hyun, J. C., and Misra, R. S.: Cerebroside glucosidase: Assay and purification. Fed. Proc., 33:1226, 1974.
46. Mapes, C. A., and Sweeley, C. C.: Preparation and properties of an affinity column adsorbent for differentiation of multiple forms of α-galactosidase activity. J. Biol. Chem., 248:2461, 1973.
47. Norden, A. G. W., and O'Brien, J. S.: Binding of human liver β-galactosidases to plant lectins insolubilized on agarose. Biochem. Biophys. Res. Commun., 56:193, 1974.
48. Desnick, R. J.: Unpublished results, 1974.
49. Wold, F.: Chemical modification of enzymes. In Desnick, R. J., Bernlohr, R. W., and Krivit, W. (eds.): Enzyme Therapy in Genetic Diseases. Birth Defects Original Articles Series. Vol. 9. Baltimore, Md., Williams & Wilkins Co., 1973, p. 46.
50. Snyder, P. D., Jr., Wold, F., Bernlohr, R. W., Dullum, C., Desnick, R. J., Krivit, W., and Condie, R. M.: Enzyme therapy II: Purified human α-galactosidase A, stabilization to heat and protease degradation by complexing with antibody and by chemical modification. Biochim. Biophys. Acta, 350:432, 1974.
51. Desnick, R. J., and Johnson, D. L.: In preparation, 1974.
52. Gregoriadis, G.: Structural requirements for the specific uptake of macromolecules and liposomes by target tissues. In Tager, J. M., Hooghwinkel, G. J. M., and Daems, W. T. (eds.): Enzyme Therapy in Lysosomal Storage Diseases. Amsterdam, North-Holland Publishing Co., 1974, p. 131.
53. Abuchawski, A., van Es, T., Palczuk, N. C., and Davis, F. F.: Preparation and properties of nonimmunogenic catalase. Fed. Proc., 33:1317, 1974.
54. Poznansky, M. J., and Chang, T. M. S.: Kinetics of semipermeable microcapsules containing catalase. Proc. Canad. Fed. Biol. Sci., 28:87, 1969.
55. Morell, A. G., Gregoriadis, G., Scheinberg, I. H., Hickman, J., and Ashwell, G.: The role of sialic acid in determining the survival of glycoproteins in the circulation. J. Biol. Chem., 246:1461, 1971.
56. Hudgin, R. L., Pricer, W. F., Ashwell, G., Stockert, R. J., and Morrell, A. G.: The isolation and properties of a rabbit liver binding protein specific for asialoglycoproteins. J. Biol. Chem., 249:5536, 1974.
57. Fiddler, M. B., Wold, F., and Desnick, R. J.: In vivo recognition of murine tissue and serum proteins. Biochem. J. (in preparation).
58. Rogers, J. C., and Kornfeld, S.: Hepatic uptake of proteins coupled to fetuin glycopeptide. Biochem. Biophys. Res. Commun., 45:622, 1971.
59. Hickman, S., Shapiro, L. S., and Neufeld, E. F.: A recognition marker required for uptake of a lysosomal enzyme by cultured fibroblasts. Biochem. Biophys. Res. Commun., 57:55, 1974.
60. Winterburn, P. J., and Phelps, C. F.: The significance of glycosylated proteins. Nature (Lond.), 236:147, 1972.
61. Zopf, D. A., and Ginsburg, V.: Studies on carbohydrate antigens of cell surfaces using oligosaccharides isolated from milk. Fed. Proc., 33:1225, 1974.
62. Shapiro, L. J., Hickman, S. G., Hall, C. W., and Neufeld, E. F.: Corrective and noncorrective forms of human α-L-iduronidase. Proceedings of the Meeting of The American Society for Human Genetics, Portland, Oregon, 79A, 1974.
63. Chang, T. M. S.: Artificial Cells. Springfield, Ill., Charles C Thomas, Publisher, 1972.
64. Gregoriadis, G., and Leathwood, P. D.: Enzyme entrapment in liposomes. Febs. Letters, 14:95, 1971.
65. Kataoka, T., Williamson, J. R., and Kinsky, S. C.: Release of macromolecular markers (enzymes) from liposomes treated with antibody and complement: An attempt at correlation with electron microscopic observations. Biochim. Biophys. Acta, 298:158, 1973.

66. Ihler, G. M., Glew, R. H., and Schnure, F. W.: Enzyme loading of erythrocytes. Proc. Natl. Acad. Sci. U.S.A., 70:2663, 1973.
67. Fiddler, M. B., Thorpe, S. R., Krivit, W., and Desnick, R. J.: Enzyme therapy III: In vivo fate of erythrocyte entrapped β-glucuronidase. *In* Tager, J. M., Hooghwinkel, G. J. M., and Daems, W. T. (eds.): Enzyme Therapy in Lysosomal Diseases. Amsterdam, North-Holland Publishing Co., 1974, p. 182.
68. Gregoriadis, G., and Ryman, B. E.: Fate of protein-containing liposomes injected into rats: An approach to the treatment of storage diseases. Eur. J. Biochem., 24:485, 1972.
69. Papahadjopoulos, G. P., and Mayhew, E.: Cellular uptake of cyclic AMP captured within phospholipid vesicles and effect on cell-growth behaviour. Biochim. Biophys. Acta, 363:404, 1974.
70. Rifkind, R. A.: Destruction of injured red cells in vivo. Am. J. Med., 41:711, 1966.
71. Chang, T. M. S.: Immobilization of enzymes, adsorbents or both within semipermeable microcapsules (artificial cells) for clinical and experimental treatment of metabolite related disorders. *In* Desnick, R. J., Bernlohr, R. W., and Krivit, W. (eds.): Enzyme Therapy in Genetic Diseases. Birth Defects: Original Articles Series. Vol. 9. Baltimore, Md., Williams & Wilkins Co., 1973, p. 66.
72. Cornelius, C. E.: Animal models—a neglected medical resource. N. Engl. J. Med., 281:934, 1969.
73. Lush, I. E.: The biochemical genetics of vertebrates except man. *In* Neuberger, A., and Tatum, E. L. (eds.): Frontiers of Biology. Vol. 3. Amsterdam, North-Holland Publishing Co., 1967.
74. Green, E. L. (ed.): Biology of the Laboratory Mouse. 2nd Ed. New York, McGraw-Hill Book Co., 1966.
75. Green, E. L. (ed.): Handbook on Genetically Standardized JAX Mice. 2nd Ed. Bar Harbor, Me., The Jackson Laboratory, Bar Harbor Times Publishing, 1968.
76. Blake, R. L., and Russell, E. S.: Hyperprolinemia and prolinuria in a new inbred strain of mice. Science, 176:809, 1972.
77. Farrell, D. F., Baker, H. J., Herndon, R. M., Lindsey, J. R., and McKhann, G. M.: Feline G_{M1} gangliosidosis: Biochemical and ultrastructural comparisons with the disease in man. J. Neuropath. Exp. Neurol., 32:1, 1973.
78. Patel, V. P., Koppang, N., Patel, B., and Zeman, W.: p-Phenylenediamine-mediated peroxidase deficiency in English setters with neuronal ceroid-lipofuscinosis. Lab. Invest., 30:366, 1974.
79. Suzuki, K., Suzuki, Y., and Fletcher, T. F.: Further studies on galactocerebroside β-galactosidase in globoid cell leukodystrophy. *In* Volk, B. W., and Aronson, S. M. (eds.): Sphingolipids, Sphingolipidoses and Allied Disorders. New York, Plenum Publishing Corp., 1972, p. 487.
80. Jolly, R. D.: Mannosidosis of children, other inherited lysosomal storage disorders. Am. J. Pathol., 74:211, 1974.
81. Feinstein, R. N., Howard, J. B., Braun, J. T., and Seaholm, J. E.: Acatalasemic and hypocatalasemic mouse mutants. Genetics, 53:923, 1966.
82. Feinstein, R. N.: Studies of acatalasemia. *In* Desnick, R. J., Bernlohr, R. W., and Krivit, W. (eds.): Enzyme Therapy in Genetic Diseases. Birth Defects: Original Articles Series. Vol. 9. Baltimore, Md., Williams & Wilkins Co., 1973, p. 55.
83. Thorpe, S., Fiddler, M. B., and Desnick, R. J.: Enzyme therapy IV: A method for determining the in vivo fate of bovine α-glucuronidase in β-glucuronidase deficient mice. Biochem. Biophys. Res. Commun., 61:1464, 1974.
84. Desnick, R. J., Fiddler, M. B., Steger, L., Dullum, C., Cumming, D., and Thorpe, S. R.: Enzyme replacement therapy: In vivo fate of native, erythrocyte and liposome entrapped β-glucuronidase. Pediatr. Res., 9:312, 1975.

CHAPTER 14

PREVENTION OF PREMATURITY AND PERINATAL MORBIDITY

Shirley Griffith Driscoll, M.D.

PRESENT STATUS

Retrospective examination of the antecedents of cerebral palsy and mental retardation implicates perinatal damage in a substantial number of cases.[3, 6, 9, 11, 19, 22, 27] As described so eloquently by Lilienfeld and Pasamanick, there appears to be "a continuum of reproductive casualty, consisting of brain damage incurred during the prenatal and paranatal periods, leading to a gradient of injury extending from fetal and neonatal death through cerebral palsy, epilepsy, behavior-disorder, and mental retardation."[16] The ultimate in perinatal damage is death, yet most victims of perinatal insults survive—many of them sustaining permanent handicaps. Thorough study of phenomena associated with perinatal mortality can thus be expected to lead to increased understanding of the causes underlying the entire spectrum of perinatal damage.

Recent decades have seen a precipitous drop in maternal deaths, with only a modest decline in fetal and neonatal mortality. Statistical reports document the large differences in perinatal mortality among communities that differ in geography, economic level, and racial and ethnic composition. Mortality of males exceeds that of females; blacks, that of whites; and poor, that of rich.[26] Specific diagnostic information has not been available to permit comparisons of such high- and low-risk categories to be made on the basis of causality. If subgroups of these populations were compared with respect to specific causes of death, it might

343

emerge that the measures for preventing many deaths—as well as near-deaths—were available to the otherwise privileged and not to the generally disadvantaged. The two most critical determinants of perinatal mortality are the general health of the mothers and the quality of their prenatal care. Overall, the sociologist seems to offer greater insights than the biologist, at the present time.

Recognition of the causes of perinatal damage often is not possible in individual cases. Certain basic premises, however, seem to apply: (1) that perinatal deaths and sublethal damage have similar pathogenesis; (2) that repetitive pregnancy wastage, of any degree of severity, tends to repeat in terms of etiology; and (3) that thorough study of all deaths continues to produce useful information relative to the overall problem of perinatal injury.

The necropsy provides significant insight as to the mode of death in most instances of neonatal deaths and in a high percentage of stillbirths. In addition to the post-mortem examination, an interpretation of the sequence of events leading toward the identification of the primary, underlying cause of death requires study of the fetal adnexa (i.e., placenta, membranes, and umbilical cord) and scrutiny of the clinical circumstances. Usually, reports focus on *fetal death* or *neonatal death,* and data are presented in relation to the locus in which death occurred. (Was death in the uterus or in the nursery?) Yet, many of the causes of fetal mortality also account for postnatal death. For example, erythroblastosis fetalis, infections, and antenatal asphyxia rank high in the hierarchy of causes of death, both *in utero* and *ex utero.* The fetus, asphyxiating prior to or during birth, may be born near death, surviving only briefly as a moribund neonate. "High risk pregnancies," such as those complicating maternal diabetes mellitus or complicated by pre-eclamptic toxemia, are at high risk for both stillbirths and neonatal deaths. Therapeutic modalities improving prenatal survival may be followed by increased neonatal mortality. On the other hand, deliberate shifts in the timing of elective delivery may forestall stillbirths, but at the price of increased neonatal mortality, owing to iatrogenic prematurity.

Cumulative experience with perinatal mortality at the Boston Hospital for Women demonstrates dramatically the overwhelming influence of antenatal circumstances on the outcome of pregnancy. Analysis of fetal deaths occurring at or after 20 weeks' gestation and of infant deaths occurring through 28 postnatal days produces a consistent series of causes, virtually all of which begin *in utero.* Antenatal asphyxia and placental insufficiency account for 40 per cent of stillbirths and contribute to 4 per cent of neonatal deaths. Some 10 to 12 per cent of perinatal deaths are attributable to malformations, and a similar percentage, to erythroblastosis fetalis; two thirds of the mortality ascribed to each of these two causes occurs prior to birth. Infections account for 8 to 10 per cent of perinatal deaths, the majority beginning *in utero.* Pulmonary hyaline membrane disease remains the principal neonatal killer – being the cause of 25 to 30 per cent of newborn deaths. Ten to 15 per cent of neonatal mortality involves tiny infants (less than 1000 grams at birth), whose prompt demise is attributed to "immaturity," in the absence of an anatomic explanation of their deaths. One sixth to one fifth of stillbirths are unexplained; in most such instances, fetal death precedes birth by days or weeks, the circumstances cannot be documented and tissue examinations are often uninformative. Thus, whereas 25 to 30 of every 1000 births do not yield a surviving infant, nearly all the perinatal deaths can be attributed to deleterious processes occurring before or as a direct consequence of birth.

The anachronistic habits of attributing unexplained fetal deaths to "asphyxia," of uncritically ascribing many stillbirths to prematurity, and of failing to carry out appropriate studies of the deceased and its afterbirth should not be tolerated. Acceptance of reproductive failure without thoughtful inquiry as to the cause partakes of the same nihilism as often attends and obstructs the evaluation of "crib deaths." The opportunity is then lost to identify causes and contributory factors, to allay parental anxiety and guilt, and to counsel families relative to future child-bearing.

Prematurity

As noted by Little in 1862, and by many others since then, premature birth is a major precursor of cerebral palsy—the probability of neurological damage increasing with the degree of prematurity, as reflected by the birth weight.[17] The lower the birth weight, the greater the neonatal mortality, and the higher the risks of brain disorders and mental retardation among survivors. The impact of prematurity is intensified by an early birth that has followed an otherwise complicated pregnancy or abnormal labor.

The single, most critical determinant of neonatal survival is birth weight; and of birth weight, gestational age. A great many factors influence mean birth weight of infants of equal gestational age. For example, at the same stage of gestation males are heavier than females; whites, than other races; and singletons, than individual members of multiple births. Usually, the interdependence of gestational age and fetal weight is such that births at less than 37 weeks' gestation yield infants weighing less than 2500 grams. These infants are at greater risk of neonatal death than are heavier infants of similar gestational age, or, conversely, of infants of greater gestational age, but weighing less than 2500 grams. Currently, the lowest neonatal mortality rate in the United States is three per 1000 livebirths—enjoyed by white infants born to mothers between the ages of 25 to 29 years and having birth weights of 3501 to 4000 grams.[25]

"Prematurity," *i.e.*, birth at less than optimal gestational age, has numerous precursors, some of them causal, and is, itself, the antecedent of postnatal hardship and a majority of neonatal deaths. Premature labor may effect the expulsion of a damaged fetus from a hostile uterus, thereby concluding a troubled pregnancy. On the other hand, because of uncertainty relative to gestational age, the obstetrician may elect early delivery, perhaps by cesarean section, abruptly converting the healthy, intact fetus into a struggling, premature infant. The quality of life *in utero* and the circumstances of its interruption influence the adaptation of any newborn to extrauterine life. The adverse effects of premature birth may be aggravated by all that goes before and eased by intensive neonatal care.[13] If the antenatal milieu is prejudicial to fetal health, corrective measures should be sought—or the fetus should be removed to a more favorable environment. Efforts to delineate specific threats to fetal integrity, to arrest or prevent undesired premature labors, and to optimize both the timing and circumstances of delivery of the mature fetus, promise to improve immeasurably both perinatal welfare and outlook.

The costs of premature birth can be estimated in terms of lives lost, productivity unrealized, funds expended for care, and personal suffering of its victims and their families. Obviously, neonatal mortality is greatest among those infants born at the youngest gestational ages—their deaths being attributable to functional immaturity of the entire organism. Placentas associated with such births tend to be abnormal—the commonest lesions being inflammatory (i.e., chorioamnionitis). Many tiny infants are found to have aspirated or swallowed acute inflammatory exudate prior to their births. It is tempting to attribute death to birth and to link the latter to placentitis.

Toward the end of the second trimester, fetal maturation seems to have progressed to a stage permitting many neonates to survive. Thereafter, the most important "disease" is the "respiratory distress syndrome" (RDS), a disorder peculiar to prematures and rare among term infants. Most observers report that one sixth to one third of affected infants die. We have found pulmonary hyaline membrane disease at autopsy in about three fourths of infants alleged to have died with RDS.[8] Farrell estimated that RDS attacked 100,000 newborns in the United States in 1968; 10,000 of these died as neonates. Of the survivors, 10,000 to 30,000 suffered mental retardation.[10] The same report indicates that 400 million dollars was expended in the primary care of infants with RDS during that year. With the increased use of "intensive care," the associated mortality will decline, but at what cost and with what residua are open questions. Increasingly, the pathology of neonatal life is complicated by iatrogenic lesions, the risks of intensive care. The cost of systematic, detailed study of every premature birth—or at least of every premature birth followed by neonatal death—would not seem exorbitant if it were to uncover causes amenable to future prevention, without substantial risk to mother or fetus.

THE FUTURE

The predominant, recognized threats to fetal and neonatal survival—and to preservation of the integrity of the liveborn—are relatively few. Antenatal asphyxia and placental insufficiency, malformations, erythroblastosis, infections, and conditions peculiar to the prematurely born account for most perinatal deaths and near-deaths. A substantial portion of stillbirths are unexplained, most of them following antepartum fetal death, with ensuing maceration. Retrospective, systematic study of all deaths should be undertaken, to seek out epidemiological associations and valid clinical correlations. Such reviews reduce the unexplained deaths to a small percentage of the whole and permit the concentration of special resources on the monitoring and care of the identifiable "high risk" gravida. In the British Perinatal Mortality Survey, a group of experienced obstetricians, working together as a team, found adequate explanation for perinatal deaths in all except less than 1 per cent of the cases, whereas the pathologist, working alone, left 17 per cent unexplained. The effort to reach a diagnosis should always be made, with tissue examination, its keystone. A review of clinical circumstances should provide necessary support for the final synthesis in every case.

Overall, the future promises an improved outlook, if the quality of life of those born can be brought close to its potential at the time of conception—that is, if prenatal and perinatal environmental injury can be minimized. Optimizing conditions at conception and protecting the conceptus thereafter are laudable, although incompletely attainable, objectives. Others have dealt with the complexities of genetic counseling, antenatal screening for birth defects, and elective, selective abortion[12, 21, 23] (see Chapters 3–9).

Prevention of damage depends on identification and elimination of environmental hazards. MacMahon and Sowa[19] summarized the evidence that mental disorders may be related to factors operative during the antecedent pregnancy or delivery, many of which are now avoidable within developed societies and their health care systems.

Direct Physical Damage

Except for accidental trauma, an infrequent cause of prenatal and perinatal injury and/or death, physical injury to the fetus is iatrogenic. Modern obstetrical skills can circumvent the risks of prolonged and difficult labors and traumatic deliveries. Direct injury to the central nervous system, with or without associated fractures, rarely occurs under present standards of care to the parturient. Exposure of the human conceptus to ionizing radiation, incidental to diagnostic study or therapy of maternal disease, should be permitted only after careful consideration of the alternatives.

Asphyxia and Anoxia

Clearly, the mode of fetal death following abruptio placentae and compression of the umbilical cord is asphyxial. The *coup de grâce* in lethal placental insufficiency is also asphyxia. Sublethal asphyxia, on these same bases, may damage the surviving fetus irreversibly. When the risks are recognized prior to fetal damage, preventive measures can often be elected. All too often, the asphyxia—causing death or near death—occurs in silence, out of reach of clinical, biochemical, and electronic surveillance.

Increased utilization of health care facilities and access to effective modalities of monitoring will improve the outlook for "high risk" pregnancies. Better means of identifying patients at risk are needed, to broaden this category to include those not recognized as being in jeopardy by methods presently in use. According to Baird and Thomson (in the British Perinatal Mortality Study),[10] antepartum hemorrhage was of uncertain origin in 60 per cent of the instances of its occurrence. Since neither the gravida nor the fetus is manifestly exposed to increased risks prior to such hemorrhage, there are seldom grounds for decisive intervention. Both the epidemiologist and the physiologist are challenged by the current dearth of relevant information bearing on this problem.

A systematic documentation of the epidemiological characteristics of pregnancies at increased risk of placental abruption could suggest avenues for future physiological and clinical evaluation. Furthermore, the detailed study of both normal and aberrant uteroplacental relationships may be enlightening. The placenta develops within the decidua, is anchored to this tissue, and functions there until parturition. Here is the interface of mother and conceptus—a site of necessary vascular and immune tranquility, if gestation is to proceed to a successful outcome. Damage to the decidua or interference with uterine circulation disrupts these relationships, thus jeopardizing fetal welfare. Essential decidual functions may prove to be more diverse and complex than those of perfusion of the placenta during gestation and its ready detachment during the third stage of labor. Protection of the decidua from injury may prevent placental lesions deleterious to the fetus, including placental hypoplasia, infarction, and premature separation. More information is needed concerning both the maintenance of the decidua and its physiological significance during gestation.

Placental Insufficiency

This rubric is often applied when circumstances do not permit a more specific diagnosis to be made, or after exclusion of other possibilities. As measures of placental function have become available for clinical use, monitoring and documentation of placental failure are now possible in many instances. Retrospectively, with the extensively infarcted placenta in hand, insufficiency can be inferred after delivery. Frequently, if the pregnancy is recognized to be "at risk," deteriorating placental function signals the need for intervention. Clinical controversy surrounding the issue of "postmaturity" may be resolved through systematic monitoring of placental functions and fetal status when pregnancy extends beyond term. Whatever the management, assessment of the status of the progeny beyond the immediate neonatal period is imperative.

Erythroblastosis Fetalis (Isoimmune Hemolytic Anemia)

The greatest progress in prevention and amelioration of perinatal damage of specific etiology has been made with reference to erythroblastosis fetalis. A common disease, isoimmune hemolytic anemia has been a killer of the fetus and neonate and a major cause of brain damage when sustained without lethal outcome. The first fetal disorder to yield to direct antenatal therapy, this is also a major perinatal disease for which effective prophylactic measures have been devised.

With widespread utilization of anti-Rh antibody to prevent primary sensitization of Rh-negative women, fewer fetuses will be exposed to the ravages of this disease. Women already sensitized, who challenge present means of early fetal treatment, will steadily decrease in number among the population of reproductive age. Means should be found to prevent or mitigate the deleterious effects of ABO incompatibility, anti-Kell antibodies of maternal origin, and other consequences of sensitization on the newborn. Erythroblastosis fetalis might then become an extinct disease among populations enjoying access to health care. It is likely that other, rarer, fetal/neonatal disorders—not overtly affecting the erythrocyte—will also be traced to fetal-maternal incompatibilities.

Malformations

Progress in understanding the primary bases of human maldevelopment has been slow and steady, recently engaging diverse resources, from epidemiology to bioengineering to molecular biology. The majority of malformations are still unanticipated and, in retrospect, unexplained. The promulgation and systematic application of new knowledge, however, promise to bring human teratology into a closer relationship with clinical medicine. Objective, detailed, and standardized recording of all malformations is essential, for both epidemiological purposes and counseling as to probable etiology. Thorough post-mortem documentation of such lesions in all instances of reproductive wastage and perinatal deaths is an obvious necessity if epidemiology and etiology are to be clarified.

Syndromes of infectious etiology, conveniently dubbed TORCH* syndromes, should yield to serological testing and vaccinating of susceptibles prior to risking pregnancy.[20] Scrupulous avoidance of unnecessary exposure of the pregnant, or possibly pregnant, woman to ionizing radiation and drugs will reduce these hazards to the unborn.

At the present time, lesions of genetic-chromosomal etiology can be prevented only by selective breeding – or by elective interruption of pregnancy. Less drastic means may be found in the future. Amelioration of the impact of hereditary disorders may become possible through manipulation of the fetal microenvironment. For example, specific dietary or drug therapy given to the gravida may protect the fetal brain, or antenatal surgery may arrest the damage that is secondary to lesser degrees of neural tube defects.

Unfortunately, the detection of embryopathies and fetal malformations prior to delivery is often impossible, or the means are impractical. Again, epidemiology and antecedent reproductive history predict some instances of maldevelopment, thus providing the opportunity for special surveillance during gestation. In addition, aberrant uterine "growth," indicating fetal growth retardation, oligo- or polyhydramnios, sometimes raises the suspicion of maldevelopment. The advent of successful fetal surgery may permit specific early diagnoses to be followed by corrective procedures, without removal from the uterus.

Infections

Most serious perinatal infections arise *in utero;* if they attack neonatally, they affect infants already burdened by other problems, notably prematurity, "birth defects," or asphyxia. Occasionally, a major neonatal infection seems to have been contracted on exposure, during delivery, to the flora of the lower genital tract.

Beyond the obvious and present objectives of prevention or optimal management of maternal infections, future progress will depend on control of transamniotic infections. Clearly, many such infections are associated with prolonged or premature rupture of membranes – recognized hazards to both mother and fetus. Choice and timing of antimicrobial therapy are keys to successful management in these cases. Interruption of the pregnancy is often elected to protect the mother, but at great risk of prematurity to the offspring. Identification of the underlying cause of membrane rupture is critical to an understanding of similar transamniotic infections with intact membranes. Does inflammation precede (and cause) premature rupture of membranes? The succeeding section will show that acute chorioamnionitis is a frequent companion of premature birth, a harbinger of perinatal mortality, as well as a key component of transamniotic fetal infections.

---

*An acronym for Toxoplasma gondii, Rubella, Cytomegalovirus, and Herpes Simplex.

Identification of the flora resident in the cervix and vagina of a gravid woman might alert attending physicians to the risks of infectious complications during a particular pregnancy. Once again, antecedent events—including past reproductive casualities—may provide the incentives for such studies and their subsequent interpretation.

Two disparate types of microorganisms, both common in the lower genital tracts of women, appear to be perpetrators of occasional intrauterine infections of clinical significance. *Candida albicans,* so frequently associated with vaginitis, is infrequently the *bona fide* pathogen, causing deciduochorioamnionitis and fetal infections. Mycoplasmas, both *M. hominis* and T strains, have been implicated in sporadic infections of the female genitalia and, occasionally, in spontaneous abortions.[15] Their ubiquity impedes efforts to clarify their role, if any, in human infertility, reproductive wastage, and premature births.[14] Recent studies document the positive association of these organisms in the maternal vagina with acute chorioamnionitis of the delivered placenta.[24] Confusion and controversy regarding these putative pathogens will be dispelled only by systematic studies of the clinical correlates of "colonization."

Prematurity and Low Birth Weight

Abramowicz and Kass, in an exhaustive, but uncritical, review of prematurity as a major public health problem, cited voluminous evidence on the relationship of a host of factors to low birth weight.[1] Clifford attempted to link both demographic attributes and complications of pregnancy to gestational wastage, with particular emphasis on prematurity and its sequelae.[7] Obviously, neonatal mortality will be lowered by any measure that increases survival of neonates of low birth weight or reduces the frequency of births of such infants. The former approach is attaining considerable success through the development of intensive care nurseries and regionalization to expedite access to such resources.[18]

Only a minority of premature births are clearly explicable on the basis of elective early delivery, cervical incompetence, or spontaneous early labor in response to polyhydramnios or multiple gestation. Other premature labors are attributed to uterine malformations, follow uterine instrumentation, or are initiated by intrauterine hemorrhage or infection. A few are iatrogenic, resulting from intervention in the face of uncertain fetal age—a dilemma which can almost always be avoided by reference to new objective tests of fetal maturity. However, most prematurity and many "small-for-dates" births are unexplained.

The incidence of neonates of low birth weight might be reduced if its causes were environmental, i.e., resulting from circumstances amenable to remedial manipulations during the pregnancy-at-jeopardy. Bergner and Susser suggest that poor maternal nutrition fulfills these criteria.[4] Acute chorioamnionitis is a frequent companion of low birth weight and a precursor of neonatal death, especially among the very smallest newborns. Clearly, the etiology of chorioamnionitis is obviously environmental, onset preceding delivery by a brief period of time. Identification of the factor(s) initiating this process may permit preventive or therapeutic measures to be taken before the inexorable sequence of premature labor, early birth, and neonatal complications of immaturity supervenes.

FINAL COMMENT

Reproduction is a continuum that should not be divided conceptually into two periods separated by birth, since birth itself is part of this continuum. The same deleterious influences affect the conceptus prior to, during, and immediately following its birth. Reproductive wastage is also a continuum, manifested in various degrees of severity—from perinatal death to minor intellectual deficits. In view of the impact of antenatal asphyxia on the unborn, of low birth weight on the neonate, and the current paucity of knowledge relative to their prevention, these two circumstances warrant intensive study. Without such scrutiny, empiricism and supportive measures—including intensive neonatal care—will continue to predominate the management of major problems at the beginning of life.

References

1. Abramowicz, M., and Kass, E. H.: Pathogenesis and prognosis of prematurity. N. Engl. J. Med., 275:878–885, 1001, 1053, 1966.
2. Baird, D., and Thomson, A. M.: The effects of obstetric and environmental factors on perinatal mortality by clinicopathological causes. *In* Butler, N. R., and Alberman, E. D. (eds.): Perinatal Problems. Edinburgh, E. & S. Livingstone, Ltd., 1969, p. 223.
3. Barron, S. L.: The epidemiology of human pregnancy. Proc. R. Soc. Lond. [Biol.], 61:1200, 1968.
4. Bergner, L., and Susser, M. W.: Low birth weight and prenatal nutrition: An interpretative review. Pediatrics, 46:946, 1970.
5. Butler, N. R., and Alberman, E. D. (eds.): Perinatal Problems: The Second Report of the 1958 British Perinatal Mortality Survey. Edinburgh, E. & S. Livingstone, Ltd., 1969.
6. Churchill, J. A., Masland, R. L., Naylor, A. F., and Ashworth, M. R.: The etiology of cerebral palsy in pre-term infants. Dev. Med. Child Neurol., 16:143, 1974.
7. Clifford, S. H.: High-risk pregnancy. I. Prevention of prematurity the *sine qua non* for reduction in mental retardation and other neurological disorders. N. Engl. J. Med., 271:243, 1964.
8. Driscoll, S. G., and Yen, S. B.: Neonatal pulmonary hyaline membrane disease: Some pathologic and epidemiologic aspects. *In* Villee, C. B., and Zuckerman, J. (eds.): Respiratory Distress Syndrome. New York, Academic Press, Inc., 1973.
9. Eastman, N. J., and DeLeon, M.: Etiology of cerebral palsy. Am. J. Obstet. Gynecol., 69:950, 1955.
10. Farrell, P.: Epidemiology of the respiratory distress syndrome. Pediatr. Res., 8:452, 1974.
11. Harper, P. A., and Wiener, G.: Sequelae of low birth weight. Annu. Rev. Med., 16:405, 1965.

12. Hirschhorn, K.: The possible use of amniocentesis for monitoring. *In* Hook, E. B., Janerich, D. T., and Porter, I. H. (eds.): Monitoring, Birth Defects, and Environment. Academic Press, Inc., 1971, p. 25.
13. Intensive care of the newborn. Lancet, 1:969, 1974.
14. Klein, J. O., Buckland, D., and Finland, M.: Colonization of newborn infants by mycoplasmas. N. Engl. J. Med., 280:1025, 1969.
15. Kundsin, R. B., and Driscoll, S. G.: The role of mycoplasmas in human reproductive failure. Ann. N.Y. Acad. Sci., 174:794, 1970.
16. Lilienfeld, A. M., and Pasamanick, B.: Association of maternal and fetal factors with development of epilepsy; abnormalities in prenatal and paranatal periods. J.A.M.A., 155:719, 1954.
17. Little, W. J.: On the influence of abnormal parturition, difficult labours, premature birth, and asphyxia neonatorum, on the mental and physical condition of the child, especially in relation to deformities. *In* Obstet. Soc. Lond., Trans., 3:293, 1861. London, Longman, Green, Longman, and Roberts, 1862, 480 pp.
18. Lucey, J. F.: Why we should regionalize perinatal care. Pediatrics, 52:488, 1973.
19. MacMahon, B., and Sowa, J. M.: Physical damage to the fetus. The Milbank Mem. Fund Q., 39:14, 1961.
20. Nahmias, A. G.: The TORCH complex. Hosp. Practice, 9:65, 1974.
21. Nelson, T., Oakley, G. P., Jr., and Shepard, T. H.: Classification and tabulation of conceptual wastage with observations on type of malformation, sex ratio, and chromosome studies: II. Methods. *In* Hook, E. B., Janerich, D. T., and Porter, I. H. (eds.): Monitoring, Birth Defects, and Environment. New York, Academic Press, Inc., 1971, p. 45.
22. Rubin, R. A., Rosenblatt, C., and Balow, B.: Psychological and educational sequelae of prematurity. Pediatrics, 52:352, 1973.
23. Shepard, T. H., Nelson, T., Oakley, G. P., Jr., and Lemire, R. J.: Collection of human embryos and fetuses: A centralized laboratory for collection of human embryos and fetuses: seven years experience: I. Methods. *In* Hook, E. B., Janerich, D. T., and Porter, I. H. (eds.): Monitoring, Birth Defects, and Environment. New York, Academic Press, Inc., 1971, p. 29.
24. Shurin, P. A., Alpert, S., Rosner, B., Driscoll, S. G., and Kass, E. H.: Genital mycoplasmas—association with chorioamnionitis. (In press, 1974.)
25. A Study of Infant Mortality from Linked Records: Comparison of Neonatal Mortality from Two Cohort Studies: U.S. Department of HEW, Vital and Health Statistics System, Series 20, No. 13, 1972.
26. A Study of Infant Mortality from Linked Records: By Birth Weight, Period of Gestation, and Other Variables. U.S. Department of HEW, Vital and Health Statistics System, Series 20, No. 12, 1972.
27. Wiener, G.: The relationship of birth weight and length of gestation to intellectual development at ages 8 to 10 years. J. Pediatr., 76:694, 1970.

CHAPTER 15

THE PREVENTION
OF LEAD POISONING

John W. Graef, M.D.

INTRODUCTION AND CURRENT STATUS

Historical Perspectives

Lead poisoning in man has been recognized since the time of the early Greeks. Moreover, the Romans used lead so extensively that more than one authority has speculated about its role in weakening their civilization. Many of the Roman aqueducts were lined with lead-containing metals. Cooking ware, food storage containers, and eating utensils were either coated with lead glazes or made of pewter. Acidic foodstuffs (wine, cider) or water stored for long periods undoubtedly increased the lead exposure of that society beyond safe limits for many citizens.

During the Dark Ages, following the fall of Rome, lead was less widely used but re-emerged in central Europe after the ninth century A.D., where it was used for many purposes. The practice of altering wine with lead became so common that it was outlawed in Germany in the fifteenth century[1]—to little avail. Industrial and other uses of lead increased, and lead colic began to be widely seen in the eighteenth century.

Until the early portion of the twentieth century, lead poisoning was thought to be a condition with acute, painful symptoms, often simulating other diseases. Neurological impairment was recognized, particularly in the peripheral nerves, but it was not considered as a sequel.[2] The disease was not frequently diagnosed among children except as an involvement of a family in an endemic or among infants as a result of the use of lead nipple shields by wet nurses. Its abortifacient effects were noted by several workers in the latter part of the nineteenth century.[3-5] There was an association not only with increased stillbirth but also with malformations noted in infants born of mothers with high lead exposure.

354

During this same period, a newer use of lead as a color-stabilizing and weather-resistant ingredient of paint became widespread. In acid or coastal climates, surfaces painted with leaded paint eventually began to crack and peel. Disturbing reports of lead poisoning resulting from ingestion of lead-laden chipped paint began to appear in the early 1920s. In 1925 Aub described the effects of lead on children,[6] and Ruddock[7] recognized the now widely accepted association between lead poisoning and the phenomenon of pica, the ingestion of nonfood substances in small children. Pica (from the Greek word for "magpie") remains a controversial and poorly understood problem in children who are continuously exposed to lead.

As the phenomenon of pica for lead paint chips became more widely recognized, the neurological sequelae of lead encephalopathy still remained relatively unnoted until the 1930s and 1940s when a number of workers began to point out this association between lead paint ingestion and conditions ranging from hydrocephalus to poor performance in school. It is important to note that until these observations were made, lead poisoning in children was generally considered an acute disease.[8-10] Other sources of exposure included lead vapor from the burning of storage battery casings and ingestion of moonshine whiskey.

After the flurry of reports of poisoning due to lead paint ingestion, restrictions were placed on the use of lead in interior paints, and there was a widespread relaxation of concern, particularly with the advent of titanium as a replacement for lead in paint.

A phenomenon then occurred that remains a tragic and continuing reminder that diseases of children do not occur out of context. What was forgotten was that lead in dangerously high concentrations had already saturated the plaster and walls of millions of homes in America, particularly in the central cities. With improved highways and the general affluence of the post-war period, many of these homes were abandoned to lower socioeconomic groups for whom adequate maintenance was financially difficult. Inadequate sanitary code enforcement, housing inspection, and maintenance, combined with large families under stress and small unsupervised children with or without pica, produced the national disaster reflected in the current estimates of 400,000 cases of childhood lead poisoning. In the United States alone there are 150 to 200 deaths per year related to this cause.[11, 12]

A far greater number of children are permanently damaged. According to some authors,[13-15] as high as 40 per cent of children with symptomatic lead intoxication suffered irreversible neurological sequelae. These include fine and gross motor defects, hyperactivity, dyslexia, memory loss, and developmental defects. Among retarded children lead poisoning has been observed as a consequence of the more pronounced pica seen in such children and the environmental hazard of lead paint coating the walls of older institutions.[16]

These sequelae have been seen most frequently following overt encephalopathy. Byers[14] in his review of 45 cases found 9 of 17 children with encephalopathy on diagnosis to have suffered permanent damage. Four died and at least three of the nine suffered new and irreversible loss of mental function. Overall, follow-up was obtained on 40 of the 45 patients, and 19 of those, including some without gross encephalopathy, suffered persistent psychological irregularities, became overtly and permanently retarded, or died.

These figures are not dissimilar from those of Perlstein and Attala,[13] who reviewed 425 cases of plumbism in Chicago over a 10-year period. Of these cases, 93 exhibited mental retardation, 85 suffered recurrent seizures, 9 demonstrated cerebral palsy, and 5 showed optic atrophy. Since some children suffered more than one sequel, the total found to have permanent damage was 39 per cent. With regard to mode of onset, those with encephalopathy on diagnosis had the greatest frequency and severity of sequelae (80 per cent). However, even among those with mild or no presenting symptoms of plumbism the incidence of mental retardation by gross measurement was appreciable (9 per cent). Unfortunately, in this as in other retrospective studies, the criteria for diagnosis were sufficiently broad and the measure of mental retardation sufficiently general to include some children for whom lead poisoning may not have been the etiological factor. Nonetheless, similar rates of sequelae have been found in other studies.[48, 54]

Table 15–1 illustrates the currently recognized effects of lead, in both symptomatic and asymptomatic children, based on the current U.S. Public Health Service guidelines and available literature. The factors that contribute to the development of symptoms remain obscure, although it is clear that blood level alone does not necessarily correlate well with intoxication except when it is very high or very low.

However, some general associations have been observed. Because lead affects rapidly growing tissues, children are most susceptible in infancy and immediately thereafter. Indeed, Scanlon has shown increased cord blood lead in infants of mothers in high lead exposure areas[25] – a disturbing finding of as yet unknown significance. Iron deficiency has been shown to increase susceptibility to intoxication in experimental animals,[17] and malnutrition, particularly dietary calcium deficiency, may also contribute.[18]

A seasonal incidence of symptoms has long been observed, with the largest number of cases seen during the summer months. Although not understood, this phenomenon has been generally attributed to vitamin D mediated alterations in calcium binding and transport and to the association of lead with calcium metabolism.[18] Many observers suspect, however, that increased incidence in summer may be due to enhanced exposure to soil and exterior paint containing lead as well as to greater family mobility during summer months.

TABLE 15–1. CLINICAL AND LABORATORY EVIDENCE OF LEAD
INTOXICATION AND ASYMPTOMATIC LEAD BURDEN*

Lead Intoxication	Asymptomatic Increased Lead Burden
Clinical	Clinical
Anorexia, constipation, irritability, clumsiness, lethargy, behavior changes, hyperactivity (sequela), abdominal pain, vomiting, fever, hepatosplenomegaly, ataxia, convulsions, coma with increased CSF pressure	History of pica Environmental lead source Positive family history
Laboratory	Laboratory
Microcytic, hypochromic anemia Basophilic erythrocyte stippling Increased δ-ALA in serum and urine Decreased δ-ALA dehydratase in erythrocytes Increased urinary coproporphyrin Increased erythrocyte protoporphyrin Decreased osmotic fragility Increased metaphyseal densities on x ray Aminoaciduria, glucosuria	Blood lead greater than 40 μg/100 ml Hair lead greater than 100 μg/100 ml in proximal segment 24-hr urinary lead excretion greater than 80–100 mg Lead mobilization test greater than 1 μg/mg EDTA ingested/24 hr Radiopacity in gastrointestinal tract on x ray

*From Graef, J. W., and Cone, T. E., Jr. (eds.): Manual of Pediatric Therapeutics. Boston, Little, Brown and Company, 1974. Reprinted with permission.

Although there is general agreement that lead encephalopathy results in permanent damage in most cases, there is considerable controversy surrounding the effect of asymptomatic lead burden.[19-22] This concept has evolved to account for the large number of children with increased absorption of lead in whom no detectable effect of lead can be measured. Such children generally have elevated concentrations of lead in their blood but are not anemic, have no radiological evidence of lead, and have no demonstrable symptoms. Because of the variability with which symptoms of lead poisoning occur, it remains unclear as to which of these children are at greatest risk for neurological damage.

Nevertheless, evidence for damage in asymptomatic lead burden continues to mount. Pueschel,[21] de la Burde,[19] David,[49] and others have shown neurological irregularities in chronic, low-lead–exposed children without overt symptoms. Silbergeld's[23] studies of lead-associated hyperactivity in mice and the demonstration by others of slow learning in sheep with borderline blood lead levels are suggestive of animal models with practical application to the human disease. Certainty is difficult to achieve because of inadequate criteria for diagnosis and insufficient tools for measurement of minor psychological irregularities. Studies in chronic low lead exposure are increasing as attention is focused on establishing the importance of this problem.

TABLE 15–2. LEAD IN ENVIRONMENT

Lead-based paint, interior and exterior
Lead-laden dust and soil
Ceramic glazes, pewter
Lead in foodstuffs—evaporated milk, drinking water
Airborne lead
Industrial lead vapors
Lead in inks, newsprint, and other sources

Current Environmental Sources

Table 15–2 shows primary sources of lead in the environment. Lead-based paint remains the major source of exposure in young children. At the present time, Massachusetts is the only state to have removed from the market exterior as well as interior lead-based paint. Because many homes contain not one, but several, layers of lead-based paint, it is not uncommon to recover paint chips containing as much lead as 40 per cent from surfaces such as back porches, windowsills, hallways, and banisters, all of which are readily accessible to a child.

The absorption range has been shown to be approximately 10 per cent in adults[26] and as high as 53 per cent in young children.[24] This helps to account for the higher degree of susceptibility to lead poisoning in young children. Certainly, the high absorption figure must be viewed with alarm in the calculation of acceptable environmental lead concentrations for the younger age group. Under conditions of absorption as high as 50 per cent, ingestion of one gram of paint containing 40 per cent lead could produce lethal concentrations of blood lead in a typical 2-year-old child. That potentially fatal ingestions of lead do not happen more often than they do is a current area of investigation.

Among other sources of lead exposure in young children, airborne lead and lead in foodstuffs are probably most important. Airborne lead either from automobile exhaust or from smelters apparently contaminates soil and house dust and is ingested, rather than inhaled, during the normal hand-to-mouth activity of small children.

Lead in foodstuffs apparently is derived from faulty canning techniques that expose the contents of the can to lead through the seams. Concentrations well above 100 μg/liter have been observed. In one study, the mean lead level of canned evaporated milk was 202 μg/liter.[27] Such levels could produce an increased lead burden in a growing infant in a matter of weeks.

The presence of lead in ceramic glazes and in paint on toys depends to some extent on the country of origin. In the United States, it is illegal to use lead paint on toys; ceramics for commercial use must be fired at temperatures greater than 1800°F or lead glazes may not be used. However, such controls do not exist elsewhere, and sporadic cases of poisoning due to these sources are reported. Of passing interest was the recent report of excessive lead in the paint coating of wooden pencils—information suggesting a widely unrecognized but common source of exposure.[28]

Drinking water continues to be reported as a source of lead poisoning in communities with lead-lined pipes or water storage facilities.[29]

Current Detection and Screening Programs

In 1970 the U.S. Congress passed the Lead Poison Prevention Act which provided funds for screening of children and inspection of housing. The funds were given to individual municipalities to establish laboratory facilities and outreach capability. Provision was not included for housing repair, medical costs, or research. Through the use of these and other locally derived funds, a number of cities and smaller communities in the United States have established screening and housing inspection programs.

Screening of Children

Although blood lead determination has been used in the majority of these programs, other indirect tests of lead intoxication, such as urinary δ-amino levulinic acid or hair lead, have been tried.[30, 31] Although blood lead remains the most reliable of screening procedures, the advent of the free erythrocyte protoporphyrin (FEP) assay shows promise in identifying children with intoxication quickly and simply.[32] Its correlation with asymptomatic increased lead burden has not yet been confirmed. Both blood lead and FEP determinations can be done on capillary blood by use of 10 to 50 λ per determination. Blood lead has the disadvantage of being susceptible to contamination from soil on the skin of the patient unless the skin is carefully cleaned or blood is obtained by venipuncture. The latter procedure, in small children, is difficult enough to have effectively frustrated attempts at mass screening until micromethods and finger-prick became available.

Results of mass screening have been remarkably consistent across the country, with a few notable exceptions. In general, the incidence of children with confirmed blood leads greater than 40 μg% whole blood is highest in the inner city, especially among Black children.[33, 34] Table 15–3 shows the results of a survey of 30 communities by the Center for Disease Control.[33] Some "lead belt" environments produce elevated blood lead levels in as high as 15 to 25 per cent of children under 6 years of age.[11] Surrounding "bedroom" suburbs generally show much lower levels, of 1 to 2 per cent of the susceptible population. An exception to this pattern is Pittsburgh, Pennsylvania where, despite extensive

TABLE 15–3. ELEVATED BLOOD LEAD RATES BY AGE AND BLACK AND NONBLACK GROUPINGS*

	<3 Yrs. Old		≥3 Yrs. Old		Total	
	No./Total	%	No./Total	%	No./Total	%
Black	34/373	9.1	50/732	6.8	84/1105	7.6
Nonblack	19/508	3.7	16/656	2.4	35/1164	3.0
Total	53/881	6.0	66/1388	4.8	119/2269	5.2

*From Anderson, D. G., and Clark, J. L.: Neighborhood screening in communities throughout the nation for children with elevated blood lead levels. Environ. Health Perspect., 7:3, 1974.

screening, lead poisoning appears to be virtually nonexistent, even in the core city.[35] This phenomenon has not been explained. Exceptions are also found in modern communities where leaded paint has not been used extensively.

Screening of Housing

Until the advent of portable x ray fluorescence, housing inspection required the chipping of paint from a surface and the analysis of the chips in the laboratory by atomic spectrophotometry or other methods. This procedure necessarily reduced the speed and the number of dwellings inspected. Now, a single inspector using a portable x ray device can inspect as many as 15 individual units per day without damaging intact surfaces. In large eastern U.S. industrial cities, 50 to 80 per cent of dwellings have interior surfaces containing potentially toxic levels of lead. Even in homes where interior paint does not contain lead, exterior paint has been used on windowsills and porches where small children have ready access. In addition, there is increasing documentation of elevated levels of lead in soil, particularly surrounding highways, near the exterior surfaces of frame housing, and in communities with lead smelters. How much this soil contributes to the elevated burden of a given child is not known, but there are many children for whom it appears to be a likely source.

Finally, the development of a technique for analysis of lead in dentine by Needleman et al.[36] suggests that determination of blood lead alone is indeed missing many children with elevated lead burden. Mean tooth lead levels in lead poisoned children were seven times those of children from suburban Boston, but those of suburban Boston were two and one-half times those of Icelandic children.[34] These findings suggest a previously undetected increased burden in the suburban children.

In summary, current detection and screening have revealed a very large number of children with high environmental lead exposure during their early formative years and an enormous number of dwellings containing dangerously high concentrations of lead on interior surfaces.

PRINCIPLES FOR DETECTION AND SCREENING

Medical

There is general agreement that with increased screening, parents and children become more aware of the dangers associated with lead ingestion. Sachs, for example, reported a decrease in cases of elevated lead levels in an area of Chicago that has been screened annually since 1966.[37] The occurrence of lead encephalopathy also appears to have decreased, although no comparable figures exist.

Therefore, the problem of screening and detection continues to fall into the area of the asymptomatic, but "at risk," child. Enough is now known about childhood lead poisoning to develop some priorities for testing.

The earliest measurable effect of lead appears to be its capacity to suppress enzyme activity by blocking sulfhydryl groups. This effect is also characteristic of other heavy metals, but in addition lead interferes specifically with heme synthesis.[38] Hence, decreased δ-amino levulinic acid (ALA) dehydrase activity can be measured in the absence of anemia, a fairly late finding in lead poisoning.[39] Lead also interferes with the ATP-ase system of the erythrocyte, causing a pronounced potassium leak with a subsequent resistance to osmotic lysis as the cell shrinks.[40] Whether this effect precedes decreased ALA dehydrase activity is not known, but it appears to be an early effect as well. Both effects correlate fairly well with blood lead levels but not necessarily with total body burden. Total body lead is made up primarily of relatively tightly bound lead salts stored in bone, and these are usually chemically inactive under normal conditions.[41]

Other steps of the heme synthesis pathway are affected, and the net result is an increase in free erythrocyte protoporphyrin, thereby reflecting an interference in the uptake of iron into the heme molecule. This increase can be measured with a fluorimeter, since erythrocytes with excess protoporphyrin fluoresce and can be quantitated.[42] This test appears to be as accurate as the ALA dehydrase assay and is so simple that a portable device can provide an accurate determination from capillary blood within minutes.

The precipitation of nucleoprotein on the red cell surface produces a pattern of basophilic stippling of the erythrocytes, easily seen on a simple smear. This may be seen in the absence of anemia.

Other screening tests include the hair lead test based on the principle that hair follicles actively excrete lead and that excess intake is reflected in differential patterns of lead concentration along the length of the strands.[31] Although the tissue is easily accessible and the test correlates well with increased burden,[43] it is a difficult determination and has not seen wide use.

Measurements of serum and urinary ALA have also been suggested, but even though the serum ALA is elevated in intoxication, it is not as sensitive an indicator of lead effect as the FEP or direct assay of ALA dehydrase activity.[44] Urinary ALA unfortunately does not accurately reflect ALA synthesis unless a 24-hour collection is obtained,[45] and this is impractical as a screening test.

Radiographs, although specific for acute or chronic lead ingestion, are expensive, require interpretation, and are generally reserved for confirmation of suspicious cases.

It has been suggested that blood lead, FEP, or a combination of the two may be the most effective way of finding children with increased burden and/or intoxication. Both these determinations can be done on capillary blood from a single finger-prick.

Confirmation of the diagnosis requires a complete history and physical examination, blood count, and x rays of the long bones, at least. In our hands, the lead mobilization test (Table 15–4)[46] is also useful, based on the principle that the excretion of lead in urine is enhanced in the presence of the chelating agent ethylenediaminotetraacetate (EDTA) if the body burden of lead is elevated.[47] A positive test is greater than 1 μg lead excreted per mg EDTA injected per 24 hours. This test, however, is indicative only of elevated burden, and not necessarily of intoxication. The decision to treat a given child must depend on a weighing of all the known parameters, including his environment.

Environmental

The factors of run-down housing, lead-based paint, small children, and a stressful family life all combine to produce childhood lead poisoning. It has been suggested that all houses be screened to determine the presence and concentration of lead in the walls. Although this is desirable, it is at the present time a formidable task, given limited manpower and resources. Therefore, priorities must be established in order to reduce the morbidity and the neurological sequelae and ultimately to eliminate the hazard.

Certainly, it appears to be a reasonable approach that any home with a child having an increased burden of lead be inspected and that repairs be initiated immediately. Satisfactory repairs and consequent protection of the child can frequently be accomplished by use of volunteer labor; wallboard or fiberboard may be used to cover over exposed surfaces. Steaming or sanding lead paint to the original wood has signifi-

TABLE 15–4. INDICATIONS FOR LEAD MOBILIZATION TEST*

1. Blood lead greater than 60 μg/100 ml whole blood
2. Blood lead 30–60 μg/100 ml whole blood *and*
   Hair lead greater than 100 μg/gm *and*
   a. Symptoms of plumbism *or*
   b. History of exposure (PICA) *or*
   c. Positive roentgenograms *or*
   d. Sibling with lead intoxication *or*
   e. Unexplained anemia and basophilic stippling *or*
   f. Positive urinary coproporphyrin or erythrocyte protoporphyrin *or*
   g. Positive urinary or serum delta ALA *or*
   h. Decreased erythrocyte delta ALA dehydratase *or*
   i. Aminoaciduria and/or glucosuria *or*
   j. Decreased osmotic fragility

*From Graef, J. W., and Cone, T. E., Jr. (eds.): Manual of Pediatric Therapeutics. Boston, Little, Brown and Company, 1974.

cant risks for the workers involved, but it is ultimately desirable if the paint is to be removed. Simply painting over lead paint has tragic consequences in children with pica, since, undeterred, they eat through the new layer. However, keeping flat wall surfaces in good repair to prevent flaking and chipping obviously reduces the hazard. Depending on the resources of the community, tax incentives to landlords can help stimulate voluntary compliance with sanitary code statutes, but stiff penalties may be required for noncompliance.

More important, and more difficult, is an overall reduction of lead in the environment. Industrial lead consumption has continued to increase, particularly for use in gasoline; in at least three major cities in the U.S. there have been significant increases in the levels of lead in air. Although the effect of inhalation of airborne lead has not been shown, the ingestion of lead-laden dust and soil by small active children has been shown to produce a chronic positive lead balance with the same effect as ingestion of lead paint.

CONCLUSIONS

As a man-made disease lead poisoning, particularly in children, continues to defy control. The combination of the ubiquitous presence of lead in the modern environment and its subtle toxicity at low levels of exposure requires that steps be taken diligently to reduce its potentially devastating effect on generations of children. Control measures are further complicated by the fact that principles of screening that are useful in the environment do not necessarily apply to the screening of children. From a number of different approaches some general guidelines have emerged:

1. Most urban areas in the U.S. have an identifiable high-risk

area or "lead belt" where pre-World War II housing has been permitted to deteriorate. Screening programs must establish these areas as a priority for the *repeated* screening of children between the ages of 1 and 6 years.

2.   All housing built prior to 1950 ultimately should be screened with an x ray fluorescence device. The lead content of the walls should be permanently recorded and the tenants *and future tenants* should be informed of the findings.

3.   Ultimately, *all* children between the ages of 1 and 6 years should be screened annually or more often, regardless of race, socioeconomic group, place of residence, and presence or absence of symptoms.

4.   Screening should be reasonably accurate, minimally invasive, available in quantity, simple to perform, possible to perform by portable devices, and inexpensive. Ideally, screening would entail a single microdetermination on capillary blood of lead (elevated burden) and free erythrocyte protoporphyrin (intoxication). When further management is required, the distinction between these two entities should be kept in mind.

5.   Adequate medical follow-up should be available. Nothing is more frustrating to a community than to initiate a screening program with no mechanism for care of the index population. All parents of children screened should be notified of the results of screening, whether they be positive or negative.

6.   Emergency measures must be instituted to reduce the lead hazard when an index case is found. Experience suggests that reinspection of a home after repairs have been initiated is prudent. No child with lead intoxication should be permitted to return to a lead-laden environment (sadly, this goal is often impractical).

7.   Any child with elevated lead levels, whether intoxicated or not, should be followed through school age. School authorities should be notified when a questionable exposure has occurred in a child's medical history.

8.   Any child with symptoms of lead intoxication should undergo a full diagnostic evaluation and should not be subjected to "screening" approaches that are designed for a different purpose.

9.   Whenever chelation therapy is contemplated, elevated excretion of lead should be demonstrated with a test dose of chelating agent before a full course of therapy is initiated, unless encephalopathy is present.

10.   Determination of lead is a laborious and exacting procedure requiring skill and practice. An inexperienced laboratory may create inaccuracies.

11.   Ideally, each community, in conjunction with its Board of Health, should establish a surveillance of its lead control program. The

medical community as well as the housing authorities should be subject to review of its efforts to reduce this hazard.

Efforts at education should include not only parents and health personnel but also legislators and jurists, since only with stiff penalties can adequate pressure be brought to bear on landlords and homeowners to repair housing.

12.   Housing in good repair, even with lead paint, is less hazardous than housing in run-down condition. Similarly, children who are in a loving environment, under adequate supervision, with adequate nutrition have reduced risk for lead poisoning.

13.   Where it is impractical to remove lead paint, covering with wallboard to a height of four feet, combined with repair of flaking and peeling areas, appears to offer satisfactory protection.

14.   Large sums of money are required on a national level to provide (a) emergency housing repair; (b) proper inspection, education, and surveillance; (c) adequate enforcement of existing and proposed environmental statutes to reduce airborne lead and permanently eliminate the sale of leaded paint; and (d) research support for studies to (i) screen less invasively and more accurately; (ii) neutralize lead in paint and other environmental sources; and (iii) develop newer, more efficient, and less costly therapy.

15.   Finally, adequate provision must be made for the victims of this disease. Special schooling and institutional care may be required, but it must also be recognized that, unlike other causes of mental retardation, this disease was caused not by nature, but by man.

THALLIUM, CADMIUM, ARSENIC, AND MERCURY

Like lead, other heavy metals, most notably thallium, cadmium, arsenic, and mercury, can inhibit enzymes and ultimately can produce damage to the central nervous system. Subacute mercury poisoning, acrodynia, is a well known entity, largely reversible when treated. It is not common in children, but it does produce marked behavioral changes if untreated. Chronic arsenic and thallium poisoning are also known to produce damage resulting in peripheral neuritis and late-onset dementia.[50] The inhalation of cadmium produces severe pulmonary symptoms, and its ingestion can produce gastroenteritis and renal disease. Neurological damage is not usually seen except in an isolated outbreak of a syndrome called itai-itai in Japan.[51]

The tragic experience, seen particularly in Japan, of severely malformed and retarded children poisoned by maternal ingestion, during pregnancy, of fish containing mercury caused widespread alarm recently.[52] However, fish diets have not been shown to have increased mercury levels except where mercury waste is washed into fish feeding grounds.

Treatment of these conditions is with chelating agents; symptoms are largely reversible. However, the association of the conditions with toxicity of other heavy metals is not well understood, particularly the suggestion that cadmium may enhance the toxicity of lead in some animals.[53]

In general, these heavy metals are found in industrial sources. Increases in ambient concentrations are found near smelters and mines where increased alertness by concerned individuals may lead to identification of high exposure rates otherwise undetected. As a cause of mental retardation in children, their effects may be more indirect than direct in enhancing toxicity of each other or of other heavy metals, since direct toxicity appears to produce more severe organic damage.

References

1. Waldron, H. A., and Stofen, D.: Sub-clinical lead poisoning. New York, Academic Press, Inc., 1974.
2. McKhann, C. F.: Lead poisoning in children: Cerebral manifestations. Arch. Neurol. Psychiat., 27:294, 1932.
3. Oliver, T.: Lead poisoning in its acute and chronic manifestations. Lancet, 1:644, 1891.
4. Hall, A.: The increasing use of lead as an abortifacient; a series of thirty cases of plumbism. Br. Med. J., 1:584, 1905.
5. Wynne, F. E.: Domestic hot water supplies as a factor in the production of lead poisoning. Br. Med. J., 2:267, 1911.
6. Aub, J. C., Fairhall, L. T., Minot, A. S. et al.: Lead poisoning. Medicine, 4:1, 1925.
7. Ruddock, J. C.: Lead poisoning in children, with special reference to pica. J.A.M.A., 82:1682, 1924.
8. Vogt, E. C.: Roentgenologic diagnosis of lead poisoning in infants and children. J.A.M.A., 98:125, 1932.
9. Levinson, A., and Zeldes, M.: Lead intoxication in children; study of 26 cases. Arch. Pediatr., 56:738, 1939.
10. Byers, R. K., and Lord, E. E.: Late effects of lead poisoning on mental development. Am. J. Dis. Child., 66:471, 1943.
11. Lin-Fu, J. S.: Childhood lead poisoning...an eradicable disease. Children, 17:2, 1970.
12. Chisholm, J. J.: Lead poisoning. Sci. Am., 224:15, 1971.
13. Perlstein, M. A., and Attala, R.: Neurologic sequelae of plumbism in children. Clin. Pediatr., 5:292, 1966.
14. Byers, R. K.: Lead poisoning: Review of the literature and report on 45 cases. Pediatrics, 23:585, 1959.
15. Berg, J. M., and Zappella, M.: Lead poisoning in childhood with particular reference to pica and mental sequelae. J. Ment. Defic. Res., 8:44, 1964.
16. Gibson, S. L. M., Lam, C. N., McCrae, W. M. et al.: Blood lead levels in normal and mentally deficient children. Arch. Dis. Child., 42:573, 1967.
17. Mahaffey, K. R., and Goyer, R. A.: The influence of iron deficiency on tissue content and toxicity of ingested lead in the rat. J. Lab. Clin. Med., 79:128, 1972.
18. Mahaffey, K. R., and Goyer, R. A.: Experimental enhancement of lead toxicity by low dietary calcium. J. Lab. Clin. Med., 76:933, 1970.

19. de la Burde, B., and Choate, M. S., Jr.: Does asymptomatic lead exposure in children have latent sequelae? J. Pediatr., 81:1088, 1972.
20. Kotok, D.: Development of children with elevated blood lead levels: A controlled study. J. Pediatr., 80:57, 1972.
21. Pueschel, S. M.: Neurological and psychomotor function in children with increased lead burden. Environ. Health Perspect., 7:13, 1974.
22. Needleman, H. L.: Lead poisoning in children: Neurologic implications of widespread subclinical intoxication. Semin. Psychiatry, 5:47, 1973.
23. Silbergeld, E. K., and Goldberg, A. M.: Hyperactivity: A lead-induced behavior disorder. Environ. Health Perspect., 7:227, 1974.
24. Alexander, F. W.: The uptake of lead by children in differing environments. Environ. Health Perspect., 7:155, 1974.
25. Scanlon, J.: Umbilical cord blood lead concentration. Am. J. Dis. Child., 121:271, 1971.
26. Kehoe, R. A.: Normal metabolism of lead. Arch. Environ. Health, 8:232, 1964.
27. Mitchell, D. G., and Aldous, K. M.: Lead contents of foodstuffs. Environ. Health Perspect., 7:59, 1974.
28. Pichirallo, J.: Lead poisoning: Risks for pencil chewers? Science, 173:509, 1971.
29. Goldberg, A.: Drinking water as a source of lead pollution. Environ. Health Perspect., 7:103, 1974.
30. Hankin, L., Hanson, K. R., Kornfeld, J. M. et al.: A dipstick test for the mass screening of children for lead poisoning based on urinary delta-aminolevulinic acid (ALA). Bull. Conn. Agricult. Exper. Station, New Haven, 716, 1970.
31. Kopito, L., Byers, R. K., and Shwachman, H.: Lead in hair of children with chronic lead poisoning. N. Engl. J. Med., 276:949, 1967.
32. Chisholm, J. J., Mellits, E. D., Keil, J. E. et al.: Variation in hematologic responses to increased lead absorption in young children. Environ. Health Perspect., 7:7, 1974.
33. Anderson, D. G., and Clark, J. L.: Neighborhood screening in communities throughout the nation for children with elevated blood lead levels. Environ. Health Perspect., 7:3, 1974.
34. Needleman, H. L., and Shapiro, I. M.: Dentine lead levels in asymptomatic Philadelphia school children: Subclinical exposure in high- and low-risk groups. Environ. Health Perspect., 7:27, 1974.
35. Moriarity, R.: Personal communication.
36. Needleman, H. L., Tuncay, O. C., and Shapiro, I. M.: Lead levels in deciduous teeth of urban and suburban American children. Nature (Lond.), 235:111, 1972.
37. Sachs, H. K.: Effect of a screening program on changing patterns of lead poisoning. Environ. Health Perspect., 7:41, 1974.
38. Gibson, K. D., Neuberger, A., and Scott, J. J.: The purification and properties of δ-aminolevulinic acid dehydrase. J. Biochem., 61:618, 1955.
39. Hernberg, S., Nikkanen, J., Mellin, G. et al.: δ-Aminolevulinic acid dehydrase as a measure of lead exposure. Arch. Environ. Health, 21:140, 1970.
40. Vincent, P. C., and Blackburn, C. R. B.: The effects of heavy metal ions on the human erythrocyte. I. Comparisons of the action of several heavy metals. Aust. J. Exper. Biol., 36:471, 1958.
41. Chisholm, J. J.: Chelation therapy in children with subclinical plumbism. Pediatrics, 53:441, 1974.
42. Piomelli, S., Davidow, B., Guinee, V. F. et al.: The FEP (free erythrocyte porphyrins) test: A screening micromethod for lead poisoning. Pediatrics, 51:254, 1973.
43. Pueschel, S. M., Kopito, L., and Shwachman, H.: Children with an increased lead burden. A screening and follow-up study. J.A.M.A., 232:462, 1972.
44. Baloh, R. W.: Laboratory diagnosis of increased lead absorption. Arch. Environ. Health, 28:198, 1974.
45. Barnes, J. R., Smith, P. E., and Drummond, C. M.: Urine osmolality and aminolevulinic acid excretion. Arch. Environ. Health, 25:450, 1972.
46. Graef, J. W., Kopito, L., and Shwachman, H.: Lead intoxication in children. Diagnosis and therapy. Postgrad. Med., 50:133, 1971.

47. Tersinger, J., and Srbova, J.: The value of mobilization of lead by calcium ethylene-diamine-tetra-acetate in the diagnosis of lead poisoning. Br. J. Indust. Med., 16:148, 1959.
48. Smith, H. D., Baehner, R. L., Carney, T. et al.: The sequelae of pica with and without lead poisoning. Am. J. Dis. Child., 105:609, 1963.
49. David, O. J.: Association between lower level lead concentrations and hyperactivity in children. Environ. Health Perspect., 7:17, 1974.
50. Chisholm, J. J., Jr.: Poisoning due to heavy metals. In Pediatric Clinics of North America. Vol. 17. Philadelphia, W. B. Saunders Company, 1970.
51. Fleischer, M., Sarofim, A. S., Fassett, D. W. et al.: Environmental impact of cadmium: A review by the panel on hazardous trace substances. Environ. Health Perspect., 7:253, 1974.
52. Snyder, R. D.: Congenital mercury poisoning. N. Engl. J. Med., 284:1014, 1971.
53. Fern, V. H.: The synteratogenic effect of lead in cadmium. Experientia, 25:56, 1969.
54. Chisholm, J. J., and Harrison, H. E.: The exposure of children to lead. Pediatrics, 18:943, 1956.
55. Graef, J. W., and Cone, T. E., Jr. (eds.): Manual of Pediatric Therapeutics. Boston, Little, Brown and Company, 1974.

CHAPTER 16

INFECTIOUS DISEASES AND THE PREVENTION OF MENTAL RETARDATION

Jerome E. Kurent, M.S., M.D.,
and John L. Sever, M.D., Ph.D.

INTRODUCTION

It has been estimated that infectious diseases are responsible for 4 to 9 per cent of institutionalized mentally retarded children.[1-4] It is not known, however, just how often an infection reduces an individual endowed with high intellect to one of only average intelligence. The infections known to cause mental retardation include the embryonal encephalopathies caused by rubella virus, cytomegalovirus, and herpesviruses, *Treponema pallidum* and *Toxoplasma gondii*. Acquired infections caused by herpesvirus, measles virus, and arboviruses during neonatal life, childhood, and adulthood may cause encephalitis, leading to severe brain damage. Bacterial meningitis may, likewise, lead to serious neurological sequelae and mental retardation. The cerebral complications of acute infantile diarrhea also may result in mental retardation. Other infectious agents have also been suggested as possible candidates for causing intellectual deficits.

It is probable that the total contribution of infections that cause mental defects is underestimated. This is particularly true of childhood infections where there is a long interval before adequate intellectual assessment can be made. The consequences of clinically inapparent and mild infections have also been difficult to assess.

In this chapter the current information on the role of infections as causes of mental retardation will be briefly summarized. Infections etiologically related to mental subnormality, as well as those merely suspect, will be described. Studies will be proposed to delineate the role of infections in causing defects of higher cerebral function. Finally, the measures now available to control infections related to mental retardation will be discussed.

INFECTIONS AND MENTAL RETARDATION

Infections Known to Cause Mental Retardation

Both congenital and acquired infections can result in mental retardation. Intrauterine infection may cause congenital brain damage; infections acquired during the neonatal period, childhood, and adulthood may also cause severe defects in intellectual function. The implicated infections that have been described include both congenital and acquired diseases.[5-10]

Congenital Infections

Rubella. Rubella virus infection of the pregnant woman may result in the birth of an infant with mental retardation, microcephaly, microphthalmia, blindness, and deafness. The child may also have cataracts, cardiovascular anomalies, anemia, petechiae, hepatosplenomegaly, pneumonitis, and lesions of the long bones. The highest risk to the fetus occurs during the first and second trimesters when up to 50 per cent of women infected very early in pregnancy may give birth to a defective infant. Undoubtedly, the most significant control measure for the prevention of rubella anomalies has been the development and utilization of live attenuated rubella virus vaccine.

Cytomegalovirus. Cytomegalovirus (CMV) is ubiquitous, but fetal infection may result in an infant with mental retardation, microcephaly, cerebral calcifications, and seizures. Anemia, petechiae, thrombocytopenia, hepatosplenomegaly and pneumonitis may also occur. The most susceptible period of gestation, the sources of infection, and the routes of inoculation are not known. There is also evidence that inapparent congenital CMV infection may be related to mild to moderate auditory and mental dysfunction in childhood.[11]

Herpesviruses. Herpesviruses occur as two distinct serotypes. Type 1 is the oral strain which causes "cold sores," or "fever blisters," whereas type 2 causes lesions of the skin and mucous membranes of the genitals. Both may cause intrauterine infections; however, 90 to 95 per cent of perinatal herpes infections are caused by herpesvirus, type 2. In most cases, infection of the infant is acquired during parturition by contact with the infected birth canal. Affected children who survive may demonstrate mental retardation, microcephaly, microphthalmia, intracerebral calcifications, and seizures. Infections of the newborn may result in generalized disease, which may include acute necrotizing encephalitis. There is no proven treatment for herpesvirus infections, although several drugs are currently being tested. These include adenine arabinoside (Ara-A), cytosine arabinoside (Ara-C), and iododeoxyuridine (IDU).[12-14]

Syphilis. *Treponema pallidum,* the spirochete causing syphilis, may infect the developing fetal brain. The consequences of the disease to the fetus may be divided into early and late effects. Early effects may include mental retardation, hydrocephalus, and seizures. Late effects occur during adolescence and may include antisocial behavior, hemiplegia, and convulsions. Congenitally acquired syphilis occurs in babies of women having primary syphilis during that pregnancy, but may also occur in infants whose mothers had untreated syphilis years before. Penicillin is the treatment of choice for both maternal and congenital infections.

Toxoplasmosis. Intrauterine infection by *Toxoplasma gondii,* an ubiquitous intracellular sporozoan, may cause CNS lesions in infected infants. These lesions may lead to mental retardation, microcephaly, hydrocephalus, periventricular intracerebral calcifications, chorioretinitis, and seizures. Most maternal infections are clinically inapparent, but lymphadenopathy may occur. The frequency of maternal toxoplasmosis in the Collaborative Perinatal Study was approximately 1 per 1000 pregnancies.[6] Severe disease generally occurs only when maternal infection takes place during the first two trimesters.[15] Mothers with antibody prior to pregnancy are not at risk.

Preventive measures include avoidance of ingestion of raw or undercooked meats, and avoidance of contact with animal excreta, particularly that of domestic cats. The natural reservoir of the pathogen is the mouse, from which the cat probably acquires the infection. Treatment may include sulfadiazine and pyrimethamine, with supplements of folinic acid and Baker's yeast.

Acquired Infections

Bacterial Meningitis. There is a characteristic age-related occurrence of the various types of bacterial meningitis. In contrast to the period preceding the development of effective gram positive chemotherapy, today gram negative bacteria cause the majority of neonatal meningitides. Most of the etiological agents are of the *Escherichia coli,* paracolon, and Klebsiella-Enterobacter groups of bacilli. Recently, however, beta-hemolytic streptococcus has been implicated in maternal and neonatal infections.[16]

In spite of major advances in antibacterial chemotherapy, the problem of neonatal bacterial meningitis has not been conquered. The difficulty of establishing an early diagnosis of meningitis during neonatal life is probably the single most important factor responsible for failure to control these infections and their serious consequences. Development of antibiotic-resistant bacterial strains, as well as difficulties in choosing the proper antibiotic, also contribute to the problem. In a large proportion of neonatal bacterial meningitides, the maternal history may provide helpful clues in recognizing infants at risk before they manifest clinical disease. The conditions of maternal urinary tract infections, premature rupture of membranes, and prolonged labor may predispose to neonatal sepsis and/or meningitis.

Hemophilus influenzae, type B, *Streptococcus pneumoniae* (pneumococcus), and *Neisseria meningitidis* (meningococcus) cause most acute bacterial meningitis in older infants and children. After the age of 5, meningitis caused by *H.* influenzae becomes less common, and the infections from pneumococcus and meningococcus become relatively more frequent. The source of most *H. influenza* and *N. meningitidis* infections is a healthy carrier. There are no effective control measures, although vaccine development is under study.[17-19] Antibiotics and intensive supportive care constitute the therapeutic mainstays. Maintenance of optimal hydration, reduction of increased intracranial pressure, and management of septic shock are the prime supportive measures.

Bacterial meningitis in adults is caused mainly by *N. meningitidis* and *S. pneumoniae.* Meningitis caused by the pneumococcus may originate from primary infectious foci, such as the middle ear, nasal sinuses, or lung. Chronically debilitated individuals are more predisposed to infections caused by the pneumococcus, as well as those caused by the gram negative bacilli. Other gram positive bacteria, such as the staphylococci and streptococci may cause meningitis. *Mycobacterium tuberculosis* is also an important cause of severe neurological sequelae of meningeal infection.

Acute Infectious Diarrhea. Infections not invading the CNS may cause severe brain damage with mental retardation. Acute infectious diarrhea may cause hypernatremia, which can result in cerebral edema, seizures, and irreversible brain damage. Prompt and appropriate fluid replacement therapy is vital to prevent damage. Two primary mechanisms have been postulated as being responsible for neurological sequelae of acute infantile gastroenteritis with electrolyte disorders:[20] (1) dehydration with hypernatremia, resulting in shrinkage of the brain; and (2) overly vigorous attempts to correct hypertonic dehydration and acidosis. It has been stated that in certain areas of the United States and other parts of the world, diarrhea is the single most important preventable cause of mental subnormality.[21]

The most common etiological agent of acute infantile diarrhea is enteropathogenic *E. coli.*[22] The infection may be acquired in nursery epidemics from sick infants, or during delivery, from the mother. Transmission is by the fecal-oral route. Sporadic cases may also occur, and a source of infection may not be identified. Viruses have also been implicated, such as members of the ECHO and Adeno groups.

Viral Encephalitis

HERPES. Acute necrotizing encephalitis caused by herpesvirus type 1 is the most common sporadic viral encephalitis. It is a highly lethal infection, although mortality estimates vary. Survivors often suffer severe neurological and intellectual deficits.[23] Host-predisposing factors are poorly understood, and there is no evidence to indicate that herpesvirus recovered from encephalitis cases differs from isolates of ordinary herpesvirus. Herpesvirus type 2 also causes acute encephalitis, primarily affecting newborns, and rarely, adults.

MEASLES. Measles encephalitis occurs as a complication of rubeola in approximately 0.1 per cent of reported cases.[24] This disorder, which may cause severe defects in mental function, is potentially preventable through employment of the commercially available attenuated live virus vaccine.

ARBOVIRUS ENCEPHALITIS. As a group, the Arboviruses include over 250 members, not all of which are known to infect man.[22] Those known to cause encephalitis in the Western Hemisphere include Western equine encephalitis, Eastern equine encephalitis, Venezuelan equine encephalitis, St. Louis encephalitis, and California encephalitis viruses. Arthropod vectors are involved in the natural transmission of all these agents, and animal reservoirs are important in maintaining a virus pool in nature. The successful prevention of Arbovirus infections may depend upon the control of mosquitoes and other arthropod vectors, since there are no vaccines available for human use.

Arboviruses may cause delayed onset of mental subnormality following childhood infection. The viruses causing Western equine encephalitis, St. Louis encephalitis and California encephalitis have been implicated in this regard.[25]

Postinfectious encephalomyelitis rarely may follow uncomplicated common viral infections, including measles, rubella, mumps, varicella zoster, and influenza. In general, accurate estimates are not available regarding the frequency of mental retardation following these infections.

Syphilis. The delayed onset sequelae of acquired syphilis are well known. Paresis with psychosis may follow untreated primary syphilis up to 20 years later. Serological tests for syphilis are valuable diagnostic tools and the tests of the Venereal Disease Research Laboratory (VDRL) and Fluorescent Treponemal Antibody Absorption (FTA-ABS) are commonly employed in this regard. Penicillin is highly effective therapy for syphilitic infections.

Infections Suspected But Not Proven to Cause Mental Retardation

Mycoplasma

Recent reports have linked mycoplasma species with abortion, stillbirth, prematurity, and small-for-date infants.[26] *Mycoplasma hominis* has also been isolated from the cerebrospinal fluid of a case of neonatal hydrocephalus,[27] and *Mycoplasma pneumoniae* has been associated with a variety of acute and subacute human neurological disorders. These include the Guillain-Barré syndrome, encephalitis, meningitis, and cerebellar ataxia.[28] However, the precise relationship of mycoplasma infection to these disorders is unknown.

Maternal Urinary Tract Infections

The possible relationship of maternal urinary tract infections that occur during pregnancy to abnormal outcomes has not been investigated. There is evidence that maternal urinary tract infections may be related to the birth of small-for-date babies.[29] The association of subnormal mental development with such infants has been suggested.

Viral Meningitis

Mumps. Mumps virus is the most frequent single cause of aseptic meningitis. Despite speculations regarding the role mumps may play in causing mental retardation, particularly as a delayed onset disease, there is inadequate data for interpretation.

Enteroviruses. The enteroviruses, as a group of agents, are responsible for most aseptic meningitis. The primary agents include Coxsackie and ECHO viruses. The long-term effects of these common CNS infections are unknown.

Viral Encephalitis

Long-term follow-up of patients with viral encephalitis is needed to evaluate the possible effects of the disease on mental function. California encephalitis and Western equine encephalitis, in particular, have been suggested as resulting in the onset of serious intellectual deficits several years after the infection occurred.

STUDIES TO PROVIDE ADDITIONAL DATA ON THE ROLE OF INFECTIONS IN CAUSING MENTAL RETARDATION

One current large investigation of infections as a cause of abnormal pregnancy outcome is included in the Collaborative Perinatal Study.[30] This is a cooperative effort between the National Institute of Neurological Diseases and Stroke and 11 University-affiliated medical centers in the United States. The program that the Infectious Diseases Branch is utilizing in this effort will be reviewed, and further studies that are needed to identify additional infectious agents responsible for causing mental retardation will be described.

The Collaborative Perinatal Study

As part of the Collaborative Perinatal Study, data from approximately 60,000 pregnancies that occurred from 1959 through 1966 are being analyzed. Children born of these pregnancies have been followed for approximately eight years after birth. Correlations are being made between the events of pregnancy, labor and delivery, and the subsequent neurological and intellectual development of the child.

The three major approaches used in the Collaborative Perinatal Study are (1) Clinically reported infections occurring during pregnancy are being correlated with findings in the children through 8 years of age. These data are compared with control pregnancies matched for age, sex, race, and institution; (2) Serological studies for antibody to infectious agents are being conducted using maternal serum specimens that were collected during pregnancy. Serum antibody levels to selected infectious agents are being determined using specimens from approximately 4000 women who had abnormal children, and from 4000 women who had normal pregnancy outcomes. The frequency of antibodies and changes in antibody concentrations are being compared for the patients and controls; and (3) Cord serum immunoglobulin (IgM) elevations are used to identify "high risk" children. This approach takes advantage of the fact that IgM levels are frequently elevated in perinatal infections. With this lead, specific IgM tests can be conducted for a variety of agents, including *Toxoplasma gondii, Treponema pallidum,* and CMV.

The recognition of clinically apparent infections during pregnancy offers an excellent opportunity to detect etiological associations between infectious agents and congenital disease, including mental retardation. In the Collaborative Perinatal Study, clinical infections occurred in 8180 of 58,828 pregnant women. The infections and their frequency of incidence include: viral, 3401; bacterial, 4539; fungal, 102; and parasitic, 138. Figure 16–1 illustrates the frequency of viral infections occurring during

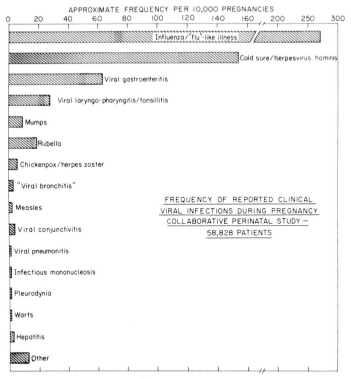

APPROXIMATE FREQUENCY PER 10,000 PREGNANCIES

Influenza/"flu"-like illness
Cold sore/herpesvirus hominis
Viral gastroenteritis
Viral laryngo-pharyngitis/tonsillitis
Mumps
Rubella
Chickenpox/herpes zoster
"Viral bronchitis"
Measles
Viral conjunctivitis
Viral pneumonitis
Infectious mononucleosis
Pleurodynia
Warts
Hepatitis
Other

FREQUENCY OF REPORTED CLINICAL
VIRAL INFECTIONS DURING PREGNANCY
COLLABORATIVE PERINATAL STUDY —
58,828 PATIENTS

Figure 16–1.

pregnancy in the Collaborative Perinatal Study. Rates varied from approximately 1 to 20 per 10,000 pregnancies for rubella, mumps, varicella-zoster, measles, warts, hepatitis, infectious mononucleosis, viral bronchitis, conjunctivitis, pneumonitis and pleurodynia. Considerably higher rates occurred for influenza-like illness, cold sores, viral gastroenteritis, and laryngopharyngitis. The data correlating these infections with findings in the children are currently undergoing detailed computer analysis.

The serological approach correlates clinical and antibody data for infections in the mother to obtain information of the effects on the fetus and child following serologically confirmed subclinical and clinical infections in the mother. Subsequent neurological and intellectual development of the children has been closely followed. The present study design involves the determination of antibody responses to 11 infectious agents using serial sets of sera obtained prospectively from approximately 4000 women with abnormal pregnancy outcomes and from a similar number of matched normal controls.

The seroepidemiological survey was also utilized to study the 1964 rubella epidemic. A total of 6161 pregnant women from 11 institutions in the Collaborative Perinatal Research Study were studied during this period.[31] Paired serum specimens obtained from 500 pregnant women before and after the 1964 epidemic demonstrated that approximately 85 per cent had prior immunity, as determined by the presence of hemagglutination-inhibiting antibody in the first specimens. In the total population, 2.4 per cent developed clinical rubella, and 1.2 per cent had inapparent infections during the epidemic. Approximately 17 per cent of first trimester infections and 10 per cent of second trimester infections resulted in congenital rubella. A similar survey conducted during 1966 demonstrated that 8 per cent of the women were susceptible to rubella.

The third major approach currently used in the Collaborative Perinatal Study is IgM determinations. Newborn cord serum IgM values are frequently elevated in congenital infections, and the serological specificity may be determined in certain infections by using the indirect fluorescent antibody method. The diagnosis of some congenital infections leading to mental retardation is thus possible by this technique. In the study, cord serum specimens were obtained from approximately 32,000 children. Cord IgM, as well as IgA and IgG levels, were determined by the radial immunodiffusion method.[32] Approximately 6 per cent of the 32,000 cord sera tested had IgM levels in excess of 20 mgm per cent, the upper limit of normal. The frequency of elevated cord serum IgM levels, however, is also inversely related to the birthweight of the child, and this must be considered before assigning pathological significance to high values.

The specificity of IgM may be determined to the agents causing cytomegalic inclusion disease, toxoplasmosis, and syphilis. Unfortunately, reproducible tests for IgM with rubella and herpesviruses are not currently available. The value of IgM determinations is limited, in that one third to two thirds of congenitally infected infants fail to have increased IgM levels. It is also not possible to prove the cause of increased IgM in all instances, and newborns may have elevated values in the absence of known intrauterine infection. These cases, however, could represent infections due to unrecognized agents and, therefore, will be studied in detail.

Proposed Studies

Detection of Congenital Infections Causing Mental Retardation by Direct Isolation Techniques

The Collaborative Perinatal Study should be extended to provide additional information regarding infections and mental retardation. The approach should emphasize attempts at direct isolation of viruses, mycoplasmas, and bacteria from all newborns (including neonatal deaths), even in the absence of obvious infection. Cultures of cord blood, nasopharynx, urine, and feces should be made. Maternal urine and cervical swabs should also be utilized in isolation attempts. Placentas should be cultured and examined histologically for evidence of inflammation, which might indicate antecedent or ongoing infection. Venous blood should be obtained from the cord and the mother at the time of delivery and again at 3 days after birth. The serum should be frozen and stored for future study if there should ever be evidence of infection or overt clinical disease in the child. Such studies would include IgM and specific antibody determinations to selected infectious agents. Specimens taken at 3 days after birth would be preferable to cord blood, since the latter may contain contaminating immunoglobulins from the maternal circulation. All infected patients and matched controls should be studied until 21 years of age and evaluated semiannually for possible psychological and learning impairments.

Detection by Serological Techniques of Acquired Infections Possibly Associated with Mental Retardation

The problem of identifying an etiological relationship between acquired but inapparent or mild, infections during childhood and adolescence in relation to mental subnormality should be investigated. The same study population previously described could provide data to test for the possible existence of such a relationship. Serum specimens would have to be obtained from approximately 10,000 randomly chosen subjects from a study population that may be reliably followed for an extended period of time. Patient populations such as those of the Kaiser Permanente Medical Group Clinics in the United States would be particularly valuable for this investigation. Patients followed by these clinics are generally available for long-term follow-up. The children should be followed semiannually until they are 21 years of age. If, during the course of routine examinations, the child demonstrated evidence of subnormal or aberrant intellectual development, a serological study of serum and cerebrospinal fluid should be performed against selected antigens. These antigens should include rubella, rubeola, herpes types 1 and 2, cytomegalovirus, Epstein-Barr virus, toxoplasma, mumps, ECHO, Coxsackie influenza, parainfluenza, and hepatitis A and B viruses. Other agents could also be studied. If the child lives in a region where specific arbovirus infections are common, antibody determinations to those agents should be performed. Additional specimens should be obtained during the acute and convalescent periods of infection, so that changes in antibody titer can be detected. Serially collected control serum specimens should be studied at the same time.

Detection of Infections Related to Mental Retardation by Clinical and Serological Techniques

The possible delayed effects of aseptic meningitis caused by mumps, Coxsackie, and ECHO viruses have not been thoroughly evaluated. To investigate this question, 5000 patients with clinically and laboratory-confirmed diagnoses should be followed for a minimum of five years or until 21 years of age, whichever is longer. These patients should be evaluated semiannually for subtle psychological and learning impairments and compared to 5000 normal subjects who have not had clinical viral meningitis.

Retrospective Study of Newly Institutionalized Patients

A careful retrospective study of new admissions to institutions for the mentally retarded should be conducted. Patients who have been institutionalized for an extended period of time should not be included, since common exposure to infectious agents within the facility would confuse the interpretation of seroepidemiological data. Serum and CSF should be obtained from approximately 1000 patients with unexplained mental retardation. These specimens should be studied for the presence of specific antibody to selected infectious agents. Aliquots of both serum and CSF should be frozen and stored for future study at such time as new etiological agents and relationships are identified or suspected. Comparison controls should be used within the large study population of mentally retarded patients. Thus, patients with microcephaly could be compared with patients having Down's syndrome regarding specific antibody patterns; other clinical comparisons could be made on a similar basis. Careful attention would have to be made to control for factors which might influence exposure to infectious agents, such as relative ability to maintain personal hygiene, and the quality of the patient's environment prior to institutionalization.

Specimens from this population of mentally retarded patients should be studied for antibody to representative members of all major virus groups. Agents that should be given top priority include the herpes viruses types 1 and 2, cytomegalovirus, Epstein-Barr virus, rubeola, rubella, mumps, influenza, parainfluenza, enteroviruses, hepatitis A and B viruses, and toxoplasma. Members of the arbovirus groups should also be studied, particularly Eastern equine encephalitis, Western equine encephalitis, Venezuelan equine encephalitis, and California encephalitis viruses.

CONTROL MEASURES FOR THE PREVENTION OF MENTAL RETARDATION DUE TO INFECTIOUS DISEASE

The prevention of mental retardation due to infectious diseases may be possible where effective and safe vaccination procedures are available. Other measures may also be used, and their application will be considered in terms of the congenital and acquired infections previously discussed.

Rubella

The rubella immunization program in the United States since 1969 has had two primary objectives: (1) the routine immunization of all children 1 to 12 years old, and (2) the selective immunization of post-pubertal girls and women of childbearing age who are not pregnant. Vaccine recipients should be advised to practice a reliable means of contraception for two months following rubella vaccination in order to minimize the chance of transplacental virus transmission. Current practice should include immunization of children at 1 year of age, and certainly before they reach school age. The first objective was intended to decrease the overall incidence of rubella and its further dissemination, whereas the second was intended to protect individuals at greatest risk. Significant decreases in the incidence of rubella have been reported since implementation of rubella vaccination programs.[33] The importance of maintaining an active vaccination program cannot be overemphasized, since recrudescence to previously high infection rates is apt to occur if a population of susceptibles is again present. There is a distinct possibility of this happening, particularly when an infection such as rubella appears to be under good control and public enthusiasm wanes.

Herpes Viruses

Although there are presently no proven control measures for members of the herpesvirus group, attempts are being made to develop a herpesvirus vaccine.[34] Antiviral drugs, such as Ara-A, Ara-C, and IDU, are being tested for efficacy against both congenital and acquired herpes virus infections and offer hope that effective antiviral chemotherapy may be achievable.

Cytomegalovirus

Live attenuated virus vaccines against cytomegalovirus are currently being tested, and it has been suggested that all adolescent girls be vaccinated.[35] All live herpes virus vaccines, however, carry the possible risk of introducing a latent infection with unrecognized effects, including oncogenesis.

Syphilis

The prevention and control of congenital syphilis ultimately depends upon the identification and treatment of maternal infection. Penicillin is highly effective therapy for *Treponema pallidum* and will prevent fetal infection when the mother is adequately treated. Both congenital and acquired syphilis should be treated with penicillin. Serological tests, such as the VDRL and FTA-ABS, are valuable detection devices for syphilis. There are no vaccines available for *T. pallidum*.

Toxoplasmosis

Toxoplasmosis gondii is usually acquired from eating uncooked meats or from contact with cat excreta. Cats acquire the infection from mice, which may be the ultimate source of the organism in nature. Since there are no available immunization measures, the only means of preventing infection is by thorough cooking or freezing of meats before eating, and by avoiding contact with cat and rodent excrement. Control by these means may be difficult to achieve, however. Vaccine development might be directed toward the immunization of domestic cats, as well as humans, to interrupt an important source of human infection.

Bacterial Meningitis

Minimizing the morbidity and mortality from bacterial meningitis depends upon prompt diagnosis and appropriate antibiotic therapy of suspected infections. Once the etiological agent is established, definitive antibiotic therapy must be vigorously pursued. Bacterial vaccines are a potentially valuable preventive measure, but require further development.

Acute Infectious Diarrhea

Acute infectious diarrhea caused by enteropathogenic *E. coli* or viral agents demands prompt but cautious correction of fluid and electrolyte imbalances, when they occur. This important measure is vital to preventing brain damage, as is antibiotic therapy, when indicated. Control of nursery epidemics involves elimination of carrier sources and interruption of fecal-oral transmission. The latter may be facilitated by hand washing by nursery personnel and isolation of infected infants.

Measles

Extensive use of measles vaccine in the United States since 1963 has demonstrated that antibodies have persisted for the 11 to 12 year follow-up period.[24] There has been an incidence of acute CNS disease in one per million persons inoculated with vaccine, but this rate must be compared with an incidence of one per thousand seen with naturally acquired measles. The pathogenesis of subacute sclerosing panencephalitis (SSPE) and its occasional association with measles immunization are not clear. A survey in the United States had identified 400 cases of SSPE, 40 of which had occurred in vaccinated individuals. It does not appear, however, that these cases are directly vaccine-related. Also, the significance of transiently abnormal EEG's in patients having either naturally acquired measles[36] or live measles vaccine[37] is unknown at this time. Based on 10 years' experience in the use of live measles vaccine, it appears to be safe and immunogenic. The highly effective immunity provided should fulfill an important role in preventing severe CNS sequelae of measles infection.

SUMMARY

The perinatal and acquired infections known to be etiologically related to mental retardation have been reviewed. Those infections that are possible candidates for causing mental subnormality have also been discussed, and the need for adequate long-term studies has been suggested. Potential studies to provide needed information regarding the role of infections in causing defects of higher cerebral function were described. Finally, the available control measures for dealing with the important infections causing mental retardation were summarized.

ACKNOWLEDGMENT: The authors gratefully acknowledge the assistance of Ms. Nelva Reckert in the preparation of this manuscript.

References

1. Rostafinski, M. J.: The incidence of preventable forms of brain damage. Va. Med. Mon., 91:22, 1964.
2. McIntire, M. S., and Adams, H. Q.: Congenital anomalies associated with mental retardation. Nebr. Med. J., 48:551, 1963.
3. U.S. President's Panel on Mental Retardation. Report of the Mission to Denmark and Sweden. Washington, D.C., U.S. Public Health Service, August, 1962, p. 5.

4. Poskanzer, D. C., and Salam, M. F.: Mental retardation related to infectious disease in patients at Walter E. Fernald State School. *In* Eichenwald, H. F. (ed.): The Prevention of Mental Retardation Through Control of Infectious Diseases. Bethesda, Maryland, U.S. Department of Health, Education, and Welfare, National Institute of Child Health and Human Development, 1966, p. 23.

5. Lundstrom, R.: Rubella during pregnancy. A follow-up study of children born after an epidemic of rubella in Sweden, 1951 with additional investigations on prophylaxis and treatment of maternal rubella. Acta Paediatr., Scand., 51(Suppl. 133):1, 1962.

6. Sever, J. L.: Perinatal infections affecting the developing fetus and newborn. *In* Eichenwald, H. F. (ed.): The Prevention of Mental Retardation Through Control of Infectious Diseases. Bethesda, Maryland, U.S. Department of Health, Education, and Welfare, National Institute of Child Health and Human Development, 1966, p. 37.

7. Sever, J. L.: Viruses and the fetus. Int. J. Gynecol. Obstet., 8:763, 1970.

8. Sever, J. L. Virus infections and malformations. Fed. Proc., 30:114, 1971.

9. Fuccillo, D. A., and Sever, J. L.: Viral Teratology. Bacteriol. Rev., 37:19, 1973.

10. Kurent, J. E., and Sever, J. L.: Pathogenesis of intrauterine infections of the brain. *In* Gaull, G. E. (ed.): Biology of Brain Dysfunction. Vol. 3. New York, Plenum Publishing Corporation, 1975. (In press.)

11. Reynold, D. W., et al.: Inapparent congenital cytomegalovirus infection with elevated cord IgM levels. N. Engl. J. Med., 290:291, 1974.

12. Will the arabinosides make it as antiviral agents? Hosp. Pract., 9:44, 1974.

13. Johnson, R. T.: Treatment of herpes simplex encephalitis. Arch. Neurol., 27:97, 1972.

14. Kern, E. R., Overall, J. C., Jr., and Glasgow, L. A.: Herpes virus hominis infection in newborn mice. I. An experimental model and therapy with iododeoxyuridine. J. Infect. Dis., 128:290, 1973.

15. Desmonts, G., and Couvreur, J.: Congenital toxoplasmosis. A prospective study of 378 pregnancies. N. Engl. J. Med., 290:1110, 1974.

16. Grossman, J., and Tompkins, R. L.: Group B beta-hemolytic streptococcal meningitis in mother and infant. N. Engl. J. Med., 290:387, 1974.

17. Smith, D. H., Johnston, R. B., Jr., Peter, G., and Anderson, P.: Studies on a vaccine for *Hemophilus influenzae B*. *In* International Congress on the Application of Vaccines Against Viral, Rickettsial, and Bacterial Diseases of Man, Scientific Publication No. 226. Washington, D.C., Pan American World Health Organization, 1971, p. 353.

18. Austrian, R.: The current status of pneumococcal disease and the potential utility of polyvalent pneumococcal vaccine. *In* International Congress on the Application of Vaccines Against Viral, Rickettsial, and Bacterial Diseases of Man, Scientific Publication No. 226. Washington, D.C., Pan American World Health Organization, 1971, p. 359.

19. Artenstein, M. S.: Polysaccharide C vaccines against meningococcal infections. *In* International Congress on the Application of Vaccines Against Viral, Rickettsial, and Bacterial Diseases of Man, Scientific Publication No. 226. Washington, D.C., Pan American World Health Organization, 1971, p. 350.

20. Dodge, P. R.: Some comments on neurological sequelae of diarrhea and electrolyte disturbances, comments by M. Z. Salam. *In* Eichenwald, H. F. (ed.): The Prevention of Mental Retardation Through Control of Infectious Diseases. Bethesda, Maryland, U.S. Department of Health, Education, and Welfare, National Institute of Child Health and Human Development, 1966, p. 290.

21. Dodge, P. R. *ibid,* comments by H. F. Eichenwald. *In* Eichenwald, H. F. (ed.): The Prevention of Mental Retardation Through Control of Infectious Diseases. Bethesda, Maryland, U.S. Department of Health, Education, and Welfare, National Institute of Child Health and Human Development, 1966, p. 292.

22. Report of the Committee on Infectious Diseases. 17th ed., Evanston, Illinois, American Academy of Pediatrics, 1974.

23. Johnson, K. P., Rosenthal, M. S., and Lerner, P. I.: Herpes simplex encephalitis. The course in five virologically proven cases. Arch. Neurol., 27:103, 1972.

24. Krugman, S., and Perkins, F. T.: Vaccination against communicable disease. Am. J. Dis. Child., 126:406, 1973.

25. Finley, K. H.: Postnatally acquired infections leading to mental subnormality. *In* Eichenwald, H. F. (ed.): The Prevention of Mental Retardation Through Control of

Infectious Diseases. Bethesda, Maryland, U.S. Department of Health, Education, and Welfare, National Institute of Child Health and Human Development, 1966, p. 101.

26. Klein, J. O., Buckland, D., and Finland, M.: Colonization of newborn infants by mycoplasma. N. Engl. J. Med., 280:1025, 1969.

27. Boe, O., Diderichsen, J., and Matre, R.: Isolation of *Mycoplasma hominis* from cerebrospinal fluid. Scand. J. Infect. Dis., 5:285, 1973.

28. Lerer, R. J., and Kalavsky, S. M.: Central nervous system disease associated with *Mycoplasma pneumoniae* infection: Report of five cases and review of the literature. Pediatrics, 52:658, 1973.

29. Norden, C. W., and Kass, E. H.: Bacteriuria of pregnancy — a critical appraisal. Annu. Rev. Med., 19:431, 1968.

30. The Women and their Pregnancies. The Collaborative Perinatal Study of the National Institute of Neurological Diseases and Stroke. DHEW Publication No. 73–379 (NIH), Washington, D.C., Superintendent of Documents, U.S. Government Printing Office, 1972.

31. Sever, J. L., Hardy, J. B., Nelson, K. B., and Gilkeson, M. R.: Rubella in the Collaborative Perinatal Research Study. II. Clinical and laboratory findings in children through 3 years of age. Am. J. Dis. Child., 118:123, 1969.

32. Sever, J. L.: Immunoglobulin determinations for the detection of perinatal infection. J. Pediatr., 75:1111, 1969.

33. Krugman, S., and Katz, S. L.: Rubella immunization. A five year progress report. N. Engl. J. Med., 290:1375, 1974.

34. Zaia, J. A.: The humoral and cellular immune response to an ether extracted antigen of herpes simplex virus. Presented at the annual meeting of the Reticuloendothelial Society at Williamsburg, Virginia. December 8, 1973.

35. Elek, S. D., and Stern, H.: Development of a vaccine against mental retardation caused by cytomegalovirus infection *in utero*. Lancet, 1:1, 1974.

36. Pampiglione, G.: Prodromal phase of measles: Some neurophysiological studies. Br. Med. J., 2:1296, 1964.

37. Pampiglione, G., Griffiths, A. H., and Bramwell, E. C.: Transient cerebral changes after vaccination against measles. Lancet, 2:5, 1971.

CHAPTER 17

EARLY INTERVENTION IN THE PREVENTION OF MENTAL RETARDATION

John H. Meier, Ph.D.

INTRODUCTION

The increasingly sophisticated technology and counseling now available for preventing the conception of children having some developmental anomalies are currently inadequately utilized. The number of children born with both absolute and relative developmental disabilities is increasing ("developmental disabilities" is a more generic, and recently popularized, term which encompasses mental retardation, epilepsy, cerebral palsy, autism, learning disabilities, and other similar conditions). Ironically, technological advances now enable more vulnerable children to survive, often in better condition than was thought possible several years ago. Unfortunately, the growing complexity of our society and the sheer survival value of higher cortical activity, which tends to suffer damage first and most in those who survive, places those damaged organisms at an even greater disadvantage. It is as though there were a kind of inflation of the commodity of intelligence, whereby it takes more intellectual prowess for the same functional and adaptive behavior, much as Toffler's *Future Shock* predicts. Many more children who have been born with various degrees of developmental disabilities are surviving, entering school, and subsequently seeking employment and the "good life," insofar as they are able. It is neither humane nor prudent—from the larger socioeconomic point of view—to assume the posture that the birth of a handicapped child is simply the parents' bad luck. It has become even more important to program even the most limited computers (human brains) to utilize their maximum capacities. To assess the capacity of a computer, one of the first questions asked deals with the size of its core storage. That is, how many bits of information can it store for later retrieval and processing? Likewise, in the human brain it is very important for maximizing the core storage of this human computer from which it gets rich data from the time of birth, when it is developing most rapidly. Where else can one find a computer containing more than ten billion flip-flop circuits, occupying less than a cubic foot of space, operating on the energy of a peanut for four hours, being totally mobile, and produced with unskilled labor?

Until genetic counseling, prenatal diagnosis, and (perhaps) genetic engineering or eugenics are more widely and efficaciously practised, as well as accommodated in public ethics, attitudes, and in legislative-governmental policies, it is obvious that there will continue to be malfunctioning and limited computers produced. Moreover, since the vast majority of the mentally retarded and developmentally disabled population has no clear-cut etiology, it is exigent to employ the best that is known in the current state of the art and science of early intervention.

Developmental disability (DD) is a relative term; compared to Einstein, most people are retarded in their mathematical comprehension. The aforementioned irony of increasing, both absolutely and relatively, the *number* of DD persons, suggests that the prevention of DD is also a relative phenomenon, in the sense that the prevention of further disability through early intervention is still prevention of otherwise inevitable further deterioration, although the total disability may not be lessened or fully compensated. The specific genetic inheritance of a DD person may establish certain ceiling limitations for growth and development, but even those lowered ceilings will probably not be reached without appropriate intervention.

This chapter is divided into two main sections. The first section addresses the rationale for instituting early intervention efforts (for this chapter, "early" means "at birth;" other chapters discuss preconceptual and prenatal intervention). This reviews the recent and relevant literature for both normally developing and deviant infants and toddlers. The second section presents and discusses a suggested schema for early intervention, with specific provision made for early and periodic developmental screening, interdisciplinary and comprehensive evaluations and treatment planning, and several representative and exemplary intervention programs.

RATIONALE FOR EARLY INTERVENTION

The Last Two Dynamic Decades

Intervention programs for young children did not develop overnight, but rather over a period of many centuries.[56] Early childhood intervention programs are therefore set upon deep and strong foundations of experience and theory from several interdisciplinary sources, which collectively have withstood faddism and have acted as a basis for more recent approaches.

As the American economy and its populace recovered from the military drain of World War II, the U.S. regained much of the domestic stability that had been shaken by the broken families in war and the baby boom shortly thereafter. Many insightful social critics again began to probe the social conscience of America. Examples were cited of children and their parents who were starving to death in the midst of unprecedented affluence and high average standards of living that had been scarcely dreamed of a generation before. Coincidentally, but not accidentally, scholars in the field of child development, early childhood education, developmental psychology, academic pediatrics, and several other related disciplines were publishing findings that underscored the plasticity and educability of the human being. These discoveries served as a cogent rationale for early intervention into the lives of children whose estimated intellectual potential would never be realized unless some compensatory experience were provided during the most moldable first five years of their lives. Since this book is concerned especially with mental retardation, the following rationale concentrates on the intellectual/cognitive domain of development, although many correlates exist for the social, emotional, linguistic, and physical domains.

Probably the most influential publication in the intellectual/cognitive domain was *Intelligence and Experience*,[37] which marshaled all of the existing evidence into an overwhelmingly compelling argument that the experiences derived from one's environment during the early years of life have a profound and irreversible impact upon one's subsequent level of intellectual functioning. The combination of the scholarly work by such Renaissance thinkers and investigators as Bloom,[4] Bruner,[7] Dennenberg,[14] Deutsch,[19] Fowler,[25] Gray,[30] Hess and Shipman,[36] Lewis,[50] Murphy,[59] and many others, and the political wisdom of providing the American public with a respectable cause tantamount to making the world safe for democracy resulted in the mid-sixties declaration of a national War on Poverty.

The Supreme Court decision of 1954 also served as a catalyst for polarizing scholars who had for years been waging battles about the relative importance of heredity and environment on human growth and development. This centuries-old nature/nurture controversy was highlighted by the recurring phenomenon of newly integrated low socioeconomic status children from Black and Chicano families into the view and lives of the more advantaged. Ethnic minority children were more readily identifiable by virtue of superficial traits of skin pigmentation and/or characteristic surnames than were their equally deprived Anglo counterparts, all of whom were failing in competition with agemates who came from more enriched and privileged backgrounds.

In spite of the failure syndrome attributed stereotypically and perhaps in a prophesy-fulfilling way to minority children, of low socio-economic status (SES), existing research data failed to support any pervasive genetic determinants to account for their persistent low quality performance when confronted with the academic demands placed upon them in middle- and upper-class school programs. Adding fuel to the perennially raging fire of the nature/nurture controversy was a report issued by Jensen.[42] This report was, in itself, a scholarly contribution, but the volatile interpretations of it generated a great deal more heat than light. Marshaling the evidence from numerous studies, Jensen cogently demonstrated that better than 50 per cent—in some cases, up to 80 per cent—of the variability in children's performance on intelligence tests was attributable to inherited capabilities. However, Jensen's overzealous disciples neglected to point out the corollary that environmental influences are accountable for 20 to 50 per cent of the variability, and that many children were being nurtured in very intellectually limiting and crippling environments. The modifiability of intelligence became clear in such intervention studies as the Milwaukee Project, discussed in a later section of this chapter.

Two Assumptions Discredited

From before the turn of the twentieth century through the 1940's there were two basic, long-entrenched assumptions that influenced both the science and the practice of childbearing: "fixed intelligence" and "predetermined development." These two assumptions suggested that intelligence is an inherited capacity which develops to a predetermined level at a fixed rate. The I.Q., accordingly, remains constant throughout life. Proponents of this point of view overlooked the corollary that deprivation of experience caused marked retardation in the rate at which the infant organism developed.

Hunt[37] presented evidence at variance with the assumption of fixed intelligence by pointing to research showing that: (1) I.Q.'s of identical twins reared apart are lower than those of identical twins reared together;[60] (2) I.Q.'s of infants obtained at successive stages show considerable variation;[1] (3) I.Q.'s of foster-home children rise with nursery school experience; and (4) children reared in orphanages score lower on tests than do children reared in foster homes.[81]

Wellman[91] conducted another study through the Iowa Child Welfare Research Station and compared the spring and fall test performance of 652 children who had enrolled for one academic year in the nursery school. After children had been in nursery school a year, they averaged 7.0 I.Q. points higher on a test in spring than they had on a test that had been given the previous fall. For 228 children who had had two years of nursery school experience, there was a gain of 4.0 points for the second year. For the 67 children who attended nursery school for a third year, the mean gain was 8.0 points for the first year, 4.0 for the second year, and 2.0 for the third year. These children were matched with a control group which showed that between the fall and spring testings the schooled group gained 7.0 points, whereas the unschooled group lost an average of 3.0 points.

The pioneer studies of Spitz[82, 83, 84] have had great influence in establishing that intelligence is not fixed, but is plastic and modifiable, and that adequate mothering during the first year of life is crucial for optimal growth and development. Spitz's studies were concerned with infants from a "foundling home," where the infants received very little attention or stimulation after being weaned from their mothers at 3 months of age. The other infants were in a "nursery" attached to a penal institution for delinquent girls where the mothers were allowed to play with and attend their children every day throughout the first year. The foundling home children came from well-adjusted mothers whose only handicap was that they could not support themselves or their children. The "nursery" children were mothered mostly by delinquent minors, some of whom were physically handicapped, psychopathic, and criminal. The mean developmental quotient for the 61 children in the "foundling home" dropped progressively during the first year of the infant's life, from a starting level of 131 to a final level of 72. On the other hand, the "nursery" children rose from 97 to 112, by months 4 and 5, remained level to months 8 and 9, then dropped to 100 for months 10 and 12. The means for the first four months was 101.5 and for the last four months, 105. For a long time these findings were, unfortunately, explained away. The belief that development is entirely predetermined and intelligence fixed was accepted by Hall,[32] who communicated the belief to his students and to the child study associations of America. The belief was thus widely held, even though compellingly contradictory evidence existed.

Importance of Early Experience

Hebb[34] concluded that experience is an essential requirement for the formation of "cell assemblies," since these are the mediators of neural connections. Thus, it is the earliest experience of primary learning input that forms much of the pattern for later information-processing capability in the system and serves as the initial software program for the human computer hardware.

Gray and Miller,[31] in reviewing the literature on cognitive development, point out that early experience has four dimensions: the nature of the experience itself, the timing in the developmental period, the duration, and the intensity. Each dimension may differentially affect cognitive structure and development.

Studies of lower animals are noteworthy, but the scope of this chapter permits only the brief mention of a few. Rats reared in darkness take longer to learn pattern discrimination than those reared in light; rats handled as pups give birth to more adaptive pups.[14] Rat pups of handled or generally more stimulated mothers are more active and alert than are offspring of less stimulated mothers.[16] Pets (cats and dogs) reared in a home with more attention and stimulation do better in learning situations than laboratory-reared animals.[76, 77] Studies by Levine,[49] Dennenberg,[15] and Harlow,[33] with various kinds of laboratory animals, have all led to the conclusion that close physical contact and stimulation are essential for adequate physiological, emotional, and adaptive development. Some other investigations[3, 45, 67] have shown that rearing in enriched environments produces anatomical, as well as biochemical, differences in the brain of rats; not only were the experimental animals more proficient on problem-solving tests, but also they had more of the enzyme acetylcholinesterase, which is associated with synaptic transmission of neural messages (data processing) in their brains. (The quantity of enzyme was determined on autopsy).

The research relevant to infant intervention and which used human subjects, has been concerned mainly with the consequences of maternal deprivation and institutionalization, and the developmental effects of handling and physical contact. The earliest carefully conducted and reported studies on the effects of maternal deprivation[5, 83] were alarmingly grim in their description of the devastating and lasting consequences of early lack of mothering. Subsequent investigations[17, 28, 66] have led to similar, less extreme, conclusions. In essence, these studies have shown that children deprived of a consistent mothering figure in early life are significantly behind normal children on almost all measures of growth and development. The most strikingly deficient areas are language behavior and social competence.

Since neither institutionalization nor mothering is a pure or unitary variable, subsequent investigations have attempted to isolate the factors in the mothering process that are crucial for the infant's development. One such factor that is now known to be basic is the handling of the infant. Brody[6] reports that visual attentiveness in infants is highly correlated with the amount of handling provided by the caretaking person. Studies by Casler,[11] Rheingold,[69] and White, Castle and Held[93] have shown that additional handling and attention of institutionalized infants facilitates their development and increases their alertness. Spitz has suggested, in his formulation of the "cradle of perception," that the infant can begin to see and learn about his environment only through his close physical relationship with his mother. Along these lines, Korner and Grobstein[44] recorded the visual scanning behavior of twelve neonates and observed that their eyes were open 90 per cent of the time when being held and only 25 per cent of the time when either left unhandled or moved to a sitting position. Clearly, this finding suggests that the development of early visual-motor schemata is facilitated by handling, since other studies show that even at this young age the child can discriminate between visual cues.[23] Yarrow[95] found a similar result in a study of children under foster care. He discovered a significant correlation between developmental test scores at 6 months of age and ratings of amount and appropriateness of maternal handling.

Rubenstein[73] reported a significant positive relationship between ratings of maternal attentiveness and measures of exploratory behavior and preference for novel stimuli in 5-month-old infants. Newborn infants may be quite susceptible to the adverse effects of currently acceptable practices shortly after delivery.

The observations of maternal behavior in 28 human mothers of full-term infants reported in this study are consistent with investigations in animals; measurable differences, lasting for as long as one year, are apparent between mothers with early and extended contact and those separated from their infants in the early hours after birth. The awareness of a special attachment period shortly after birth — during which brief periods of partial or complete separation may drastically distort a mother animal's feeding and caring for her infant — would lead a caretaker or naturalist to be extremely cautious about any intervention in the period after birth.

In the human mother, the disproportionately high percentage of mothering disturbances, such as child abuse and deprivation–failure-to-thrive, which occur after a mother has been separated from her sick newborn infant, force a thorough review and evaluation of our present perinatal practices.[42a]

High Risk Infants and Toddlers

There is no doubt that a high-risk situation ultimately poses its greatest threat to the central nervous system of a developing infant. If brain damage occurs, it is this system that bears the brunt of the insult on a long-term basis. The more recently acquired higher cortical functions, such as language, refined visual and auditory perception and discrimination, and abstract problem-solving seem to be the most vulnerable to such insults. These subtle, distinctly human capabilities are typically the first to be impaired and, unfortunately, the last to be detected in infant assessment. The belief that early brain damage could be more readily compensated, both physiologically and experientially, because of the greater plasticity of the developing brain, has now been disputed by Isaacson.[41] In reconsidering his data and lengthy experience with brain damaged infants, he concluded that the brain damage sustained by very young infants has more pervasive, diffuse, and pernicious effects on higher cortical functions than originally believed. Furthermore, Purpura[67] has determined that the human brain is a good deal more mature at birth than had been previously believed, rendering it even less amenable to intervention intended to capitalize on the plasticity phenomenon.

Most of the babies considered to be of high risk are those of low birth weight; hence, most follow-up studies deal with outcome of this group. Less information is available about follow-up of babies of normal birth weight, but with perinatal complications that place them at significant risk. Although this latter group of newborns represents a relatively small percentage of the intensive care nursery population, they must be considered at significantly high risk and are usually found to be suffering from disorders similar to the complications of prematurity that do lasting damage. They are frequently babies with birth injury and hypoxia, respiratory distress, and transient metabolical, hematological, and/or infectious problems — all of which can permanently insult the nervous system.

Much more specific information is available about the outcome of the low birth weight baby. Virtually all follow-up studies have shown a striking number of dull and retarded children in low birth weight follow-up groups.[20, 21] Even in the best environment, the proportion of children considered to be developmentally below average at 2 and 4 years of age ordinarily increases as birth weight decreases. Because the general outcome for the larger low birth weight baby differs characteristically from the smaller babies — i.e., those weighing less than 1580 grams (3 lb 8 oz) — it is discussed separately.

When one considers the very small premature — less than 1580 grams — one finds that literally all statistics agree on a poor prognosis regarding future development.[21, 51, 64] Within this group of prematures Lubchenco reports an incidence and severity of retardation that is inversely related to birth weight. Her findings are corroborated by Drillien,[21] Knobloch,[43] and others, all of whom report an overall incidence of mild to moderate retardation in approximately 65 per cent of the sample studies. In Lubchenco's population, 48 per cent of the retarded children had associated findings of cerebral palsy. Of additional interest was the fact that, of the 35 children followed with normal intelligence in this very low birth weight group, 20 were experiencing difficulty in school characterized by significant underachievement, problems in speech, reading, mathematics, and behavior.

In low birth weight babies with birth weights greater than 1580 grams (3 lb 8 oz), less severe but significant defects in intellectual development are also shown. This is particularly true of such babies in the lower social classes. The incidence of actual mental retardation in these larger premature infants in good homes is only slightly above that in the full-term population, whereas the mental retardation incidence doubles when the matched low birth weight population is born into lower social classes. However, when social class is disregarded, it has been demonstrated that most of these low birth weight babies performed significantly below their siblings from identical genetic and environmental milieus.

From the findings reported so far, it seems likely that a combination of inferior genetic endowment, a poor environment, and restricted opportunity has a more marked effect on development in those who were small at birth than on maturely born children from a similar background.[21]

An I.Q. of 100 in a child of a middle- or upper-class home, where parents and siblings are of above average or superior intelligence, may represent mental retardation to that family. Again, of the small babies who did score within the average range of intelligence at 2 and 4 years of age, there is subsequently an increased incidence of learning disability and school maladjustment.[21]

Maturation and Critical Periods

Piaget and Inhelder[38] take an essentially maturational position and argue that specific levels of cognitive development must be achieved before certain conceptual strategies can be learned. Piaget's view is that the more new things a child has seen and heard, the more he wants to see and hear; the broader the child's repertoire (developed by experience), the more new relationships he can discover. Intelligence incorporates all the given data of experience within its framework, thus the greater variety of experience to which the child has been exposed, the greater his capacity for coping with new experiences. This would be true of cognition as well as of sensorimotor intelligence. Piaget feels that in every case intellectual adaptation involves an element of assimilation; however, "assimilation can never be pure, because by incorporating new elements into its early schemata the intelligence constantly modifies the latter in order to adjust them to new elements."[65] The elements in the environment to which the child pays attention depend on his prior experience, and he learns only from that experience to which he attends.

Piaget, in stressing the importance of introducing learnings based on the natural stages of the child's interaction with the environment, alludes to the problem of "appropriate match" between the new encounters and the earlier schemata already assimilated into the individual's repertoire. This principle of the match seems to be of great significance in both motivational and intellectual development, since a teacher or parent must match the learning encounter to the child's critical period of development in which there is maximum capability for and interest in the learning of that particular material.

Social/Cultural and Parental/Environmental Influences

Bruner,[7] using Piaget's work as a cornerstone, has used a wide variety of experimental techniques and has taken as one of his major themes the impact of culture in the nurture and shaping of growth. His view is that cognitive growth occurs as much from the outside in as from the inside out — a belief which reinforces the previous statements regarding the nature/nurture issue.

Freiberg and Payne,[26] in a review of parental influence on cognitive development in early childhood, report mounting evidence for the potency of early environment in shaping later cognitive abilities. They note that early learning effects in children and "cumulative deficit," resulting from deprived environment, have been reviewed by many researchers. They cite Zingg's study[96] to point out that there appear to be extremes of social and cultural deprivation resulting in a cumulative deficit for which compensatory training provides only limited benefit. They conclude that "the formation of cognitive and intellectual skills can be reasonably conceived of as developmental in nature and modifiable by variation in the environment."[26] Further, they raise the question, "If such is granted, then how might changes be effected in early intellectual development through the use of appropriate child-rearing and educational practices?"[26]

It is a question that has prompted considerable further investigation. If human cognitive development is to be enhanced through early environmental enrichment or other intervention, then the home or other preschool program — not the traditional school — is to be recognized as the beginning educational unit. Gray[30] stated that the mother plays a key role in providing environmental control conducive to learning in the child's early years. Rolcik[71] found a significant relationship between scholastic achievement and parental interest in the child and in his education. Shaw and White[78] concluded that the child-parent identification bore a relationship to school performance, whereas Norman[62] found that the parent value system influenced academic achievement. Hess and Shipman,[36] who assessed interaction patterns of mothers and children from several socioeconomic classes, argue that cognitive growth is dependent on the cognitive meaning in the parent-child communication system, and that the mother's pattern of interaction and communication with the child tends to determine whether the child later will actively participate in formal educative processes and other basic institutions of society. Renewed interest is being focused on the interaction between the family and the infant at risk.[2] Siqueland,[80] for example, found that intervention with premature infants in the first few weeks of life affected the babies in such a way that the infants evoked more responsive interaction from the mothers. Similarly, Falender[22] found that intervention produced more talkative children who asked more questions and made more requests, which in turn provoked more verbal interaction from the mother. This view suggests that the timing of various accomplishments might be important insofar as it influenced the mother's response to her child. Moreover, major federal projects, such as the Office of Child Development (OCD) Project Homestart, as well as private efforts, such as the Meeting Street School efforts,[13] manifest the emphasis again being placed on intervening with the parents and families of handicapped children.

Fowler[25] summarized a large body of research done by educators and psychologists who have evaluated techniques applied to children during their preschool years in order to accelerate intellectual growth. Fowler felt that Gesell[27] and McGraw,[53] who attempted to support a maturational point of view, often de-emphasized the fact that "specific training invariably has produced large gains regardless of whether training came early or late in development." In this same review, studies on verbal memory and language improvement point to the advantages of early verbal stimulation provided by oral, written, and dictorial material, as well as to the general experience gained in making observations and learning to discriminate between objects.

Bloom[4] studied all of the available data published from a number of longitudinal studies carried out over the last half-century. He concluded that "the introduction of the environment as a variable makes a major difference in our ability to predict the mature status of a human characteristic."[4] Furthermore, he suggested that "in terms of intelligence measured at age 17, about 50 per cent of the development takes place between conception and age 4, 30 per cent between ages 4 and 8, and about 20 per cent between ages 8 and 17."[4]

Caldwell,[10] who performed a systematic study of the environments of infants of varying socioeconomic status (SES) backgrounds, stressed that it is as important to measure the environment in which development is occurring as it is to measure the developmental processes themselves. Bayley studied the relationships between infant development and SES. In 1965, she published standardization data on a sample of 1409 children. At all assessment points up to 15 months of age there were no significant differences as a function of sex, birth order, parental education, and other similar parameters. Coleman[12] found that as early as first grade most groups of children from lower SES backgrounds and most children representing minority groups tended to score significantly lower than the national average on most measures of school achievement and, thus, lower than children from high SES backgrounds. These deficits increased as the children progressed through the typical school experience. Caldwell noted that lower SES homes varied greatly in their stimulation value and that SES alone failed to discriminate low SES children with favorable environments from those with unfavorable environments.

The heterogeneity of home stimulation among lower SES families is also supported by Werner, Bierman, and French,[92] who found that their educational stimulation scale was much more predictive of later school success than was any SES variable. Their study, *The Children of Kauai*, was published in 1971 by the University of Hawaii Press. It was a longitudinal study of children from the perinatal period through age 10, from a wide range of ethnic and socioeconomic backgrounds, and all living on the Island of Kauai, one of the Hawaiian Islands. It was discovered that for each 1000 live births an estimated 1311 pregnancies were required because of early and late fetal deaths. Of the 1000 live births, about 850 children remained free of any observable physical defect or serious developmental disability at 2 years of age. By the time they were 10 years old, only two thirds of the original 1000 children were considered to be functioning satisfactorily in school and to be having no recognizable physical, intellectual, or behavioral problem. During the period of conception to 10 years of age, about one half of the original 1311 subjects either died or developed some handicap. Many of the handicaps were the more diffuse, ill-defined, minimal cerebral dysfunction manifestations of the sort seen in the deaf-blind rubella child or the child who has sensory impairment problems for other reasons. This study analyzes a great deal of data and demonstrates some very meaningful relationships between various environmental predictions of handicap. For example, the structured family interview item dealing with an educational-stimulation rating predicted the I.Q. a child might have ten years later better than did parental socioeconomic status, education, occupation, or intelligence. Environmental casualties are shown to take a much greater toll than hereditary factors—a fact underscoring the urgent need for prevention and intervention programs that begin at, or before, conception.

A SUGGESTED MODEL

Screening and Assessment

This section is drawn from a much more extensive monograph[56] that addresses itself to the state of the art and science of operational or proposed methods and materials for the early developmental screening and assessment of young children. Table 17–1 presents existing techniques and instruments for the early identification of children who have various developmental disorders or who are at considerable risk of later experiencing such disorders. It lists a representative sample of tests and procedures for identifying, evaluating, and classifying the developmental status of young children. To obviate a hardening of the categories by not going beyond such a paralysis of analysis, an intervention/prevention scheme is presented in Table 17–2.

To categorize neatly and consistently everything germane to the aforementioned is virtually an impossible task; nevertheless, to establish a point of departure for future more comprehensive and sophisticated endeavors is attempted in this model. To stay within the scope and focus of this chapter, discussion of the following considerations has been omit-

Developmental Domain	Page[1]	Test or Procedure	Developer(s) Author(s)	Age Range[2]	Reliability	Validity	Time[5]	Cost per Child[6]	Administration[7]	Recommended Stage[8]
PHYSICAL	30	Automated Multiphasic Health Testing Services	Collen & Cooper	Over 4 yr.	A[3]	A	70	30[5]	Mix	Ter.
	34	Biochemistry & Cytogenetics	Guthrie	5-3 mo.	A	A	U[4]	<1	LT&EE	Sec.
	38	Amniocentesis	O'Brien	C-B	A	A	60	20	P	Sec.
	40	Metabolic	Howell, Holtzman & Thomas	B-3 mo.	A	A	<30	2	LT&EE	Sec.
	41	Ultra-Micro Automated System	Ambrose	B-3 mo.	A	A	60	1	Mix	Sec.
	42	Nutritional Status	Fomen	B-30 mo.	A	U	20	1	PP	Sec.
	44	Gestational Age	Lubchenco	B-1 mo.	A	A	5	2	PP	Sec.
	49	Statistical Mortality Morbidity	MCH	B-12 mo.	A	A	Neg[9]	Neg	P	Pre-Pri
	51	Statistical Epidemiology	Tarjan, et al.	Pre-B	A	A	Neg	Neg	P	Pre-Pri
	55	Data System	Scurletis, et al.	Pre-B	A	A	Neg	Neg	PP	Pri
	57	Prevention	de la Cruz & LaVeck	Pre-C	U[4]	U	U	U	P	Pri & Pre-Pri
	58	Apgar Rating	Apgar	B	A	A	6	1	P	Pri
	59	Vision	Press & Austin	Over 30 mo.	U	U	Neg	1	PP	Pri
	61	Eye Screening	Barker & Hayes	B-5 yr.	A	U	Neg	<1	PP	Pri
	62	Electro-Oculograph	Petre-Quadens	1-6 yr.	A	U	120	10	LT&EE	Ter.
	64	Hearing High-Risk Register	Hardy	C-3 yr.	A	U	Neg	Neg	PP	Pri
	67	Hearing Screening	Young; Downs & Silver	9-12 mo.	A	A	5	2	PP	Pri
INTELLECTUAL/ COGNITIVE	71	Potential Battered Children	Kempe & Helfer; Walworth & Metz; Gil	C-2 yr.	U	U	U	U	P	Sec.
	73	Vocalization Analysis	Fillippi & Rousey	B-12 mo.	A	U	40	20	PP	Sec.
	75	Behavioral & Neurological Assessment Scale (I)	Brazelton	B-3 yr.	A	U	40	30	P	Ter.
	75	Neuro-Developmental Observation	Ozer & Richardson	Over 5 yr.	U	U	20	15	PP	Sec.
	80	Attention to Discrepancy	Kagan	B-12 mo.	A	U	30	20	LT,EE	Ter.
	83	Ordinal Scales of Cognitive Dev.	Uzgiris & Hunt	B-3 yr.	U	U	60	30	P	Sec.
	86	Infant Intelligence Scale (CIIS)	Cattell	B-30 mo.	A	A	25	15	P	Sec.
	86	Bayley Scale of Infant Dev.	Bayley	B-30 mo.	A	A	45	25	P	Sec.
	88	Kuhlmann-Binet Infant Scale	Kuhlmann	B-30 mo.	A	A	30	15	P	Sec.
	88	Griffiths Mental Dev. Scale	Griffiths	B-4 yr.	A	A	30	15	P	Sec.
	89	Gesell Developmental Scale (Revised Scale)	Gesell, et al.	B-5 yr.	A	A	40	30	P	Sec.
	92	Ivanov-Smolensky	Luria	B-24 mo.	A	U	20	15	LT	Ter.
	93	Habituation	Lewis, et al.	B-18 mo.	A	A	30	15	PP	Sec.
	93	Psychophysiological	Crowell	B-3 mo.	A	A	80	50	Mix	Ter.
LANGUAGE	98	Playtest	Friedlander	3-12 mo.	A	A	50	25	LT,EE	Ter.
	99	Infant Cry Analysis	Ostwald, et al.	B-3 mo.	A	U	30	15	LT,EE	Ter.
	104	Expressive Language	Reyes, et al.	2-4 yr.	A	A	40	20	PP	Sec.
	108	Receptive Language	Marmor	1-3 yr.	A	U	30	15	PP	Sec.
	108	Early Language Assessment Scale	Honig & Caldwell	3-48 mo.	A	U	30	15	PP	Sec.
SOCIAL/ EMOTIONAL	114	Behavioral & Neurological Assessment Scale (II)	Brazelton, et al.	B-3 yr.	A	U	30	15	PP	Sec.
	114	Behavior Problem Checklist	Quay & Peterson	B-4 yr.	U	U	30	20	P	Sec.
	116	Rimland Diagnostic Check List	Albert & Davis	B-4 yr.	U	U	30	20	P	Sec.
	116	Behavior Checklist	Ogilvie & Shapiro	3-6 yr.	A	U	45	30	P	Sec.
	117	Quantitative Analysis of Tasks	White & Kaban	1-6 yr.	A	A	60	30	PP	Sec.
	118	Behavior Management Observation Scales	Terdal, et al.	B-4 yr.	U	U	60	20	PP	Sec.
	118	Vineland Soc. Maturity Scale	Doll	B-18 yr.	A	U	25	10	PP	Pri/Sec.
	118	Preschool Attainment Record	Doll	B-7 yr.	A	U	30	15	PP	Pri/Sec.
	119	Behavioral Categorical System	DeMyer & Churchill	2-5 yr.	A	U	30	20	P	Sec.
	125	Psychological Assessment: Functional Analysis	Bijou & Peterson	B-Adult	A	A	U	U	P or PP	Ter.
COMPREHENSIVE SYSTEMS	128	First Identification of Neonatal Disabilities (FIND)	Wulkan	B-12 mo.	U	U	U	U	U	All
	128	System of Comprehensive Health Care Screening & Service	Scurletis & Headrick	C-4 yr.	A	U	U	U	Mix	All
	132	Preschool Multiphasic Program	Belleville & Green	B-4 yr.	A	U	U	U	Mix	All
	136	Pluralistic Assessment Project	Mercer	5-11 yr.	U	U	U	U	U	Sec.
	140	Pediatric Multiphasic Program	Allen & Shinefield	Over 4 yr.	A	A	120	30	Mix	All
	143	Rapid Developmental Screening Checklist	Giannini, et al.	B-5 yr.	A	A	5	1	PP,P	Pri
	143	Guide to Normal Milestones of Development	Haynes	B-3 yr.	A	A	15	5	PP,P	Pri
	150	Developmental Screen, Inventory	Knobloch, et al.	5-18 mo.	A	A	20	10	PP,P	Pri
	153	CCD Develop. Progress Scale	Boyd	B-8 yr.	A	A	30	15	PP	Pri
	156	Denver Develop. Screening Test	Frankenburg & Dodds	B-6 yr.	A	A	30	15	PP	Pri
	16	At Risk Register	Alberman & Goldstein, Sheridan; Oppe; Walker	Pre-C	A	A	Neg	Neg	PP,LT	Pre-Pri.
	19	Risk Factors (Kauai Study)	Werner, Bierman & French	Pre-C to 12 yr.	A	A	Neg	Neg	PP	Pre-Pri.

NOTES: 1. Number of first page discussing topic in *Screening and Assessment of Young Children at Developmental Risk* (by Meier, J. H., Wash., D.C., Gov't. Printing Office, 1973).
2. C=Conception; B=Birth.
3. A=Adequate, i.e., >.75, when reported or estimated (only concurrent and face validity — not predictive).
4. U=Unknown — in any category indicates that data are either unavailable, too variable, or sparse.
5. Minutes required for administration and interpretation — estimated average with normally developing child.
6. Estimated total in dollars including time and materials under optimum conditions.
7. P=Professional trained to administer test(s); PP=ParaProfessional, properly trained; LT=Laboratory Technician; EE=Elaborate Equipment (in laboratory and usually not portable); Mix=Combination of preceding. A trained professional is required to interpret test results.
8. Recommended Stage in Screening System — Pri=Primary; Sec.=Secondary; Ter.=Tertiary; Pre-Before.
9. Neg.=Negligible amount of time or cost per child.

TABLE 17-1.

Age	Satisfactory Progress If not ➤	Screening and Risk Assessment If screening results or risk factors are positive ➤	Evaluation, Close Observation and Diagnosis to ➤	Intervention and Follow-Along ➤
PRE-CONCEPTUAL	Intent to Conceive Adaptive & Physiological Readiness (Normal Maternal & Family History)	**Presence of One or More Maternal Risk Factors:** Physical/Medical 1. Malnutrition 2. Age < 16 or > 35 3. Poor Reproductive History 4. Suspect Metabolic and/or Genetic Disease Social/Behavioral 1. Low SES 2. Sixth Grade Education 3. Functionally Illiterate 4. Low Adaptive Behavior Rating	Nutritional/Metabolic Tests Derive Genetic Pedigree Literacy/Educ. Tests Adult Adaptive Behavior Rating (Nihira)	Genetic Counseling (Sterilization) Diet Therapy Contraceptive Counseling (Planned Parenthood) Maternal Training (Jr. & Sr. High School)
PREGNANCY (first 3 mo.)	Request for Service (Suspected Pregnancy Confirmed) Regular OB/GYN Checks Normal Progress	**Complications During Pregnancy:** 1. Infections 2. Rubella 3. Toxemia 4. Drug Overuse 5. Radiation 6. Blood Incompatibility 7. Malnutrition 8. Maternal Psychosis 9. Unwanted Pregnancy	Appropriate Medical Tests to Evaluate Maternal & Embryo Condition Amniocentesis Social/Behavioral Tests of Maternal Ability and Attitudes	Counseling Therapeutic Abortion Psychotherapy
PREGNANCY (last 3 mo)	Regular OB/GYN Checks Normal Progress	Above First Request for OB/GYN Services	Evaluation of Maternal and Fetal Condition	Counseling Positive Attitude (Natural Childbirth)
NEWBORN (first month)	Hospital Admission Normal History of Pregnancy and Routine OB/GYN Checks Uneventful Delivery	**Complications During Delivery:** 1. Hemorrhage 2. Dystocia 3. Excessive Anesthesia 4. Trauma 5. Placental Damage 6. Cesarean 7. Premature (SGA) 8. Postmature 9. Hospital Admission with no prior OB/GYN checks	Appropriate Medical Tests to Evaluate Maternal & Infant Condition	Necessary Procedures to Insure Maternal & Infant Viability
INFANCY	Normal Neonatal Growth & Development	**Pediatric Physical and Developmental Exam:** 1. Apgar (5 min.) 2. Metabolic/Genetic Screens (e.g., PKU) 3. Trauma 4. Infection 5. Malnutrition 6. Head Circumference 7. Guide to Normal Milestones of Development (@ 1 mo.)	Behavioral & Neurological Assessment Scale (Brazelton & Harowitz, @ 1 mo.) Environmental Quality Maternal Attitude & Aptitude (Below for Specifics)	Sensory Stimulation Behavior Modification Environmental Enrichment Maternal Training (Below for Specifics)

*Developed by John H. Meier for Governor's Conference on Prevention of Developmental Disabilities, held at Newporter Inn and University of California (Irvine), December 1973. The author gratefully acknowledges the suggestions from Tadashi Mayeda regarding the format of this matrix.

TABLE 17–2.

	Satisfactory Progress — If not →	Screening and Risk Assessment — If screening results or risk factors are positive →	Evaluation, Close Observation and Diagnosis to →	Intervention and Follow-Along →
INFANCY	Monthly Well-Baby Physical & Developmental Checks (1st year) Normal Progress Bi-Monthly Physical & Developmental Checks (2nd year)	Physical: 1. Trauma 2. Infection 3. Diseases 4. Malnutrition 5. Vision 6. Hearing 7. Maternal Postnatal Depression/Rejection, Neglect and/or Abuse 8. Prolonged Separation of Infant from Mother Developmental: 1. Rapid Developmental Screening Checklist (@ 6 mo. and 1 yr.) 2. Developmental Screening Inventory (@ 18 mo.) 3. Developmental Progress Scale (@ 12, 18, 24 mo.) 4. Denver Developmental Screening Test (@ 12, 18, 24 mo.) 5. Behavior Problem Checklist (@ 24 mo.)	**DQ** { Albert Einstein Scales of Sensori-Motor Development; Fantz-Nevis Visual Preference Test; White-Held Visually-Directed Prehension Test; Gesell Developmental Scale; Bayley Scale of Infant Development **Cognitive Q** { Ordinal Scales of Cognitive Development; Griffiths Mental Development Scale; Kahn Intelligence Tests; Infant Rating Scales (Hoopes); Kuhlman-Binet Infant Scale; Infant Intelligence Scale **Environ./Parent Q** { Caldwell (A Procedure for Patterning Responses of Adults and Children — APPROACH); Parental Attitude Research Instruction; Parents' Attitude Scale; Wechsler Adult Intelligence Scale (WAIS) **Language Q** { Irwin Speech Sound Development Test; Prelinguistic Infant Vocalization Analysis (Ringwell, et al.); Shield Speech and Language Development Scale; Early Language Assessment Scale (Honig); Receptive-Expressive Emergent Language (REEL, Bzoch) **Ach. Q** { Preschool Attainment Record (Caldwell) **P/N Q** { Psychophysiological/Neurological Maturity (Brazelton, Crowell) **Soc./Behav. Q** { Vineland Social Maturity Scale (Doll); Emotional Maturity Adaptive Behavior Scales (Nihira); Pluralistic Assessment (Mercer)	Bobath & Ayres (Physical Therapy) Gordon (Home Learning Center — Florida) Gray, et al. (DARCEE) Heber & Garber (Milwaukee Project) Keister (North Carolina Infant Day Care) Lally & Honig (Syracuse Infant Project) Levenstein (Mother-Child Home Program) Parent-Child Center Programs (Costello, Holmes) Meier, et al. (Education System for High-Risk Infants) Robinson (Frank Porter Graham Infant Project) Weikart & Lambie (Ypsilanti-Carnegie Infant Education Project) White & Kaban (Brookline)
TODDLER – EARLY CHILDHOOD	Periodic Physical & Developmental Checks Approximately every 6 mos. Normal Progress	Pediatric Physical Exams (See above considerations) Developmental Screens (Nos. 4, 5, & 6 above) Peabody Picture Vocabulary Test Goodenough-Harris Draw-A-Person	Preschool Inventory (Caldwell) Leiter International Performance Scale Slosson Intelligence Test Raven's Coloured Progressive Matrices Stanford-Binet Intelligence Scale Developmental Articulation Test (Hejna) Illinois Test of Psycholinguistic Abilities (Kirk & McCarthy) Verbal Language Development Scale (Mecham) Developmental Test of Visual-Motor Integration (Beery) Developmental Test of Visual Perception (Frostig) Detroit Tests of Learning Aptitude Minnesota Preschool Scale IPAT Test of G-Culture Fair (Cattell) Arthur Point Scale of Performance Tests California Tests of Mental Maturity and Personality Metropolitan Readiness Test Oseretsky Tests of Motor Proficiency Weoman Auditory Discrimination Test	Model Preschool Programs (by last names of developers — for description, see SOURCES below). Anderson & Bereiter Blank Hooper Kamii Karnes, Zehrbach, & Teska Meier Miller & Camp Montessori Nedler Nimnicht Palmer Robison Shaeffer & Aaronson Weikart Whitney & Parker

SOURCES: Battle, C. U. & Ackerman, N. C. *Early Identification and Intervention Programs for Infants with Developmental Delay and Their Families — A Summary and Directory.* Chicago: Nat'l. Easter Seal Society for Crippled Children & Adults, 1973. Guthrie, P. D. with Horne, E. V. *Measures of Infant Development — An Annotated Bibliography.* Washington, D.C.: Head Start Test Collection, Educational Testing Service, December 1971. Hoepfner, R., Stern, C. & Nummedal, S. G. (Eds.), *CSE-ECRC Preschool/Kindergarten Test Evaluations.* Los Angeles, Calif.: UCLA Graduate School of Education, 1971. Meier, J. H. *System For Open Learning, Facilitator's Handbook I: SOL Foundations and Rationale.* Denver, Colo.: Publisher's Press Inc./Monitor Publications, 1973. Meier, J. H. *Screening and Assessment of Young Children at Developmental Risk.* Washington, D.C.: U.S. Gov't. Printing Office, 1973. Parker, R. K. (Ed.), *The Preschool in Action — Exploring Early Childhood Programs.* Boston, Mass.: Allyn and Bacon, Inc., 1972. Williams, T. M. *Infant Care — Abstracts of the Literature.* Washington, D.C.: Consortium in Early Childbearing and Childrearing, August 1972.

TABLE 17–2. *Continued.*

ted: (1) federal mandates for early and periodic screening, diagnosis, and treatment of developmental disabilities; (2) definition(s) of developmental anomalies; (3) controversial issues regarding at-risk registers and populations; and (4) some of the broader societal, ethnic, ethical, and legal considerations relative to any massive identification/intervention programs. Since any complete and balanced presentation should deal with these areas, each factor is considered in detail elsewhere in this book.

Significant advances have been made beyond both the Procrustean notion of making all individuals fit a predetermined ideal human mold and the Spartan notion of gross screening by throwing infants into cold water and keeping only those who can save themselves. This chapter is concerned with the reverse side of the coin, that is, with the early identification of those individuals who are likely to sink. Fortunately, these persons either can be taught to swim or can be protected from a fatal total immersion. At least they can be enabled to crawl, walk, or run. The massive screening program that is envisioned promises to reveal those factors that contribute to developmental risks in varying degrees and thereby to allow them to be weighted in terms of their relative contribution to handicapping conditions. This screening and identification, in turn, will facilitate the prevention of more serious developmental disabilities, which the earlier sections of this chapter indicate can be attenuated by early detection and appropriate intervention. Moreover, the appropriateness and efficacy of various intervention/prevention methods and materials can be determined. Especially germane to the state of the art and science of early screening and assessment is the statement in Footnote 3 of Table 17–1, which points out the inadequacy or absence of predictive validation for most extant tests and procedures used in isolation. This potentially serious deficiency is lessened by the fact that undoubtedly predictive screening will be accomplished with a battery of screening procedures from which most valid portions are extracted and in which the composite will be more accurately predictive than the sum of the parts.

Beyond the Paralysis of Analysis

When a satisfactory comprehensive developmental screening system has been thoroughly field-tested, it is useful only if it is matched with practical intervention programs. Table 17–2 presents a matrix designed to orchestrate the multiple screening, evaluation, and intervention considerations. Several appropriate intervention procedures and programs have been reported in the literature. Column V of Table 17–2 contains a representative, but far from exhaustive, list of such "Intervention and Follow-Along" efforts. Since much of this book is concerned with genetic considerations, which are represented in both the preconceptual and pregnancy sections of Table 17–2, no discussion of these will be included in this chapter. Nevertheless, it should be readily apparent from Column III, "Screening and Risk Assessment," that there are many preconceptual factors, in addition to genetics, which singularly or in combination are salient in the determination of risk for probable developmental disabilities in the potential offspring of procreative partners. To the extent that these factors are present, one should then proceed to Column IV, "Evaluation, Close Observation, and Diagnosis," to delineate more clearly the likelihood of conceiving and giving birth to a developmentally disabled child.

It is rare, of course, that such an opportunity for preconceptual analysis presents itself. This is a strong argument for courses and other educational programs aimed at adolescents in junior and senior high school to inform them about the various risk factors and the means either for preventing pregnancy in the first place or for insuring as many positive conditions as possible for a planned pregnancy. One promising movement in recognition of this need is the National Consortium on Early Childbearing and Childrearing, which is sponsoring state-wide conferences throughout the United States to deal with these critical issues. The Consortium is working in conjunction not only with school boards, school administrations, health care centers, and representatives from other such institutions, but also with the pregnant and frequently unwed teenage mothers (and occasionally fathers). Historically, these young patients have been systematically discriminated against, ridiculed, and maintained in a state of ignorance and shamefulness by a punitive puritanical system, all of which serves to exacerbate the debilitating features of the typically unwanted pregnancy and attendant responsibilities.

It is obviously beyond the purview of this chapter to elaborate upon the various procedures, instruments, and model programs indicated at various strategic points in the matrix. Needless to say, it is most desirable for any potential subject in this system to begin and remain normal, and thereby progress down Column I. However, for those who yield positive screening results and are subsequently found to have *bona fide* developmental delays or disabilities, the sooner they are identified and placed in properly matched remediation/prevention programs, *e.g.*, Column V, the better it is for the child, the professional, and the society. Since individual subjects and individual professionals and paraprofessionals bring various requirements to each case, several options are mentioned in the evaluation and intervention columns.

Some Intervention Models

Specific instructional techniques used by parents to assist the young child in cognitive skill acquisition are not numerous. McCandless[52] reported a study by Irwin indicating that working-class mothers who spent 10 minutes per day reading to their child from 12 months to about 20 months of age achieved improvement in their youngsters in all phases of language acquisition. Fowler,[25] in his surveys of gifted children — one of whom was his daughter — indicates that these children were generally exposed to instructional techniques developed by a parent, and many learned to read by 3 years of age. Moore,[58] Meier,[54] and Nimnicht[61] reported encouraging results from using a nonautomated "Talking Type-writer" system, among many other accelerated "academic" achievements, for enabling very young preschoolers to learn to read. These procedures and concepts were then formulated into an autotelic responsive environment for widespread implementation.[54, 61]

Weikart and Lambie[90] stated that the proliferation of preschools may be attacking the problem of enhancing educational opportunity of the disadvantaged child in the wrong way. Perhaps retraining the parents might ameliorate his learning deficits. The Perry Project in Ypsilanti, Michigan, uses weekly home visits to provide direct instruction to the child, to share with the mother information about the educative process, and to encourage her to participate in the actual teaching of her child. The home visitor's demonstration of child-management techniques indirectly teaches the mother the most effective ways of handling her children.

Another current project that is attempting to meet the needs of children from homes lacking cultural and economic advantages is a research and demonstration unit at Upstate Medical Center in Syracuse, New York. Richmond and Caldwell have designed an educationally oriented day care program for children from 6 months to 5 years of age. The program goals are the following: to create an atmosphere in which infants and children can grow happily, to be a bridge between whatever culture the parents offer and the culture of a larger world, and to provide specific learning experiences for stimulating cognitive growth of the infants and children. The staff of the unit, which is part of a larger research project on patterns of learning during the first years of life, includes educators, nurses, social workers, researchers, and pediatricians. In the first year, the I.Q.'s of the 29 children who attended the center for 3 months or more showed average increases of 5.5 points, whereas the control group showed a downward drift. In general, when the child's environment, including the parents, is seriously deficient, the I.Q. of a potentially bright child (I.Q. of 120) can drop twenty points in the period between the ages of 6 months and 24 months.

Children from more stimulating home environments gained about twice as much in I.Q. points as children from less stimulating homes. This environmental difference in response to day care may indicate the limitations of day care in counteracting the effects of an unstimulating home environment.[88] Recent efforts at Syracuse are directed toward increased intervention efforts delivered in the home.[94]

The curriculum of many intervention programs is highly structured and theoretically oriented, as was the case with the Piaget-based curriculum in Syracuse. The home-based intervention studies of Gordon,[29] Palmer and Rees,[63] and Weikart and Lambie,[90] however, showed that the specific curriculum may not be as important to the success of the program as is the method by which any well-conceived guideline to stimulation is carried out. When the stimulation is parent-child centered, with emphasis on playfulness and on the existing strengths of the child, mothers tend to continue along the stimulation guidelines between sessions and after the termination of home visits. Both the Gordon and Weikart programs showed gains prior to 12 months of age. Palmer and Rees showed substantial gains between 24 and 36 months of age.

The Washington, D.C. Infant Education Research Project[75] supported the guideline approach to curriculum, but emphasized the need for careful training of intervention personnel. Schaeffer and Aaronson documented the lack of response of some mother-infant dyads to the intervention program. Mothers who scored high on hostility were associated with hostile infants and with depressed mental test scores at 36 months. Enrichment infants who showed the greatest gains had mothers who had "flexible, spontaneous, and pleasant" relationships with them. The importance of the quality of the caretaker/child interaction has gained renewed emphasis in the studies reported by Beckwith[2] and Lewis and Rosenberg.[49a]

The home-based program of Levenstein[47] has shown some of the most impressive gains in I.Q. scores during and following the two-year program that begins at 20 months of age. The explicit purpose of the home visits is for the "toy demonstrator" to model verbal stimulation techniques for the mother through the medium of a prescribed sequence of books and toys. The mother is encouraged to take over the demonstrating of the toy as she becomes able. Recent work by Levenstein[48] has shown a cost/benefit advantage over other home-based programs. Volunteers and mothers who have been through the two-year program with their own child have become effective toy demonstrators themselves; recent reports show that gains made by the children taking part in this early intervention program have been maintained into the first grade.

In other studies, when the experimental subjects have "regressed" toward the control group, they have done so because the program terminated too early[10, 29, 47, 90] and failed to involve the parents to a degree sufficient to alter the home environment.[85] Levenstein attributes the unusual holding power of her program's effects to having influenced the mother's child-rearing habits, starting as early as 20 months, as well as to having continued the intervention intensively for two years during the most critical period for language development. Lane[46] states that the mother is a small child's most important teacher. To influence the mother's style of teaching, the staff has devised different tasks—games, puzzles, walking boards, and other similar devices—and these are left in the home for one week and their use explained in detail to the mother: how she can lift out learnings, influence attitudes, and stimulate curiosity, by the use of these tasks.

The next noteworthy study is the Milwaukee Project.[35] Besides being mentioned in numerous technical and research-oriented journals, it has received acclaim in several more popular magazines such as *American Education,* which recently ran a feature article entitled, "Can Slum Children Learn?" The answer is a very strong *yes.* As a result of a survey that revealed that children whose mothers are retarded have a fourteen times greater chance of being retarded themselves, an experimental intervention program was launched to determine whether or not some of the retardation can be prevented. The highly structured program was designed to enrich greatly the early environment of these children.

The Milwaukee Project started with newborn infants and brought them into a center for five days a week with structured and unstructured experiences. Unfortunately, the specific curriculum used with each infant was not carefully documented, although considerable information does exist about the overall program. It is now exigent that replication studies document very carefully every part of the intervention program and the differential impact of every phase of the program on each child, because these children in the Milwaukee Project made phenomenal progress (mean I.Q. of experimentals, independently assessed in second grade, is 120, compared to mean I.Q. of 80 in controls, whose I.Q. is still dropping).

The Office of Child Development has sponsored a series of experimental Parent-Child Centers across the country. These are each designed to provide comprehensive services to approximately 100 children, from birth to 3 years of age, who are euphemistically described as disadvantaged, environmentally-deprived, culturally different, of low SES, and belonging to ethnic minority groups of the culture of poverty. The 36 programs around the country bring these children into centers for various and sundry services—including cognitive stimulation. Two booklets of infant curriculum, *How Babies Grow and Learn* and the *H.O.M.E. Infant Stimulation Program,* have been forthcoming from these collaborative efforts. These publications describe the kinds of things a mothering person might do with the infant, including illustrations, rationales, written procedures, and lists of necessary equipment. The equipment is very simple and readily available. To ensure easy and correct use, the pieces of equipment for each learning episode have been placed in cloth bags on which the instructions for use have been outlined. The Portage Project,[79] which serves 75 preschool multiply-handicapped children from infancy to 6 years of age is conducted exclusively in the children's homes and has developed a series of curriculum cards for a large series of sequential learning episodes.

Many experienced nonworking mothers acknowledge and complain that although they had plenty of time for their first child and showered all kinds of attention on him or her, with each subsequent child they had correspondingly less time to give to the younger ones. It is easier to neglect the little one than it is to ignore the ones who can now talk and make their demands known, so the younger ones consequently tend to get progressively less adequate interaction with adults. Studies on superior intelligence, including Terman's *Studies of Genius*[86] indicate that it is the firstborn who typically tends to grow and develop in the cognitive area most fully and rapidly. Some of the recent birth order studies also support this notion.[95a] The various apparatus for cribs are not designed to substitute for human mothering, but rather to supplement the interaction which an infant has with his environment during those many alert moments he would otherwise spend staring into space in a relatively sterile and unresponsive environment.

A truly responsive environment for infants probably facilitates their developing a sense of control over their destiny; this control is closely linked to competence. Such early experience may enable the child in later years to move easily into new situations, to deal efficiently with vast amounts of new tentative data, to be comfortable in complex situations, and to have a relatively high tolerance for ambiguity in situations — in short, to have effective coping behavior.

For the past decade, Meier has been interested and actively involved in creating a responsive nursery school and even an autotelic responsive crib environment. One experimental version[56a] allows the infant as young as 3 months old to: (1) summon the adults by sounding a buzzer or bell; (2) operate a videotape playback unit (at this time Sesame Street was just coming on television so some of the appropriate portions from it, as well as various face schema and other audio-visual materials, were included); (3) turn on a continuous loop film projector for movies on the ceiling; (4) control flashing lights strung around the periphery of the ceiling; (5) turn on an audio tape playback unit with music and/or language programs on it; (6) turn on a vibrator that helps expel intestinal gases or a heartbeat simulator, both of which have a very quieting tactile-kinesthetic effect; and (7) control the room illumination with an easily manipulated dimmer.

Several of the items are on time-delay switches, so they go off automatically and must be reactivated for another cycle. Other apparatuses including heaters, fans, and similar devices can be controlled to alter the features of the environment. A long arm supports a manipulandum that the baby can grab or just hit with a fisted swiping and cause a motorized mobile hung up above on another arm to rotate. The experimental version also has several event and time unit counters to record the number of times and duration of operation of various items to give some index to their autotelic properties, that is, to determine how intrinsically interesting they are to an infant.

It must, of course, be understood that the hardware just described is in no way intended to supplant the mother or other caretaking adults, but rather to supplement in the caretaking process. Ironically, such apparatuses frequently reveal capabilities in infants that mothers otherwise would never have suspected, much less fostered. From all of the foregoing, it should be evident that it is the enhancement of adult/infant interaction that infant education programs must accomplish for any real long-term results.

References

1. Bayley, N.: Mental growth in young children. *In* Yearbook of the National Society for the Study of Education. Vol. 39, Chicago, University of Chicago Press, 1940, pp. 11–47.
2. Beckwith, L.: "Caregiver-Infant Interaction and the Development of the Risk Infant." Los Angeles, University of California at Los Angeles, 1974 (Mimeographed).
3. Bennett, E. L. et al.: Chemical and anatomical plasticity of the brain. Science, 146:610, 1964.
4. Bloom, R. S.: Stability and Change in Human Characteristics. New York, John Wiley & Sons, 1964.
5. Bowlby, J.: Maternal Care and Mental Health. World Health Organization, 1951.
6. Brody, S.: Patterns of Mothering. New York, International Universities Press, Inc., 1951.
7. Bruner, J. S.: Toward a Theory of Instruction. Cambridge, Massachusetts, Harvard University Press, 1966.
8. Caldwell, B. M.: Descriptive evaluations of child development and of developmental settings. Pediatrics, 40:46, 1967.
9. Caldwell, B. M., and Richmond, J. B.: The Children's Center in Syracuse, N.Y. *In* Dittmann, L. L. (ed.): Early Child Care: The New Perspectives. New York, Atherton, 1968, pp. 326–358.
10. Caldwell, B. M., and Richmond, J. B.: Social class level and stimulation potential of the home. *In* Hellmuth, J. (ed.): Exceptional Infant. Seattle, Special Child Publications, 1967, pp. 455–466.
11. Casler, L.: The effects of extra tactile stimulation on a group of institutionalized infants. Genet. Psychol. Monogr., 1965.
12. Coleman, J. S.: Equality of Educational Opportunity. Washington, D.C., U.S. Government Printing Office, 1966.
13. Denhoff, E. et al.: "Parent Programs for Developmental Management." Providence, R.I., Meeting Street School, 1974 (Mimeographed).
14. Dennenberg, V. H.: Early experience and emotional development. Sci. Am., 208:138, 1963.
15. Dennenberg, V. H.: Critical periods, stimulus input, and emotional reactivity: A theory of infantile stimulation. Psychol. Rev., 71:335, 1964.
16. Dennenberg, V. H., and Thoman, E. B.: From Animal to Infant Research. Paper presented at the National Conference on Early Intervention with High Risk Infants and Young Children, University of North Carolina at Chapel Hill, May 5–8, 1974.
17. Dennis, W., and Najarian, F.: Infant development under environmental handicap. Psychol. Monogr., 71(7), 1957.
18. Deutsch, M.: The disadvantaged child and the learning process. *In* Passow, A. H. (ed.): Education in Depressed Areas. New York, Bur. Pub., Teachers College Press, Columbia University, 1963, pp. 163–179.
19. Deutsch, M. et al.: *The Disadvantaged Child.* New York, Basic Books, Inc., 1967.
20. Drillien, C.: A longitudinal study of the growth and development of prematurely and maturely born children. Arch. Dis. Child, 36:233, 1961.
21. Drillien, C.: The Growth and Development of the Prematurely Born Infant. Baltimore, Williams & Wilkins Co., 1964.

22. Falender, C.: Mother-child interaction and the child's participation in the Milwaukee project: An experiment in the prevention of cultural-familial mental retardation. Paper presented at Society for Research in Child Development Meeting, Philadelphia, 1973.

23. Fantz, R. L., and Nevis, S.: Pattern preferences in perceptual cognitive development in early infancy. *Merrill-Palmer Q.,* 13:77, 1967.

24. Findley, W.: Programmatic Research and Development in Early Educational Stimulation. Symposium of Amer. Educ. Res. Ass'n., Session 8.5, Chicago, February, 1968, (Unpublished).

25. Fowler, W.: Cognitive learning in infancy and early childhood. *Psychol. Bull.,* 59:116, 1962.

26. Freiberg, S., and Payne, D. T.: A survey of parental practices related to cognitive development in young children. Child Dev., 38:65, 1967.

27. Gesell, A : The ontogenesis of infant behavior. *In* Carmichael, C. (ed.): Manual of Child Psychology. New York, John Wiley & Sons, 1954.

28. Goldfarb, W.: Emotional and intellectual consequences of psychological deprivation in infancy: A re-evaluation. *In* Hock, P., and Zubin, J. (eds.): Psychopathology of Childhood. New York, Grune & Stratton, 1955, pp. 105–119.

29. Gordon, I.: Early Child Stimulation Through Parent Education: A Final Report to the Children's Bureau. Gainesville, Fla., University of Florida Press, 1969.

30. Gray, S.: Before First Grade. New York, Teacher's College Press, 1966.

31. Gray, S., and Miller, J. O.: Early experience in relation to cognitive development. Rev. Educ. Res., 37:475, 1967.

32. Hall, G. S.: Founders of Modern Psychology. New York, Appleton, 1912.

33. Harlow, H. F.: The maternal affectional system. *In* Foss, B. M. (ed.): The Determinants of Infant Behavior. New York, John Wiley & Sons, 1963, pp. 3–33.

34. Hebb, D. O.: The Organization of Behavior. London, Chapman & Hall, Ltd., 1949.

35. Heber, R. et al.: Rehabilitation of Families at Risk of Mental Retardation, A Progress Report. Madison, Wisconsin, University of Wisconsin Press, 1971.

36. Hess, R. D., and Shipman, V. C.: Early experience and socialization of cognitive modes in children. Child Dev., 36:869, 1965.

37. Hunt, J. McV.: Intelligence and Experience. New York, Ronald Press Company, 1961.

38. Inhelder, B., and Piaget, J.: The Growth of Logical Thinking. New York, Basic Books, Inc., 1958.

39. Irwin, O. C.: Infant speech: Effect of systematic reading of stories. J. Speech Hear. Res., 3:187, 1960.

40. Isaacson, R. L., and Nonneman, A. J.: Early brain damage and later development. *In* Satz, P., and Ross, J. J. (eds.): The Disabled Learner. The Netherlands, Rotterdam University Press, 1973.

41. Isaacson, R. L.: Recovery"?" from Early Brain Damage. Paper prepared for the National Conference on Early Intervention with High Risk Infants and Young Children, University of North Carolina at Chapel Hill, May 5–8, 1974.

42. Jensen, A.: How much can we boost I.Q. and scholastic achievement? Harvard Educ. Rev., 39:1, 1969.

42a. Kennell, J. H. et al.: Maternal behavior one year after early and extended postpartum contact. Dev. Med. Child Neurol., 16:172, 1974.

43. Knobloch, H., et al.: Neuropsychiatric sequelae of prematurity: A longitudinal study. J.A.M.A., 161:581, 1956.

44. Korner, A., and Grobstein, R.: Visual alertness as related to soothing in neonates: Implications for maternal stimulation and early deprivation. Child Dev. 31:867, 1967.

45. Krech, D., Rosenzweig, M. R., and Bennett, E.: Relations between brain chemistry and problem-solving among rats raised in enriched and impoverished environments. J. Comp. Physiol. Psychol., 55:801, 1962.

46. Lane, M. B.: "Consultation to the Campus Planning Committee of the Parent-Child Educational Centers of Litchfield Park Area, Arizona." Tempe, Arizona, Arizona State University, February, 1968 (Unpublished).

47. Levenstein, P.: Cognitive growth in preschoolers through stimulation of verbal interaction with mothers. Am. J. Orthopsychiatry, 40:426, 1970.

48. Levenstein, P.: Manual for replication of the mother-child home program. New York, Family Service Ass'n. of Nassau County, Inc., 1971 (Mimeographed).
49. Levine, S.: Stimulation in infancy. Sci. Am., 1960.
49a. Lewis, M., and Rosenburg, L. A. (eds.): The Effect of the Infant on Its Caregiver. New York, John Wiley & Sons, Inc., 1974.
50. Lewis, O.: Pedro Martinez: A Mexican Peasant and His Family. New York, Random House, 1964.
51. Lubchenco, L.: High Risk Infant Follow-Up Study. Progress Report, (Unpublished).
52. McCandless, B.: Children and Adolescent Behavior and Development. New York, Holt, Rinehart, & Winston, 1961.
53. McGraw, M. B.: Behavior of the newborn infant and early neuro-muscular development. Res. Publ. Assoc. Res. Nerv. Ment. Dis., 19:244, 1939.
54. Meier, J. H.: An autotelic nursery for deprived children. In Masserman, J. H. (ed.): Current Psychiatric Therapies, Vol. 10. New York, Grune & Stratton, 1970, pp. 30–45.
55. Meier, J. H.: The evaluation of remediation/prevention methods and materials. In Satz, P. (ed.): The Disabled Learner: Early Detection and Intervention. Rotterdam, The Netherlands, Rotterdam University Press, 1973, pp. 187–207.
56. Meier, J. H.: Screening and Assessment of Young Children at Developmental Risk. Washington, D.C., U.S. Government Printing office (DHEW Pub. No. OS 73–90), 1973.
56a. Meier, J. H., Segner, L. L., and Greuter, B. B.: An education system for high-risk infants: A preventive approach to developmental and learning disabilities. In Hellmuth, J. (ed.): The Disadvantaged Child. Vol. 3. New York, Brunner/Mazel, Inc., 1970, pp. 405–444.
57. Montes, F., and Risley, T. R.: Evaluating traditional day care practices: An empirical approach. Child Care Q., 1974. (Submitted for publication).
58. Moore, O. K.: Autotelic responsive environments and exceptional children. In Hellmuth, J. (ed.): The Special Child in Century 21. Seattle, Special Child Publications, 1964, pp. 87–138.
59. Murphy, L. B.: Child development–then and now. Children Educ. 44:302, 1968.
60. Newman, H. H. et al.: Twins: A Study of Heredity and Environment. Chicago, University of Chicago Press, 1937, pp. 325–327.
61. Nimnicht, G.: Low-cost typewriter approach helps preschoolers type words and stories. Nation's Schools, 80:34, 1967.
62. Norman, R. D.: The interpersonal values of parents of achieving and non-achieving gifted children. J. Psychol. 64:49, 1966.
63. Palmer, F. H., and Rees, A. H.: Concept training in two year olds: Procedures and results. Paper presented at the meeting of the Society for Research in Child Development, Santa Monica, March 1969.
64. Parmelee, A. H., and Hober, A.: Who is the 'Risk Infant'? Clin. Obstet. Gynecol., 16:376, 1973.
65. Piaget, J.: Play, Dreams, and Imitation in Childhood. New York, W. W. Norton & Co., 1962.
66. Provence, S., and Lipton, R. C.: Infants in Institutions: A Comparison of Their Development with Family-Reared Infants During the First Year of Life. New York, International University Press, 1962.
67. Purpura, D. P.: Discussants' Comments to First Plenary Session, National Conference on Early Intervention with High Risk Infants and Young Children, University of North Carolina at Chapel Hill, May 5–8, 1974.
68. Ramey, C. T. et al.: The Carolina Abecedarian Project: A Longitudinal and Multidisciplinary Approach to the Prevention of Developmental Retardation. Chapel Hill, N.C.: University of North Carolina, 1974 (Mimeographed).
69. Rheingold, H. L.: The effect of environmental stimulation upon social and exploratory behavior in the human infant. In Foss, B. (ed.): Determinants of Infant Behavior. New York, John Wiley & Sons, 1961.
70. Rheingold, H. L.: Independent behavior of the human infant. In Pick, A. D. (ed.): Minnesota Symposium on Child Psychology. Vol. 7, 1973, pp. 175–203.
71. Rolcik, J. W.: Scholastic Achievement of Teenagers and Parental Attitudes Toward and Interest in Schoolwork. Family Life Coordinator, 14:158, 1962.

72. Rosenthal, R., and Jacobson, F.: Teacher expectations for the disadvantaged. Sci. Am., 218:19, 1968.
73. Rubenstein, J.: Maternal attentiveness and subsequent exploratory behavior in the infant. Child Dev., 38:1089, 1967.
74. Sameroff, A. et al.: Infant casualty and the continuum of infant caretaking. *In* Horowitz, F. D., Hetherington, E. M., Seigel, M., and Salapatek, S. C. (eds.): Review of Child Development Research, Vol. 4, 1974. (In press.)
75. Schaeffer, E. S., and Aaronson, M.: Infant education research project: Implementation and implications of a home tutoring program. *In* Parker, R. K. (ed.): Conceptualizations of Preschool Curricula. Boston, Allyn & Bacon, Inc., 1970.
76. Scott, J. P.: Critical periods in the development of social behavior in puppies. Psychosom. Med., 20:42, 1958.
71. Scott, J. P., Frederickson, E., and Fuller, J. L.: Experimental exploration of the critical period hypothesis. *Personality,* 1:268, 1951.
78. Shaw, M. C., and White, D. L.: The relationship between child-parent identification and academic underachievement. J. Clin. Psychol., 21:10, 1964.
79. Shearer, M. S., and Shearer, D. E.: The Portage Project: A model for early childhood education. Except. Child., 39:210, 1972.
80. Siqueland, E. R.: Biological and experimental determinants of exploration in infancy. *In* Stone, L. J., Smith, H. T., and Murphy, L. B. (eds.): The Competent Infant. New York, Research Commentary Basic Books, 1973.
81. Skeels, H. M. et al.: A study of environmental stimulation: An orphanage preschool project. University of Iowa Study of Child Welfare, Vol. 15, 1938, p. 264.
82. Spitz, R. A.: Hospitalism: An inquiry into the genesis of psychiatric conditions in early childhood. Psychoanal. Study Child, 1:53, 1945.
83. Spitz, R. A.: Anaclitic depression. Psychoanal. Study Child, 2:313, 1946.
84. Spitz, R. A.: Hospitalism: A follow-up report. Psychoanal. Study Child, 2:113, 1946.
85. Starr, R. H., Jr.: Cognitive development in infancy: Assessment, acceleration, and actualization. Merrill-Palmer, 17:153, 1971.
86. Terman, L. M.: Genetic Studies of Genius, Vol. 1, The Mental and Physical Traits of a Thousand Gifted Children. Stanford, Calif., Stanford University Press, 1925.
87. Toffler, A.: Future Shock. New York, Random House, Inc., 1970.
88. Wachs, T. D. et al.: Cognitive development in infants of different age levels and from different environmental backgrounds: An exploratory investigation. Merrill-Palmer, 17:283, 1971.
89. Weikart, D., and Lambie, D. J.: Preschool intervention through a home teaching program. *In* Hellmuth, J. (ed.): Disadvantaged Child, Vol. II. New York, Brunner/Mazel, Inc., 1968, pp. 435–501.
90. Weikart, D. P., and Lambie, D. Z.: Early Enrichment in Infants. Paper presented at the meeting of the American Association for the Advancement of Science, Boston, Mass., December, 1969.
91. Wellman, B. L.: Iowa's studies on the effects of schooling. Natl. Soc. Study Educ., 39:377, 1940.
92. Werner, E. E., Bierman, J. M., and French, F. E.: The Children of Kauai. Honolulu, Hawaii, University of Hawaii Press, 1971.
93. White, B. L., Castle, P., and Held, R.: Observations on the development of visually directed teaching. Child Dev., 35:349, 1964.
94. Wright, C., Lally, J. R., and Dibble, M.: Prenatal–postnatal intervention: A description and discussion of preliminary findings of a home visit program supplying cognitive, nutritional, and health information to disadvantaged homes. Paper presented at the meeting of the American Psychological Association, Miami, September, 1970.
95. Yarrow, L. J.: Research in dimensions of early maternal care. Merrill-Palmer Q., 1963, 101–114.
95a. Zajonc, R. B.: Birth order and intelligence: Dumber by the dozen. Psychol. Today, 8:37, 1975.
96. Zingg, R. M.: Feral man and extreme cases of isolation. Am. J. Psychol., 53:487, 1940.

CHAPTER 18

REGIONAL GENETIC COUNSELING PROGRAMS

Vincent M. Riccardi, M.D.*

INTRODUCTION

Many instances of genetic disorders and mental retardation can be prevented, either by minimizing the effects of the disorder (*e.g.,* phenylketonuria) or by avoiding the birth of an affected child. In both cases the preventive measures depend on the widespread routine availability of genetic counseling services *and* on a relatively good understanding of genetics by health professionals and the general public. Merely to provide genetic counseling clinics, without associated educational programs, would be self-limited. A combined approach has been utilized to maximize the preventive impact of genetic counseling over the 201,000 square miles of Colorado and Wyoming. A comprehensive program to educate and provide clinical services was established by the Department of Biophysics and Genetics, University of Colorado Medical Center in Denver. This Colorado-Wyoming Regional Genetic Counseling Program (CW-RGCP), now in its third year, has filled a serious void.[1] The need for the program in the Rocky Mountain Region is particularly great in view of the relative isolation of many communities. Travel to the region's major medical referral center at Denver is often limited, unless there is an unusual or emergency situation. This isolation means that the 1.54 million residents of Colorado and Wyoming beyond Denver had not had genetic counseling directly available, as had the 1 million residents of Denver. To overcome this restriction and to make

*From the Department of Biophysics and Genetics. No. 590, University of Colorado Medical Center, Denver, Colorado. Supported by the Colorado-Wyoming Regional Medical Program, the Kaiser Foundation, and the National Foundation—March of Dimes.

410

clinical genetics a routine part of the medical care for the entire 2.54 million residents, the CW-RGCP was established. Although geographic and climatic conditions, as well as population densities, will vary from one region to another, the regional approach to health care needs can certainly be projected directly to many other areas.

Programs like CW-RGCP have been made feasible and appropriate by recent developments in medical genetics and general health care: (1) Increase in importance of the morbidity and mortality of genetic disorders and congenital malformations as some diseases due to other causes—particularly infections—have decreased in prominence;[2] (2) Improved techniques for earlier and more specific diagnosis, including amniocentesis for intrauterine diagnosis;[3] (3) Enhanced dietary, medical, and surgical therapy;[3, 4] (4) Improved methods of genetic counseling;[5] (5) Larger numbers of specially trained medical geneticists and allied personnel; (6) Acknowledgment of the need for direct involvement of university medical center staff in the health care of local communities; and (7) Increased professional and public awareness of medical genetics and its implications at various clinical, social, ethical, and moral levels.[6]

The CW-RGCP has thus combined three approaches: (1) Establishment of regional genetic counseling clinics; (2) Training of health professionals; and (3) Education of the general public. Others have acknowledged the general need for genetic counseling programs to provide entire populations of large regions with both educational and clinical services.[3, 7] The CW-RGCP, however, represents the first comprehensive program to serve a large area by bringing these services directly to each outlying community. Although satellite clinics *per se* are no longer unique, the structured, comprehensive approach of the CW-RGCP does represent an innovation in primary health care. The conviction that the incidence and impact of genetic disease ultimately can be minimized by the uniform availability of Regional Genetic Counseling Programs has been basic to our efforts. The purpose of this chapter is to delineate the procedures for establishing similar programs elsewhere.

REGIONAL GENETIC COUNSELING PROGRAMS: EXPERIENCE AND GENERAL PRINCIPLES

In the first 15 months of operation (beginning January 1973) the CW-RGCP provided 61 clinics in eight separate communities. A total of 369 counseling sessions were provided to 268 families. The rate of referral and requests for expansion of the program's activities have increased steadily. Table 18–1 indicates the number and percentage of patients referred according to the genetic nature of their disorder. This classification[1] has proved to be helpful, both for record-keeping purposes and for emphasizing recurrence risks as a fundamental aspect of medical care

TABLE 18–1. CLASSIFICATION OF COLORADO-WYOMING REGIONAL GENETIC COUNSELING PROGRAM (CW-RGCP) PATIENTS ACCORDING TO DIAGNOSIS AND REPRESENTING THE FIRST 15 MONTHS OF OPERATION *

	Number	Per Cent
A. Genetic		
1. Chromosomal	27	10.1
Preamniocentesis†	6	2.2
2. Single gene	106	39.5
Preamniocentesis‡	1	0.4
3. Polygenic	45	16.8
Preamniocentesis§	1	0.4
	186	69.4
B. Probably genetic	7	2.6
C. Questionably genetic	16	6.0
D. Developmental	44	16.4
E. Environmental	3	1.1
F. Other, nongenetic	12	4.5
	82	30.6
Total	268	100

*"Genetic" disorders have defined recurrence risks; "Probably genetic" and "Questionably genetic" disorders have questionable recurrence risks; "Developmental," "Environmental," and "Other, nongenetic" disorders have essentially no excess recurrence risks. Preamniocentesis counseling was provided eight times (3%).
  †Indications included maternal age.
  ‡Indications included a previous child with cystinosis.
  §Indications included a previous child with anencephaly.

(see legend of Table 18–1). The emphasis on both the patient-care aspect of the program and the long-term goal of prevention, has been very helpful in establishing genetic counseling as a routine aspect of local medical care. It cannot be stressed too strongly that, currently, genetic counselors will have only limited, short-term influence on the frequency and burden of genetic disease unless the relevance of genetic counseling is demonstrated simultaneously. This relevance is conveyed by example, as well as by training and education programs. Such programs, directed both at professionals and the general public, have been very effective, as judged by enthusiastic responses and increases in patient referrals and follow-ups.

The actual cost of implementing a Regional Genetic Counseling Program is determined mainly by salaries, travel expenses, secretarial assistance, and office supplies. Using the CW-RGCP as a model, one sees clearly that the minimum requirement for full-time salaried personnel is a physician geneticist, nurse-geneticist, secretary, and cytogenetics technician. Travel expenses of $10,000 per year is a minimum estimate; this figure includes a travel cost of $18.00 per family visit. The costs for telephone calls, postage, photocopying of clinic reports, stationery, and other supplies are substantial. The point is that the energy, time, and material needed to establish properly and operate an RGCP of any magnitude require considerable financial support. Costs for any given RGCP would of course vary with the individual needs and resources of each region. In any case, it would not be possible to charge patients or families themselves what it actually costs to provide patient care, in particular when a large number of patients seen are from low-income families.[1] For example, the minimum average total cost at CW-RGCP has been about $200 per family visit. Operating costs have thus been deferred by several specific grants.

In view of the uniqueness of Regional Genetic Counseling Programs, there is need to educate as well as to provide patient care in order to minimize the many misunderstandings about medical genetics. There is a considerable inertia to be overcome before full impact can be realized. Thus, there must be caution against the premature application of cost-effectiveness formulae. Ultimately, however, Regional Genetic Counseling Programs will essentially pay for themselves by offsetting the cost of caring for affected individuals. For example, for each Down's syndrome birth that is prevented, an estimated $150,000 could be saved.[1] Both private and government sources are aware of these benefits. The state would be an obvious and appropriate source of funding since state institutions often become responsible for the care and support of many individuals with congenital or inherited defects.

Before actually implementing an RGCP, one must consider organization in terms of the needs and resources of each program and the population it will serve. There are three particularly important sets of considerations:

1. Personnel and administrative organization. The CW-RGCP functions as a semiautonomous unit within the Department of Biophysics and Genetics of the University of Colorado Medical Center. It is also an integral part of the Medical Center's interdepartmental Genetics Unit. The program's full-time physician-geneticist (Vincent M. Riccardi) is responsible for organizing and implementing its activities. He establishes each clinic site, attends most clinics, supervises and coordinates ongoing patient care, and arranges the training and teaching efforts. He is assisted by the other physician-geneticists of the Genetics Unit and by the program's nurse-geneticist.

2. Logistics.   Travel to clinic sites is a critical consideration, since it is relatively expensive, very time-consuming, and fatiguing. Travel time often dictates clinic sites, the frequency with which each site is visited, and patient loads. The minimum length of time spent traveling is about four hours; overnight stays are sometimes necessary. It should be remembered, however, that the limitations of travel are one of the reasons that many patients are not seen at the medical center. New techniques in laboratory diagnoses are made more immediately accessible to local communities through the RGCP, since samples obtained locally

can be analyzed at the medical center. However, when charges are involved it has been found that many families could not afford them. This situation has been most obvious with chromosome analyses, which are frequently required and relatively expensive.[1] Thus, a cytogenetics technician, whose salary can offset the cost of at least some of the analyses, is an important element of the program. Communication with patients, their families, and physicians regarding the genetic counseling and consequent recommendations is particularly important. On one hand, the referring physician must be adequately informed of the counseling report without undue delay. On the other hand, the patient's family must be given a written summary of the counseling for it to be effective.[5,8] Thus, copies of the report and/or summary letters are provided to both the referring physician and the family in each case. Finally, because of the long distances involved, telephone bills for the program are large.

3. Manner and intensity of interaction with local health personnel. Interaction with local health professionals in each local community has been predicated on the following principles: (a) Desire to become part of the community's routine primary health care resources, both as clinical consultants and as teachers; (b) Establishment of a regional clinic would be only with the explicit sanction of local physicians; (c) Involvement of local Public Health Nurses was considered to be an important, if not critical, facet of the program; and (d) Emphasis on simplicity, with minimum effort and cost to each community.

REGIONAL GENETIC COUNSELING PROGRAMS: ESTABLISHMENT AND OPERATION

What follows is a detailed description of the means and mechanisms by which the CW-RGCP was established. This history and the summary outline of Table 18-2 are meant to provide guidelines for the development of other RGCPs. Two features of such programs must be consistently emphasized: simplicity and service. This emphasis takes into account the described inertia of clinical genetics and the idiosyncrasies of private practice in relatively small communities. Thus, because the initial cost in time and money to the community was minimal (simplicity) and because patient care and teaching were to be provided locally (service) the program had a natural and sensible appeal. These two considerations are probably critical to the successful establishment of RGCP clinics.

Preclinic Organization

A series of three meetings at the University of Colorado Medical Center with physicians representing various communities throughout

TABLE 18–2. OUTLINE OF CONSIDERATIONS FOR THE ESTABLISHMENT AND OPERATION OF A REGIONAL GENETIC COUNSELING PROGRAM (RGCP)

  I. Preclinic organization
    A. Preliminary meetings
    B. Local community meetings
  II. Clinic organization
    A. Clinic facilities
    B. Patient scheduling
    C. Precounseling data collection
    D. Genetic counseling sessions; dictation
    E. Laboratory samples
    F. Amniocentesis
    G. Travel considerations
    H. RGCP personnel
  III. Equipment
    A. Data forms
    B. Dictaphone
    C. Physical examination
    D. Laboratory specimen collection
    E. Photography
    F. Reference books
  IV. Nurse-geneticist
  V. Secretarial assistance
  VI. Patient records
    A. Disbursement of clinic reports
    B. Genetics unit folders
  VII. Case review conferences
  VIII. Public Health Nurses
  IX. Training and education
    A. Professional
    B. General public

Colorado and Wyoming was held to determine receptivity to the program and to decide where to establish the first clinics. Communities selected for clinics were determined largely as a function of the enthusiasm of the local representative, who was designated the local coordinator. Through this representative, one or two local meetings with other physicians in the community were held and the approval of this small group was obtained. It was made clear frequently that there were only two requirements of the community: (1) a place for patient interviews, physical examinations, and dictation, and (2) someone to schedule appointments and forward the list of names and diagnoses to the RGCP staff a week or so prior to the clinic. All other equipment for operating the clinic would be brought with the staff. The local coordinator would continue to serve as liaison. About a week prior to the first clinic in each community, spot announcements indicating the time, place, and means of scheduling appointments were released to local newspapers and radio stations. There were no restrictions on sources of referrals, and self-referrals were encouraged. The sole criterion for referral was the concern, Will it happen again? (*i.e.,* what is the risk of occurrence or recurrence of a disorder). It was not necessary for a diagnosis to have been established beforehand. The two most frequent sources of referral were physicians and Public Health Nurses.[1]

Clinic Organization

Clinics have been located in various types of facilities, including Health Department offices, hospital examining rooms and offices, and even a hospital's small auditorium. In any case, two rooms are needed — one for the counseling, and one for dictation. All clinics are held monthly — except two, which are held every other month. Each meets on a specified day of the month. A minimum of 90 minutes is required for each new family, and 45 minutes, for each follow-up. Six to 10 families are scheduled for each clinic, with the anticipation that there will be one cancellation. This scheduling has worked well. An average of 61 family visits are provided for each 10 clinics, including 44 initial visits and 17 follow-ups. Occasionally, patients are seen as inpatients in local hospitals or other institutions. A list containing each proband's name, reason for referral, and available medical data or its source is provided about one week prior to each clinic. If a Public Health Nurse is involved, frequently a pedigree will be completed before the clinic meets. Otherwise, it is constructed by the Program's nurse-geneticist, who also obtains the preliminary medical history. This material is then presented to the physician-geneticist who supplements the history as necessary, examines the proband and/or family members and then proceeds with the counseling; sometimes counseling *per se* is deferred until additional data is obtained (*e.g.,* confirmation of previous diagnoses, laboratory determinations, and other similar information). If appropriate, samples for laboratory analysis are obtained or suggested and photographs are taken. The counselor can then dictate his report on the family just seen, while the next family is being interviewed by the nurse-geneticist. Follow-up visits are scheduled if a final disposition has not been reached, if the counseling is not straightforward, if there is obvious emotional turmoil, or if the counselor will become involved in the patient's subsequent management. Unfortunately, it has not been possible for every family to have a formal follow-up session; 27 per cent of family visits represent follow-ups. However, many more receive follow-up visits by their Public Health Nurse, who often sits in at the counseling session.

If an amniocentesis for prenatal diagnosis is to be performed, preamniocentesis counseling is provided. At this session indications are clarified, family and other pertinent histories are obtained, and details of the procedure are discussed. Occasionally, the amniocentesis is then performed locally and the liquor brought back to the medical center. At other times, the patient is referred to the medical center for the procedure to be performed at a later date. With time, increasing numbers of patients have been referred for this reason, with varying indications (Table 18–1). This fact alone demonstrates that substantial progress is being made.

Patient loads and clinic hours are interrelated with travel considerations, especially when travel is by airplane. Thus, it is sometimes necessary to travel on two consecutive days, either to facilitate a normal caseload or to accommodate a larger one. Ordinarily, the program's full-time geneticist attends each clinic in the company of the nurse-geneticist. When one of the other medical center's genetic counselors attends, the nurse-geneticist provides the necessary continuity.

Clinic Equipment

All the materials necessary for patient records, physical examinations, laboratory samples, photography, and reference reading are brought to each clinic. At each initial family visit an RGCP folder is made up for each family, designated by the name of the consultand or proband, and by an assigned "family number." This and additional pertinent information, such as addresses and referring physicians, are noted on a specially designed worksheet. This form has proved useful as a source of data for communications on the patient's behalf, and for collection of statistical data, such as the source of referral. Other necessary forms are those for drawing pedigrees and those for preparing releases to obtain previous medical records.

The physical examination equipment is standard for any genetics clinic. Venipuncture equipment is very important, especially for collecting specimens for chromosome analysis, amino acid determinations, and linkage studies. In addition, equipment should be available for obtaining buccal smears, skin biopsies for tissue culture, saliva, and urine specimens. A camera to record physical features of patients is a necessity and greatly enhances the ability of other physicians at the medical center to contribute diagnostic considerations.

Nurse-Geneticist

This member of the RGCP team is indispensable for operations of the magnitude described here. These professional roles as clinician, coordinator, liaison, and teacher have been clarified through this program. She directly assists the genetic counselor by constructing a family pedigree and taking a preliminary history. She also collects and collates data for each family and provides continuity when more than one counselor is involved, or there is a long interval between the initial visit and final disposition. She is an important link between the program and local nurses, both for patient care and for professional training.

Secretarial Assistance

The role of the secretary in the RGCP is singled out here because this staff member is so vital to the maintenance of patient records and communication with patients and health professionals in the outlying communities.

Patient Records

RGCP folders are established for each consultand or proband. Each folder contains the worksheet as described, the family pedigree, a copy of each genetic counseling report, and all other relevant correspondence and accumulated data (laboratory reports, old records, etc.). Each genetic counseling report leads to or provides clarification of three specific considerations: diagnosis, prognosis and natural history of the disorder, and recurrence risks for various family members, both current and future. The original report is provided to the local clinic. Copies are also sent to the referring physician, to other physicians as directed by the family, to the patient's medical center hospital chart, and to the consultand. A summary letter may be sent to the family. Notation is made of everyone to whom each report is sent. Assigned "family numbers" are very useful for assimilating all data applicable to individual families accruing under different names. Before a final disposition can be made and communicated, the data for each folder must be reviewed by other genetic counselors.

Case Review

Each case is presented to other members of the Genetics Unit at biweekly conferences convened precisely for that purpose. All cases are reviewed by at least four clinical geneticists. Each final disposition, therefore, represents a consensus. The importance of this approach to patient care cannot be overemphasized.

Public Health Nurses

Patient referral, clarification of the family history, follow-up reinforcement of the genetic counseling, and liaison with a referring physician and/or genetic counselor are obvious and frequent ways that Public Health Nurses contribute to patient care through genetic counseling.[6] In addition, when a nurse—as opposed to a secretary—has been responsible for scheduling patients, clinic case loads are consistently greater, and the cancellations, fewer. Moreover, in all nine current clinics of CW-RGCP, Public Health Nurse involvement is critical to the continuation and local success of the program. In Wyoming especially, the Public Health Nurses are the cornerstone of the program. Their roles in genetic counseling care are being expanded and enhanced. Keenly aware of this, they all have participated with enthusiasm in the training and educational programs for nurses.

Training and Education

Initial encounters with physicians in their own communities have always involved a staff conference and a lecture on genetic counseling. Additional lectures on specific topics of medical genetics have been provided regularly, thereafter. In one community, an accredited five-hour course in clinical genetics was given. The primary goals were to emphasize the relevance of genetic counseling by clarifying general principles and to enhance diagnostic acumen and patient management. These efforts have always been followed by increased numbers of patient referrals and a more intimate involvement in local health care. For example, there has been a definite association with increased requests for inpatient consultations.

For nursing personnel, in addition to didactic lectures, considerable training has been provided to develop specific skills. In particular, patient interviewing, pedigree construction, and techniques for home follow-up have been stressed. Professionally, Public Health Nurses are important as sources of referrals and as providers of follow-up support. In general, they have contributed a great deal to the care of patients through genetic counseling—especially in the more rural communities. An average of six to eight hours per month of in-service training programs is being provided for nurses. In addition, several yearly workshops, lasting from one-half to three full days, have been conducted. These workshops, for all the Public Health Nurses in Wyoming, were very successful and led directly to the subsequent establishment of two more clinics. On several occasions, a nurse has been brought to the medical center for more intensive training when, for example, she was particularly skillful in running the clinic in her community. The responsibilities for the nurses' training have been shared by the program's counselors and the nurse-geneticist, who has proved particularly valuable, in this respect.

Educational efforts on behalf of the general public have been directed toward lectures, pamphlets, radio, television, and newspapers. Special effort has been made to utilize regularly these vehicles of communication. Positive results have been indicated by the many consequent self-referrals and telephone inquiries.

There is no doubt that these efforts to train and educate are vitally important in establishing genetic counseling as an effective means of patient care and preventive medicine.

PROBLEMS

The primary problem is, of course, financing. For an RGCP, however, it is not merely a matter of competing for health care funds; there is an associated problem: Many people still have to be convinced that genetic counseling is a critical part of modern medical care. As stressed before, a directed effort must be made to demonstrate its relevance and application in day-to-day medical practice. As the effectiveness of genetic counseling becomes more obvious, financial resources will, in turn, become less limited. This is another reason why educational efforts have had such high priority in the CW-RGCP. To date, support from federal and private institutions has been the most readily available to this program, but for the long-run, state support would be the most appropriate. Specific considerations for estimating operating costs have been dealt with previously, but the need for offsetting the cost of chromosome analyses for at least a portion of the population serviced must be emphasized. The time, effort, and cost of travel can be limiting factors in the location, size, and scheduling of individual clinics. The special need for accurate communication over long distances requires special attention, as does a well-organized patient-record system and an effective means of collating and dispensing data.

SUMMARY

Genetic counseling is an essential part of today's preventive medicine. The Colorado-Wyoming Regional Genetic Counseling Program has demonstrated that projects to optimize the preventive impact of genetic counseling are both feasible and worthwhile. Thus, by simultaneously providing educational and clinical services on a routine basis over a widespread area (201,000 square miles), genetic counseling has become part of the primary medical care of many families who, otherwise, would not have received it. The details of establishing and operating such a program, including financial considerations, are presented and discussed. Patient data derived from the first 15 months of operation of the program are also presented.

References

1. Riccardi, V. M., and Robinson, A.: Preventive Medicine Through Genetic Counseling: A Regional Program. Prev. Med. (in press).
2. Gordon, H.: Genetic counseling: Considerations for talking to parents and prospective parents. J.A.M.A., 217:1215, 1971.
3. Clow, C. L., Fraser, F. C., Laberge, C., and Scriver, C. R.: On the application of knowledge to the patient with genetic disease. Progr. Med. Genet., 9:159, 1973.
4. Howell, R. R.: Genetic disease: The present status of treatment. *In* McKusick, V. A., and Claiborne, R. (eds.): Medical Genetics. New York, HP Publishing Co, 1973, pp. 271–280.
5. Reynolds, B. D., Puck, M. H., and Robinson, A.: Genetic counseling: An appraisal. Clin. Genet., 5:177, 1974.
6. Institute of Society, Ethics, and the Life Sciences: Ethical and social issues in screening for genetic disease. N. Engl. J. Med., 286:1129, 1972.
7. Wendt, G. G., and Theile, U.: A pilot scheme for a genetic clinic. Humangenetik, 21:145, 1974.
8. Leonard, C., Chase, G. A., and Childs, B.: Genetic counseling: A consumer's view. N. Engl. J. Med., 287:433, 1972.

CHAPTER 19

THE ROLE OF LAW IN THE PREVENTION OF GENETIC DISEASE

Philip Reilly, J.D.*

THE LEGAL FRAMEWORK

Only recently have we begun to challenge the notion that technological developments that offer benefit to humanity must be incorporated into society. The art of anticipating the unforeseeable must be mastered if we are to avoid exacerbating the problems that we attempt to solve. Advances in the biochemical sciences constitute a special class of developments to be scrutinized — so immediate is their impact on man.

An expanding number of genetic disorders can be identified with a simple, albeit refined, series of assays. Unfortunately, the technology of detection of persons at risk for having offspring with genetic disease has significantly outdistanced the ability to cure children with these disorders. A new era of "preventive genetic medicine" is emerging. Currently, the reduction of numbers of persons suffering from most genetic diseases can be accomplished only by reducing the number of births of such persons. Either mating choice itself or post-marital reproductive decisions can be influenced to reach this goal. This chapter considers the role of law in effecting the reduction of genetic disease through genetic screening legislation.

What is the reach of permissible legal activity in the service of public health? The state can regulate the public health pursuant to its "police power," yet a power that authorizes not only public health regulations, but also economic controls, does not lend itself to concise definition. It is commonly characterized as plenary and inherent, since it refers to the state's general authority to protect itself and its members by

*This paper supported in part by USPHS Grant GM 19513.

regulating the activities of its people. Thus, it is "an indispensable prerogative of sovereignty."[1]

The United States is a federal system of government, which divides authority between the nation and the states. The Constitution specifies those powers that the national government may exercise. The public health power is not enumerated. Certainly, the eighteenth century draftsmen of that document thought that the public health would be largely the responsibility of the states. The Supreme Court has consistently broadly construed federal powers to tax, spend, and regulate commerce. Pursuant to these powers the federal government can wield great influence over state public health programs. In practice, it appears that this is a shared power; the states may claim it, but it is enmeshed in the fabric of national government. Of course, in nonfederal systems (most nations) the power is within the repository of the national government.

Although the police power of the states and the enumerated powers of the federal government are broad, they are limited by constitutional restraints that protect individual rights. In all nations the state's exercise of a power must be balanced against the infringements upon personal interests that such exercise engenders.

The health of citizens is of course a special concern of government. When the threat posed by a contagious disease is significant, it has been established that the state may compel preventive treatment of its citizens. In 1905 the Supreme Court held that the state could compel a person to undergo smallpox vaccination.[2] For the individual the threat of smallpox was sufficiently great, and the risks from vaccination sufficiently small, that this determination was warranted. Indeed, many nations have statutory programs to provide for the control of contagious diseases. The Public Health Act of 1961 of Great Britain offers one example of comprehensive efforts in this area.[3]

It may be appropriate for the state to force treatment of a person even if there is no discernible risk to society in the illness. Thus, a woman may be ordered to receive a blood transfusion to save the child that she is carrying (although recent legalization of abortion may negate this precedent).[4] The state may compel treatment of an infant despite the religious objections of the parents.[5] These examples reflect the special concern of the state for the well-being of persons who cannot care for themselves and are an exercise of "parens patriae" power. The scope of activity in which the state can engage to guide the health of an adult without dependents may be much more strictly limited. To be sure, the health of society is a function of the health of its citizens, yet traditionally we have honored the individual's rights to make decisions affecting himself. State action to save the life of a citizen is a special case. The boundaries of permissible government activity to regulate more generally the health of society as a whole will be investigated. Such an inquiry requires a consideration of several constitutional constraints on the public health power. The law of the United States will be looked to in the delineation of these constraints. The underlying principles are honored by other nations as well.

In the United States there is a constitutional guarantee of the freedom of a citizen to his religious convictions. On occasion, exercise of the public health power has conflicted with this freedom. For example, the fluoridation of public water supplies infringed upon the rights of those who are opposed for religious reasons to medication in any form. In court tests it has been held that despite the real infringement, fluoridation was a reasonable activity designed to benefit public health.[6] In assessing the conflict between religions and state programs the courts have distinguished freedom to believe from freedom to act pursuant to a particular belief. It is possible to argue that fluoridation of water did not compel people to violate their religious convictions (use of public water is not required), but even an unmistakable act of compulsion could be justified "under the pressure of great dangers."[2] What constitutes such dangers remains unclear.

A second constitutional protection is the *Due Process* clause of the Fifth and Fourteenth Amendments. It states that no person may be deprived "of life, liberty, or property without due process of law." The contours of these words are vague, particularly their import in contexts other than procedural due process, which requires that a person have the right to challenge a taking. Whereas due process may be a potent weapon against poorly drafted legislation or cavalier administrative activity, against carefully drafted laws that provide for adequate mechanisms for protection of personal rights, the utility of invoking the clause is reduced. If in the future the Supreme Court chooses to ground "fundamental interests" on the due process clause, then the importance of that clause will increase for public health legislation.

A third important restraint on public health laws that are aimed at a target population is found in the *Equal Protection* clause of the Fourteenth Amendment. "No state . . . shall deny to any person within its jurisdiction the equal protection of the law." Until recently, that clause was read narrowly to require only that a statutory classification be rationally related to a legitimate governmental objective. Today, however, some statutory classifications are required to satisfy a more rigorous demonstration: that compelling state interests support them. Racial classifications are "suspect" and must satisfy this higher test. This has obvious implications for certain forms of genetic screening laws. The compelling state interest test may also be applied to any classification if the result of the classification affects a fundamental right. Voting and procreation are two examples of such rights. Classifications that are measured by the compelling state interest test must demonstrate a special relevance of classification to purpose. The presumption of the validity of the law is reversed. "The burden of justification is on the state rather than on the challenging party."[7] The court will demand a greater quantum of support to approve the legislation,[8] and will inquire if alternative means can accomplish the same purpose.[9]

Finally, there is the still-forming doctrine of *Fundamental Interests*. It holds that certain human activities, although not specifically mentioned in the Bill of Rights, deserve special judicial protection. The Ninth Amendment and the Due Process clause offer the framework for this judicial stance. Privacy, originally the right to be let alone, is slowly being expanded to include rights of personal decision-making. Recently, several important decisions have been written that have significantly reduced the power of the state to regulate the private use of birth control technology. These decisions have helped to revolutionize voluntary family planning in the United States. Nine years ago the Supreme Court invalidated a law that effectively prohibited the use of contraceptives in Connecticut.[10] In 1973 it found that women had a right to abort their pregnancies in the first two trimesters.[11]

Despite clear theoretical limitations on the exercise of the public health power, the state traditionally has had little difficulty in justifying compulsory requirements under its rubric. Indeed, since the days of the first quarantine laws, compulsion has been common to laws in this field. A brief catalogue of compulsory laws should illustrate the point. Numerous statutes in a variety of jurisdictions require that persons be vaccinated against smallpox,[12] measles,[13] and poliomyelitis.[14] The state requires testing for venereal disease[15] and Rh disease.[16] It may demand treatment of venereal disease[17] and tuberculosis.[18] It demands preventive treatment for neonatal ophthalmia.[19] It reserves the right to incarcerate mental incompetents.[20]

A variety of eugenics laws have been passed. Every state in the United States and many other nations prohibit some consanguineous marriages. A common statute is one that prohibits all matings of persons as close as or closer than first cousin.[21] Perhaps the most potent examples of eugenics legislation currently in force are the compulsory sterilization laws. A significant number of jurisdictions retain legal authority to sterilize institutionalized mental defectives.[22] A few state laws permit the sterilization of certain convicted felons.[22] Under the rubric of eugenics, at least one wave of insidious laws did a disservice to society until an embarrassingly recent time. The antimiscegenation statutes were just invalidated in 1967.

State action to influence procreative behavior may be discerned in many countries. For example, Article 4 of the Marriage Law of 1950 in the People's Republic of China reads, "A marriage may be contracted only after the man has reached twenty years and the woman eighteen years of age."[23] Thus, marriage is delayed longer in China than in the United States. Unofficially, the government stance in China is to suggest that optimal marital age is between 25 and 30 for a man, and between

22 and 28 for a woman.[23] Such a policy has contributed significantly to the successful Chinese birth control program.

Although American case law is limited, it indicates consistent judicial deference to a wide variety of public health laws. Compulsory vaccination,[2] sterilization of institutionalized mental defectives,[24] and premarital blood tests[25] have been validated. It has been decided that a tuberculosis chest film is a proper prerequisite to attendance at a state university, despite religious objections.[26] No clear delineation of the outer limits of state action pursuant to the public health power can be made. It has been wisely noted that "the power to order an individual to undergo a medical procedure such as immunization . . . is a potentially far-reaching one. . . . The potential of this power for evil as well as for good is evident, and proposals for the extension of the power deserve careful scrutiny."[1] Genetic screening legislation represents an extension of the power. Against a tradition of validation of state law, it is appropriate to examine this novel activity.

EARLY GENETIC SCREENING LAWS

Genetic screening legislation has been written to provide for the detection of both treatable disease and carrier status. The discovery that persons with phenylketonuria who are placed early on a low-phenylalanine diet avoid severe mental retardation combined with the refinement of simple serum phenylalanine testing methods initiated a demand for state support of neonatal screening. Between 1962 and 1971, forty-three states and the District of Columbia passed laws mandating such a screening program.[27] Controversy engulfed the legislation aimed at detecting this treatable disorder only after most states had acted. However, criticism by Bessman and others may have served to halt federal legislation.[28] Despite important unanswered questions asked several years ago about the nature of the hyperphenylalaninemias that had ramifications both for testing and treatment of neonates, today it appears settled that the mandated testing programs have successfully combated retardation in most affected children.

Phenylketonuria screening programs have been initiated in at least 20 nations. Over 13 million children have been tested for the disorder.[29] It is interesting to note that most of these nations did not legislate screening programs. Some programs were initiated by proclamation from the national department of health. However, researchers reporting on several national screening programs point with pride to the high percentage of children tested without need for compulsory legislation. The province of Manitoba, screening since 1965, has tested, without any legal mandate, about 98.5 per cent of children born there.[30] Since early 1966 Ireland has had an "entirely voluntary" program.[31] The experiences in these nations suggest that compulsory testing laws are unnecessary to accomplish the purpose. Because precedent is an important concept in legal reasoning, it is advisable to be wary of needless compulsory laws. In fact, little public discussion preceded the passage of these laws in the United States. The relatively low cost of the programs, the appeal of a direct dividend on biomedical research funding, and the effective lobbying by special interest groups were among the important factors that helped the movement of the law from committee to code book.

Sickle cell anemia legislation followed hard on the heels of phenylketonuria laws. It is a more complicated story. With the development of a rapid inexpensive blood test for sickle cell heterozygosity there appeared a capability to inform the at-risk population of the distribution of an allele that in the homozygous state indicated a grave incurable genetic disease. By late 1972, 12 states and the District of Columbia had legislated sickle cell anemia screening laws—that is, laws that provide for a search for heterozygotes.[32] A brief scrutiny of those original laws would indicate how *not* to write genetic legislation. In them, scientific error was often compounded by ignorance of the dangers implicit in such laws. Concern for confidentiality of test results and adequate genetic counseling were sorely lacking. At a time when the need for programs of prophylaxis in genetic disease is beginning to be realized, it is well to be familiar with the dangers inherent in misuse of technology.

Sickle cell anemia screening legislation may be properly called eugenic. The intended purpose of these laws can be easily discerned. Heterozygote screening informs a person whether or not he or she is at risk for having a child with a genetic disorder. Knowledge of carrier status is of little specific benefit to the health of the heterozygote, who is (for the most part) functionally normal. It is, however, directly important to procreative choice. Thus, the obvious purpose of the laws was to influence childbearing decisions by heterozygous couples. The hope was to reduce the number of births of affected children.

Genetic screening laws were of two general types: mandatory premarital blood tests for at-risk couples and mandatory preschool tests for children. Because of the association of sickle cell anemia with only one race screening laws confronted an unusual equal protection problem. The distribution of the allele suggests that only black persons be tested. A legislative classification based on race, however, is constitutionally suspect. The state must demonstrate a compelling argument for the classification if a constitutional challenge is initiated. It is amusing to note the contortions some law writers went through to avoid an obvious racial classification. The New York preschool testing law *required* that children attending urban schools include a record of a sickle cell test with their school physical examination certificate; for rural children, this requirement was *optional* with the discretion of the school physician.[33] This of course reflects the demographic distribution of blacks. The language of the New York premarital screening law was ludicrous. It required persons who were "not of the Caucasian, Indian or Oriental races" to be tested for sickle cell trait prior to obtaining a marriage license.[34]

A more typical legislative approach was to delegate the decision to require screening for the trait to a public health official. For example, the Arizona law reads, "The state department of health may require that a test be given for sickle cell anemia to any identifiable segment of the population which the department determines is susceptible to sickle cell anemia at a disproportionately higher ratio than is the balance of the population."[35] At least one state confronted the classification problem squarely. Kentucky required all "Negro" marriage license applicants to be tested for sickle cell trait.[36] Violation of the law carried the threat of a fine of one hundred dollars.[37]

Let us briefly examine one of the possible constitutional objections to a compulsory sickle cell anemia screening law: the equal protection test. Does the state have a compelling interest in identifying phenotypic normal carriers for a particular autosomal recessive disease? The benefits to family and society from the reduction in births of persons with sickle cell disease due to sensible procreative decisions based upon screening test results must be weighed against risks inherent in data acquisition that has a potential for wide scale discrimination. Sporadic, yet serious, discrimination against sickle cell heterozygotes has been documented.[32] Individuals carrying the trait have been at risk for foreclosure of certain job categories and payment of higher life insurance premiums. Unfortunately, it is insidiously easy to translate "deleterious" gene to "bad" gene, with a concomitant shift in attitude toward the possessor of that genome.

The original sickle cell screening laws failed to provide for confidentiality of test information, availability of free, competent, genetic counseling services to explain the implications of a positive result in heterozygous testing, or programs designed to educate the general public about the meaning of genetic disease. As such, these laws were woefully inadequate. Given the severe racial tension in our society, the functional health of the carrier, and the potential for misuse of genetic information, these laws were probably unconstitutional. This question has been mooted by passage of a well-funded federal law that supports only voluntary state sickle cell screening and research programs.[38] The lure of federal funds has helped to stimulate revision of these laws in several states and the passage of new legislation in other states. Nevertheless, it is possible that a compulsory genetic screening law could survive a constitutional test.

Genetic screening legislation continues to be written, despite the storm of controversy that broke around those first laws. In the past year New Mexico, Kansas, and North Carolina have joined the ranks of states with sickle cell screening legislation. The coverage of genetic screening laws is expanding. Virginia's new law is entitled a "Voluntary Program for Control of Genetic Diseases."[39] North Carolina has provided for screening programs to detect all the hemoglobinopathies.[40] Montana, a quantum leap ahead of its sisters, now mandates the screening of neonates for all the "inborn metabolic errors."[41] It seems imperative that guidelines be developed to reduce the dangers of hastily written legislation.

One disturbing trend is already apparent. Since the phenylketonuria laws were first written, there has been a spate of *disease-specific* legislation. California now provides by law that a couple be informed of the availability of premarital testing for Tay-Sachs disease.[42] In the wake of the federal sickle cell anemia law, passage of a national Cooley's Anemia bill was secured.[43] Disease-specific laws appear to be a fragmented way to attack genetic disorders. From such legislation one can expect an unnecessary multiplication of administrative activity, which might siphon off research funds to project management. Perhaps more important is the confusion that can be caused in the lay public's conception of our health care priorities. Does galactosemia really pose a threat to the public health sufficient to warrant a legislature specifically providing for neonatal screening?

GENETIC SCREENING LEGISLATION: A PERSPECTIVE ON THE ROLE OF LAW

Genetic screening programs encompass one of two goals: treatment and reduction of the number of births of affected persons. Treatment is the traditional aim of screening. For a growing number of genetic diseases, treatment—often in the form of dietary management—is beginning to prove effective. Besides phenylketonuria and galactosemia, maple syrup urine disease and argininosuccinic aciduria may be mitigated by careful control of diet. The development of therapeutic protocols provides impetus for mass neonatal screening. An issue which lurks behind the decision to screen for a treatable disorder is what role the frequency of newborns afflicted with the malady should play in implementation of a program. This is primarily a problem in allocation of resources. A decision to legislate a search for galactosemic neonates may provoke criticism from those who question the position of that disorder in the hierarchy of health care needs.

The decision to allocate tax dollars to screen neonates for untreatable disorders is perhaps subject to more severe criticism. Of course, it may be argued that no condition is completely recalcitrant to treatment. Nevertheless, it is preferable that the law should not become involved in mandating tests for specific disorders when such an act will eventually run the gamut of those who question spending priorities.

For many genetic disorders the current posture of medicine is to reduce the number of births of affected individuals through genetic screening and to counsel persons who threaten their offspring with the disease. Chromosomal abnormalities, which are usually not hereditary, are among these disorders. Three distinct kinds of persons pose risk to offspring. Some persons threaten progeny regardless of whom they marry. This includes persons affected with dominant disorders (retinoblastoma), carriers of X-linked diseases (hemophilia), and persons with balanced chromosomal translocations. Other persons pose a high risk of conceiving progeny with genetic disorders only because of mating choice. Couples heterozygous for the same autosomal recessive disease comprise this category. Finally, particular pregnancies may be threatened because of maternal age or exposure to clastogenic agents. The age-morbidity curve for Down's syndrome is well established. Those persons who carry a detectable genetic disease trait may be identified before procreation. Amniotic monitoring is the only means to detect most chromosomal abnormalities after procreation.

Genetic diagnostic technology has grown rapidly. Significant advances in automation and concomitant economics of scale promise an era of multiphasic genetic screening. Programs that screen neonates for a wide assortment of inborn errors of metabolism—both treatable and untreatable—are already under way. Routine karyotyping could easily be incorporated into such efforts. Undoubtedly, a nation-wide screening program could be geared up if funds were appropriated for training needed technicians. Introduction of legislation that would provide support for training is imminent.[44] Realistically, a crucial purpose of genetic screening legislation is to acquire and disperse data that indicates whether or not an individual poses a genetic risk to anticipated offspring. If the state is to consider the implementation of such data gathering it must carefully ascertain the magnitude of benefit and detriment implicit in such an activity.

In the abstract, it is not difficult to suggest a plan whereby the technology of screening is made freely available to all persons at their own discretion. Yet, a truly passive voluntary public health screening program accomplishes little. At the very least, it must be advertised. On a continuum stretching from pure voluntarism to pure compulsion where is the demarcation line of governmental intervention? In our technological society genetic screening may have an alluring quality, yet the information acquired could be used insidiously and coercively against the individual confronted with it. A fetal diagnosis of trisomy 21 carries a powerful impetus for abortions. Is it proper to characterize genetic screening as mere information gathering? How shall its tremendous potential for influencing procreative choices be channeled?

What is the spectrum of possible benefits that might justify invocation of the public health power? Mass screening for genetic disease, when coupled with appropriate mating prohibitions, could permit the reduction of all identifiable disorders to mutation level in a single generation. The most far-reaching kind of program would screen all persons at birth and forbid all heterozygotes and homozygotes for important genetic diseases from becoming natural parents. Artificial insemination and adoption could still provide such couples with a normal family life. The social havoc that such a program could create needs no delineation.

The elimination of births of persons affected with genetic disease may be a more palatable goal for a public health program than the eradication of selective alleles. An adequate technological capability will soon be available. The state could identify and forbid marriage between persons heterozygous for the same disorder. It could forbid natural parenthood to individuals affected with a dominant disorder. It could implement a program requiring that all "at risk" pregnancies be monitored. Compulsory abortion of affected fetuses could be incorporated into such a program. The program would be both distasteful and impossible to enforce in a manner consistent with a free society. A most realistic aim of the state would be to act to reduce the number of births of affected persons without the use of compulsory measures. This is the implicit position of public health authority today. It is the *raison d'être* for heterozygote screening programs.

There are a variety of models for state action guiding technological implementation. It is possible that the state could decide that the technology under consideration promised more disadvantages than advantages. By refusing to fund a project it could sound its death knell. Such is the history of the Supersonic Transport (SST) controversy. A more absolute approach would be to ban *some* activity. One remarkable action recently taken by the federal government has been the suggestion of a moratorium on studies of human fertilization *in vitro*.[45]

A realistic policy for government to adopt is one of *laissez-faire*. For many biomedical programs, such a posture is a death sentence. Eugenic technology, however, particularly amniocentesis and automated screening programs, could be expected to develop despite lack of tax revenue support. In such a situation the course of development is determined by the health professionals and the insurance establishment. For example, it is interesting to speculate upon the impetus given to the development of amniocentesis as a routine diagnostic tool when it was placed on the Blue Shield fee schedule. Another important parameter is the role played by biomedical industry. Refinement and cost reduction in automated karyotype analysis equipment will be translated into increased sales and utilization. Government inaction appears merely to shift decision making to more specialized forums.

Currently, the model that government seems to favor in the field of eugenic legislation is "pure voluntarism." Such a value is emphatically claimed by the National Sickle Cell Anemia Control Act and the

Cooley's Anemia Act. Both of these laws aim for a comprehensive attack on the diseases. Provisions for research for cures accompany provisions for heterozygote screening. State and private programs receive support only upon demonstration of voluntary models of screening, yet it remains uncertain whether the concept of "pure voluntarism" is accurate. I would argue that the availability of very low-cost or free screening programs, coupled with public advertising of the virtues of screening, is actually a type of positive incentive.

Obvious positive incentives to action are myriad in government policy. The oil-depletion allowance is an incentive to enter the field of primary oil production. Tax deductions of any kind fit this model. Special mortgage assistance to low-cost housing programs is another example. Federal health insurance supporting the needs of the impoverished is an incentive to equalize distribution of health care services in our society. Amniocentesis is also listed on the Medicaid fee schedule.

Negative incentives are usually framed so as to penalize, *de facto,* certain behavior. An important example of such an effect may be discerned from a recent Supreme Court interpretation of the validity of a state action under a section of the Social Security Act. In *Dandridge v. Williams,*[46] the Supreme Court held that the state of Maryland could place a ceiling on its payment to families supported under the Federal Aid to Families with Dependent Children program. The plaintiff argued that the state discriminated against children in large families. A family of seven children received less per capita support than a family of four if both families received the same monthly assistance. The Court, however, chose to look at the child support as a problem in "the social and economic field" rather than to perceive it as a fundamental interest which would require more rigorous justification of the state funding policy. Interestingly, although Maryland argued that incentives to family planning were a legitimate state interest, the Supreme Court evaded comment on that position.

A final possible mode of state action is compulsion. Numerous state health programs carried out in this manner have already been listed. The limits defining compulsory programs are uncertain. The infrequency of constitutional adjudication in the field of public health law, combined with the novelty of the idea of mass eugenic legislation, makes prophecy here a fool's endeavor. Realistically, some constitutional questions are greatly influenced by public opinon. A striking example was the decision on abortion. The sense of public opinion could greatly influence adjudication of eugenic legislation.

Is it possible to draft and implement sensible genetic screening legislation? Even the most foresighted law cannot anticipate all problems arising under its domain of influence, yet prolonged reflection on the proper uses and means of such legislation could be rewarded with a rationale for implementation. The primary question is, should the state act? State action is optimal when society and individual citizens benefit and no person suffers. This is a rarely realized ideal. Given that few, if any, programs could meet such criteria, it is necessary to investigate if there is sufficient benefit in a plan to warrant its trial. From a societal perspective, the value of a program designed to reduce drastically the births of persons affected with incurable genetic disease appears positive. Our willingness to tolerate intervention into human procreative decisions must be weighed against the value of such a program. The response to that intervention is a function of the manner of intervention. Can we design a program that maximizes the reduction of disease while it sufficiently minimizes the violation of the individual citizen's rights?

GUIDING PRINCIPLES FOR GENETIC SCREENING LEGISLATION

Initial efforts to discern guiding principles for mass genetic screening appear to have been an *ad hoc* response to complications emerging from early programs. One of the first well articulated sets of principles grew out of the Tay-Sachs screening undertaken in Washington, D.C. in 1971.[47] A year later a comprehensive effort to identify principles for the design and operation of programs was made by a research group associated with the Institute of Society, Ethics, and the Life Sciences.[48] In 1973 the state of Maryland enacted the first genetic screening law that showed evidence of perceptive attention to subtle consequences of that activity.[49] The following principles are offered in anticipation of future legislative activity. Because they focus on the language of implementing legislation they are somewhat different from principles of program operation.

Among the major principles that a genetic screening law should embrace are *scientific rationality, paramount concern for the physical and psychic health of the individuals screened, assurance of high quality genetic counseling for all persons with significant test results, recognition that a program of public education is integral to the dual goals of reducing genetic disease and eradicating genetic discrimination, and special attention to the problem of maintaining strict confidentiality of any screening records that are stored.*

An insistence that the laws meet the test of scientific rationality is not a requirement too obvious to bear stating. Two of the many incidences of the failure of genetic screening laws to satisfy this standard are illustrative. The title of a Georgia session law that provided for screening school children for sickle cell trait reads, "Education—Immunization for Sickle Cell Anemia Required for Admission to Public Schools."[50] Fortunately, this ludicrous requirement was not incorporated into the body of the law. The National Sickle Cell Control Act contained the following declaration of purpose clause that confused trait with disease: "The Congress finds and declares — (1) That sickle cell anemia is a debilitatory, inheritable disease that afflicts approximately two million American citizens and has been largely neglected;... "[51] Perhaps two million Americans have the trait, but only about fifty thousand have the disease. State legislators rarely are budgeted to afford the services of scientific consultants. Nevertheless, a policy decision to implement biomedical technology that is not preceded by scientific critique builds a social program on a shaky foundation. Poor drafting of national legislation for which there is easy access to first-rate scientific advice is inexcusable.

Screening legislation must be written (or rejected) with a clear understanding of immediate potential risks and benefits and a recognition of policy implications. For the committee that reviews the law, the spectrum of detectable diseases (treatable and untreatable) and of carrier states, the cost of screening for various disorders, and the accuracy of available testing methods should be among the subjects investigated. It is neither possible nor desirable that categories of diseases to be screened for and protocols to be followed be written into law. Such an effort would produce inflexible and obsolescent language. However, recognizing a concern for the well-being of its citizens, the legislature should set limits on screening activity. For example, the test methods employed could be required to satisfy a certain threshold accuracy.

No state should legislate a program of genetic screening without first securing funding adequate to support a competent staff, first-rate laboratories, and effective educational programs. Currently, state laws evidence little concern for the fact that diagnostic testing represents a small fraction of the total cost of a well-administered broad spectrum program.

Scientific rationale for certain genetic screening surveys raises a delicate constitutional question. Because of the skewed distribution of the alleles, it is sensible to test only certain groups of persons at risk for certain genetic disorders: whites, for cystic fibrosis; blacks, for sickle cell anemia; Ashkenazic Jews, for Tay-Sachs disease; and women over forty, for fetuses with chromosomal abnormalities. This selectivity of testing is sensible in terms of allocation of resources, but it raises the aforementioned issue of classification by race, which requires special or "compelling" arguments for justification. Arguably, the compelling interest of the state is the reduction of suffering in the citizenry, maximized by the widest distribution of services. In support of selective screening perhaps the concept of "benign racial classifications" could be adopted.[7] This concept, born in civil rights legislation and designed to redress racial imbalance in our society, argues that clear benefit adheres to the group so classified by a particular law. This position is vulnerable to attack, based upon the threat of discrimination implicit in gathering genetic information.

A second major principle of a genetic screening law should be concern for the integrity of the individuals who will be tested. The cornerstone of integrity is voluntary participation. As Justice Harlan noted, "There is . . . a sphere within which the individual may assert the supremacy of his own will and rightfully dispute the authority of any human government, especially of any free government existing under a written constitution, to interfere with the exercise of that will."[2] Only rarely can a government justify this kind of intrusion. Mandatory screening laws might achieve a more rapid reduction of genetic disease than might voluntary legislation. Nevertheless, the unique potential that genetic information holds for subtle social discrimination is a risk that outweighs the envisioned benefit.

The willing participation of an individual in a public screening program requires that he or she know why the test is offered, what test results could mean for personal health or procreative decisions, significant risks associated with the test itself, and availability of counseling services and treatment, where it is indicated. Because tests are overwhelmingly designed to directly benefit the individual being screened, it is appropriate that parents be allowed to give consent for the testing of their children. Informed consent is a much unsettled doctrine, with complexities beyond the scope of this paper. However, the *principle* is both clear and applicable to screening situations. Further, it is broad enough to require that the screened individual receive full disclosure of all test results.

A genetic screening law should guarantee that persons recording a positive test result be given immediate access to free, genetic counseling administered by a highly qualified staff. The genetic counselor may strongly influence the future of the person screened. His function — assisting in very crucial decisions — is so sensitive that some attempt to delineate the boundaries of this activity is warranted.

The ideal of nondirective counseling is not one that can be standardized, yet the principle is clear (see Chapter 3). Genetic counselors should not impose personal concepts of health and disease upon those seeking their service. The state must employ competent, experienced professionals who, hopefully, will bring to their work the requisite deference to the integrity of those who invoke their aid. Licensing laws rarely serve to police a profession, but educational standards and on-the-job training demanded by a license offer some degree of self-regulation.

Dynamic counseling could be instrumental in the reduction of live births of persons with incurable genetic disease. Lay persons confronted with arcane medical data frequently defer to the advice of experts.[52] It is theoretically possible that an amorphous set of eugenic guidelines could emerge from the counseling experience of persons employed by the state. If counselors were to suggest uniformly that a given situation required a specific response (sterilization, abortion?), that response would be frequently forthcoming. Social attitudes toward specific anomalies would be shaped. Anticipating this ethical problem, Maryland law requires that "counseling be nondirective; and that such counseling emphasize informing the client and not require restriction of childbearing."[49]

Genetic counselors working in accordance with a state supported screening program should be discouraged from disclosing any confidential information to third parties. Despite limited case law that suggests a counselor could contact a blood relative or a spouse because of the consultand's genetic profile, it is preferable that the person tested decide to whom the information should be transmitted. Any other procedure could result in serious emotional injury to the consultand.

The espousal of value-free counseling raises at least one unwieldy question. In certain situations a screening test result might imply adultery (or an infrequent spontaneous point mutation in the parental germ line). Should the counselor withhold test information, convey the information without explanation, or suggest the implications of the test result? The law traditionally has acknowledged the right of a physician to withhold information from his patient if it is in the patient's best interests.[53] The requirement of informed consent is reducing the physician's right to modify or withhold medical facts. The concept of a proper withholding has been rooted in the preoperative situation where the physician could demonstrate a realistic concern for the patient's anxiety. Genetic counseling does not usually precede medical intervention. It may be that the genetic counselor should be able to convey or withhold facts in the best interests of his consultand. At the moment, the law has not commented on that right. If the counselor is to be given this latitude, the possibility of abuse must be recognized.

A program of general public education, designed to provide information about the nature of the human genome and the problem of genetic disease, offers a valuable long-range weapon to avert the threat of discrimination against persons with a particular genetic status. This is indeed part of the long-term plan. For the immediate future specialized information campaigns must precede the invocation of any large-scale screening. While accentuating the implications of genetic disease for the health of present and future generations, the program must avoid dramatization. Good citizen response necessitates that fear of the unfamiliar be diminished. The most recent genetic screening laws recognize the value of genetic education. For example, recent North Carolina legislation requires that "any program for voluntary testing shall begin no sooner than 60 days after the implementation of an adequate and effective educational program."[54]

Confidentiality of genetic screening results is a complex problem that is intertwined with the decision of when to screen. Based on the assumption that the integrity of the individual is paramount, what is the most appropriate screening time? The answer depends on the goal of the program. Clearly, screening designed to prevent retardation from phenylketonuria demands that the neonate be tested, despite the inability of the newborn to voice consent to the procedure or to understand its results. At that stage of a person's life, parental consent is operative. An interesting parenthetical question is whether the state should tolerate religious objection to blood-taking for a therapeutic test of an infant. Currently, states have taken different views on this question. One may argue that the possibility of the birth of an afflicted child to parents who object to blood testing is so remote as to be of less importance to the state than its concern for deference to freedom of religion.

Screening neonates in hospital nurseries is a logistically sound procedure. However, the state that decides to do this screening must develop a mechanism to insure that the infant receives the test result and adequate counseling at an age when he will be able to understand the implications of the test. Informing parents is not enough. There is little certainty that they will retain the meaning of the test over a long period of time. It is also possible that a misguided sense of shame or guilt would cause them to withhold the information from their child.

Evolving such a mechanism is neither an easy nor an inexpensive task. Certain possibilities present themselves. A coded alert could be placed on the official birth certificate (for example, representing heterozygosity for Hurler's syndrome). This alert would prompt the contacting of such a person during his sixteenth year by public health officials who would counsel him about this disorder. The ominous nature of such a meeting is disturbing. Perhaps the better solution is to insure that each public school district contain one competent genetic counselor who would be responsible for explaining carrier status for genetic disorders to children as the need to know arose.

The difficulty of insuring accurate information transmission suggests that neonatal screening should be reserved only for treatable disorders. Screening for carrier status for a variety of disorders (both the sex-linked and autosomal recessive conditions) could properly be done in adolescence. For example, states could require a public health examination prior to the third year of high school. Such an examination could include an offer of genetic screening for certain conditions. This offer could vary with sex and ethnicity of each student. Such a program could be nicely integrated with public education about human genetics. It has the advantage of directly conveying information to a person prior to choosing a marriage partner. Obviously, the receipt of such information should occur early enough to influence mating choice. The emotional stability of an adolescent population and the potential dangers of social stigmatization of carriers are two drawbacks to such a plan.

Premarital screening offers certain benefits. Prior to obtaining a marriage license a couple can be informed of the risks of genetic disease and of the availability of free test and counseling facilities. Because these persons are in most cases anticipating parenthood, the offer of test services can be expected to be frequently accepted. For those rare couples that prove to be at high risk for conceiving offspring with genetic disease, options are available that can help them. Artificial insemination, adoption, or pregnancy monitoring could prove palatable. It is possible that test results could alter marriage plans. If the test is freely taken, however, it would seem that the misfortune of a terribly diseased child is of more importance than the risk posed to a few marriages. Testing at this time has the advantage of guaranteeing privacy (results need not be stored) and being transmitted at an optimal age for comprehension. A very important drawback to premarital screening is its failure to alert women who are likely to bear children out of wedlock.

The principle of confidentiality requires that test results be transmitted only to the person tested or to his legal guardian. Employers, insurance companies, and the government need not have access to the results. Of course, test information could be valuable to biomedical research and genetic counseling. Thus, it may be appropriate to keep anonymous records which would permit the compilation of accurate frequency figures for certain disorders. If it is calculated that resources to accomplish these ends are unavailable, broad spectrum genetic screening should not be undertaken.

Current legislative trends indicate a predisposition toward multiphasic screening of neonates for treatable disorders and nontreatable disorders, as well as for carrier status. Sixteen of the phenylketonuria laws mandate screening for "other metabolic disorders."[27] Massachusetts screens newborns for a variety of disorders without legal mandate. The New York legislature recently passed a law that specifically provides for neonatal screening for "homozygous sickle cell disease, branched chain ketonuria, galactosemia, homocystinuria, adenine deaminase deficiency, histidinemia, and other such diseases and conditions as may from time to time be designated."[55]

Available legislation does not refer directly to data storage. Inevitably, serious discussion will develop around the concept of a "state genetic registry." Certainly, the concept is not novel. New York requires that "birth defects and allied diseases" be reported to state officials.[56] Minnesota authorizes the state board of health to collect data on hereditary disease, to prepare information to aid counselors, and to maintain a cytogenetics laboratory that will offer diagnostic services to the public.[57]

Federal legislation creating a Genetics Data Bank could be a positive feature of any national plan to reduce genetic disease. Such data acquisition could refine current crude empirical risk figures for certain genetic diseases and congenital anomalies. Variance of incidence by race could be ascertained. Unusual alterations in disease frequency could provide confirmation that a dangerous mutagenic or clastogenic agent had been introduced in some section of the environment. The puzzling issue of heterozygote advantage might receive new information. The extent and meaning of isozyme variants could be further explored.

Of course, the potential for misuse of such data banks poses a real obstacle to their development. What constitutes misuse? Genetic records could theoretically be utilized to *positively* identify a male in a paternity suit or a suspect whose blood had been discovered at the scene of the crime. Would such activity be acceptable? Relevant case law answers in the affirmative.[58] It is important to note that in civil proceedings (paternity suits) blood tests may be ordered at the discretion of the court.[59]

The critical question to ask is, can a system be devised to protect a large computer health record data bank from invasion? Employers, life insurance companies, even a person's neighbors might have an interest in obtaining such data. At least one state has been willing to act upon the belief that it can protect stored data of a sensitive nature. Recently enacted New York legislation attempts to safeguard a computer data bank that contains the patient records of psychiatric hospitals located in several other jurisdictions.[60] Remote access terminals permit multistate data storage. Among important features of the law are those that shield data from subpoena, render it inadmissible in court, and protect the information of one state from perusal by another. Provision for annual system security review by the Commissioner of Mental Hygiene is included. Publication of data to benefit mental health researches must be anonymous. It remains to be seen whether such a system will be inviolable.

Based on the assumption that government-supported genetic registries will evolve it is important to develop guidelines on the use of such information. Employers should be denied the right to request genetic information during a pre-employment physical.[61] For those rare situations where it is justifiable to request the information, regulatory exceptions can be developed. It may be that sickle cell heterozygotes can be legally excluded from piloting commercial aircraft. However, such exceptions will be exceedingly rare. Until mortality records prove the argument, persons heterozygous for a genetic disease should not be penalized with high life or health insurance premiums. Perhaps heterozygosity testing should be precluded from insurance physicals. Clearly, access provisions to information must be carefully drafted. The data should be restricted to the person screened or to a legal guardian. Neither spouses nor next of kin need be informed. A heavy penalty should deter persons from obtaining this data by illegal methods.

It may be possible to draft compulsory genetic screening legislation that would survive constitutional challenge. However, the potential for abuse of information elicited by such a law, once well understood, should curb its invocation. Indisputably, genetic screening offers benefit to individual families and society as a whole. The law best serves a program designed to reduce genetic disease by guiding technological implementation. Successful phenylketonuria screening, accomplished without legal mandate in many countries, proves that compulsion is an unnecessary feature. For now, perhaps the best course would be to continue neonatal screening programs for those disorders that are amenable to treatment and to develop laws that inform couples seeking a marriage license that free genetic testing is available if they wish to utilize it. The elimination of genetic disease must proceed with paramount concern for the procreative rights of the individual.

ACKNOWLEDGMENT: The author wishes to express his thanks to Harold Edgar, Professor of Law, Columbia University School of Law, and Aubrey Milunsky for reading the manuscript.

References*

1. Grad, F. P.: Public Health Law Manual. New York, The American Public Health Association, Inc., 1970, p. 5.
2. *Jacobson v. Massachusetts,* 197 U.S. 12 (1905).
3. Public Health Act of 1961, 9 & 10 Eliz. 2, c. 64, §38.
4. *Raleigh Fitken-Paul Morgan Memorial Hospital v. Anderson,* 42 N.J. 421, 201 A.2d 537 (1964), *cert. denied,* 377 U.S. 985 (1964).
5. *State v. Perricone,* 37 N.J. 463, 181 A.2d 751 (1962), *cert. denied,* 371 U.S. 890 (1962).
6. *De Aryan v. Butler,* 119 Cal. App. 2d 674, 260 P.2d 98 (1953), *cert. denied,* 347 U.S. 1012 (1954). *Froncek v. Milwaukee,* 269 Wis. 276, 69 N.W.2d 242 (1955).
7. Note, Developments in the law: Equal protection, 82 Harv. L. Rev. 1067 (1969).
8. *Loving v. Virginia,* 388 U.S. 1 (1967).
9. *McLaughlin v. Florida,* 379 U.S. 184 (1964).
10. *Griswold v. Connecticut,* 381 U.S. 479 (1965).
11. *Roe v. Wade,* 410 U.S. 113 (1973).
12. N.Y. Public Health Law §2164 (McKinney 1972).
13. Cal. Health and Safety Code §3400 (West Supp. 1973).
14. Cal. Health and Safety Code §3380 (West Supp. 1973).
15. Cal. Health and Safety Code §4300 (West Supp. 1973).
16. Rev. Code Mont. 69-6702 (1973 Supp.).
17. Cal. Health and Safety Code §3185 (West 1972).
18. Cal. Health and Safety Code §3285(d) (West 1972).
19. Cal. Business and Professional Code §551 (West 1962).
20. Cal. Welfare and Institutions Code §5150 (West Supp. 1974).
21. Keezer, F. H.: *In* Morland, J. W. (ed.): On the Law of Marriage and Divorce. Indianapolis, The Bobbs-Merrill Co., Inc. 1946.
22. Fersterr, E.: Eliminating the unfit—Is sterilization the answer? 27 Ohio St. L.J. 591 (1964).
23. Chen, P.: Population planning: policy evaluation and action program. *In* Wegman, M., Lin, T.-Y., and Purcell, E. (eds.): Public Health in the People's Republic of China. New York, Josiah Macy Jr. Foundation, 1973, p. 246.

*Please refer to *A Uniform System of Citation,* 11th ed., 1967, published by The Harvard Law Review Association, Gannett House, Cambridge and U.S. Government Office Style Manual, 1967 for legal references.

24. *Buck v. Bell*, 274 U.S. 200 (1927).
25. *Gould v. Gould*, 78 Conn. 242, 61 A. 604 (1905).
26. *Holcomb v. Armstrong*, 39 Wash. 2d 860, 239 P.2d 545 (1952).
27. Swazey, J.: Phenylketonuria: a case study in biomedical legislation. 48 J. Urban Law 833 (1971).
28. Bessman, S. P.: PKU laws, a model for the future? Md. State Med. J. 15:145, 1966.
29. Levy, H. L.: Genetic screening. *In* Harris, H., and Hirschhorn, K. (eds.): Advances in Human Genetics. Vol. 4. New York, Plenum Publishing Corp., 1974, p. 26.
30. Fox, J. G., Hall, D. L., Haworth, J. C., et al.: Newborn screening for hereditary metabolic disorders in Manitoba, 1965–70. Can. Med. Assoc. J., 104:1085, 1971.
31. Cahalane, S. F.: Phenylketonuria: Mass screening of newborns in Ireland. Arch. Dis. Child., 43:141, 1968.
32. Reilly, P. R.: Sickle cell anemia legislation. J. Leg. Med., 1:39, 1973.
33. N.Y. Education Law §903 (McKinney 1972).
34. N.Y. Domestic Relations Law §13aa (McKinney 1972).
35. Ariz. Stat. §36-797.42.B. (1973).
36. Ky. R.S. 402.310-340 (1973).
37. Ky. R.S. 402.990 (16) (1973).
38. 42 U.S.C. 300b (1973).
39. Va. Code §32-112.20–23 (1972).
40. N.C. Gen. Stat. §143B-188–196 (1973).
41. Mont. Rev. Stat. 69-6711–13 (Supp. 1973).
42. Cal. Civil Code §4302 (West 1973).
43. 42 U.S.C. 300c (1973).
44. A bill sponsored by Senator Javits, The National Genetic Disease Control Act, will be introduced into committee.
45. Proposed HEW Regs.: Protection of human subjects – policies and procedures. Fed. Reg., 38:31738, 1973.
46. 397 U.S. 471 (1970).
47. Kaback, M. M., and O'Brien, J. S.: Tay-Sachs: Prototype for prevention of genetic disease. Hospital Practice, 8:107, 1973.
48. Lappé, M., Gustafson, J., et al.: Ethical and social issues in screening for genetic disease. N. Engl. J. Med., 286:1129, 1972.
49. Md. Stat. 43 §814 (1973).
50. Georgia Laws, 1972 Session, p. 962.
51. 86 Stat. 137.
52. Leonard, C., Chase, G., and Childs, B.: Genetic counseling: A consumer's view. N. Engl. J. Med., 287:433, 1972.
53. *Salgo v. Leland Stanford Jr. University*, 154 Cal. App. 2d 560, 317 P.2d 170 (1957).
54. N.C. Gen. Stat., §143B-196 (1973 Supp.).
55. N.Y. Public Health Law §2105 (McKinney 1974 Supp.).
56. N.Y. Public Health Law §2730–33 (McKinney 1972).
57. Minn. Stat. Health §144.94 (1973).
58. *Schmerber v. California*, 384 U.S. 7 (1966).
59. Annot., *Blood Grouping Tests*, 46 Am. L. Rep. 2d 994 (1954).
60. N.Y. Civil Rights Law §79(j) (McKinney 1973 Supp.).
61. *Griggs v. Duke Power Co.*, 381 U.S. 424, 1972.

CHAPTER 20

THE ECONOMICS OF PRENATAL GENETIC DIAGNOSIS

Ronald Conley, Ph.D., and Aubrey Milunsky, MB.B.Ch., M.R.C.P., D.C.H.

In 1975 over 20,000 infants with chromosomal abnormalities will be born in the United States, and over 700,000 worldwide[1] (Table 20–1). The prodigious economic implications of such births provide cause for serious concern, especially with the advent of prevention through carrier detection and prenatal genetic diagnosis. There can be no measure of the emotional costs. Indeed, there are those who find odious the need to recognize in monetary terms the medical costs of these tragedies. The realities of the situation, however, demand that society study the cost-benefit aspects of preventive programs.

TABLE 20–1. ESTIMATED INCIDENCE OF CHROMOSOMAL ABNORMALITIES IN LIVEBORN INFANTS (1975)*

	United States	World
Total population	216,553,000	4,000,000,000
Live births	4,000,000	136,000,000
Autosomal abnormalities (1.00 per 1000)	4,000	136,000
Sex-chromosome abnormalities (2.28 per 1000)	9,120	310,080
Structural autosomal rearrangements (1.80 per 1000)	7,200	244,800
Total chromosomal abnormalities (5.2 per 1000)	20,800	707,200

*From Milunsky, Aubrey: A PRENATAL DIAGNOSIS OF HEREDITARY DISORDERS, 1973. Courtesy of Charles C Thomas, Publisher, Springfield, Illinois.

442

TABLE 20-2. SUMMARY OF PRENATAL GENETIC STUDIES IN
526 CONSECUTIVE CASES

Indications for Amniocentesis
CHROMOSOMAL ANALYSIS
Translocation carrier	9
Maternal age 35–39	132
Maternal age 40 or over	137
Family history of Down's syndrome	
Prior birth	84
Other close relative	35
Miscellaneous	89
SEX-LINKED DISORDERS	12
METABOLIC DISORDERS	28
Total Cases Studied	526

Affected Fetuses

	Number	Live Birth	Therapeutic Abortion	Diagnosis Confirmed at Birth or Abortion
Chromosomal Disorders				
Down's syndrome	4	—	4	4
Klinefelter's syndrome	1	1	—	1
Trisomy 18	1	—	1	1
X-linked disorders	4	2	2	4 (males)
Metabolic disorders				
Tay-Sachs disease	4	—	4	4
Hunter's syndrome	1	—	1	1
Other				
Rubella	1	—	1	1
Total	16	3	13	16

Each year, with the birth of hundreds of thousands of chromosomally defective offspring, a future commitment to care by society in excess of two billion dollars is created.[1] Over 20 years, without consideration of the costs of inflation, the commitment will have grown to about forty billion dollars[1]—a figure not very dissimilar from the recent calculations of Swanson.[2]

Rather than entertain theoretical estimates for benefit-cost analyses, in this chapter we have considered the specific consecutive diagnostic case experience in one author's (A.M.) laboratory.

In the summary of this consecutive case experience (Table 20–2), "affected" fetuses were diagnosed in 16 of 526 cases (studied between January 1970 and December 1973). Chromosomal analysis for advanced maternal age is seen to be the commonest indication by far for prenatal genetic studies. All abnormalities diagnosed prenatally have been confirmed after abortion or birth. There have been no errors. In one case in which the parents elected to continue the pregnancy even in the face of a prenatal diagnosis of Klinefelter's syndrome, a twin was later and unexpectedly found and was normal at delivery. The parents elected to terminate those pregnancies in which the fetus had Down's syndrome, trisomy 18, fetal rubella, or a metabolic disorder. In these cases there is little doubt as to the sequelae of the disorders: moderate to severe retardation in all the cases and early death in the last group (four cases of Tay-Sachs disease and one of Hunter's syndrome).

Five fetuses were identified as having a high risk of defect: one with rubella and four who were carried by mothers with sex-linked disorders. Two of the latter were therapeutically aborted. The two carried to term were normal when examined after delivery.

BENEFIT-COST THEORY

Benefit-cost analysis consists of comparing social benefits (the total increase in social well-being that results from a given action) with social costs (the value of the resources that would have been available for alternative uses had the action not been taken). Social benefits take many forms. Among the more important are increases in productive capacity (earnings, homemaking, other unpaid work at home and away from home); decreases in special medical and custodial costs (institutional care, clinical care); decreases in special training costs (special education classes, other similar training programs); increases in life expectancies; improvements in mobility; and decreases in special housing needs. Two points must be made: (1) social benefits will be realized by many people, not just those directly affected; and (2) social benefits are not restricted to variables that can be measured in monetary form. The purpose of benefit-cost analysis is to influence the allocation of resources; all effects of a given activity that increase social well-being must be considered.

Measuring the benefits of preventing prenatal defects is a formidable task because of a highly inadequate data base and several conceptually unresolved problems in the definition of benefits. Three situations can be envisioned after the detection of a defective fetus: (1) the fetus is aborted and in effect is replaced by a subsequent infant; the total number of children born to the family remains unchanged; (2) the fetus is aborted, and for one reason or another the parents decide to reduce family size, possibly by more than the single birth; and (3) no abortion is performed, but early prenatal or postnatal treatment (in methylmalonic aciduria, galactosemia) reduces the damage to the fetus.

The third situation, which may be termed the "repair" case, is relatively straightforward. Social benefits are measured by the increase in well-being (measured by the many variables that affect well-being) that result because the effects of prenatal conditions are minimized or eliminated. In the first situation, which may be termed the "replacement" case, social benefits should probably be measured by the increase in well-being that results from the "replacement" person being more productive and more capable of enjoying life than the original fetus would have been if born. In addition, the individual will require less care. It is reasonable to assume that the "replacement" infant will be average with respect to mental and physical capacity. Thus, the two cases have identical measures of benefits if, in the "repair" case, damage can be completely prevented.

It is the second, or "nonreplacement," situation that presents the greatest confusion. In the other two situations, substantial benefits are received by a "repaired" or "replacement" fetus. In the "nonreplacement" case, however, the only social benefits realized are those received by persons other than the aborted fetus. This has major effects on the definition of social benefits.

In the "repair" or "replacement" cases, benefits are measured by the increase in productivity, the decrease in special medical, custodial, and training costs (and other special supportive costs such as counseling), and the many other changes that affect individual well-being and that are realized by the surviving fetus as compared to that which would have happened had the fetus not been repaired or replaced.

In the "nonreplacement" case, however, social benefits are measured by the savings in medical, educational, custodial, and other costs to society (society itself having been diminished by one person). This differs from the "repair" and "replacement" cases in that: (1) *all* educational, medical, and custodial costs that would have been expended are considered a social benefit and not simply the difference in these costs between an "average" person and what would have been expended had the fetus not been replaced or repaired; (2) there are obviously no incremental earnings to count as a benefit; and (3) reductions in income maintenance costs (public assistance, social security) are counted as social benefits, whereas in the other two cases they are regarded as transfer payments with the cost to the taxpayer being offset by the benefit to the recipient.

It seems inconsistent to consider increased earnings as a benefit in the "repair" or "replacement" situations, but not to consider lost earnings (what the productivity of the aborted fetus would have been) as a social loss in the nonreplacement case. Similarly, it is inconsistent to count reductions in transfer payments as a benefit in the latter case, but not in the former ones. Nevertheless, this is the logical conclusion of a model that places zero value on the life of an unborn fetus and measures increased well-being only to the surviving population, which may include a replacement fetus, may be diminished by one person, or perhaps by more (if some families reduce family size by more than one).

Measuring Benefits

Our benefit-cost comparisons will be limited to an evaluation of the prevention of chromosomal abnormalities and the metabolic disorders (four cases of Tay-Sachs disease and one of Hunter's disease).

One immediate benefit, in the replacement case, is that the surviving infants will have longer life expectancies. On the average, a person born in 1971 has a life expectancy of 71 years—a figure that developing medical technology may lengthen. Infants with Tay-Sachs disease, on the other hand, usually live less than four years. Children with Hunter's syndrome or Down's syndrome survive longer, but they still have life expectancies considerably below average.

A second benefit, also applicable only in the "repair" or "replacement" cases, is increased productivity. The extreme, but not atypical, situation is when the damaged fetus would have been incapable of any productive activity during its lifetime, and the repaired or replacement fetus is expected to have normal productivity. An indication of the potential size of the benefits is derived by estimating lifetime earnings of workers, *i.e.*, wage, salary, and self-employment income. Under normal conditions, an employer could not pay a worker appreciably less than the value of what he produces, since this situation would provide other employers with an incentive to hire the worker away; it would not pay the employer to pay a worker more than the value of what he produces.

A male infant born in 1972 can be expected to receive about 1.5 million during his lifetime. This figure is based on the assumption that there will be an average annual growth in output of 2.5 per cent, is adjusted for mortality and the possibility of having no income at various age levels, and makes no allowance for inflation. Thus, these estimates are expressed in terms of the price levels prevailing in 1972. For a female infant, the comparable figure is a little over half a million dollars.

TABLE 20-3.   PRESENT AND FUTURE INCOME DATA*

Age	Income in 1972 of Persons with Income		Estimated Income of 1972 Infants as They Move Into Age Group†		Estimated Income Adjusted for Mortality and Discounted to Present Value	
	MALE	FEMALE	MALE	FEMALE	MALE	FEMALE
14–15	$ 550	$ 384	$ 248	$ 155	$ 180	$113
16–19	1,636	1,217	1,803	1,132	535	339
20–24	5,058	3,225	8,201	4,337	1,780	956
25–34	9,628	4,278	19,738	5,676	2,575	759
35–44	12,172	4,482	32,166	7,666	2,070	487
45–54	12,092	4,715	40,826	10,347	1,251	339
55–64	10,320	4,125	44,520	12,451	593	192
65+	8,635	3,577	34,388	15,158	192	127

*Mortality data were derived from U.S. Department of Health, Education, and Welfare, Public Health Service: Vital Statistics of the United States, 1971. Vol. 2, Sect. 5, Life Tables. All income data were derived from U.S. Department of Commerce: Current Population Reports. Series P-60. No. 90. Money Income in 1972 of Families and Persons in the United States, December, 1973, p. 114.

†These income estimates represent the average for all persons surviving to a given age group.

These astoundingly large estimates are placed in better perspective if discounted to present value, *i.e.,* the amount that people would have been willing to pay in 1972 for these future dollars. For males, the present value of lifetime income, in terms of 1972 dollars, discounted at 7 per cent, is $78,000. For women, the comparable figure is $26,000.

Our choice of a 7 per cent discount rate may appear low, in view of current market interest rates, but current rates are distorted by an unusually high inflation, a phenomenon not reflected in the above estimates.

These estimates were derived as follows: Cross-sectional estimates of income by age and sex for 1972 were assumed to represent the expected lifetime profile of income for an infant born in 1972, after adjustment for productivity growth. The adjustment for productivity change was made by increasing the earnings for each age group by the factor $(1.025)^{y-72}$, where $y-72$ represents the person's age during the year y. These estimated future incomes were then multiplied by: (1) the probability of survival through the given age group, (2) the percentage of persons in that age group actually receiving income in 1972, and (3) the application of a 7 per cent discount factor to these future earnings. In view of the continuing increase in the percentage of women working and the rapidly expanding employment opportunities for women, the lifetime earnings estimate for female infants is undoubtedly low.

These estimates overstate the actual anticipated *earnings* of persons born in 1972, since they include dividends, rents, interest, social security, public assistance, and other unearned income received by individuals. About 85 per cent of the income on which the above estimates were based was earned income. This is a conservative estimate of the percentage of estimated lifetime income that should be attributed to the productive activity of people because: (1) some nonearned income, such as social security and unemployment compensation, is a form of deferred income;* (2) some dividends, rents, and interest are paid on personal savings that would not have been realized had the person not been born to earn the income from which these savings were derived; and (3) a disproportionate share of unearned income is received by the aged and would have a minuscule effect when discounted to present value.

There were five cases with biochemical diseases among the therapeutically aborted fetuses. Had these fetuses survived birth, there is almost no possibility that any would have had any gainful employment. Therefore, the output benefits of a therapeutic abortion (in the replacement case) would be $67,000 for men and $22,000 for women (85 per cent of the present value of estimated lifetime income).

---

*Another way of describing this problem is that social security, in part, and unemployment compensation, in total, are financed by taxes on the employer and do not enter into anyone's income until received as an income maintenance benefit. Public assistance, however, is financed by income taxes, and these dollars are counted twice in income figures—once by the taxpayer and once by the public assistance recipient.

There were four therapeutic abortions because of Down's syndrome or trisomy 18. Had these fetuses survived, significant retardation would have existed with I.Q. levels probably < 50. Existing follow-up studies indicate that the productivity of persons at this level of retardation is very limited, probably being no more than 20 per cent of the norm for their age and sex counterparts.[2] Although it is arguable whether this percentage will be higher (because of improved services) or lower in the future, it provides the most reasonable basis on which to estimate the output benefits of therapeutic abortions in these cases—$53,000 and $18,000 (0.8 × $67,000 and $22,000) for men and women, respectively.

Two other components of lost productivity must be considered: homemaking services and other unpaid work. It will be assumed that unpaid work represents about 25 per cent of gross earnings in the case of males, *i.e.,* that they devote about 10 hours per week to home maintenance and other volunteer work.[4]

The valuation of homemaking services is an unsettled and controversial area. In the case of *full-time* homemakers, one approach is to assume that the value of homemaking services must be worth at least as much as the income, net of taxes, that women could earn if employed. Although the question may be raised as to whether women who remain home could earn as much as those who remain employed, it is generally considered that this is a conservative approach to estimating the value of a homemaker's time, considering the diversity of skills required to be a housekeeper, financial manager, and mother.[3]

But what of working women? Several studies have indicated that work does not relieve women of housekeeping responsibilities. In one study,[5] it was reported that working women spent 54 per cent as much time on housekeeping as full-time homemakers; in another study, the comparable percentage was 78 per cent. On the basis of the lower of these estimates, we calculate that the value of homemaking services produced by working women is half as large as that produced by full-time housewives or 37.5 per cent (50 per cent of 75 per cent) of their gross earnings (taxes are estimated at 25 per cent of women's earnings).

On the basis of these assumptions, total lifetime productivity for male infants is valued at $83,000 ($67,000 × 1.25) in 1972. For women, the comparable lifetime estimate of productivity is $47,000, which is derived as follows: On the average, the present value of the earnings of female infants born in 1972 is $22,000. However, since about half of the women of working age are engaged in full-time homemaking, $17,000 (75 per cent of $22,000) must be added to this figure to allow for the homemaking services women provide while they are full-time homemakers. Finally, another $8,000 is added to account for homemaking services produced by women while gainfully employed.

Our estimate of the total gain in productivity of preventing a case of Tay-Sachs disease or Hunter's disease is, therefore, $83,000 in the case of males and $47,000 in the case of females. The comparable estimates for preventing cases of Down's syndrome or trisomy 18 are $66,000 and $38,000 (assuming the total productive capacity of these infants would have been 20 per cent of the norm).

A third major benefit of preventing genetic disease is the savings in institutional care costs. During fiscal 1971, the latest year for which data are available, the average annual cost per resident in public institutions for the mentally retarded was $5,537.[6] However, these costs are increasing at a rapid rate. Between 1971 and 1972, costs in public mental hospitals rose by 17.6 per cent.[7] If we make the reasonable assumption that costs in institutions for the retarded rose at the same rate, per resident cost in 1972 would have been $6,509.

Even if there were no inflation, institutional costs would be expected to rise over time, reflecting the increased wages paid to employees. If institution employees are to share equally in productivity growth with the rest of the population, then, given our previous assumption, their wages must rise by 2.5 per cent per year. On this basis, institutional costs for the next 70 years, in terms of 1972 dollars, are estimated in Table 20–4.

If a genetically defective infant were institutionalized in 1972 and remained institutionalized for the next 70 years, the total cost would be $1,206,000, which would have a present value of $149,000. In most cases, however, the cost would be far lower because many seriously defective infants will not survive 70 years and many will spend a substantial portion of their lives outside institutions.

In fact, except in extreme cases, public institutions are reluctant to admit infants. It is unlikely that the fetuses with Tay-Sachs disease would have survived long enough to be institutionalized. The fetuses with Down's syndrome and Hunter's syndrome were obviously at substantial risk of institutionalization. Moreover, institutionalization may have been extremely costly, since many patient's with Down's syndrome now live to be 50 or 60 years old.

If we assume, conservatively, that persons with Down's and Hunter's syndrome will, on the average, spend 20 years in institutional care, and

TABLE 20–4. INSTITUTIONAL COSTS

Years After 1972	Estimated Annual Cost in 1972 Dollars	Estimated Annual Costs Discounted at 7%
0–9	$ 7,292	$5,480
10–19	9,335	3,566
20–29	11,949	2,320
30–39	15,296	1,510
40–49	19,580	983
50–59	25,064	639
60–69	32,084	417

that this care will occur between the ages of 10 and 30, then the present value of their care would be $59,000. In the nonreplacement case, this entire amount would be considered a benefit of prevention. In the replacement case, however, normal consumption and normal educational expenditures should be removed from the estimate of institutional costs, since these costs would have to be met for the nonretarded replacement and would not represent a net social saving. It has been estimated that only about 20 per cent of institutional costs can be considered normal consumption.[3] During fiscal 1972, average annual current expenditures per pupil in public schools in the United States were $934, which is less than 15 per cent of average annual institutional costs.[8] Reducing the estimates of institutional costs by 20 per cent overall, and by 15 per cent during the first 10 of the estimated 20 years in institutions, should adequately allow for normal consumption and education costs. The benefits of preventing institutionalization in the replacement case are, therefore, $42,000.

These estimates do not include capital costs. The reasonable assumption that capital costs are 10 per cent of current operating costs would raise the estimates of the savings in institutional costs due to preventing cases of Down's syndrome, trisomy 18, and Hunter's syndrome to $65,000 and $48,000 for the nonreplacement and replacement cases, respectively.

Another important benefit of prevention is the savings in medical costs for the two fetuses that would have had Tay-Sachs disease. These costs are estimated at $30,000 each for their projected lifespan, which is usually 2 to 4 years.

It is not possible at present to estimate all of the benefits of prevention. There may be substantial costs for special schooling incurred for severely retarded persons living in the community (especially since the recent right-to-education court decisions) or for daycare. In addition, there is frequent need for medical and dental attention; often, major cardiac, orthopedic, or abdominal surgery is required. Prevention, of course, eliminates these costs.

Perhaps the most important of the unmeasured benefits of prevention is the reduction of stress, bitterness, frustration, and other psychic reactions by the families. These feelings may affect behavior and may result in divorced or separated parents, alcoholism, psychiatric illness, and, among siblings, school problems and delinquency.

The sum of our estimates of increased productivity and reduced institutional and medical costs is a close approximation of the actual financial gain per prevented case. A part of institutional costs is used for services that would otherwise be provided in the community, and our period of residential care encompasses the age range in which most of these costs would be incurred.

These benefits are summarized in Table 20–5. The benefits of

TABLE 20–5. PRESENT VALUE OF BENEFITS OF PREVENTING
GENETIC DISEASE IN 1972

Type of Condition Prevented	Replacement Case	Nonreplacement Case
Tay-Sachs disease	$ 95,000	$30,000
Hunter's syndrome	113,000	65,000
Down's syndrome	100,000	65,000
Trisomy 18	100,000	65,000

preventing cases among males and females are averaged together to provide a basis for comparison with the costs of prevention.

Measuring Costs

The cost of preventing the birth of a prenatally defective fetus depends upon: (1) the number of cases that must be screened to identify a prenatally defective fetus, (2) the cost of the screening procedure, and (3) the cost of the subsequent abortion.

Amniocentesis and the subsequent genetic studies on the fluid cost about $250 in 1972. This figure reflects the cost of consumable supplies and technician and physician effort, but not capital allowances for buildings or equipment. However, a second amniocentesis was needed in 8.6 per cent of the cases screened and a third amniocentesis in 0.6 per cent of the cases. The average cost of a completed genetic study, therefore, was about $275 ($250 × 1.092).

Prenatal genetic studies for *chromosomal* disorders were done in 486 cases (Table 20–2). Six cases with abnormalities were detected. The parents elected therapeutic abortion in four. One was stillborn (trisomy 18) and one was born live. The single live birth had Klinefelter's syndrome, a condition characterized by physical and "behavioral" defects, including mental retardation in 10 to 15 per cent of cases. In the four cases that were therapeutically aborted, however, serious mental defect would unquestionably have occurred.

Four cases with major congenital defects were averted—a case finding rate of one in every 121.5 screenings. If it is assumed that the cost per therapeutic abortion is $400 (the known range for outpatient/inpatient saline induction is $100 to $400, and for hysterotomy, $800), then the cost per case averted is estimated at about $34,000* (121.5 × $275 + $400).

Amniocenteses are performed to detect a large number of different, though rare, metabolic disorders. Thus, the case finding rate cannot be

---

*See also Chapter 9.

used to determine the cost of detecting Tay-Sachs disease or Hunter's syndrome since screening may be for other metabolic disorders.

However, in cases where both parents are known to be carriers of the recessive gene that causes the disorder, presumably because of the prior birth of an infant with a metabolic disorder, the probability of a subsequent child suffering from the same disorder is 1 in 4. If screening is limited to this high-risk group, then the cost per case detected and aborted would be about $1500 (4 × $275 + $400).

Benefit-Cost Comparison

Table 20–6 summarizes the results of dividing the earlier measures of benefits by the estimate of the cost of detecting and aborting each case of a genetically defective fetus. The following points are made:

1.   In all cases, the benefit-cost ratios exceeded 1.0 so that even without a "complete" measure of benefits the prevention of genetic defect is shown to be economically worthwhile.

2.   The benefit-cost ratio is considerably higher in the replacement than in the nonreplacement case. In a world where, increasingly, family size is planned the replacement case is the most relevant.

3.   The benefit-cost ratio is spectacularly high among groups in which the risk of a genetically defective child is known to be high, as in the case of parents who have already given birth to a child with a metabolic disorder.

SPECIAL CONSIDERATIONS

Several major issues must be addressed when prevention procedures that require abortion are evaluated. One issue, already discussed, is the appropriateness of ascribing a zero value to the potential life of an unborn defective fetus. Perhaps most people would accept this valuation in the cases of fetuses who are certain to be born with severe mental and physical handicaps.

But what of cases in which nondefective fetuses are aborted? Such situations may occur inadvertently if the test for a defective fetus is in

TABLE 20–6.   BENEFIT PER DOLLAR-OF-COST OF PREVENTING SELECTED GENETIC DEFECTS

Type of Condition Prevented	Replacement Case	Nonreplacement Case
Tay-Sachs disease	$63.3	$20.0
Hunter's syndrome	75.3	43.3
Down's syndrome	2.9	1.9
Trisomy 18	2.9	1.9

error. This has happened several times in prenatal diagnosis[1] although, in general, the analyses have proven correct. Nondefective fetuses may be aborted deliberately in cases in which there are twins, one defective and one nondefective. Such a case occurred in our series, although no abortion took place. Statistically, the abortion of nondefective fetuses must result if abortion takes place because there is an unacceptably high risk of having a genetically deficient infant, *e.g.*, in the case of exposure to rubella, and in males in the case of sex-linked conditions. In these situations, one must argue either that there is no value to the life of an unborn fetus, *whether or not defective, or that the financial and psychic burden* of caring for a seriously defective child is so overwhelming that it takes precedence over other considerations.

What of cases in which the genetic defect, although detected, is treatable (*e.g.*, phenylketonuria, hemophilia)? The productivity benefits or institutional savings may be minimal in these cases. Abortion must be justified on the ground of medical expense or, more likely, the desire of parents to have as nearly normal a child as possible.

Prevention Possibilities

It has been estimated that the prevalence of mental retardation would decline by almost 80 per cent if all demographic groups in society had the same rate of mental retardation as middle- and upper-class whites.[3]* This would be a conservative estimate of the possibilities of preventing mental retardation. It is conservative because middle- and upper-class whites by no means avail themselves of all prevention possibilities. Moreover, in recent years new preventive techniques have been developed (*e.g.*, rubella vaccines, amniocentesis) that could bring about even larger reductions in the prevalence of mental retardation.

Of course, part of the reduction in mental retardation will be achieved by environmental changes to expand the mind-developing experiences of children living in deprived situations. Nevertheless, a significant part of the reduction in retardation will come as a result of preventing damaging diseases and accidents before birth, during birth, and during the childhood years. Because not all diseases and accidents lead to mental retardation, the total reduction in childhood disabilities, both mental and physical, as a result of new preventive efforts could be quite large.

The effect of new preventive techniques on childhood disabilities can be striking. In the past, over half of all offspring with Down's syndrome were born to women over 35. Figures from a national survey of 1633 amniocenteses show that women between 35 and 39 years of age

---

*Nonwhites are six to seven times more likely to have I.Q.'s below 70 than whites, and the children of the poor are thirteen times more likely to be retarded than the children of the middle and upper classes.

appear to have a risk of about 1:60 of bearing a chromosomally abnormal child.* It has been calculated that women over 40 years of age have a risk of roughly 1:40 for bearing chromosomally abnormal offspring.[9] At a risk factor of 1:60, it would cost about $17,000 for each genetically defective case averted—a figure well below any of our earlier calculations of benefits. In short, it is economically feasible, on the basis of the limited benefits that were measured, to prevent the births of over half of all cases of Down's syndrome.

The case of the effectiveness and value of amniocentesis in preventing chromosomal abnormalities is striking and clear. Even if this prevention method were applied widely throughout the country, the total annual reduction in the incidence of retardation would only be about 6000 to 7000 cases. There is no single road to prevention. Many separate avenues must be sought, each of which will affect a relatively small part of the total problem. Collectively, however, they will have a major effect on social well-being.

Evaluating Preventive Methods

Mental retardation will never be completely eradicated. In fact, it is unlikely that all cases of preventable retardation will be eliminated, at least in the foreseeable future. Inevitably, a cost barrier will be confronted that will cause many to question, and deny, the advisability of preventing additional cases of retardation. For example, it would cost over $250,000 to detect a single case of Down's syndrome among women under 35 (assuming an incidence of 1:1000). Such prevention is probably not justifiable on the basis of increased earnings and reduced institutionalization costs. It may well be justified on the basis of preventing the anguish and emotional chaos resulting from the birth of a severely retarded child, but this is an unquantifiable variable that requires public understanding if it is to be actionable.

Despite the importance of the emotional costs involved in the birth of retarded children, appeals for expanded preventive efforts are more readily received and acted upon when they are supported by firm benefit-cost and cost-effectiveness analyses. Even if they do not always show favorable ratios, they may indicate that the sacrifice being requested of society is not as great as first appears.

If the needed data are known, it is not difficult to extend the benefit-cost comparisons described above to other types of preventive efforts. If the condition being prevented would have led to total dependency, then

---

*This figure, stemming from consecutive amniocenteses, is more than three times the risk figures previously used for this age range. It would be wise to regard the matter as yet unsettled and these figures as a tentative guide, pending the publication from the Collaborative Amniocentesis Registry Project.

one can compare an estimate of cost with the above estimate of lifetime productivity (adjusted for price level changes between 1972 and the year in which the cost data are collected), the anticipated cost of institutionalization, which can be easily calculated on the basis of the table given in the text, and any other estimated savings in medical, schooling, or other costs.

If the disability is less profound, then the difficult task must be confronted of determining what the expected level of productivity would be if the infant were born. If productivity is expected to be "a per cent" of the norm, then productivity benefits of prevention (in the replacement case) can be calculated by multiplying our estimates of lifetime productivity by $(1 - a)$ per cent. These estimates will be as serviceable as almost any that could be calculated.

References

1. Milunsky, A.: The Prenatal Diagnosis of Hereditary Disorders. Springfield, Ill., Charles C Thomas, Publisher, 1973.
2. Swanson, T. E.: Economics of mongolism. Ann. N.Y. Acad. Sci., 171:679, 1971.
3. Conley, R.: The Economics of Mental Retardation. Baltimore, Md., Johns Hopkins Press, 1973.
4. Saenger, G.: The Adjustment of Severely Retarded Adults in the Community. Report to the N.Y. State Interdepartmental Health Resources Board, Albany, N.Y., 1957.
5. Hall, F. T., and Schroeder, M. P.: Time spent on household tasks. J. Home Econ., 62:23, 1970.
6. U.S. Department of Health, Education, and Welfare, Social and Rehabilitation Service, Rehabilitation Services Administration, Division of Developmental Disabilities: Unpublished data.
7. U.S. Department of Health, Education, and Welfare, Public Health Service, National Institute of Mental Health: Statistical Note 106, Provisional Patient Movement and Administrative Data—State and County Mental Hospital Inpatient Services, July 1, 1972 to June 30, 1973.
8. U.S. Department of Health, Education, and Welfare, Office of Eduction: Digest of Educational Statistics. 1973, p. 65.
9. Milunsky, A., Littlefield, J. W., Kanfer, J. N., et al.: Prenatal genetic diagnosis. N. Engl. J. Med., 283:1370, 1441, 1498, 1970.

CHAPTER 21

CAN EUGENIC POLICY BE JUST?

Marc Lappé, Ph.D.

UNKNOWN QUANTITIES AND QUALITIES IN EUGENICS

Eugenics is a largely undeveloped and unevaluated science that aims to improve the genetic quality of whole populations over time. It is "undeveloped" because until recently there have been no feasible mechanisms to implement a meaningful eugenic program. Among the opportunities currently available for genetic intervention are restrictions (both implicit and explicit) on reproduction through agencies like Family Planning; mass population screening for the carrier status of specific genes; burgeoning use of amniocentesis for detection of fetal abnormalities; abortion in cases of highly probable genetic disease; and a widespread network of genetic counseling units. It is "unevaluated" largely because the evaluation of eugenics requires a depth and breadth of understanding, which few have mastered, of science and the ultimate purposes it should serve.

Only one aspect of all the factors that go into a comprehensive analysis of eugenic policy will be addressed here—namely, whether *any* stratagems that could have eugenic effects should be implemented. This question has never appeared to have merited analysis before, largely because the apparently unmitigated goods of improvement—or even warding off genetic deterioration—appear to have an imperative of their own. Only Paul Ramsey has addressed this "self-evident truth." He observes that a doctrine of divine providence does not adjure a Christian or a Jew *"to succeed* in preventing genetic deterioration . . . nor is that person under an *absolute command* to intervene in a population's genetic fate."[1] Ramsey believes that it is essential to begin *without* eugenics

Supported by NIH Grant No. GM-19922–02.

in view, so that the end does not *dictate* (at great peril of ethical omissions) the acts that must be done to accomplish it. Moreover, the likelihood that a given policy of genetic intervention would be a highly efficient means of achieving given ends does not, in Ramsey's view, authenticate those means. I take this same neutral posture with regard to eugenic ends to avoid prejudging the merits (or demerits) of genetic policies.

To illustrate that eugenic policies do raise ethical questions, note that *any* stratagem that is going to meaningfully affect the genetic composition of future generations must affect the reproductive behavior of a large number of persons. Gone is the era in which small breeding populations, subjected to extreme selective forces or isolated in microreproducing units, pass through one of the classic genetic "bottlenecks" which permit a few individuals to contribute markedly to the total genetic composition of several future generations. For example, the late William Boyd noted in his *Textbook of Biology* that virtually all of some 962 individuals with Huntington's chorea could be traced to six immigrants to the United States in the seventeenth century. Had the present knowledge of this dread disease been known then, would these persons have been sterilized? What may be said of the more than 1000 individuals who now are at risk for transmitting this disease to one half their offspring?

The population has changed so dramatically in its demographic characteristics (age at childbearing, mean number of offspring, etc.) that the selective advantage which may previously have existed between families carrying the trait for Huntington's chorea and the rest of the population has largely disappeared. Does this fact make a difference in an analysis? Resolving this or any other question about eugenics depends only partly on one's ability to answer questions of fact. Questions of values—that is, *ethical questions*—are often inextricably bound up with what appear to be scientifically ascertainable facts. What relevant genetic "facts" *mean* and how they are to be used constitute the major ethical issues addressed in this analysis.

Among the questions of fact that are relevant to eugenics are the following:

1. How will current demographic trends affect the composition of the gene pool in the future?

2. To what extent will intentionally disrupting these trends change the gene pool's structure?

3. What are the genes that can now be detected in the carrier state? What is the current reproductive behavior of the carrier population?

4. To what extent will altering the reproductive behavior of only such carriers affect the genetic composition of the population?

Dudley Kirk has examined some of these questions in a general way, and has emphasized the importance of the relaxation of selection.[2] More technical treatments of these issues will be discussed below.

These questions of scientific fact raise an almost mirror image set of questions of values:

1.   What constitutes a "good" or "ideal" demographic profile, in a genetic sense, for a country? How might attempting to attain this profile conflict with the needs and values of the populace? If it were inconsistent with a group's or a society's population policy, for example, how would one equitably alter an "undesirable" demographic characteristic?

2.   What human traits are considered sufficiently important (and how are criteria set for assessing these traits) to warrant a eugenic policy that favors them? Should the proportion with undesirable characteristics be reduced in number or should those with desirable ones be augmented? What are the human consequences of each policy?

3.   Are there variations in degree in defining desirable or undesirable genotypes? What constitutes a "deleterious" gene for an individual? an extended family? an ethnic group? an entire population? How may "desirable" genes be defined when the environments of different groups differ with regard to the manifestation of "deleterious" or possibly beneficial genes?

The question of how there may be intervention in the reproductive behavior of individuals who harbor desirable or undesirable genes subsumes the ethical question of intervening in the reproductive behavior of social classes or larger reproductive units. This is an especially vexing problem where the interpretations given by individuals and society to the term "desirable" differ radically (as has proved to be the case in at least a few jurisdictions for sickle cell trait). I am acutely conscious that ethical analysis of this sort often grates most stridently against the imperatives of science where the needs of individuals and society come into conflict. I hope to avoid, or at least to minimize, this bias in this chapter. In doing so, I have relied in part on the thinking and approaches of those who have thought about this problem longer and harder than I have.*

Definitions

Eugenic programs traditionally have been cast as if there were two seemingly discrete objectives: (1) the reduction of the frequency of presumably "deleterious" genes (negative eugenics), and (2) the improvement of the genetic status of the population (positive eugenics). The first objective is usually presumed to be attainable primarily through restricting the procreation of couples at risk for having offspring with deleterious genes, and the second, by encouraging the procreation of those

---

*Especially James Neel, James Gustafson, Sumner Twiss, Daniel Callahan, and most importantly, John Rawls.

individuals who carry the greatest proportion of "desirable" genes. In practice, eugenic objectives can rarely be so easily separated. More importantly, the attainment of such objectives will be substantially more complex than this model suggests, since it involves profound social and ethical questions.

Historical Roots

It is likely that a "eugenic" program in the United States today could not survive under the onus imposed by the history of the Eugenics Movement. If (as some have asserted) the opportunity to justify "Eugenic Programs" in this country has been lost, it is partly because of a continued heritage of cross-contamination of human genetic ideals with political and racial motives.[3] The much publicized sterilization of indigent "retarded" girls in Alabama, or the proliferation of compulsory sickle cell laws, has again raised the specter of genetic programs run by the privileged at the expense of the deprived.

The first task then would be to attempt to rescue eugenics from its historical associations. This can be done only by extricating it from its adherent racial and class overtones. Like many others, I have been convinced that it was more than an accidental confluence of historical circumstances that led scientists to promulgate ideas of social policy which were intrinsically racist and class-oriented at a time when scientific concepts of race and genetics were still in their infancy. In late nineteenth century Europe science was the exclusive domain of the upper class. "Darwinism" did not become "Social Darwinism" in the hands of Machiavellian politicians; it was "social" by its nature.[4]

The lesson of Darwin's "like perpetuating like" was not lost on those who believed in the primacy of maintaining a class society. It is clear that among the progenitors of eugenic policy in this country and England were not only geneticists and educators, but also members of the upper class intent on perpetuation of a social order that they believed was crucial to the survival of mankind.

It is one of those strange turns of history that the person who was to write the most widely read articles on eugenics in 1966, Frederick Osborn, was almost alone in his ardent opposition to the racist and class appeal of the eugenic literature of 30 years before. Osborn's intention to advocate fairness is clear in his support of the American Eugenics Society's position statement of 1961:[5]

One course open to eugenics is to find, and to further, those institutional arrangements, those social, economic and psychological pressures, which would result in a tendency for individuals to have more or fewer children in proportion to their success or achievement in their particular environments. . . . There is no conflict between those desiring to raise the level of our genetic inheritance and those desiring to raise the level of our social inheritance.

It is difficult to imagine how one could "raise our genetic inheritance" without compromising the moral structure of society. The history of the eugenics movement then (and perhaps even now) is replete with contradictions to Osborn's expectations. In a class society, one cannot, in one breath, use the criterion of *achievement* for rewarding genetic gifts of "more children," and in another, speak of the need to "raise the level of our social inheritance." I find it difficult, eugenically and ethically, to reconcile the paradox that among those who would be rewarded with the opportunity of perpetuating their kind because of their "success" would be those whose success had been assured by their high social standing alone, and *not* by their genes.

This historical background should not be taken to mean that I believe eugenic programs are intrinsically wrong or could not be justified under some system of equity and justice. Indeed, my purpose in writing this chapter is to examine how eugenic ideals might be implemented ethically, albeit at the sacrifice of some of the pseudogenetic goals of the early advocates of the movement.

A key ethical dilemma of eugenic policy in a class society, then, is to reconcile the fact that those who advocate eugenic programs would be the ones having the most to gain by their implementation. To be just, eugenic policy should counter, rather than reinforce, this tendency.

Resolution of this apparent paradox requires a demonstration (1) that at least some eugenic objectives contain intrinsic "goods"; (2) that one or more of these goods is of sufficient merit to warrant potential inequities in distributive justice (as in the number of offspring permitted a couple); and (3) that over time, any such inequity would be rectified (*e.g.*, through compensation of the disadvantaged in kind or quality). What follows is an attempt to analyze these criteria and to determine if a eugenic policy is justified.

ASSESSING EUGENIC OBJECTIVES

Step 1:   Define the Genetic Status Quo

All objectives of eugenic policy are ultimately measured against some real or imaginary genetic status quo. There is disagreement over a fundamental question that substantially dictates the ethical imperatives of eugenic policy, namely, is or is not the genetic composition of society deteriorating to such an extent that remedial actions are mandated simply to maintain the status quo?[6] Some, like J. H. Sang, believe that the apparent increase in human heterozygosity for deleterious genes "will lead to their spread through the population" and conclude that "we may now be starting a phase of 'genetic pollution' roughly equivalent to the beginnings of environmental pollution a generation ago."[7] This fear of a genetic apocalypse has been with us for over 100 years, beginning with

Darwin himself, who was disturbed about the potential genetic consequences of society sustaining the "weak in body and mind" through medical interventions.

Others, like P. B. Medawar, challenge this view of humankind's genetic fate and consider that although genetic dangers confronting us are "genuine" and deserve our attention, they "are not of great urgency or gravity."[8]

The truth lies somewhere between these poles and requires explication through an analysis of the consequences of dynamical trends in populations and an honest statement about our ignorance of the meaning of the vast degree of genetic heterogeneity revealed by study of human polymorphisms. For example, it is uncertain whether selective forces or purely neutral events, like genetic drift, have generated the degree and extent of genetic diversity in human populations. The ethical weight one would give to sustaining the current level of diversity, reducing it, or increasing it, depends on an understanding of its evolutionary significance.

A key demographic variable which defines the genetic status quo is the average maternal (and to a lesser extent, paternal) age at childbearing. In addition to the well-known effects of increasing maternal age on the frequency of Down's syndrome, increasing parental age will also affect the mutation rate (since mutations accumulate, with age, in some as yet unknown fashion). The risk of birth defects generally increases only with grand multiparity as shown by the increased frequency of congenital disorders in children with birth orders of 5 or more.[9] The relationships between lower birth order, maternal age, and congenital malformations and Down's syndrome are shown in Table 21–1.

The net effect of changing the demographic characteristics of a population on the incidence of malformations and Down's syndrome can be seen in the result of lowering the birth rate and age-at-last-birth in the Japanese population. By instituting widespread contraceptive and abortion policies *without an implicit eugenic purpose*, Japan was able to reduce its incidence of fatal birth defects by 10 per cent and its incidence of Down's syndrome by 40 per cent.[10] Currently, 94 per cent of births in Japan are in the 20 to 34 age group, further reinforcing this trend.

The genetic status quo of a given population can only be defined dynamically by its demographic characteristics—its average fertility rate, its differential mortality, and its age characteristics. These in turn have important genetic consequences. For example, if the population is growing rapidly, it will tend to have a lower mutation rate (and hence lower "genetic load" over time) than will a stationary or declining population, largely as a result of earlier reproductive ages in the fast-growing model. An important eugenic consequence of this basic demographic characteristic is that selection against some deleterious traits that are expressed with increasing probability with age (*e.g.*, Huntington's chorea or multiple polyposis) will be much weaker in fast-growing than in slow-

TABLE 21-1. STATISTICAL CORRELATIONS BETWEEN BIRTH INCIDENCE OF CONGENITAL MALFORMATIONS, DOWN'S SYNDROME AND MATERNAL AGE AND BIRTH ORDER*

Congenital Malformation per 1000 Births†

	MATERNAL AGE GROUP					
	Under 20	20–24	25–29	30–34	35–39	40–44
Birth order						
1	44.2	46.0	48.7	53.3	62.0	—
2	38.7	40.3	42.0	44.7	50.6	—
3	—	41.8	37.0	39.3	49.2	—
4	—	37.9	39.8	42.8	47.3	65.8

Down's Syndrome per 1000 Live Births‡

		MATERNAL AGE GROUP					
	AVERAGE RATE	Under 20	20–24	25–29	30–34	35–39	40–44
Birth order							
1	0.567	0.469	0.426	0.533	1.03	2.71	8.51
2	0.685	0.360	0.471	0.513	1.01	2.99	7.54
3	0.814	—	0.402	0.517	0.841	2.42	8.37
4	1.144	—	0.393	0.486	0.877	3.00	9.40

*Reprinted with the permission of The Population Council from "Parental Age as a Factor in Pregnancy Outcome and Child Development," by Dorothy Nortman. REPORTS ON POPULATION/FAMILY PLANNING, no. 16 (August 1974): 40–41.

†United States data 1961–69, compiled by U.S. Obstetrical Cooperative, Downstate Medical Center, Brooklyn, N.Y., from approximately 500,000 birth records. Adapted from Nortman, D.: op. cit., Table 11B.

‡Michigan data, 1960–64. Adapted from Nortman, D.: Ibid., Table 11A.

growing populations—especially in those populations where rapid population growth is accompanied by a high early mortality.[11] These dynamic effects, operating largely on dominant conditions, and the currently high gene frequencies for recessive conditions like cystic fibrosis and Tay-Sachs disease, suggest to some population geneticists that the present genetic "status quo" is a relict of selective forces (*e.g.*, those favoring heterozygotes or carriers for the responsible genes), genetic drift through founder effects, and/or genetic bottlenecks operating in the past. For any discussion of eugenics, therefore, it is crucial to recognize that the current population is *not* in genetic equilibrium and that sustaining or changing demographic patterns will have potentially major effects on the frequency of "beneficial" genes. At least one prominent population geneticist has noted that these demographic and other trends in human evolution leave in doubt the relevance of our definitions of "favorable" or "unfavorable" genes.[12]

Since eugenic changes require changes in the relative frequencies of human genes, such a statement compounds the ambiguity of any ethical analysis. But, there is agreement that demographic factors are important in determining whether eugenic changes are possible. For example, although a slow-growing population will have increasingly less opportunity for selection, reduction of family size *per se* does not mean that changes in gene frequency cannot still occur. However, as the amount of variability in family size decreases ("variance") to the point that it is lower than the mean family size, the effective "breeding population" will become smaller than the effective population size. The immediate consequence of such a shift (as has occurred in Japan) is to give the *appearance* of a eugenic effect, in that the frequency of recessive genes declines initially. However, as Imaizumi and his colleagues have demonstrated, the long-term effect of this demographic trend is to eventually increase the rate of appearance of such genes to a higher level than before the population shift occurred.[13]

Another demographic phenomenon with profound implications for ethical interpretation includes the well-known existence of assortative (nonrandom) mating in the human population. For example, one consequence of nonrandom mating is to decrease, rather than increase, the spread of genetic variability in the population.[14] If genetic diversity is indeed desirable (or even a *summum bonum*, as many human geneticists maintain), then this is a dysgenic trend in the population. If, however, the ultimate end of eugenics is to improve the proportion of individuals with particularly "desirable" genotypes, then assortative mating is virtually an imperative—hence, the dilemma.

There remain significant questions regarding the definition of "good" or "deleterious" genes. Perhaps even greater questions of value remain in determining how one is to assign ethical weight to such genes, especially when one realizes that these genes are part of a much larger genetic universe of individuals, and, ultimately, of the genetic pool of the species. For the purposes of this chapter, it must be assumed that enough features of the factors maintaining genetic polymorphisms, generating chromosome abnormalities, and sustaining complex polygenic characters that typify human populations can be described to allow some general statements to be made about eugenic policy.

Step 2: Justify Changing the Status Quo

The first objective is to determine if a eugenic policy to change the status quo is scientifically warranted. It is necessary to take account of the ambiguities of our understanding of the mechanisms underlying the forces of population genetics—in particular, the largely unknown con-

TABLE 21-2.   SIMPLIFIED ILLUSTRATION OF SOME OF THE
COMPONENTS OF EUGENIC POLICY*

The Genetic Consequences of Each of the Following Are to Be Known:

CRITICAL DEMOGRAPHIC FACTORS AFFECTING GENE FREQUENCY, INCIDENCE OF
  CHROMOSOMAL ANEUPLOIDY, AND OPPORTUNITY FOR SELECTION
    Assortative mating
    Mean and variance of live births
    Age at parenthood
    Differential mortality
SOCIOECONOMIC FACTORS THAT MAY MODIFY EXPRESSION OF GENETIC COMPOSITION
    Exposure to mutagens or teratogens
    Perinatal care
    Environmental components affecting mortality
    Nutrition
OPPORTUNITIES FOR GENETIC INTERVENTION†
    Population screening
      for affected, homozygous individuals (newborns)
      for carrier, heterozygous individuals (usually adults)
    Genetic counseling
      directive
      nondirective
    Amniocentesis
      for chromosomal aneuploidy (trisomy 21)
      for single gene diseases
      for polygenic disorders (?)

The Ethical Implications of Each of the Following Are to be Known:‡

POLICY ALTERNATIVES FOR EFFECTING EUGENIC ENDS (HYPOTHETICAL)
    Incentive programs; e.g., procreative inducements in the form of tax credits
    Penalize dysgenic behavior; e.g., proscribe further reproduction through enforced
      sterilization of retarded minors
    Encourage differential fertility among socioeconomic groups; e.g., tax child-bear-
      ing on a flat-rate basis
CHOOSE AMONG POLICIES THOSE THAT ARE THE MOST EFFICIENT
    Offer inducements for nonprocreation of carriers of deleterious genes detected in
      adult screening programs
    Use amniocentesis to identify heterozygotes and abort these as well as homo-
      zygotes (or in sex-linked disease, carriers)
    Make successful treatment of homozygote-affected individuals contingent on
      their restricting their childbearing
UTILIZE COST-BENEFIT ANALYSIS TO ARRIVE AT POLICY DETERMINATIONS
    Assess social worth/societal cost of individuals against some economic standard
    Make acceptability of policy contingent on financial remuneration in contrast to
      other reciprocal benefits
    Rank potential for improvement on the basis of net societal gain

---

*This tabulation supplements the limited analysis presented in this chapter. It
suggests the extent and complexity of the items to be considered in a comprehensive
analysis of eugenic policy considerations, rather than provides a definitive list.

†See references 13, 29–31 for examples and explication.

‡A model for how such an analysis can effectively be performed for many of these
factors is exhaustively treated in *Ethics, Population and the American Tradition*, The
Task Force on Ethics and Population of the Institute of Society, Ethics, and Life
Sciences.[16]

sequences of nonrandom mating, the factors that maintain polymorphisms and restrict homozygosity (especially for histocompatibility alleles),[15] the existence of modifier genes, and the consequences of position effects and genetic heterogeneity on the phenotypic expression of genetic composition. Some of the scientific parameters that must be resolved are shown in Table 21–2.

The second objective is to determine if a change in the status quo is ethically warranted. To take a specific example, a thorough understanding of the cultural forces which affect assortative mating, migration, and other demographically important forces would be relevant to the ethics of restricting or encouraging new childbearing patterns for implementing eugenic policy.

As an operating principle, the justification for a eugenic program should be congruent with the justifications given for other public policies which affect individuals' reproductive behavior (for example, population control). A cursory glance at the literature, however, shows a complete spectrum of attitudes—from support of compulsory policies to advocacy of policies that would voluntarily compensate individuals who restrict childbearing.[16]

The population model most relevant to eugenics is sterilization for the mentally retarded. Reed and Elving make a strong "eugenic" case for restricting the childbearing of the mentally retarded.[17] They have calculated that if no retarded person reproduced, there would be 17 per cent fewer retarded individuals in the next generation. Although the net reduction in the number of retarded persons would in fact be modest (from 2 per cent to 1.7 per cent), it would obviously be a significant "improvement." However, the customary means and assumptions underlying implementation of such a program (*i.e.,* sterilization) raise serious legal and ethical questions. Although under court order to the Department of Health, Education, and Welfare (HEW), it is currently illegal to use governmental funds to order the sterilization of a mentally incompetent adult or a competent minor (see *National Welfare Rights Organization v. Weinberger et al.,* Civil Action No. 74–243 USDC, D.C.). *The Rights of Mental Patients*[18] lists 16 states where the institutionalized retarded may be sterilized: Arkansas, California, Delaware, Washington, D.C., Idaho, Maine, Michigan, Mississippi, Nevada, North Carolina, Oklahoma, South Carolina, South Dakota, Vermont, Virginia, and Wisconsin. It is imperative to note, however, that the assumption that all retarded persons cannot control their sexual motivations or are incapable of meaningful marriage is largely without merit.[19] Moreover, average family size among retarded couples remains substantially below the mean. Justification for the necessity of exclusive and aggressive restriction of *their* reproduction to effect eugenic change, and not the reproduction of others, is as yet unsubstantiated.

Step 3: Justify Changing the Status Quo In a Particular Direction

In answering the question, What factors of the human condition warrant a eugenic policy? I would begin by examining the premise that eugenics is intrinsic to a concept of human progress. Galton's vision of eugenics (like Darwin's) was inspired by the ostensible genetic improvement possible through breeding — a phenomenon which appeared empirically confirmed in the domestication of plants and animals. Both cousins saw in man's growing knowledge of the hereditary forces that shaped his nature the opportunity not only to "improve the inborn qualities of a race" but also to "develop them to the utmost advantage" (Galton's words). Darwin himself noted the "bad effects" of the "weak surviving and propagating their kind," but he optimistically hoped that at some point society would be able to discourage the propagation of such individuals.[6]

When Galton defined eugenics as "the science which deals with all the influences that improve the inborn qualities of a race . . . " he derived this justification from a natural law argument. He stated that since man is instinctively motivated ("with pity and other kindly feelings"), he has an obligation to prevent suffering. To do this, Galton said, man should begin by replacing natural selection with "other processes that are more merciful and not less effective." Galton failed to recognize that "effective" natural selection was potentially doomed because it could not compensate for the absence of a fixed relationship between genetic potential and a specified direction to the evolutionary process. All things are not possible to all beings.

Darwin recognized this fallacy; instead of natural selection, he visualized the improvement of the human stock by utilizing the mechanisms used for improving livestock. He noted that "Man scans with scrupulous care the character and pedigree of his horses, cattle and dogs before he matches them; but when he comes to his own marriage he rarely, or never, takes any such care."[20]

The argument for eugenics, which is based on the ethical imperative to improve man, must rest on stronger argumentation than the appeal to man's ascendant place in the order of nature. Yet, a renowned scientist, Roger Sperry, who claims that such a rationale can and must be derived from science, ultimately resorts to similar reasoning. Sperry has stated that "Social values are necessarily built in large part around inherent traits in human nature written into the species by evolution." By implication, Sperry would advocate eugenic programs as one of these values because he believes that we are impelled to follow the direction indicated by evolution:

The upward thrust (sic) of evolution . . . becomes something to preserve and revere. This would imply a commitment to progress and improvement . . . in terms of furthering the advancement of the evolutionary trend towards greater complexity, diversity, and improvement in the quality and dimension of life and the life experience.[21]

Unfortunately, as Stent and others have pointed out, there is nothing in the evolutionary "program" which would vindicate a natural law justification for improvement, be it through eugenics or any other system.[22] Simpson has debunked this rationale in a classic article, "The Concept of Progress in Organic Evolution," in which he stated that "There is no innate tendency toward evolutionary progress and no one, overall sort of such progress."[23] Responsibility for rationalizing eugenics rests with man, not nature.

Evolutionary biologists, like Simpson and Dobzhansky, would deny that science can serve "as an arbiter of values and belief systems." It can serve as the fabric of reality against which ethical hypotheses are tested, but it cannot serve, as so many Natural Law theorists have maintained, as the wellspring for them.

I do not believe that eugenics can be justified on the assumption that there is an ethical imperative to, in Shinn's words, "build upon and direct the processes of nature."[24] In matters of human genetics it is a dangerous course to commence one's journey with *any* injunction, much less the one which Shinn cites: "To be human is to modify both nonhuman and human nature." Such a starting point rules out the possibility that genetic equilibria yet to be reached by existing forces would bring about the "best" balance of potentials and buffering qualities of the system. In many aspects of human genetics, selective forces beyond current understanding operate to assure a modicum of equitable and adaptive distribution of genetic materials. (Neel, for example, appears to have been increasingly impressed with the necessity of recognizing the "wisdom" of genetic patterns to be found in preliterate societies).[25]

In rejecting an implicit moral imperative to eugenic "improvement," I have argued more from the position that we do not know enough about what constitutes "improvement" genetically to warrant such policies. This leaves for consideration two alternatives to eugenic policy: act to maintain the status quo (*i.e.*, prevent any deterioration), or do nothing (*laissez faire*). Both will be considered in Step 4.

Step 4: Justify Maintaining the Status Quo

A reasonable starting point to a just eugenic policy is to base it on the rationale that there is an implicit obligation to future generations to leave them no worse off genetically than the present generation is now. Such a minimal definition is consistent with contemporary understandings of the science of eugenics and congruent with some ethicists' interpretation of intergenerational responsibility. For example, in the *International Encyclopedia of the Social Sciences for 1968* Allen defines eugenics as "an applied science that seeks to maintain *or* [italics added] improve the genetic potentialities of the human species."

Gustafson has noted that our contemporary knowledge of genetics — patterns of inheritance and recurrence rates, for example — "creates the conditions under which the dimensions of responsibility. are enlarged; present generations *are* [italics added] 'causally' responsible to some extent for the genetic health of future generations, and thus it can be argued that they also have a 'moral' responsibility to them."[26] Some theologians, like Rabbi Narot, would go further and assert that we have an implicit obligation to "not knowingly multiply the individual or collective griefs of mankind" by taking genetic risks that are even "mild" or "moderate."[27] Obviously, the definition of risk itself is both a fact and a value determination.

Increased genetic knowledge and the enhancement of our ability to predict the consequences of reproductive behavior with an unprecedented accuracy based on either empirical estimates *or* "precise" determinations via amniocentesis have rather abruptly brought two American traditions into conflict: assigning a high value to autonomy and self-determination in the right to bear children, and instilling in some individuals a sense of duty to act in the common good by restricting their procreation.

When faced with these alternatives, ethicists differ as to where they place the weight of moral imperatives. Some ethicists believe, along with Twiss, that society does not have an "unmitigated right"[28] to intervene in parenthood and reproductive behavior, whereas others would allow that right and would expand it to include extensive educational measures regarding genetics. Gustafson states that it is "fitting to raise the consciousness" of persons about the future, including the "immediate social consequences of their reproductive behavior."[26] Genetic counseling is potentially a vehicle for implementing such a conclusion, but close scrutiny reveals practical and ethical difficulties.

Fraser has documented the degree to which different kinds of genetic counseling could affect the population frequencies of the single genes responsible for autosomal recessive and X-linked genetic disease.[29] He emphasizes that retrospective counseling (*i.e.*, after a proband is found) is increasingly ineffectual in reducing gene frequency as population growth declines. Since Imaizumi *et al.*[13] have shown that recessive gene frequency will probably *increase* under these demographic conditions, and others[2] have shown that selection pressures, in general, have been relaxed (especially against late-manifesting conditions), it seems clear that *some interventionist policy is needed.* Among the alternatives frequently cited and reviewed are amniocentesis, carrier screening, and genetic counseling, all of which raise ethical problems in their implementation as tools for achieving eugenic ends.[30,31] The purpose of this chapter is to suggest some of the ethical considerations that would universally apply to implementation of any policies involving genetic intervention.

Step 5: Assure That the Policy Is Just

Attainment of eugenic aims—whatever their rationale—might appear to conflict strongly with traditional values of liberty, justice, and autonomy, since their implementation would seem to require selective and coercive restrictions on some, and not other, members of society. The problem of justice is exacerbated by the fact that, by definition, negative aspects of eugenic policy (*e.g.*, restrictions on procreation of those who bear the greatest "genetic burden") would appear to have to be directed at the least, rather than the most, fortunate members of society.

Gottesman and Erlenmeyer-Kimling would rectify this inequality by basing any allotment of an individual's genetic contribution to the next generation on his or her social worth.[32] Their "Index of Social Value" fails in the same way that virtually all other such systems have failed: they assign a genetic weight to traits which have little or no demonstrated genetic basis and ignore the environmental deprivations that may overshadow an individual's manifest "worth." Certainly, we know enough about population genetics to be skeptical of policies aimed at assessing individuals' "genetic" worth, and enough about ethics and justice to question the use of constraints on childbearing as punishment for individuals with "high social cost." To be ethical, a eugenic policy must recognize at least some of the ends of justice. Paradoxically, it is those features of justice that would appear most difficult to reconcile with eugenic policy which are the most important to incorporate.

Several position statements by Rawls, in his classic book, *Principles of Justice,* would serve as useful baselines against which to judge proposed eugenic policy. Perhaps the most difficult principle to rectify in eugenic policy is the concept of fairness, which in Rawls' definition embodies the requirement that "all inequalities be justified to the least advantaged" and that "undeserved inequalities call for redress."

Taking as a working assumption* based on statistical treatments such as those shown in Table 21–3 that the "genetically least advantaged" are concentrated in the lower socioeconomic stratum of society, a just eugenic policy would have to return more to these individuals than it would take away. Since Rawls (and others) assign the highest

---

*I am aware that environmental variables play an undoubtedly large role in determining the socioeconomic gradient for congenital defects, and that neither the narrow heritability nor the actual genetic basis for most congenital abnormalities has been established. Moreover, there are grave political risks in taking this unproven assumption as proven.

TABLE 21-3. STATISTICAL CORRELATIONS BETWEEN CAUSE OF DEATH DUE TO CONGENITAL MALFORMATIONS AND EDUCATION OR INCOME (1964-1966)*†

Educational level of parents:

Father—white; less than 8 years education 5.5/1000

9–11 years 4.7/1000

12 years or more 3.1/1000

Mother—white; less than 8 years education 4.9/1000

9–11 years 4.2/1000

12 years or more 3.4/1000

Income level of parents:

Income level under $3000; white 4.4/1000

Income level $3000–$4999; white 3.8/1000

Income level over $5000; white 3.6/1000

*From Infant Mortality Rates: Socioeconomic Factors. U.S. Dept. HEW Series 22, No. 14, DHEW No. (HSM) 72-1045, Tables 14 and 15, pp. 34–35.
†These same associations do not appear to apply to blacks.

value to liberty, and since it is precisely among the poor that the liberty to bear children is restricted in classic eugenic programs, society is immediately confronted with the parodox it sought to avoid. Indeed, in Rawls' view, the precedence of liberty over justice implies that liberty can only be restricted "for the sake of liberty itself."

There are several ways to rectify this situation in Rawls' view: When there is consensual agreement to restrict liberty according to certain rules such that some restrict their liberty, those who have submitted have a claim on those who have benefited from their behavior. In the case of the mentally retarded, for example, it might suffice if the agreement to restrict their procreation for the common good were compensated by better conditions for their (fewer) offspring than they themselves received. This seems problematical, for society would probably be asking them to undergo such restrictions *non*voluntarily, and would be returning goods neither in kind nor in proportion to their sacrifice.

Rawls points out that it is just as problematical to assume that the "less fortunate" would benefit by policies which would reduce the talents of others to their level. Rawls would have the less fortunate accept societal encouragement from those with greater abilities, since presumably the development of their own traits would be "a social asset to be used for the common advantage." He goes on to say, however, that "it is also in the interest *of each* [italics added] to have greater natural assets" and hence all would "want to insure for *their* [italics added] descendants the best genetic endowment (assuming their own to be fixed)."

Rawls concludes that "the pursuit of reasonable policies . . . is something that earlier generations owe to later ones." As a minimum expression of eugenic policy, I concur with his conclusion that over time a society should take steps at least to preserve the general level of natural abilities and to prevent the diffusion of serious defects.* His statement that "These measures are to be guided by principles that the parties would be willing to consent to for the sake of their successors" should be carefully considered.

Something which has been implicit in all of the previous discussion but which *is by no means proven* is that the attributes being discussed are, in fact, genetic — or at least heritable to some significant degree. If, for example, programs intended to enhance some arbitrary trait (such as I.Q.) were to operate on the assumption that the genetic component of intelligence was "0.8" (a fact not strictly ascertainable by heritability estimates or I.Q. scores), such programs *would be* largely eugenic (assuming it had been agreed that enhanced I.Q. was desirable). If such estimates were incorrect, however, noneugenic programs to increase intelligence, such as more optimal perinatal nutrition, would probably be more efficient.

Perhaps the most important unfinished task in validating eugenic inquiry is separating the genetic from the environmental causes of important human traits. Until valid estimates are made of the genetic component for characteristics like intelligence at one extreme and anencephaly at the other, crude "eugenic" programs, based on broad estimates of heritability or empiric recurrence rates, will be ethically risk-laden, and politically highly questionable. For example, a recent note on heritability challenges the validity of the very equation that forms the basis of the I.Q./heredity dispute raging in educational circles. Moran[34] probably was the first to point out that since the contributions of genotype and the environment to a phenotype or heritability *themselves covary,* a coefficient of heritability for something like human intelligence cannot at this time be defined at all. Layzer more recently concludes that there are no suitable data for estimating the narrow heritability of I.Q.; hence, he discounts the inferences which have been drawn regarding I.Q. differentials among socioeconomic and/or ethnic groups.[35]

Nevertheless, the most consistent and ubiquitous finding of very real and often very steep socioeconomic gradients for I.Q. (documented extensively by Jensen),[36] congenital malformations (see Table 21–2) and anencephaly[37] call out for resolution. It would seem that differentials for *any* human attribute in which there are *both* genetic and environmental components would most equitably be treated by considering *first* the environmental hypothesis and *then* the genetic, thereby affording the disadvantaged immediate compensation.[38] This view has been attacked by Jensen as unworkable in practice, and by Lowe as unreachable in reality, but it still holds an ethical imperative as yet untested in the real world.

---

*How one would define the point at which society begins to suffer appreciably from neglect of Rawls' maxim to such a degree that it would be imperative to implement policies to offset genetic deterioration is again unanswered.

A first step toward eugenic policy that obviates the dilemma of doing the right thing for the wrong reasons (*e.g.,* improving I.Q. scores via eugenic policy) or not doing anything would be simply to encourage those demographic forces which have eugenic effects.

Purely external and demographic factors could have a far-reaching effect on the incidence of genetic and congenital abnormalities in the population. Reducing the environmental sources of mutagens has long been advocated as a first step toward reducing the society's genetic load. Demographic changes can effect similar shifts. As I have documented (see Table 21–1), lowering the average age of childbearing can reduce both the mutation load (especially the paternal component) and the incidence of major birth defects. Mothers who bear children between the ages of 19 and 28 comprise on the average an optimum childbearing group, in contrast to older and younger women.

In England, for example, over 45 per cent of all children born with Down's syndrome are born to women over the age of 35; yet, they comprise only 10 per cent of all births. Reducing the proportion of multiparous women (part of reducing the overall variance in births) would also achieve a radical reduction in the incidence of certain congenital abnormalities, such as spina bifida or anencephaly.[39] Encouraging other demographic policies that already have some public and legal sanction, such as restricting consanguineous marriages or encouraging prospective genetic counseling, coupled with amniocentesis and selective abortion could also effect a further reduction in the frequency of rare genetic *diseases*, but would not appreciably affect genetic load.

In general, those population-wide policies whose fundamental effect is to shift childbearing or other demographic characteristics through provision of public information, counseling, voluntary screening, and other noncoercive public devices would appear to be more consistent with equity and justice than are policies directed at compulsory control of individuals or groups.[40] Ultimately, however, such programs will falter if they fail to incorporate the growing evidence that the *primary* means of improving the quality of life of individuals still lies in eliminating environmental, not genetic, deprivations.

CONCLUSIONS

This analysis has led me to conclude that a eugenic program today would be premature for several reasons:

1. We lack a sufficient understanding of the critical demographic factors that affect gene frequency, the incidence of chromosomal aneuploidy, and the opportunity for selection in human populations. Thus, we do not know with reasonable certainty whether the current genetic status of the population (as it is affected by shifts in demographic vari-

ables) is undergoing a dysgenic trend, or whether it is stabilizing in terms of the genetic load.

2. We have not yet systematically studied the socioeconomic factors that interact strongly with environment to produce disparate incidences of congenital malformations, many of which have appreciable genetic components.

3. We have not sufficiently evaluated the desirability of utilizing existing methodologies for genetic intervention, such as population screening for carrier status, genetic counseling, or amniocentesis—all of which are now of value to *individuals*—as means of effecting population-wide change.

The incompleteness of our knowledge does *not* mean, however, that eugenic programs could not be justifiable *per se*.

A representative "just" eugenic policy would be one that promised real or potential improvements in meaningful social factors to benefit a maximum number and to compensate those who would be deprived by its implementation. A representative program would include at least the following:

A. Eugenic policy determination that gives proportional representation to the individuals who would be most affected, so that they had a significant voice in setting priorities and goals.

B. Policy that provides a set of benefits to the disadvantaged; *e.g.*, one which would compensate them proportionately to the degree that they voluntarily restricted their reproductive behavior.

C. The policy would protect fundamental rights:
   —to have at least one child;
   —to have privacy protected;
   —to make minimum incursions on liberties other than child-bearing;
   —to afford continuing opportunities toward the "pursuit of happiness."

D. Identity groups would be maintained.

E. Abuse would be minimized.

In our current state of ignorance of the benefits and consequences of these alternatives, I would rank the following objectives in order of their feasibility and ethical acceptability:

1. Improve those environmental components that most undermine the expression of human genetic potential (poverty *per se* would rank highest).

2. Stabilize the number of persons in a given area so that equitable distribution of resources remains possible.

3. Preserve the status quo of the gene pool, particularly by minimizing the contribution of novel mutations.

4. Act in ways that are "just" in the sense described in this chapter to avoid the deterioration of the gene pool.

References

1. Ramsey, P.: Fabricated Man: The Ethics of Genetic Control. New Haven, Yale University Press, 1970, pp. 29–30.
2. Kirk, D.: Patterns of survival and reproduction in the United States. Proc. Natl. Acad. Sci. U.S.A., 59:662, 1968.
3. Haller, M. H.: Eugenics: Hereditarian Attitudes in American Thought. New Brunswick, N.J., Rutgers University Press, 1963.
4. Wilson, R. J.: Darwinism and social ethics. In Darwinism and the American Intellectual. Chap. 3. Homewood, Ill., The Dorsey Press, 1967.
5. Osborn, F.: Eugenics (Encyclopedia Britannica). Eugenics Q., 13:155, 1966.
6. Lappé, M.: Moral obligations and the fallacies of "genetic control." Theol. Studies, 33:411, 1972.
7. Sang, J. H.: Nature, nurture, and eugenics. Postgrad. Med. J., 48:227, 1972.
8. Medawar, P.: The genetical control of medicine. Ann. Intern. Med., 67:28, 1967.
9. Newcombe, H. B.: Screening for the effects of maternal age and birth order in a registry of handicapped children. Ann. Hum. Genet., 27:367, 1964.
10. Lerner, M.: Heredity, Evolution, and Society. San Francisco, W. H. Freeman & Co., 1968, p. 268.
11. Charlesworth, B., and Charlesworth, D.: The measurement of fitness and mutation rate in human populations. Ann. Hum. Genet., 37:175, 1973.
12. Fraser, G. R.: The implications of prevention and treatment of inherited disease for the genetic future of mankind. J. de Génétique Humaine, 20:185, 1972.
13. Imaizumi, Y., Nei, M., and Furusho, T.: Variability and heritability of human fertility. Ann. Hum. Gen., 33:251, 1970.
14. Wilson, S. R.: The correlation between relatives under the multifactorial model with assortative mating. I. Ann. Hum. Genet., 37:189, 1973.
15. Degos, L., Colombani, J. et al.: Selective pressure on HL-A polymorphism. Nature [New Biol.], 249:62, 1974.
16. Task Force on Ethics and Population of the Institute of Society, Ethics, and the Life Sciences: Ethics, Population, and the American Tradition. In Parke, R., Jr., and Westoff, C. (eds.): Research Reports Volume VI, Aspects of Population Growth Policy. Chap. 1. Washington, D.C., U.S. Government Printing Office, 1972.
17. Reed, S., and Elving, V.: Effects of changing sexuality on the gene pool. In de la Cruz, T. F., and La Veck, G. D. (eds.): Conference on Human Sexuality and the Mentally Retarded, Hot Springs, Arkansas. New York, Brunner/Mazel, Inc., 1973, pp. 111–137.
18. Ennis, B., and Siegel, L.: The Rights of Mental Patients. New York, Avon Books, 1973.
19. Anonymous: Medical Aspects of Human Sexuality, 8:48, 1974.
20. Darwin, C.: The Variation of Animals and Plants Under Domestication. Vol. 2, London, John Murray, 1868, 405ff.
21. Sperry, R. W.: Science and the problem of values. Zygon, 9:7, 1974.
22. Stent, G. S.: Molecular biology and metaphysics. Nature [New Biol.], 248:779, 1974.
23. Simpson, G. G.: The concept of progress in organic evolution. Social Res., 41:28, 1974.
24. Shinn, R. L.: Perilous progress in genetics. Social Res., 41:83, 1974.
25. Neel, J. V.: Lessons from a "primitive people." Science, 70:815, 1970.
26. Gustafson, J.: Genetic screening and human values, an analysis. In Bergsma, D., Lappé, M., Gustafson, J., and Roblin, R. (eds.): Ethical, Social and Legal Dimensions of Screening for Human Genetic Disease. New York, Stratton Intercontinental Medical Book Corp., 1974, pp. 201–223.
27. Narot, J. R.: In de la Cruz, T. F., and La Veck, G. D. (eds.): Human Sexuality and the Mentally Retarded. New York, Brunner/Mazel, Inc., 1973, p. 204.

28. Twiss, B.: Parental responsibility for genetic health. Hastings Center Report, 4:2, 1974.
29. Fraser, G. R.: The short-term reduction in birth incidence of recessive diseases as a result of genetic counseling after the birth of an affected child. Hum. Hered., 22:1, 1972.
30. Motulsky, A. G., Fraser, G. R., and Felsenstein, J.: Public health and long-term genetic implications of intrauterine diagnosis and selective abortion. Birth Defects, 7:22, 1971.
31. Turner, J. K. G.: How does treating congenital diseases affect the genetic load? Eugenics Q., 15:191, 1968.
32. Gottesman, I. I., and Erlenmeyer-Kimling, L.: Prologue: A foundation for informed eugenics. Social Biol., 18:51, 1971.
33. Rawls, J.: A Theory of Justice. Cambridge, Mass., Belknap Press, 1971.
34. Moran, P. A. P.: A note on heritability and the correlation between relatives. Ann. Hum. Genet., 37:217, 1973.
35. Layzer, D.: Heritability analyses of I.Q. scores: Science or numerology. Science, 183:1259, 1974.
36. Jensen, A. R.: Genetics and Education. New York, Harper & Row Pubs., Inc., 1972.
37. Lowe, C. R.: Congenital malformations and the problem of their control. Br. Med. J., 2:515, 1972.
38. Lappé, M.: Censoring the hereditarians. Commonweal, 50:183, 1974.
39. Frederick, J.: Anencephalus: Variation with maternal age, parity, social class, and region in England, Scotland, and Wales. Ann. Hum. Genet., 34:31, 1970.
40. Lappé, M.: Human Genetics. Ann. N.Y. Acad. Sci., 216:152, 1973.

APPENDIX

Heterozygote Detection in Selected Autosomal Recessive Biochemical Disorders of Metabolism

Observations in heterozygotes not listed in the table are, for the most part, dealt with in the text (use index). Disorders have not been listed when no abnormalities in heterozygotes have been noted. Reliable heterozygote detection frequently is not possible in the disorders listed, despite observations made in some or all heterozygotes studied to date.

Autosomal Recessive "Biochemical Disorders" (Check index if not listed)	Observations of Abnormalities in Heterozygotes	References
Abetalipoproteinemia	Possible decreased serum lipoprotein concentration	1
Acatalasemia	Intermediate levels of catalase in rbc's* and cultured skin fibroblasts	2, 6
Adenylate kinase deficiency	Intermediate levels of adenylate kinase in rbc's	3
Afibrinogenemia, congenital	Partial deficiency of fibrogen in some	4
Albinism	Abnormal iris translucency in some	5
Amaurotic family idiocy (Batten's disease)	Cytoplasmic granules in wbc's†	6
Argininosuccinicaciduria	Intermediate activity of argininosuccinase in rbc's and fibroblasts	7
Ataxia-telangiectasia	Possible chromosomal breakage and impaired responsiveness to phytohemagglutinin by wbc's	8
Atransferrinemia	Low/normal transferrin levels	9
Chediak-Higashi syndrome	Abnormal cytoplasmic granules in cultured fibroblasts and lymphocytes	10, 11

Autosomal Recessive "Biochemical Disorders" (Check index if not listed)	Observations of Abnormalities in Heterozygotes	References
Cholesterol ester storage disease	Intermediate activity of acid lipase in cultured skin fibroblasts	12
Cystathioninuria	Intermediate responses (plasma and urine) to methionine load	13
Cystinosis types I–III	Abnormal free cystine retention in wbc's and cultured skin fibroblasts	14, 21
Cystinuria	Moderate aminoaciduria (mainly cystine and lysine) in type II	9
Diphosphoglycerate mutase deficiency	Intermediate activity of rbc 2,3-diphosphoglycerate mutase	15
Disaccharide intolerance I	Intermediate intestinal disaccharidase values	16, 9
Dystonia musculorum deformans	Phenothiazine-induced symptoms in some asymptomatic heterozygotes	18
Factor V deficiency	Lower than normal levels of clotting factor V	17
Factor VII deficiency	Lower than normal levels of clotting factor VII	19
Factor X	Low to intermediate levels of plasma factor X	20
Factor XI deficiency	Partial plasma thromboplastin antecedent deficiency	21
Factor XIII deficiency	Partial clotting factor XIII deficiency	22
Familial dysautonomia	Decreased plasma dopamine-β-hydroxylase activity in some	23
Galactokinase deficiency	Reduced rbc galactokinase activity	24
Glutathione peroxidase deficiency	Intermediate levels of glutathione peroxidase in rbc's	25
Glutathione reductase deficiency	Reduced levels of glutathione reductase in rbc's	26
Glutathione synthetase deficiency	Intermediate levels of rbc glutathione synthetase	27
Glycogen storage disease type II	Intermediate activity of α-glucosidase in wbc's and cultured skin fibroblasts	28
Glycogen storage disease type III	Intermediate activity of amylo-1,6-glucosidase in wbc's	29
Glycogen storage disease type IV	Intermediate activity of debrancher enzyme in cultured skin fibroblasts	30
Glycogen storage disease type V	Painful muscle cramps during exercise	31
Glycogen storage disease type VI	Decreased wbc phosphorylase activity	32
Glycogen storage disease type VII	Partial deficiency of phosphofructokinase in rbc's	33
β-2-Glycoprotein I deficiency	Intermediate levels of glycoprotein	34
Glyoxalase II deficiency	Intermediate values for rbc hydroxyacyl-glutathione hydrolase	35
Hageman factor deficiency	Partial deficiency of this factor in some	36
Hartnup's disease	Various but inconsistent findings with oral tryptophan loading; urinary indoles; photosensitivity	37
Hexokinase deficiency	Low levels of rbc hexokinase	38
Hexosephosphate isomerase deficiency	Partial deficiency of glucose phosphate isomerase in rbc's and wbc's	39
Histidinemia	Variable results on blood and urine histidine, L-histidine, oral loading tests, and histidase assays on skin and liver	37
Hunter's syndrome	Decreased activity of sulfoiduronate sulfatase in cultured fibroblasts; cell selection in culture following storage in liquid nitrogen	40, 41
Hydroxykynureninuria	Markedly increased excretion of xanthurenic acid after oral load of L-tryptophan	42
Hyperlipoproteinemia I	Possible slight hyperlipidemia and decreased plasma postheparin-lipolytic activity	43

Autosomal Recessive "Biochemical Disorders" *(Check index if not listed)*	Observations of Abnormalities in Heterozygotes	References
Hyperlysinemia	Possible reduced activity of lysine: α-keto-glutarate reductase	44
Hyperprolinemia type I	Modest hyperprolinemia in some	37
Hypogammaglobulinemia	Impaired incorporation of labeled precursors into DNA and RNA by PHA-stimulated lymphocytes	45
Hypophosphatasia	Low serum levels of alkaline phosphatase	46
Hypoprothrombinemia	Intermediate levels of plasma prothrombin	63
Iminoglycinuria	Hyperglycinuria with glycine loading in some	47
Intrahepatic cholestasis	Cholestasis of pregnancy a possible manifestation of heterozygosity	48
Isovaleric acidemia	Impaired $^{14}CO_2$ production from 2-^{14}C leucine by cultured fibroblasts	49
Kartagener's syndrome	Low serum levels of IgA globulin in some	50
Leigh's encephalopathy	Intermediate concentrations in urine of an inhibitor of thiamine synthesis	51
Lysyl-protocollagen hydroxylase deficiency	Intermediate activity of lysyl-protocollagen hydroxylase in cultured skin fibroblasts	62
Methemoglobin reductase	Intermediate levels of diaphorase activity; possible impaired reduction of methemo-globin during drug exposure	5
β-Methylcrotonylglycinuria	Slightly increased amounts of urinary β-methylcrotonylglycine and β-hydroxyisovaleric acid	37
Methylmalonicaciduria types I and II	Intermediate activity of methylmalonic CoA isomerase or CoA mutase in wbc's and cultured skin fibroblasts	52, 53
Mucolipidosis II (I-cell disease)	Intermediate elevation of selected serum lysosomal enzymes, with similar observa-tions in the culture medium of fibroblasts from heterozygotes	54, 55
Myeloperoxidase deficiency	Decreased activity of myeloperoxidase in wbc's	56
Pentosuria	Decreased serum L-xylulose	5
Phenylketonuria	Discriminatory analysis of plasma phenylalanine/tyrosine ratio indicating heterozygosity with $\pm$ 95% certainty	37
Porphyria, congenital erythropoietic	Intermediate activity of cosynthetase in hemolysates and cultured skin fibroblasts	57
Propionyl CoA carboxylase deficiency (ketotic hyperglycinemia)	Intermediate activity of propionyl CoA carboxylase in wbc's and cultured skin fibroblasts	5
Pyruvate kinase deficiency	Intermediate activity of pyruvate kinase in rbc's	5
Suxamethonium sensitivity	Partial deficiency of serum pseudo-cholinesterase in some	58
Thrombasthenia	Impaired clot retraction	59
Triosephosphate isomerase deficiency	Intermediate activity or triosephosphate isomerase in rbc's and wbc's	60
Wilson's disease	Decreased ceruloplasmin levels in some; intermediate concentration of hepatic copper	5
Xeroderma pigmentosum	Intermediate activity of ultraviolet endo-nuclease in cultured skin fibroblasts	61

*rbc — erythrocyte
†wbc — leukocyte

REFERENCES

1. Salt, H. B., Wolff, O. H., Lloyd, J. K., et al.: On having no beta-lipoprotein. A syndrome comprising a-beta-lipoproteinaemia, acanthocytosis, and steatorrhoea. Lancet, 2:325, 1960.
2. Krooth, R. S., Howell, R. R., and Hamilton, H. B.: Properties of acatalasic cells growing in vitro. J. Exp. Med., 115:313, 1962.
3. Szeinberg, A., Kahana, D., Gavendo, S. et al.: Hereditary deficiency of adenylate kinase in red blood cells. Acta Haematol., 42:111, 1969.
4. Prichard, R. W., and Vann, R. L.: Congenital afibrinogenemia; report on a child without fibrinogen and review of the literature. Am. J. Dis. Child., 88:703, 1954.
5. Stanbury, J. B., Wyngaarden, J. B., and Fredrickson, D. S.: The metabolic basis of inherited disease. New York, McGraw-Hill Book Company, 1972.
6. Strouth, J. C., Zeman, W., and Merritt, A. D.: Leukocyte abnormalities in familial amaurotic idiocy. N. Engl. J. Med., 274:36, 1966.
7. Jacoby, L. B., Littlefield, J. W., Milunsky, A. et al.: Argininosuccinase deficiency in cultured human cells. Am. J. Hum. Genet., 24:321, 1972.
8. Hecht, F., Koler, R. D., Rigas, D. A., et al.: Leukemia and lymphocytes in ataxia-telangiectasia. Lancet, 2:1193, 1966.
9. McKusick, V. A.: Mendelian Inheritance in Man: Catalogs of Autosomal Dominant, Autosomal Recessive, and X-Linked Phenotypes. 3rd ed. Baltimore, Md., Johns Hopkins Press, 1971.
10. Danes, B. S., and Bearn, A. G.: Cell culture and the Chediak-Higashi syndrome. Lancet, 2:65, 1967.
11. Blume, R. S., Glade, P. R., Gralnick, H. R., et al.: The Chediak-Higashi syndrome: Continuous suspension cultures derived from peripheral blood. Blood, 33:821, 1969.
12. Beaudet, A. L., Lipson, M. H., Ferry, G. D. et al.: Acid lipase in cultured fibroblasts: Cholesterol ester storage disease. J. Lab. Clin. Med., 84:54, 1974.
13. Perry, T. L., Hardwick, D. F., Hansen, S. et al.: Cystathioninuria in two healthy siblings. N. Engl. J. Med., 278:590, 1968.
14. Schulman, J. D., Wong, V. G., Bradley, K. H. et al.: Cystinosis: Biochemical, morphological and clinical studies. Pediatr. Res., 4:379, 1970.
15. Schröter, W.: Kongenitale nichtsphärocytare hämolytische Anämie bei 2,3-Diphospho-glyceratmutase-mangel der Erythrocyten im frühen Säuglings-alter. Klin. Wochenschr., 43:1147, 1965.
16. Dahlqvist, A.: Localization of the small-intestinal disaccharidases. Am. J. Clin. Nutr., 20:81, 1967.
17. Friedman, I. A., Quick, A. J., Higgins, F. et al.: Hereditary labile factor (factor V) deficiency. J.A.M.A., 175:370, 1961.
18. Eldridge, R.: The torsion dystonias: Literature review and genetic and clinical studies. Neurology, 20:1, 1970.
19. Kupfer, H. G., Hanna, B. L., and Kinne, D. R.: Congenital factor VII deficiency with

normal Stuart activity: Clinical, genetic and experimental observations. Blood, 15:146, 1960.

20. Graham, J. B., Barrow, E. M., and Hougie, C.: Stuart factor defect. II. Genetic aspects of a "new" hemorrhagic state. J. Clin. Invest., 36:497, 1957.

21. Rapaport, S. I., Proctor, R. R., Patch, M. J. et al.: The mode of inheritance of PTA deficiency: Evidence for the existence of major PTA deficiency and minor PTA deficiency. Blood, 18:149, 1961.

22. Duckert, F.: Factor XIII deficiency. Proc. 10th Intern. Congr. Soc. Haematol., Stockholm, 1964.

23. Weinshilboum, R. M., and Axelrod, J.: Reduced plasma dopamine-β-hydroxylase activity in familial dysautonomia. N. Engl. J. Med., 285:938, 1971.

24. Cotton, J. B.: Déficit héréditaire en galactokinase. Pediatrie, 22:609, 1967.

25. Necheles, T. F., Maldonado, N., Barquet-Chediak, A. et al.: Homozygous erythrocyte glutathione-peroxidase deficiency: Clinical and biochemical studies. Blood, 33:164, 1969.

26. Blume, K. G., Gottwik, M., Lohr, G. W. et al.: Familienuntersuchungen zum Glutathionreduktäsemangel menschlicher Erythrocyten. Humangenetik, 6:163, 1968.

27. Prins, H. K., Oort, M., Loos, J. A. et al.: Congenital nonspherocytic hemolytic anemia, associated with glutathione deficiency of the erythrocytes. Hematologic, biochemical and genetic studies. Blood, 27:145, 1966.

28. Nitowsky, H. M., and Grunefeld, A.: Lysosomal α-glucosidase in type II glycogenosis; activity in leukocytes and cell cultures in relation to genotype. J. Lab. Clin. Med., 69:472, 1967.

29. Chayoth, R., Moses, R. W., and Steinitz, K.: Debrancher enzyme activity in blood cells of families with type III glycogen storage disease: A method for diagnosis of heterozygotes. Isr. J. Med. Sci., 3:433, 1967.

30. Howell, R. R., Kaback, M. M., and Brown, B. I.: Glycogen storage disease type IV. Branching enzyme deficiency in skin fibroblasts and possible heterozygote detection. J. Pediatr., 78:638, 1971.

31. Dawson, D. M., Spong, F. L., and Harrington, J. F.: McArdle's disease: Lack of muscle phosphorylase. Ann. Intern. Med., 69:229, 1968.

32. Wallis, P. G., Sidbury, J. B., Jr., and Harris, R. C.: Hepatic phosphorylase defect. Studies on peripheral blood. Am. J. Dis. Child., 111:278, 1966.

33. Layzer, R. B., Rowland, L. P., and Ranney, H. M.: Muscle phosphofructokinase deficiency. Arch. Neurol., 17:512, 1967.

34. Haupt, H., Schwick, H. G., and Storiko, K.: Über einen erblichen beta(2)-glycoprotein I-mangel. Humangenetik, 5:291, 1968.

35. Valentine, W. N., Paglia, D. E., Neerhout, R. C. et al.: Erythrocyte glyoxalase II deficiency with coincidental hereditary elliptocytosis. Blood, 36:797, 1970.

36. Ratnoff, O. D., and Steinberg, A. G.: Further studies on the inheritance of Hageman trait. J. Lab. Clin. Med., 59:980, 1962.

37. Scriver, C. R., and Rosenberg, L. E.: Amino acid metabolism and its disorders. Philadelphia, W. B. Saunders Company, 1973.

38. Valentine, W. N., Oski, F. A., Paglia, D. E. et al.: Hereditary hemolytic anemia with hexokinase deficiency. Role of hexokinase in erythrocyte aging. N. Engl. J. Med., 276:1, 1967.

39. Paglia, D. E., Holland, P., Baughan, M. A. et al.: Occurrence of defective hexosephosphate isomerization in human erythrocytes and leukocytes. N. Engl. J. Med., 280:66, 1969.

40. Bach, G., Eisenberg, F., Jr., Cantz, M. et al.: The defect in the Hunter syndrome: Deficiency of sulfoiduronate sulfatase. Proc. Natl. Acad. Sci. U.S.A., 70:2134, 1973.

41. Booth, C. W., and Nadler, H. R.: In vitro selection for the Hunter gene. N. Engl. J. Med., 288:636, 1973.

42. Komrower, G. M., and Westall, R.: Hydroxykynureninuria. Am. J. Dis. Child., 113:77, 1967.

43. Boggs, J. D., Hsia, D. Y. Y., Mais, R. F. et al.: The genetic mechanism of idiopathic hyperlipidemia. N. Engl. J. Med., 257:1101, 1957.

44. Dancis, J., Hutzler, J., Cox, R. P. et al.: Familial hyperlysinemia with lysine-ketoglutarate reductase deficiency. J. Clin. Invest., 48:1447, 1969.

45. Kamin, R. M., Fudenberg, H. H., and Douglas, S. D.: A genetic defect in "acquired" agammaglobulinemia. Proc. Natl. Acad. Sci. U.S.A., 60:881, 1968.
46. Rathbun, J. C., MacDonald, J. W., Robinson, H. M. C. et al.: Hypophosphatasia: A genetic study. Arch. Dis. Child., 36:540, 1961.
47. Scriver, C. R.: Renal tubular transport of proline, hydroxyproline, and glycine. III. Genetic basis for more than one mode of transport in human kidney. J. Clin. Invest., 47:823, 1968.
48. McKusick, V. A., and Clayton, R. J.: Cholestasis of pregnancy. N. Engl. J. Med., 278:566, 1968.
49. Shih, V. E., Mandell, R., and Tanaka, K.: Diagnosis of isovaleric acidemia in cultured fibroblasts. Clin. Chim. Acta, 48:437, 1973.
50. Holmes, L. B., Blennerhassett, J. B., and Austen, K. F.: A reappraisal of Kartagener's syndrome. Am. J. Med. Sci., 255:13, 1968.
51. Murphy, J. V., Diven, W. F., and Craig, L.: Detection of Leigh's disease in fibroblasts. Pediatr. Res., 8:393, 1974.
52. Morrow, G., III, Mellman, W. J., Barness, L. A. et al.: Propionate metabolism in cells cultured from a patient with methylmalonic acidemia. Pediatr. Res., 3:217, 1969.
53. Mahoney, M. J., and Rosenberg, L. E.: Defective metabolism of vitamin B_{12} in fibroblasts from children with methylmalonic aciduria. Biochem. Biophys. Res. Commun., 44:375, 1971.
54. Leroy, J. G. and van Elsen, A. F.: I-cell disease (mucolipidosis type II) serum hydrolases in obligate heterozygotes. Humangenetik, 20:119, 1973.
55. Wiesmann, U. N., and Herschkowitz, N. N.: Studies on the pathogenetic mechanism of I-cell disease in cultured fibroblasts. Pediatr. Res., 8:865, 1974.
56. Lehrer, R. I., and Cline, M. J.: Leukocyte myeloperoxidase deficiency and disseminated candidiasis: The role of myeloperoxidase in resistance to candida infection. J. Clin. Invest., 48:1478, 1969.
57. Romeo, G., Kaback, M. M., and Levin, E. Y.: Uroporphyrinogen III. Cosynthetase activity in fibroblasts from patients with congenital erythropoietic porphyria. Biochem. Genet., 4:659, 1970.
56. Scott, E. M., Weaver, D. D., and Wright, R. C.: Discrimination of phenotypes in human serum cholinesterase deficiency. Am. J. Hum. Genet., 22:363, 1970.
59. Papayannis, A. G., and Israels, M. C. G.: Glanzmann's disease and trait. Lancet, 2:44, 1970.
60. Schneider, A. S., Valentine, W. N., Hattori, M. et al.: Hereditary hemolytic anemia with triosephosphate isomerase deficiency. N. Engl. J. Med., 272:229, 1965.
61. Regan, J. D., Setlow, R. B., Kaback, M. M. et al.: Xeroderma pigmentosum: A rapid sensitive method for prenatal diagnosis. Science, 174:147, 1971.
62. Krane, S. M., Pinnell, S. R., and Erbe, R. W.: Lysyl-protocollagen hydroxylase deficiency in fibroblasts from siblings with hydroxylysine-deficient collagen. N. Engl. J. Med., 69:2899, 1972.
63. Pina-Cabral, J. M., and Justica, B.: Congenital hypoprothrombinemia in a Portuguese family. Thromb. Diath. Haemorrh., 30:415, 1973.

INDEX

485

Consultand, 64–89
Contraception, 73, 76, 106, 107, 141, 169, 195, 218, 295, 380, 398, 425, 461
Copper, 161, 284, 288, 291, 479
Coproporphyrin, 357
Cornea, 164, 168
  micro-, 171
  opacity of, 172, 184, 186
Coronary disease, 1, 306–315
Corpus callosum, agenesis of, 160, 168
Cost(s), 359, 427, 442. See also *Economics.*
  analysis of, 232
  /benefit, 102, 254, 403, 442–455, 464
  /effectiveness, 14, 59, 102, 412
  of counseling, 412, 420
  of prematurity, 346
  of screening, 434
Cosynthetase, 479
Counseling, ethical considerations in, 13, 69
Counselor. See *Genetic counselor.*
Coxsackie virus, 374–382
Craniofacial, 35
Craniosynostosis, 223
Cranium. See *Skull, skeletal.*
Creatine phosphokinase, 96, 141, 150, 152, 169
Cretinism, 39. See also *Hypothyroidism.*
Crib death, 345
Cri-du-chat syndrome, 32
Crouzon's disease, 35
Cryptorchidism, 37, 164
Cultural considerations, 393
Cyanate, 209, 296
Cyclic AMP, 159, 160
Cystathionine, synthase, 116, 120, 121, 124, 239, 265, 267
Cystathioninuria, 122, 124, 239, 265, 290, 478
Cysteine, 292
Cystic fibrosis, 10, 94, 98, 108, 109, 117, 205, 223, 242, 289, 291, 434, 462
  ciliary inhibition factor in, 242
Cystic hygroma, 236
Cystine, 118, 283, 478
Cystinosis, 159, 240, 289, 290, 412, 478
Cystinuria, 118, 239, 265, 267, 283, 292, 478
Cytomegalic inclusion disease, 20, 27, 350, 369–382
  frequency of, 28
Cytomegalovirus, 20, 27, 28, 350, 369–382
Cytoplasmic inheritance, 22
Cytosine arabinoside, 371

Data bank, genetic, 438
Deafness, 21, 36, 165, 172, 281, 370, 396, 399

Deafness (*Continued*)
  conductive, 161
  neural, 161
  relationship of, to albinism, 161
Decidua, 348
Decision-making, 64–89
Degeneration, macular, 171, 172, 183–186
Dementia, 185–187
Demography, 461–475
Dentinogenesis imperfecta, 281
Deprivation, 20, 21, 31, 41, 54, 388–406, 453, 459, 472
Dermal hypoplasia, 164, 168
Developmental disability, 3, 22, 27, 41, 343–352, 355–366, 369–382, 385–406
  definition of, 4
  frequency of, 26
  screening, 385–406
Dextrocardia, 162, 168
Diabetes insipidus, 142, 159, 165, 288, 291
  nephrogenic, 134, 158, 289
  neurohypophyseal, 158
Diabetes mellitus, 1, 40, 195, 224, 237, 324, 344
Dialysis, 292
Diaphorase activity, 479
Diarrhea, acute infantile, 369–382
Diet, 33, 54, 57, 74, 116, 118, 312, 350, 398, 411, 429
  augmentation of, 283
  carbohydrate, 284
  control of, 286, 299
  during pregnancy, 287
  exclusion, 283
  fish, 365
  low-fat, 308
  low phenylalanine, 39, 266, 285, 426
  low protein, 241
  restriction(s) in, 284, 285, 318
  soy-protein, 268
  substitution, 291
  supervision of, 286
  supplementation of, 288
  synthetic, 285, 291
  therapeutic principles of, 282–302
5-α-Dihydrotestosterone, 166
Dihydroxyacetone phosphate, 157
L-Dihydroxyphenylalanine, 294
Diphenylhydantoin, 40
2,3-Diphosphoglycerate, 157, 478
Diphtheria toxoid, 334
Disaccharidase, 478
Disaccharide intolerance, 479
Disease(s). See specific names.
Divorce, 85, 450
DNA, 479
Dopamine-β-hydroxylase, 478
Drugs, 22, 85, 101, 195, 293, 308–315, 350, 371, 398, 478, 479
  antiviral, 380